Soil Biology

Volume 22

Series Editor
Ajit Varma, Amity Institute of Microbial Technology, Amity University Uttar Pradesh, Noida, UP, India

For further volumes:
http://www.springer.com/series/5138

Girish Shukla • Ajit Varma
Editors

Soil Enzymology

Springer

Editors
Dr. Girish Shukla
Biological, Geological & Environmental Sciences
Cleveland State University 2121
Euclid Avenue
Cleveland, OH 44115-2214, USA
g.shukla@csuohio.edu

Prof. Dr. Ajit Varma
Director General
Amity Institute of Microbial Technology
Amity University Uttar Pradesh
& Vice Chairman
Amity Science, Technology & Innovation Foundation
Block A, Amity Campus, Sector 125
Noida, UP 201303
India
ajitvarma@aihmr.amity.edu

ISSN 1613-3382
ISBN 978-3-642-14224-6 e-ISBN 978-3-642-14225-3
DOI 10.1007/978-3-642-14225-3
Springer Heidelberg Dordrecht London New York

Library of Congress Control Number: 2010935854

Cover design: SPi Publisher Services

Printed on acid-free paper

Springer is part of Springer Science+Business Media (www.springer.com)

Preface

Current food and agriculture production appears to be sufficient to meet the present and near future demand of the world. However, due to population growth and environmental alterations that are intimately connected to global warming trends, continuous food production to meet the future demand remained uncertain. To meet additional food demand, it is crucial that we must focus on the healthy state of earth and its well being. This brings us to our relationship with the soil to which we are intimately bonded. Organic compounds, microorganisms, enzymes, and soil constitute a healthy composition for soil microecosystems. Decomposition of organic matter is central to recycling and balancing of the soil nutrients. Soil enzymes and a number of associated factors as key mediators are vital to soil macrosystem. Soil enzymes are involved in the aspects of nutrient recycling, maintaining the environmental quality, thus maintaining the soil fertility, and enhancing the productivity of economically important crops. Based on our teaching and research experience, it was imperative that a book is needed, which encompasses various facets of soil enzymes and their function in maintaining the soil quality, carbon sequestration, nutrient recycling, and variety of bioremediation of contaminated soil.

In this book, we have attempted to compile broader aspects of soil enzymes including their biochemical and microbiological properties, environmental nutrients, microorganisms' and enzymes' mechanistic roles in maintaining a healthy soil state. The authors have presented various existing and potential environmental challenges to soil enzymes and the soil and have provided knowledge to deal with it.

Numerous soil enzymes are very important to carry out basic catalytic activities and abundant biological processes in various types of soils. These processes in turn are connected to soil's well being. The book is composed of 20 chapters encompassing various aspects of soil enzymology. The first chapter provides an overview of soil enzymes and general mechanistic aspects. Chapter 2 provides a broad and comprehensive account of the role of soil enzymes in maintaining soil health. Soil carbon sequestration and nutrient cycling is controlled by the decomposition of organic matter in the soil. Chapter 3 covers the aspects of soil enzymes, which facilitate soil carbon and nutrient balance. Chapter 4 highlights the enzymes that are found in forest soils, their activities, and factors affecting the activities of soil enzymes. Chapter 5 describes phosphohydrolases and their importance in organic phosphorus cycling. Hydrolytic enzymes and their role in the fast and prudent recovery of microbial cells and reestablishment of microbial communities in dry-wet cyclic has been presented elegantly in Chap. 6. Due to heavy machinery usage, pollution, and constant insults to environment by human interactions, the soil quality is affected in many cases irreversibly. Soil enzymes are one indicator, which corresponds to the quality and health of the soil. Chapter 7 illustrates the distinct types of soil modifications and provides relevant sources, classification, and properties of soil enzymes that make them excellent indicators. Rhizosphere is directly affected by various root secretions and its association with the soil microbes. In Chap. 8, authors present a comprehensive account of enzymes activities and factors affecting them in Rhizosphere of plant.

Lignocellulose-degrading soil enzymes, phenol oxidases, and fungal oxidoreductases have been covered in Chaps. 9 through 11. Evolutionary economic principles, which modulate the production of soil enzymes, are elegantly portrayed in Chap. 12. Activity of enzymes is directly affected by temperature as determined by numerous investigations in purified laboratory systems. However, relatively fewer studies have touched upon the effect of temperature on activities of soil enzymes in its native environment. Chapter 13 fulfils this promise by providing a detailed aspect of enzyme structure–function and the effect of temperature in enzyme kinetics under field conditions. Soil enzymes, which are derived from bacterial sources including Keratinases, Pectinases, Xylanases, and Lipases, and their properties are illustrated in Chap. 14. Chapter 15 outlines the broad range of enzyme–organo-mineral interactions that occurs in the soil. Mechanistic aspects, which influence soil enzyme activity, have also been covered in the Chapter.

Throughout the human history, man continued to discover methods to protect his crop from pests. Development and use of a range of pesticide has prevented total decimation of food producing crop; however, it also has caused irreversible damage to soil. While in soil, these pesticides influence the activity of soil enzymes. Chapter 16 outlines an historical and scientific perspective of the interaction of diverse pesticides with the soil and their influence on soil enzyme activity. Chapter 17 provides a unique perspective on the behavior of soil enzyme activity on volcanic soils. The chapter covers the detailed properties of volcanic ashes-derived soils and its resident soil enzymes' activity. Screening, characterization, and optimization of microbial Pectianase have been covered in the Chap. 18. Appropriate molecular approaches in order to study polymorphism in closely related microorganisms with respect to protein phosphatase are covered in Chap. 19. In addition to use of fossil fuel and its derivative, various other pollutants including chlorinated compounds, synthetic dyes, and aromatic hydrocarbons have contaminated large areas of productive crop land. Bioremediation appears to be a logical and environmentally sustainable method to counter the effect of contaminated soil. Chapter 20 provides an in depth coverage of the production and the use of a number of white rot fungi for decontamination of oil polluted soil.

In this volume, we attempted to cover many aspects of soil enzymology. We hope that this volume would be an essential resource for teachers, students, and research professional who are interested in basic and applied aspects of soil enzymology. In addition to extraordinary contribution by the authors of the series, the volume would have been a dream without the help of a large number of volunteers for their selfless efforts.

We thank Dr. Jutta Lindenborn, Springer, Heidelberg, Germany, for her admirable patience and valuable suggestions. GCS is thankful to Dr. Bakshi for her important contribution and providing an outstanding support in fine-tuned editing of the book. Finally, we like to thank our contributors who dedicated their valuable time and expertise, and without their contribution, the volume would still be a distant dream.

Cleveland, OH, USA — Girish Shukla
Noida, Uttar Pradesh, India — Ajit Varma

Contents

Contributors

Ahn, Mi-Youn Soil and Water Science Department, University of Florida, PO Box 110290, Gainesville, FL 32611, USA, mahn@ufl.edu

Allison, Steven D. University of California, Irvine, 321 Steinhaus, Irvine, CA 92697, USA, allisons@uci.edu

Bakshi, Madhunita Amity Institute of Microbial Technology, Amity University, Sector 125, Noida, Uttar Pradesh 201303, India, bakshi.madhunita@gmail.com

Baldrian, Petr Laboratory of Environmental Microbiology, Institute of Microbiology of the ASCR, v.v.i., Vídeňská 1083, 14220 Praha 4, Czech Republic, baldrian@biomed.cas.cz

Cetin, Sema Camci Agricultural Faculty, Soil Science Department, Gaziosmanpasa University, Tokat, Turkey, semacamcicetin@hotmail.com

Dao, Thanh H. US Department of agriculture, Agriculture Research Service, Beltsville, MD, USA, thanh.dao@ars.usda.gov

Das, Shonkor Kumar Department of Applied Chemistry and Biotechnology, Graduate School of Engineering, University of Fukui, 3-9-1Bunkyo, Fukui 910-8507, Japan, skdas76@yahoo.com

Dubrovskaya, Ekaterina V. Institute of Biochemistry and Physiology of Plants and Microorganisms, Russian Academy of Sciences, 13 Prospect Entuziastov, Saratov 410049, Russia, ecbio@ibppm.sgu.ru

Egamberdieva, Dilfuza Faculty of Biology and Soil Sciences, The National University of Uzbekistan, Tashkent, 100175, Uzbekistan, dilfuza_egamberdiyeva @yahoo.com

Ernakovich, Jessica Natural Resource Ecology Laboratory, Colorado State University, Fort Collins, CO 80523, USA; Graduate Degree Program in Ecology, Colorado State University, Fort Collins, CO 80523, USA, jagilman@nrel.colostate.edu;

Gartner, Tracy B. Department of Biology and the Environmental Science Program, Carthage College, Kenosha, WI 53140, USA, tgartner@carthage.edu

Gianfreda, Liliana Dipartimento di Scienze del Suolo della Pianta, dell'Ambiente e delle Produzioni Animali, Università di Napoli Federico II, Via Università 100, 80055 Portici, Napoli, Italy, liliana.gianfreda@unina.it

Islam, Rafiq Soil and Water Resources, Ohio State University South Centers, Piketon, OH, USA, islam.27@osu.edu

Karaca, Ayten Faculty of Agriculture, Department of Soil Science, Ankara University, Diskapi, Ankara, Turkey, akaraca@agri.ankara.edu.tr

Khan, Zaved Ahmed School of Biotechnology, Chemical and Biomedical engineering, VIT University, Vellore, Tamil Nadu 632014, India, zavedkhan@vit.ac.in, khan.zaved@gmail.com

Kizilkaya, Ridvan Agricultural Faculty, Soil Science Department, Ondokuz Mayis University, Samsun, Turkey, ridvank@omu.edu.tr

Leontievsky, A.A. Institute of Biochemistry and Physiology of Microorganisms, Russian Academy of Sciences, Pushchino, Moscow Region 142292, Russia, leont@ibpm.pushchino.ru

Lisov, A.A. Institute of Biochemistry and Physiology of Microorganisms, Russian Academy of Sciences, Pushchino, Moscow Region, 142292, Russia, alex-lisov@rambler.ru

López, Ada Instituto de Biología Vegetal y Biotecnología, Universidad de Talca, Curico, Chile, adlopez@utalca.cl

López, Roxana Facultad de Ciencias Agrarias, Departamento de Producción Agrícola, Universidad de Talca, Curico, Chile, rolopez@utalca.cl

Makarov, Oleg E. Institute of Biochemistry and Physiology of Plants and Microorganisms, Russian Academy of Sciences, 13 Prospect Entuziastov, Saratov 410049, Russia, room107@ibppm.sgu.ru

Malla, Rajani Central Department of Biotechnology, Tribhuvan University, Kathmandu, Nepal, rajanimalla2000@yahoo.com

Marxsen, Jürgen Limnologische Fluss-Station des Max-Planck-Instituts für Limnologie, 36110 Schlitz, Germany; Institut für Allgemeine und Spezielle Zoologie, Tierökologie, Justus-Liebig-Universität Gießen, Heinrich-Buff-Ring 26-32, 35392 Gießen, Germany, Juergen.Marxsen@allzool.bio.uni-giessen.de

Nikitina, Valentina E. Institute of Biochemistry and Physiology of Plants and Microorganisms, Russian Academy of Sciences, 13 Prospect Entuziastov, Saratov 410049, Russia, nikitina@ibppm.sgu.ru

Pokharel, Utprekshya Department of Microbiology, Punjab University, Chandigarh, India, utp_pok@yahoo.com.in

Pozdnyakova, Natalia N. Institute of Biochemistry and Physiology of Plants and Microorganisms Russian Academy of Sciences, 13 Prospekt Entuziastov, Saratov 410049, Russia, nataliapozdnyakova@yahoo.com

Prasad, Ram Amity Institute of Microbial Technology, Amity University Uttar Pradesh, Sector-125, New Super Express Highway, Uttar Pradesh, Noida, India, rprasad@amity.edu, rpjnu2001@gmail.com

Rao, Maria A. Dipartimento di Scienze del Suolo della Pianta dell'Ambiente e delle Produzioni Animali, Università di Napoli Federico II, Via Università 100, 80055 Portici, Napoli, Italy, mariarao@unina.it

Renella, Giancarlo Plant Nutrition and Soil Science, University of Florence, Florence, Italy, giancarlo.renella@unifi.it

Rosas, Anali Universidad de Concepción, Facultad de Agronomía, Departmento de Ciencia del Suelo y Recursos Naturales, P.O. Box 537, Av. Vicente Méndez 595, Chillan, Chile, arosas@udec.cl

Shi, Wei Department of Soil Science, North Carolina State University, Raleigh, NC 27695-7619, USA, wei_shi@ncsu.edu

Sinsabaugh, Robert Department of Biology, University of New Mexico, Aluquerque, NM 87131, USA, rlsinsab@unm.edu

Šnajdr, Jaroslav Laboratory of Environmental Microbiology, Institute of Microbiology ASCR, Videnska 1083, 14220 Praha 4, Czech Republic, snajdr@biomed.cas.cz

Steinweg, J. Megan Natural Resource Ecology Laboratory, Colorado State University, Fort Collins, CO 80523, USA; Graduate Degree Program in Ecology, Colorado State University, Fort Collins, CO 80523, USA, steinweg@warnercnr.colostate.edu

Štursová, Martina Laboratory of Environmental Microbiology, Institute of Microbiology ASCR, Videnska 1083, 14220 Praha 4, Czech Republic, stursova@-biomed.cas.cz

Suneetha, V. School of Biotechnology, Chemical and Biomedical engineering, VIT University, Vellore, Tamil Nadu 632014, India, vsuneetha@vit.ac.in

Turgay, Oguz Can Agricultural Faculty, Soil Science Department, Ankara University, Ankara, Turkey, turgay@agri.ankara.edu.tr

Turkovskaya, Olga V. Institute of Biochemistry and Physiology of Plants and Microorganisms, Russian Academy of Sciences, 13 Prospect Entuziastov, Saratov 410049, Russia, ovtur@ibppm.sgu.ru

Varma, Ajit Amity Institute of Microbial Technology, Amity University, Sector 125, Noida 201303, Uttar Pradesh, India, ajitvarma@aihmr.amity.edu

Waldrop, Mark P. U.S. Geological Survey, 345 Middlefield Rd, MS 962, Menlo Park, CA 94025, USA, mwaldrop@usgs.gov

Wallenstein, Matthew Natural Resource Ecology Laboratory, Colorado State University, Fort Collins, CO 80523, USA; Graduate Degree Program in Ecology, Colorado State University, Fort Collins, CO 80523, USA, mawallen@nrel. colostate.edu;

Weintraub, Michael N. Department of Environmental Sciences, University of Toledo, Toledo, OH 43606, USA, michael.weintraub@utoledo.edu

Wirth, Stephan Leibniz-Centre for Agricultural Landscape Research, Institute of Landscape Matter Dynamics, Müncheberg, Germany, swirth@zalf.de

Zavarzin, A.A. Faculty of Biology and Soil Sciences, St-Petersburg State University, 199034 St. Petersburg, Russia, az@bio.pu.r

Zavarzina, A.G. Faculty of Soil Science, Moscow State University, 119991 Moscow, Russia, zavarzina@mail.ru

Zimmerman, Andrew R. Department of Geological Sciences, University of Florida, 241 Williamson Hall, PO Box 112120, Gainesville, FL 32611, USA, aximmer@ufl. edu

Zoppini, Annamaria Istituto di Ricerca Sulle Acque, Consiglio Nazionale delle Ricerche, Area della Ricerca di Roma 1-Montelibretti, via Salaria, CP10, 00016 Monterotondo, Rome, Italy, zoppini@irsa.cnr.it

Chapter 1
Soil Enzyme: The State-of-Art

Madhunita Bakshi and Ajit Varma

1.1 Introduction

Soil may be defined as a thin layer of earth's crust which serves as a natural medium for growth of plants. It is the unconsolidated mineral matter that has been subjected to and influenced by, genetic and environmental factors-parent material, climate, organisms, and topography all acting over a period of time. Soil is a natural body consisting of layers (soil horizons) of mineral constituents of variable thicknesses, which differ from the parent materials in their morphological, physical, chemical, and mineralogical characteristics. It is composed of particles of broken rock that have been altered by chemical and environmental processes that include weathering and erosion. Soil differs from its parent rock due to interactions between the lithosphere, hydrosphere, atmosphere, and the biosphere. It is a mixture of mineral and organic constituents that are in solid, gaseous, and aqueous states.

Soils differ among themselves in some or all the properties, depending on the differences in the genetic and environmental factors. Thus some soils are red, some are black, some are deep and some are shallow, some are coarse textured and some are fine-textured. They serve as a reservoir of nutrients and water for crops, provide mechanical anchorage and favorable tilth. The components of soil are mineral matter, organic matter, water, and air, the proportions of which vary and which together form a system for plant growth; hence the need to study the soils in perspective.

M. Bakshi and A. Varma (✉)
Amity Institute of Microbial Technology, Amity University Uttar Pradesh, Sector 125, Noida 201303, India
e-mail: ajitvarma@aihmr.amity.edu

G. Shukla and A. Varma (eds.), *Soil Enzymology*, Soil Biology 22,
DOI 10.1007/978-3-642-14225-3_1, © Springer-Verlag Berlin Heidelberg 2011

1.2 Soil: A Physical Point of View

The physical properties of a soil like the water holding capacity, permeability to water, aeration, plasticity, and nutrient-supplying ability, are influenced by the size, proportion, arrangement, and mineral composition of the soil particles.

1.2.1 Particle Shape and Size

The soil particles vary in shape from spherical to angular. They differ in size from gravel and sand to fine clay (Table 1.1)

1.2.2 Textural Classes

The varying proportion of particles of different size groups in a soil constitute what is termed as a textural class. The principal textural classes are: clay, sandy clay, silty clay, clay loam, sandy clay loam, silty clay loam, loam, sandy loam, silt loam, sand, loamy sand, and silt.

The textural classes differ not only in the particle size analysis, but also in their bearing on some of the important factors affecting plant growth, such as

1. The moveability and availability of water
2. Aeration
3. Workability
4. The contents of plant nutrients

Sandy soils are very permeable and well drained but are less water retentive and hence need more frequent irrigation for successful crop growth than fine textured soils. The clayey soils can hold more moisture, but they have high wilting percentage. The rate of water intake of these soils is low. They are subject to water-logging resulting in poor aeration and workability. The moderately fine-textured soils for

Table 1.1 Classification of Soil according to their particle size

Types	International system of classification (Diameter in mm)
Gravel	2 and more
Very coarse sand	2–1
Coarse sand	1.0–0.5
Medium sand	0.5–0.25
Fine sand	0.25–1
Very fine sand	0.10–0.05
Silt	0.05–0.02
Clay	less than 0.002

example, loams, clay loams or silt loams are by far the excellent soils for plant growth, since they have the advantages of both sands and clays.

1.2.3 Colors

Colors of a soil are indicator of soil conditions and some important properties. It is due either to mineral or organic matter or mostly to both. Red, yellow, or brown colors are usually related to the different degrees of oxidation, hydration, and diffusion of iron oxides in the soil. Dark colors of a soil are associated with one or a combination of several factors, including impeded drainage conditions, content, and state of decomposition of organic matter.

1.2.4 Density

Soils having larger particles are usually heavier in weight per unit volume than those having smaller particles. True density of a soil is based on the individual densities of soil constituents and according to their proportionate contribution.

1.2.5 Pore Space

The pore space of soil is the portion occupied by air and water and it is determined largely by structural conditions. Sands have low pore space of about 30%, whereas clays may have as much as 50–60%. Although clays possess greater total porosity than the sands the pore spaces in the latter being individually larger are more conducive to good drainage and aeration.

1.2.6 Plasticity and Cohesion

Plasticity and cohesion reflect the soil consistency and workability of the soils. Plasticity is the property that enables a moist soil to change shape on the application of force and retain this shape even when the force is withdrawn. Cohesion is the tendency of the particles to stick to one another. Plastic soils are cohesive. On this basis, sandy soils may be considered to be non-plastic and clayey soils to be plastic.

1.2.7 Soil Temperature and Heat

Soil temperature is one of the important factors that control the microbiological activity and all the processes involved in the growth of plants. The temperature needed for germination and root growth varies with crops and varieties. Crops, for example, wheat, barley, and peas grown in India during winter germinate at relatively low temperatures as compared with maize, and those at which groundnut or cotton germinate. Heat is necessary for seed germination, root growth, and other biological activities.

1.2.8 Soil Water

Water has perhaps the greatest influence on the growth and yield of a crop. It is needed in much larger quantity than that of any other substance that contributes to growth or yield. Water serves the following functions in relation to plant life.

1. It is an essential part of plant food. It constitutes nearly 90% of plant tissues.
2. It serves as a solvent and carrier of plant nutrients.
3. It maintains cell turgidity and regulates temperature.

Water is held in soil in the following forms:

1. *Hygroscopic Water:* It occurs as a thin film (4–5 μm) and is held tenaciously with a tension of 31 atmospheres or more. It is not available to plants.
2. *Capillary Water:* It forms a continuous film around soil particles (outside the film of hygroscopic water) and in the micropore spaces. It is held by surface tension. Capillary water is held at a tension ranging from 1/3 to 31 atmospheres.
3. *Gravitational Water:* It is free water held at a tension below 1/3 atmospheres. It saturates the soil and percolates downwards under the influence of gravity.

1.3 Soil: A Chemical Point of View

1.3.1 Organic Matter

Organic matter, though forming a small part of mineral soils, plays a vital role in the productivity and conditioning of soils. It serves as a source of food for soil bacteria and fungi, which are responsible for converting complex organic materials into simple substances readily used by the plants. The intermediate products of decomposition of fresh organic matter help to increase the physical condition of the soil. The addition of organic matter also improves the working quality or friability of the soil. In association with clay and calcium, it helps to form the aggregates of soil particles to produce the "crumb structure".

1.3.2 Humus

The organic matter in the soil consists largely of plant remains, the residues of soil-microorganisms feeding on them, and several products of their decomposition. In consonance with its origin, organic matter contains most of the mineral elements found in the plants. The plant remains may occur in recognizable form but more commonly they are found as a fairly stable, dark, amorphous complex colloidal substance called humus. Chemically, humus represents a mixture of decomposed or altered products of carbohydrates, proteins, fats, resins, wax, and other similar substances. These complex compounds are gradually decomposed by soil organisms into simple mineral salts, carbon dioxide, water, organic acids, ammonia, methane, and free nitrogen, depending upon the initial composition of the organic matter. The average composition of humus is as follows Table 1.2.

1.3.3 Soil Colloids

The most active portions of the soil are those which are in the colloidal state. The colloidal state implies a two-phase system in which the material called the dispersed phase (fine clay and humus) is dispersed in the dispersed medium (water). In soils, the mineral and organic colloids exist in intimate and heterogeneous admixture. The mineral colloid is present almost exclusively as the clay of various kinds, whereas the organic colloid is present as humus. It is generally believed that clay particles less than one micron in diameter possess colloidal properties and these properties increase with a decrease in the size of the particles. The most distinctive colloidal properties are:

1. The large specific surface or interface and
2. The capacity to hold solids, gases, salts, and ions.

1.4 Soil: A Biological Point of View

Soil is a natural medium in which microbes live, multiply, and die. Microbial diversity in the soil is a critical environmental topic that concerns people from all

Table 1.2 Chemical composition of humus

	Percentage
Carbon	50
Oxygen	35
Nitrogen	5
Hydrogen	5
Ash (containing phosphorus, potassium, sulfur and other elements)	5

walks of life. Raw organic matter in the soil is not directly used by the plants as food. It must be broken down first into humus and then into simpler products before it can be utilized. This work is done by different kinds of microorganisms in the soil. The decomposition of organic matter forms part of the feeding and growth process of these microscopic plants and animals. However, not all the soil organisms are beneficial. There are certain bacteria which under anaerobic conditions of water-logged soil cause denitrification releasing free nitrogen which gets lost into the air.

1.4.1 Kind of Soil Organisms

Soil contains a wide range of microorganisms described as "black box" (Paul and Clark 1989). Microorganisms in soil are closely associated with soil particles, mainly clay-organic matter complexes (Foster 1988). Their activity and interaction with other microbes and larger organisms and with soil particles depend largely on conditions at the microhabitat level that may differ among microhabitats even over very small distances (Wieland et al. 2001). Increasing attention is being drawn to microorganisms because the fertility of soil depends not only on its chemical composition, but also on the qualitative and quantitative nature of microorganisms inhabiting it (Fig. 1.1; Giri et al. 2005).

Microorganisms are generally divided into five major taxonomic categories: Algae, Eubacteria, Fungi, Protists, and Viruses. The more complex cells constitute eukaryotes, which include Algae, Fungi, and Protists. The less complex cell constitutes prokaryotes comprising two microbial groups: Eubacteria (including Cyanobacteria) and the Archaebacteria.

1.4.1.1 Eubacteria

Eubacteria also known as true bacteria are recognized as the most dominant group of microorganisms among the various kinds of soil (Liesack and Stackebrandt 1992; Visscher et al. 1992; Borneman et al. 1996).They are present in all types of soil, but their population decreases as the depth of soil increases.

Bacteria live in soil as cocci (sphere, 0.5 μm), bacilli (rod, 0.5–0.3 μm), or spiral. The bacilli are common in soil, whereas spirilli are very rare in natural environments (Giri et al 2004). The most common bacteria belong to the genera *Pseudomonas, Arthrobacter, Clostridium, Achromobacte, Bacillus, Micrococcus, Flavobacterium, Coryne bacterium, Sarcina, Azospirillum,* and *Mycobacteria* (Loper et al. 1985; Bruck 1987; Lynch 1987a, b). *Escherichia* is encountered rarely in soil except as a contaminant from sewage, whereas *Aerobacter* is frequently encountered and is probably a normal inhabitant of certain soils (Subba Rao 1997).

Cyanobacteria are organisms that are very important in the formation of biological soil crusts (Fig.1.2). They are photosynthetic and live within the first

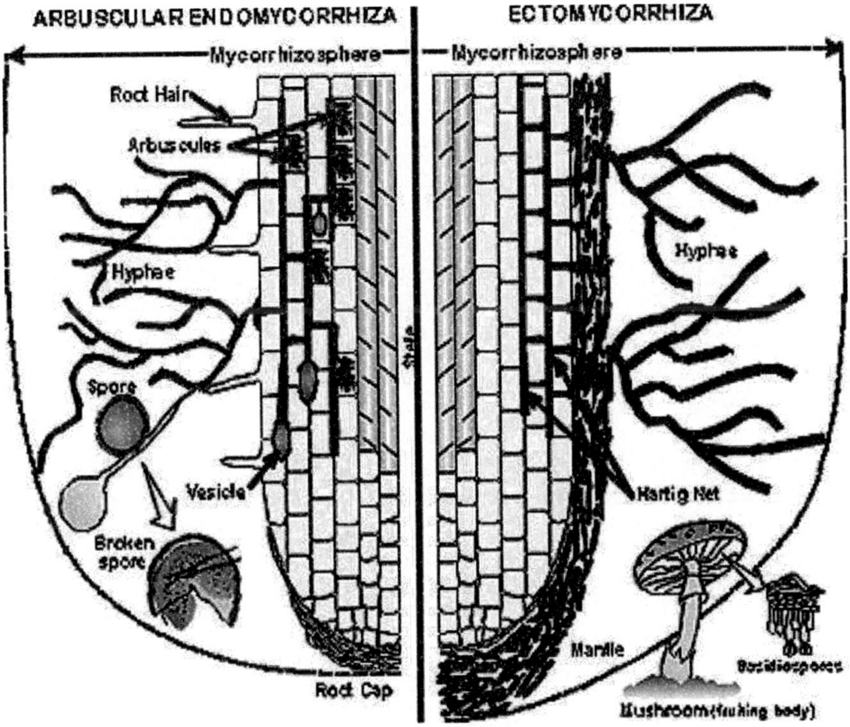

Fig. 1.1 A typical view of mycorrhizosphere (**a**) an enlarged view of the root system containing the dynamic root system embedded with nanoparticles (**b**)

Fig. 1.2 Cyanobacteria present on soil surface

ten inches of topsoil. Cyanobacteria help to reduce erosion by helping bind the particles of soil together. When the filaments of cyanobacteria become wet, they absorb water and swell up to ten times their original size which helps to store moisture within the upper layer of soil where many plants root systems and other organisms live. Cyanobacteria also play a more direct role in aiding plant survival and growth. Cyanobacteria live in the roots of some plants viz., Bryophytes (species of *Anthoceros*, *Riccia*, and *Marchantia*), Pteridophytes (*Azolla* spp) and coralloid roots in Cycas spp. (Gymnosperms). These symbioses provide dinitrogen directly for the hosts use. They have also formed symbiotic relationships with fungus (Schussler and Kluge 2001) and lichens (Kranner et al. 2002).

1.4.1.2 Fungi

Soil fungi are microscopic plant-like cells that grow in long threadlike structures or hyphae that make a mass called mycelium. The mycelium absorbs nutrients from the roots it has colonized, surface organic matter or the soil. Fungi perform important services related to water dynamics, nutrient cycling, and disease suppression. There are three functional groups of fungi found in soil:

Decomposers

Decomposers or saprophytic fungi convert dead organic matter into fungal biomass (i.e., their own bodies), carbon dioxide, and organic acids. They are essential for the decomposition of hard woody organic matter. By consuming the nutrients in the organic matter, they play an important role in immobilizing and retaining nutrients in the soil. The organic acids they produce as by products help create organic matter that is resistant to degradation.

Mutualists

These fungi develop mutually beneficial relationships with plants, for example, Mycorrhiza – *P. indica*, which colonize plant roots where they help the plant to obtain nutrients such as phosphorus from the soil. Their mass hides the roots from pests and pathogens and provides a greater root area, through which the plant can obtain nutrients.

Pathogens

This group includes the well known fungi such as *Verticillium*, *Phytophthora*, *Rhizoctonia*, and *Pythium*. These organisms penetrate the plant and decompose the living tissue, creating a weakened nutrient deficient plant or death. The pathogenic fungi are usually the dominant organism in the soil.

1.4.1.3 Algae

Algae are generally found on the surface of moist soils, where there is sufficient light for their photosynthetic reactions. The major types present are green algae Chlorophyceae, and the diatoms (Bacillariophyceae).

Microalgae are tiny plants which are active converters of solar energy, carbon dioxide, and other nutrients into sugars, proteins, and other complex organic compounds beneficial to the nutrient cycling and soil structure of croplands. Some species are able to fix atmospheric nitrogen; others form compounds conducive to the growth of associated plants. Some blue-green algae, along with a few other species (such as those of the genus Azotobacter), are rather unique in their ability to perform the nitrogen-fixing function in the presence of oxygen at the soil/ air interface, and excrete substances which further serve to glue soil particles into aggregates. This provides more aerated soil area for further growth of – these and associated beneficial microorganisms. Additionally, Anabaena has been shown to help release soil-bound phosphorus, although the exact ways in which this happens are not known at this time. Some of the common green algae occurring in most soils belong to the genera *Chlorella*, *Chlamydomonas*, *Chlrococcum*, *Oedogonium*, *Chlorochytrium*, and *Protosiphone*.

1.4.2 Population Density

The soil organisms vary in number from a few per hectare to many millions per gram of soil. The density of population is determined by food supply, moisture, temperature, physical condition, and the reaction of the soil. In neutral soils, bacteria dominate over other types of microscopic life. If the soil is acidic and rich in organic matter, fungi predominate. Algae abound on the soil in constantly moist or shady situations.

1.4.3 Bacterial Activity

The soil microflora typically produces ammonia from organic compounds when they set free more of nitrogen than they can assimilate and convert into their own protoplasm. Ammonia so released is converted into nitrites by one group of organisms called *Nitrosomonas*:

$$NH4^{+} + 11/2O_2 \rightarrow NO_2^{-} - +H_2O + 2H^{+}$$

The nitrite can then be further oxidized by *Nitrobacter* and *Nitrococcus* to yield nitrate.

$$NO_2^- - +1/2O_2 \rightarrow NO_3$$

When two genera work together, ammonia in the soil is oxidized to nitrate in a process called nitrification (Prescott's Microbiology, 3rd ed, 1996).

1.4.3.1 Energetics of Nitrification

Energy released upon the oxidation of both ammonia and nitrite is used to make ATP by oxidative phosphorylation. Since electron moves spontaneously from molecules of more negative reduction potential to acceptors with more positive reduction potential, so molecules like ammonia and nitrite which have more positive reduction potentials than NAD^+, cannot directly donate their electrons to form the required NADH and NADPH. Nitrifying bacteria solve this problem by using proton motive force or ATP to reverse the flow of electrons in their electron transport chains and reduce NAD^+ with electrons from nitrogen. Since energy is used to generate NADH as well as ATP, the net yield is fairly low (Fig. 1.3).

The nitrate-forming bacteria are generally confined to the top 25–30 cm of the soil, where the content of organic matter is also more. These organisms are most active between 25 and 38°C and under favorable conditions of tillage, aeration,

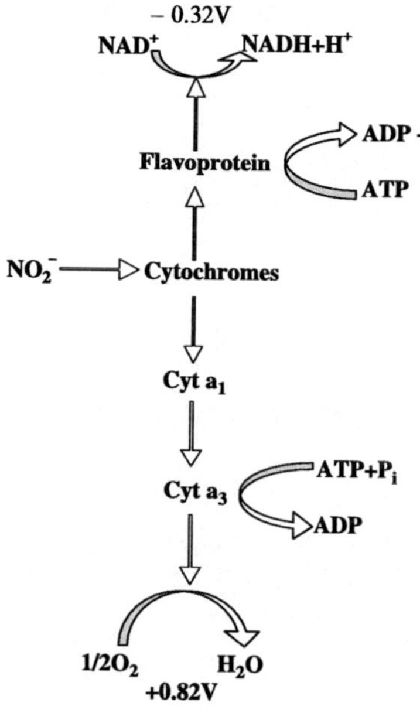

Fig. 1.3 Reverse electron flow. The flow of electrons in the transport chain of Nitrobacter (Prescott et al., 5ed)

neutral soil reaction, and moisture content at field capacity. The bacteria cease or reduce their activity when the pH value of the soil falls below 5.0.

1.5 Enzymes

Enzymes are biologically produced proteinic substances, having specific activation in which they combine with their substrates in such a stereoscopic position that they cause changes in the electronic configuration around certain susceptible bonds. Their significance in all spheres including soil is worth tested and reported. In plant nutrition, their role cannot be substituted by any other substance and their function is quite pragmatic in solubilizing and dissolving the much needed food in ionic forms for the very survival of the animal and plant kingdom.

Enzymes are the key to understanding below-ground biochemistry and the role of soil in the global carbon cycle. These biological mediators of change are active both within living soil organisms and independently as extracellular proteins that are actively secreted into the soil by roots and fungi, or released as prokaryotic and eukaryotic cells that die and decompose. These enzymes can persist in the soil for weeks while maintaining a ghostly after life activity.

1.6 Soil Enzymes

Soil enzymes play key biochemical functions in the overall process of organic matter decomposition in the soil system (Burns 1983; Sinsabaugh et al. 1991). They are important in catalyzing several important reactions necessary for the life processes of micro-organisms in soils and the stabilization of soil structure, the decomposition of organic wastes, organic matter formation, and nutrient cycling (Dick et al. 1994). These enzymes are constantly being synthesized, accumulated, inactivated, and/or decomposed in the soil, hence playing an important role in agriculture and particularly in nutrients cycling (Tabatabai 1994; Dick 1997). A better understanding of the role of these soil enzymes activities in the ecosystem will potentially provide a unique opportunity for an integrated biological assessment of soils due to their crucial role in several soil biological activities, their ease of measurement, and their rapid response to changes in soil management practices (Dick 1994, 1997; Bandick and Dick 1999).

The activity of these enzymes in soil undergoes complex biochemical processes consisting of integrated and ecologically-connected synthetic processes, and in the immobilization and enzyme stability (Khaziyev and Gulke 1991). In this regard, all soils contain a group of enzymes that determine soil metabolic processes (McLaren 1975) which, in turn, depend on its physical, chemical, microbiological, and biochemical properties.

Soil enzyme activities are often used as indices of microbial growth and activity in soils. Quantitative information concerning which soil enzymes most accurately reflect microbial growth and activity is lacking (Department of Soil and Environmental Sciences, University of California, USA).

Categories of soil enzymes include:

1. Enzymes associated with living, metabolically active cells in soil; found in cell's cytoplasm, bound to cell wall or as extracellular enzymes that have been recently produced by the cell (Fig. 1.4).
2. Enzymes associated with viable but non-proliferating cells (such as spores)
3. Enzymes that are attached to dead cells or to cell debris, or which have diffused away from dead/dying cells that originally produced them.
4. Enzymes that are "permanently" immobilized on soil clay and humic colloids. Such enzymes can remain active for long periods of time. Such immobilized soil enzymes can arise from either eukaryotic or prokaryotic cells.

Binding of enzymes to soil surfaces (especially clay and humic materials) may take place *via* Ionic interactions, Covalent bonds, Hydrogen bonding, Entrapment of enzyme by soil colloids, and other mechanisms (Fig. 1.5).

These enzymes may include amylase, arylsulphatases, glucosidase, cellulose, chitinase, dehydrogenase, phosphatase, protease, and urease released from plants (Miwa et al. 1937), animals (Kanfer et al. 1974), organic compounds and micro-organisms (Dick and Tabatabai 1984; James et al. 1991; Richmond 1991; Hans and Snivasan 1969; Shawale and Sadana 1981), and soils (Cooper 1972; Gupta et al. 1993; Ganeshamurthy et al. 1995).

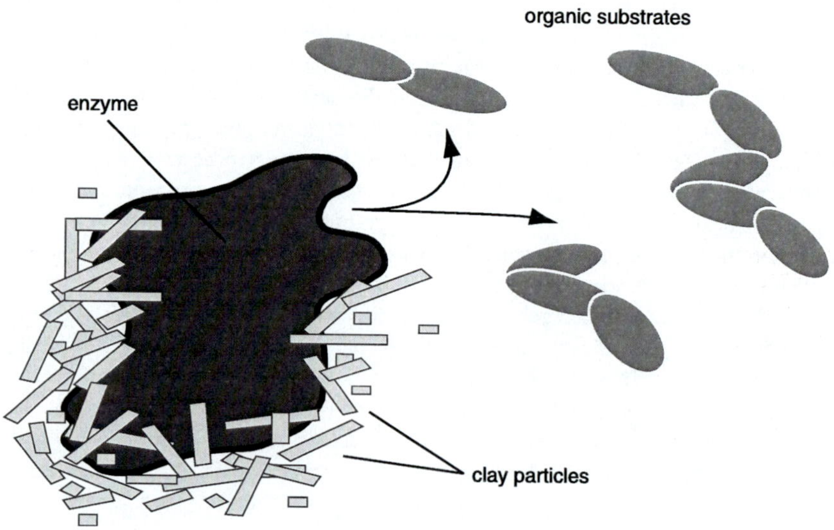

Fig. 1.4 A soil extra cellular enzyme physically protected by a clay particle sheath

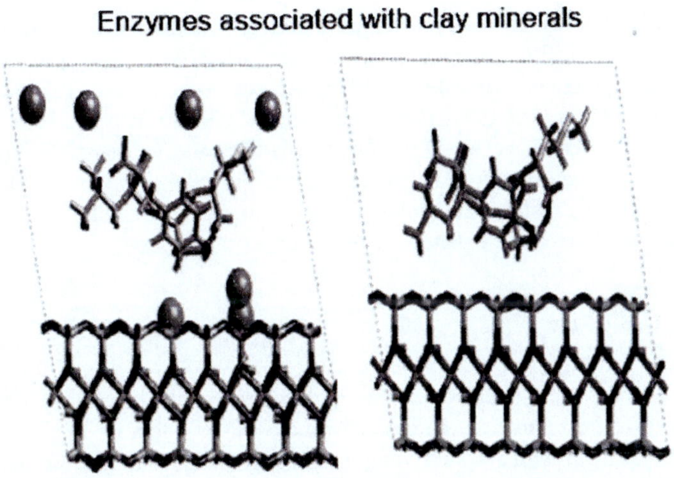

Fig. 1.5 Enzymes associated with clay minerals via different types of bond

1.6.1 Amylase

Amylase is a starch hydrolyzing enzyme (Ross 1976). It is known to be constituted by amylase and amylase (Pazur 1965; King 1967; Thoma et al. 1971). Studies have shown that amylases are synthesized by plants, animals, and micro-organisms, whereas amylase is mainly synthesized by plants (Pazur 1965; Thoma et al. 1971). This enzyme is widely distributed in plants and soils so it plays a significant role in the breakdown of starch.

The roles and activities of amylase and amylase enzymes may be influenced by different factors ranging from cultural practices, type of vegetation, environment, and soil types (Ross 1968; Ross and Roberts 1970; Pancholy and Rice 1973; Ross 1975). For example, plants may influence the amylase enzyme activities of soil by directly supplying enzymes from their residues or excreted compounds, or indirectly providing substrates for the synthetic activities of micro-organisms.

1.6.2 β-Glucosidase

β-Glucosidase is a common and predominant enzyme in soils (Eivazi and Tabatabai 1988; Tabatabai 1994). It is named according to the type of bond that it hydrolyses. This enzyme plays an important role in soils because it is involved in catalyzing the hydrolysis and biodegradation of various glucosides present in plant debris decomposing in the ecosystem (Ajwa and Tabatabai 1994; Martinez and Tabatabai 1997). Its final product is glucose, an important C energy source of life to microbes in the soil (Esen 1993).

β-Glucosidase is characteristically useful as a soil quality indicator, and may give a reflection of past biological activity, the capacity of soil to stabilize the soil organic matter, and can be used to detect management effect on soils (Bandick and Dick 1999; Ndiaye et al. 2000). β-Glucosidase enzyme is very sensitive to changes in pH, and soil management practices (Dick et al. 1996; Acosta-Martínez and Tabatabai 2000; Kuperman and Carreiro 1997; Bergstrom et al. 1998; Leirós et al. 1999; Bandick and Dick 1999; Madejón et al. 2001). Acosta-Martínez and Tabatabai (2000) reported β-glucosidase as sensitive to pH changes. β-Glucosidase enzyme is also known to be inhibited by heavy metal contamination such as Cu and several others (Haanstra and Doelman 1991; Deng and Tabatabai 1995; Wenzel et al. 1995).

1.6.3 Cellulase

Cellulose is the most abundant organic compound in the biosphere, comprising almost 50% of the biomass synthesized by photosynthetic fixation of CO_2 (Eriksson et al. 1990). The growth and survival of micro-organisms that are important in most agricultural soils depend on the carbon source contained in the cellulose occurring in the soils (Deng and Tabatabai 1994).

Cellulose in plant debris has to be degraded into glucose, cellobiose and high molecular weight oligosaccharides by cellulase enzymes (White 1982). Cellulases are a group of enzymes that catalyze the degradation of cellulose; polysaccharides build up of four linked glucose units (Deng and Tabatabai 1994). Currently, it is accepted that the cellulase system comprises three major types of enzymes. They are: endo-1,4-β-glucanase which attacks the cellulose chains at random, exo-1,4-β-glucanase which removes glucose or cellobiose from the non-reducing end of the cellulose chains, and β-D-glucosidase which hydrolyses cellobiose and other water soluble cellodextrins to glucose.

Studies have shown that activities of cellulases in agricultural soils are affected by several factors. These include temperature, soil pH, water, and oxygen contents (abiotic conditions), the chemical structure of organic matter and its location in the soil profile horizon (Rubidge 1977; Gomah 1980; Tabatabai 1982; Klein 1989; Deng and Tabatabai 1994; Alf and Nannipieri 1995), quality of organic matter/plant debris and soil mineral elements (Burns 1978; Hope and Burns 1987; Klein 1989; Sinsabaugh and Linkins 1989; Deng and Tabatabai 1994) and the trace elements from fungicides (Deng and Tabatabai 1994; Petker and Rai 1992; Arinze and Yubedee 2000; Atlas et al. 1978; Vincent and Sisler 1968).

Since cellulases play an important role in global recycling of the most abundant polymer, cellulose in nature, it would be of critical importance to understand this enzyme better so that it may be used more regularly as a predictive tool in our soil fertility programmes.

1.6.4 Laccase

Laccases (EC. 1.10.3.2 *p*-benzenediol: oxygen oxidoreductase) belong to a family of multi-copper oxidases. Laccases are widely distributed enzymes in higher plants, fungi, some insects, and bacteria. They are characterized by low substrate specificity, oxidizing various substrates, including diphenols, polyphenols, different substituted phenols, diamines, aromatic amines, and even inorganic compounds like iodine. Laccases oxidize their substrates by a one-electron oxidation mechanism, and they use molecular oxygen as an electron acceptor. The enzyme having pH optimum within 5–8 and working at temperatures between 30 and 80°C, is well suited to industrial applications requiring high pH and temperature conditions whereas the majority of known fungal laccases function in an acidic pH range and are not very thermostable.

The exo-oxidative enzyme laccase has been detected in a large number of basidiomycete ectomycorrhizal fungi, in a few ectomycorrhizal ascomycetes and only in one endomycorrhizal species. Most fungi tested showed a strong laccase activity. *Piriformospora indica* and *Sebacina vermifera* ss. Warcup and Talbot also showed a positive reaction to the ABTS test i.e., presence of laccase activity (Pham et al. 2004).

1.6.5 Chitinase

Chitinase or chitinolytic enzymes are key enzymes responsible for the degradation and hydrolysis of chitin (poly β-1-4-(2-ncetamido-2-deoxy)-D-glucoside). They are also considered as the major structural component of many fungal cell walls that use the hyper parasitism mechanisms against pests/pathogen attack, (Bartinicki-Garcia 1968; Chet and Henis 1969; Chet and Henis 1975; Chet 1987). For example, in plants, the chitinase enzyme is induced and accumulated in response to microbial infections and it is thought to be involved in the defense of plants against pathogen infections (Boiler et al. 1983; Boiler 1985).

As biological control of most pathogenic diseases is increasingly gaining popularity in recent times due to their environmental friendliness, a better understanding of the chitinolytic enzymes is likely to uncover more application avenues for this enzyme in agricultural systems and, consequently, increase plant growth and final yields.

1.6.6 Dehydrogenase

The dehydrogenase enzyme activity is commonly used as an indicator of biological activity in soils (Burns 1978).This enzyme is considered to exist as an integral part of intact cells but does not accumulate extracellularly in the soil. Dehydrogenase enzyme is known to oxidize soil organic matter by transferring protons and electrons from substrates to acceptors. These processes are part of respiration pathways of soil microorganisms and are closely related to the type of soil and soil air-water conditions (Doelman and Haanstra 1979; Kandeler 1996; Glinski and Stepniewski 1985).

Studies on the activity of dehydrogenase enzyme in the soil are very important as it may give indications of the potential of the soil to support biochemical processes which are essential for maintaining soil fertility. A study by Brzezinska et al. (1998) suggested that soil water content and temperature influence dehydrogenase activity indirectly by affecting the soil redox status. The relationship between dehydrogenase activity and redox potential (Eh) as well as Fe^{2+} content may also be used to illustrate the reactions of soil microorganisms to the changes in soil environment. Additionally, dehydrogenase enzyme is often used as a measure of any disruption caused by pesticides, trace elements, or management practices to the soil (Reddy and Faza 1989; Wilke 1991; Frank and Malkomes 1993), as well as a direct measure of soil microbial activity (Skujins 1978; Trevors 1984; Garcia and Hernández 1997). It can also indicate the type and significance of pollution in soils. For example, dehydrogenase enzyme is high in soils polluted with pulp and paper mill effluents (McCarthy et al. 1994) but low in soils polluted with fly ash (Pitchel and Hayes 1990).

1.6.7 Phosphatase

Phosphatases are a broad group of enzymes that are capable of catalyzing hydrolysis of esters and anhydrides of phosphoric acid (Schmidt and Laskowski 1961). Apart from being good indicators of soil fertility, phosphatase enzymes play key roles in the soil system (Dick and Tabatai 1992; Dick et al. 2000). In fungi, these phosphatases may be located in the periplasmic space, cell wall, vacuoles, and culture medium. Acid and alkaline phosphatases are the two forms of active phosphatase. Alkaline phosphatase, occurs in roots mainly after mycorrhizal colonization and has been proposed as a marker for analyzing the symbiotic efficiency of root colonization (Tisserant et al. 1993).A newly discovered fungus *Piriformospora indica* is known to produce one form of intracellular ACPase irrespective of the phosphate concentration. The enzyme is possibly a constitutive enzyme showing a molecular mass of 66 kDa (Malla et al. 2007).

In soil ecosystems, these enzymes are believed to play critical roles in P cycles (Speir and Ross 1978) as evidence shows that they are correlated to P stress and plant growth.

The amount of acid phosphatase exuded by plant roots has been shown to differ between crop species and varieties, (Ndakidemi 2006; Izaguirre-Mayoral et al. 2002) as well as crop management practices (Ndakidemi 2006; Patra et al. 1990; Staddon et al. 1998; Wright and Reddy 2001). For instance, research has shown that legumes secrete more phosphatase enzymes than cereals (Yadav and Tarafdar 2001). This may probably be due to a higher requirement of P by legumes in the symbiotic nitrogen fixation process as compared to cereals. In their studies, Li et al. (2004) reported that chickpea roots were also able to secrete greater amounts of acid phosphatase than maize.

1.6.8 Protease

Proteases in soil play a significant role in N mineralization (Ladd and Jackson 1982), an important process regulating the amount of plant available N (Stevenson 1986) and plant growth. This enzyme in the soil is generally associated with inorganic and organic colloids (Burns 1982; Nannipieri et al. 1996). Protease activities have been reported to occur partly in soil as a humocarbohydrate complex (Mayaudon et al. 1975; Batistic et al. 1980).

The amount of this extracellular enzyme activity may be indicative not only of the biological capacity of soil for the enzymatic conversion of the substrate, which is independent of the extent of microbial activity, but might also have an important role in the ecology of micro-organisms in the ecosystem (Burns 1982).

1.6.9 Urease

Urease enzyme is responsible for the hydrolysis of urea fertilizer applied to the soil into NH_3 and CO_2 with the concomitant rise in soil pH (Andrews et al. 1989; Byrnes and Amberger 1989). Due to this role, urease activities in soils have received a lot of attention since it was first reported by Rotini (1935), a process considered vital in the regulation of N supply to plants after urea fertilization. Urease extracted from plants or micro-organisms is rapidly degraded in soil by proteolytic enzymes (Burns et al. 1972; Pettit et al. 1976; Zantua and Bremner 1977). Urease activity in soils is influenced by many factors. These include cropping history, organic matter content of the soil, soil depth, soil amendments, heavy metals, and environmental factors such as temperatures (Tabatabai 1977; Bremner and Mulvaney 1978; Yang et al. 2006). Generally, urease activity increases with increasing temperature. It is suggested that higher temperatures increase the activity coefficient of this enzyme. Therefore, it is recommended that urea be applied at times of the day when temperatures are low. This is because during such times the activation energy is low, thus, resulting in minimum loss of N by the volatilization process. A better understanding of this enzyme would provide more effective ways of managing urea fertilizer especially in high rainfall areas, flooded soils, and irrigated lands as well as where urea fertilizer is vulnerable to urease enzyme.

1.6.10 Arylsulphatase

Sulfur uptake in plants is in the form of inorganic sulfate (SO_4) and its availability depends on its mineralization or mobilization (Williams 1975; Fitzgerald 1976) from aromatic sulfate esters ($RO–SO_3$). This is due to the fact that certain proportions of sulfur in different soil profiles are bound into organic compounds and are indirectly available to plants.

In this regard, its availability will depend on the extracellular hydrolysis of these aromatic sulphate esters or intracellular oxidation of soluble organic matter absorbed by the micro-organisms to yield energy and carbon skeletons for biosynthesis by which some SO_4–S are released as a by-product (Dodgson et al. 1982).

1.7 Conclusion

Physical, chemical, and biological properties of a soil are indicators of soil quality, while soil fertility can be determined by its biological activity. Soil provides natural environment for the survival of microorganism and they need favorable physical and chemical conditions for their optimal function. An imbalance of soil microorganisms, nutrient deficiency, and change in physicochemical properties, like a decrease in pH, can result in decreased soil fertility and crop productivity.

Understanding other possible roles of soil enzymes is vital to soil health and fertility management in ecosystems. These enzymes may have significant effects on soil biology, environmental management, growth, and nutrient uptake in plants growing in ecosystems.

References

Acosta-Martínez V, Tabatabai MA (2000) Enzyme activities in a limed agricultural soil. Biol Fertil Soils 31:85–91

Ajwa HA, Tabatabai MA (1994) Decomposition of different organic materials in soils. Biol Fertil Soils 18:175–182

Alf K, Nannipieri P (1995) Cellulase activity, Methods in Applied Soil Microbiology and Biochemistry. Academic Press, London

Andrews RK, Blakeley RL, Zerner B (1989) Urease: a Ni (II) metalloenzyme. In: Lancaster JR (ed) The bioinorganic chemistry of nickel. VCH Publishers, New York, pp 141–166

Arinze AE, Yubedee AG (2000) Effect of fungicides on *Fusarium* grain rot and enzyme production in maize (*Zea mays* L.). Glob J Appl Sci 6:629–634

Atlas RM, Pramer D, Bartha R (1978) Assessment of pesticide effects on non-target soil microorganisms. Soil Biol Biochem 10:231–239

Bandick AK, Dick RP (1999) Field management effects on soil enzyme activities. Soil Biol Biochem 31:1471–1479

Bartinicki-Garcia S (1968) Cell wall chemistry, morphogenesis and taxonomy of fungi. Annu Rev Microbiol 144:346–349

Batistic L, Sarkar JM, Mayaudon J (1980) Extraction, purification and properties of soil hydrolases. Soil Biol Biochem 12:59–63

Bergstrom DW, Monreal CM, King DJ (1998) Sensitivity of soil enzyme activities to conservation practices. Soil Sci Soc Am J 62:1286–1295

Boiler T (1985) Induction of hydrolases as a defense reaction against pathogens. In: Key JL, Kosuge T (eds) Cellular and molecular biology of plum stress. Liss, New York, pp 247–262

Boiler T, Gehri A, Mauch F, Vogeli U (1983) Chitinase in bean leaves: induction by ethylene, purification, properties and possible function. Planta 157:22–31

Borneman J, Skroach PW, O'Sullivian EW, Palus JA, Rumjanek NG, Jansen JL, Nienhuis J, Triplett EW (1996) Molecular microbial diversity of an agricultural soil in Wisconsin. Appl Environ Microbiol 62:1935–1943

Bremner JM, Mulvaney RL (1978) Urease activity in soils. In: Bums RG (ed) Soil enzymes. Academic, London, pp 149–196

Bruck TD (1987) The study of microorganisms in situ: progress and problems. In: Fletcher M, Gray TRG, Jones JG (eds) Ecology of microbial communities. SGM symposium 41. Cambridge University Press, Cambridge, pp 455–493

Brzezinska M, Stepniewska Z, Stepniewski W (1998) Soil oxygen status and dehydrogenase activity. Soil Biol Biochem 30(13):1783–1790

Burns RG (1978) Enzyme activity in soil: some theoretical and practical considerations. In: Bums RG (ed) Soil enzymes. Academic, London, pp 295–340

Burns RG (1982) Enzyme activity in soil: location and possible role in microbial ecology. Soil Biol Biochem 14:423–427

Burns RG (1983) Extra cellular enzyme–substrate interactions in soil. In: Slater JH, Wittenbury R, Wimpenny JWT (eds) Microbes in their natural environment. Cambridge University Press, London, pp 249–298

Burns RG, Pukite AH, McLaren AD (1972) Concerning the location and persistence of soil urease. Soil Sci Soc Am Proc 36:308–311

Byrnes BH, Amberger A (1989) Fate of broadcast urea in a flooded soil when treated with N-(n-butyl) thiophospheric triamide, a urease inhibitor. Fertil Res 18:221–231

Chet I (1987) *Trichoderma*-application, mode of action, and potential as biocontrol agent of soil borne pathogenic fungi. In: Chet I (ed) Innovative approaches to plant disease control. Wiley, New York, pp 137–349

Chet I, Henis Y (1969) Effect of catechol and disodium EDTA on melanin content of hyphal and sclerotial walls of *Sclerotium rolfsii* Sacc. and the role of melanin in the susceptibility of these walls to _-l-3 glucanase and chitinase. Soil Biol Biochem 1:131–138

Chet I, Henis Y (1975) Sclerotial morphogenesis in fungi. Annu Rev Phytopathol 13:169–192

Cooper PJM (1972) Arylsulphatase activity in Northern Nigerian soils. Soil Biol Biochem 4:333–337

Deng SP, Tabatabai MA (1994) Cellulase activity of soils. Soil Biol Biochem 26:1347–1354

Deng SP, Tabatabai MA (1995) Cellulase activity of soils: effect of trace elements. Soil Biol Biochem 27:977–979

Dick RP (1994) Soil enzyme activities as indicators of soil quality. In: Doran JV, Coleman DC, Bezdicek DF, Stewart BA (eds) Defining soil quality for a sustainable environment, soil science society of America. American Society of Agriculture, Madison, pp 107–124

Dick RP (1997) Soil enzyme activities as integrative indicators of soil health. In: Pankhurst CE, Doube BM, Gupta VVSR (eds) Biological indicators of soil health. CAB International, Wellingford, pp 121–156

Dick WA, Tabatabai MA (1984) Kinetic parameters of phosphatase in soils and organic waste materials. Soil Sci 137:7–15

Dick WA, Tabatai MA (1992) Potential uses of soil enzymes. In: Metting FB Jr (ed) Soil microbial ecology: applications in agricultural and environmental management. Marcel Dekker, New York, pp 95–127

Dick RP, Sandor JA, Eash NS (1994) Soil enzyme activities after 1500 years of terrace agriculture in the Colca Valley, Peru. Agric Ecosyst Environ 50:123–131

Dick RP, Breakwell DP, Turco RF (1996) Soil enzyme activities and biodiversity measurements as integrative microbiological indicators. In: Methods for assessing soil quality. Soil Science Society of America. Madison, WI, pp 9–17

Dick WA, Cheng L, Wang P (2000) Soil acid and alkaline phosphatase activity as pH adjustment indicators. Soil Biol Biochem 32:1915–1919

Dodgson KS, White G, Fitzgerald JW (1982) Sulphatase enzyme of microbial origin, vol I. CRC, Boca Raton, FL

Doelman P, Haanstra L (1979) Effect of lead on soil respiration and dehydrogenase activity. Soil Biol Biochem 11:475–479

Eivazi F, Tabatabai MA (1988) Glucosidases and galactosidases in soils. Soil Biol Biochem 20:601–606

Eriksson KEL, Blancbette RA, Ander P (1990) Biodegration of cellulose. In: Eriksson KEL, Blanchette RA, Ander P (eds) Microbial and enzymatic degradation of wood and wood components. Springer, New York, pp 89–180

Esen A (1993) β-Glucosidases-biochemistry and molecular biology, ACS symposium series, 533. American Chemical Society, Washington, DC

Foster RC (1988) Micro environment of Soil Microorganism. Biology and Fertility of Soils 6:189–203

Fitzgerald JW (1976) Sulphate ester formation and hydrolysis: a potentially important yet often ignored aspect of the sulphur cycle of aerobic soils. Bacteriol Rev 40:628–721

Frank T, Malkomes HP (1993) Influence of temperature on microbial activities and their reaction to the herbicide Goltix in different soils under laboratory conditions. Zentralblatt für Mikrobiol 148:403–412

Ganeshamurthy AM, Singh G, Singh NT (1995) Sulphur status and response of rice to sulphur on some soils of Andaman and Nicobar Islands. J Indian Soc Soil Sci 43:637–641

Garcia C, Hernández T (1997) Biological and biochemical indicators in derelict soils subject to erosion. Soil Biol Biochem 29:171–177

Giri V, Anuradha, Nandhini A, Geetha M, Gautham P (2004) A novel medium for the enhanced cell growth and production of prodigiosin from *Serratia marcescens* isolated from soil. BMC Microbiol 4:1–10

Giri B, Pham HG, Rina K, Ram P, Minu S, Garg A, Oelmuller R, Varma A (2005) Mycorrhizosphere: strategies and functions. In: Buscot F, Varma A (eds) Microorganisms in soils: roles in genesis and function, 3. Springer, Heidelberg, pp 213–252

Glinski J, Stepniewski W (1985) Soil aeration and its role for plants. CRC, Boca Raton, FL

Gomah AM (1980) CM-cellulase activity in soil as affected by addition of organic material, temperature, storage and drying and wetting cycles. Zeitschrift fuer Pflanzenernaehrung und Bodenkunde 143:349–356

Gupta VVSR, Farrell RE, Germida JJ (1993) Activity of arylsuphatases in Saskatchewan soils. Can J Soil Sci 73:341–347

Haanstra L, Doelman P (1991) An ecological dose–response model approach to short- and long-term effects of heavy metals on arylsulphatase activity in soil. Biol Fertil Soils 1:18–23

Hans YW, Snivasan VR (1969) Purification and characterization of β-glucosidases of *Alcaligenes faecalis*. J Bacteriol 100:1355–1363

Hope CFA, Burns RG (1987) Activity, origins and location of cellulases in a silt loam soil. Biol Fert Soils 5:164–170

Izaguirre-Mayoral ML, Flores S, Carballo O (2002) Determination of acid phosphatases and dehydrogenase activities in the rhizosphere of nodulated legume species native to two contrasting savannah sites in Venezuela. Biol Fertil Soils 35:470–472

James ES, Russel LW, Mitrick A (1991) Phosphate stress response in hydroponically grown maize. Plant Soil 132:85–90

Kandeler E (1996) Nitrate. In: Schinner F, Öhlinger R, Kandeler E, Margesin R (eds) Methods in soil biology. Springer, Berlin, pp 408–410

Kanfer JN, Mumford RA, Raghavan SS, Byrd J (1974) Purification of β-glucosidase activities from bovine spleen affinity chromatography. Anal Biochem 60:200–205

Khaziyev FK, Gulke AY (1991) Enzymatic activity of soils under agrocenosesa: status and problems. Pochvovedenie 8:88–103

King NJ (1967) Glucoamylase of *Coniophora cerebella*. Biochem J 105:577–583

Klein DA (1989) Cellulose functions in arid soil development. Arid Soil Res Rehab 3(1):85–198

Kranner I, Beckett RP, Varma AK (2002) Protocols in lichenology, culturing, biochemistry, ecophysiology and use in biomonitoring. Springer Lab Manual. Springer, Heidelberg

Kuperman RG, Carreiro MM (1997) Soil heavy metal concentrations, microbial biomass and enzyme activities in a contaminated grassland ecosystem. Soil Biol Biochem 29:179–190
Ladd JN, Jackson RB (1982) In: Stevenson FJ (Ed) Nitrogen in agricultural soils. American Society of Agronomy, WI, pp 173–228
Leirós MC, Trasar-Cepeda C, Garcíá-Fernández F, Gil-Sotrés F (1999) Defining the validity of a biochemical index of soil quality. Biol Fertil Soils 30:140–146
Liesack W, Stackebrandt E (1992) Occurrence of novel groups of the domain bacteria as revealed by analysis of genetic material isolated from an Australian Terrestrial Environment. J Bacteriol 174:5072–5078
Li Y, Guohua M, Fanjun C, Jianhua Z, Fusuo Z (2004) Rhizosphere effect and root growth of two maize (*Zea mays* L.) genotypes with contrasting P efficiency at low P availability. Plant Sci 167:217–223
Loper JE, Hack C, Schroth MN (1985) Population dynamics of soil Pseudomonads in the rhizosphere of potato. Appl Environ Microbiol 60:2394–2399
Lynch JM (1987a) Microbial interactions in the rhizosphere. Soil Microorg 30:33–41
Lynch JM (1987b) Soil biology-accomplishments and potential. Soil Sci Soc Am J 51:1409–1412
Madejón E, Burgos P, López R, Cabrera F (2001) Soil enzymatic response to addition of heavy metals with organic residues. Biol Fertil Soils 34:144–150
Malla R, Tanaka Y, Mori K and Totawat KL (2007) Effect of Short-term Sewage Irrigation on Chemical Build Up in Soils and Vegetables. The Agricultural Engineering International: The CIGR Ejournal. Manuscript LW 07 006 Vol. IX
Martinez CE, Tabatabai MA (1997) Decomposition of biotechnology byproducts in soils. J Environ Qual 26:625–632
Mayaudon J, Batistic L, Sarkar JM (1975) Proprietés des activités proteolitiques extraites des sols frais. Soil Biol Biochem 7:281–286
McCarthy GW, Siddaramappa R, Reight RJ, Coddling EE, Gao G (1994) Evaluation of coal combustion byproducts as soil liming materials: their influence on soil pH and enzyme activities. Biol Fertil Soils 17:167–172
McLaren AD (1975) Soil as a system of humus and clay immobilized enzymes. Chem Scripta 8:97–99
Miwa T, Ceng CT, Fujisaki M, Toishi A (1937) Zur Frage der Spezifitat der Glykosidasen. I. Verhalted von β-d-glucosidases verschiedener Herkunft gegenuberden β-d-Glucosiden mit verschiedenen Aglykonen. Acta Phytochim (Tokyo) 10:155–170
Nannipieri P, Sequi P, Fusi P (1996) Humus and enzyme activity. In: Piccolo A (ed) Humic substances in terrestrial ecosystems. Elsevier, New York, pp 293–328
Ndakidemi PA (2006) Manipulating legume/cereal mixtures to optimize the above and below ground interactions in the traditional African cropping systems. Afr J Biotechnol 5:2526–2533
Ndiaye EL, Sandeno JM, McGrath D, Dick RP (2000) Integrative biological indicators for detecting change in soil quality. Am J Altern Agric 15:26–36
Paul EP, Clark FE (1989) Soil Microbiology and Biochemistry. Academic Press, Sandiego
Pancholy SK, Rice EL (1973) Soil enzymes in relation to old field succession; amylase, cellulose, invertase, dehydrogenase and urease. Soil Sci Soc Am J 37:47–50
Patra DD, Brookes PC, Coleman K, Jenkinson DS (1990) Seasonal changes of soil microbial biomass in an arable and a grassland soil which have been under uniform management for many years. Soil Biol Biochem 22:739–742
Pazur JH (1965) Enzymes in the synthesis and hydrolysis of starch. In: Whistler R, Paschall EF (eds) Starch: chemistry and technology, vol 1, Fundamental aspects. Academic, New York, pp 133–175
Petker AS, Rai PK (1992) Effect of fungicides on activity, secretion of some extra cellular enzymes and growth of *Alternaria alternata*. Indian J Appl Pure Biol 7:57–59
Pettit NM, Smith ARJ, Freedman RB, Burns RG (1976) Soil urease: activity, stability and kinetic properties. Soil Biol Biochem 8:479–484

Pham GH, Kumari R, Singh A, Sachdeva M, Prasad R, Kaldorf M, Buscot F, Oelmuller R, Tatjana P, Weiss M, Hampp R, Varma A (2004) Axenic culture of *Piriformospora indica*. In: Plant Surface Microbiology (eds A Varma, L Abbott, D Werner and R Hampp). Springer-Verlag, Germany 2004:593–616

Pitchel JR, Hayes JM (1990) Influence of fly ash on soil microbial activity and populations. J Environ Qual 19:593–597

Reddy GB, Faza A (1989) Dehydrogenase activity in sludge amended soil. Soil Biol Biochem 21:327

Richmond PA (1991) Occurrence and functions of native cellulose. In: Haigler CH, Weimer PJ (eds) Biosynthesis and biodegradation of cellulose. Dekker, New York, pp 5–23

Ross DJ (1968) Activities of enzymes hydrolysing sucrose and starch in some grassland soils. Trans 9th International Congress Soil-Science 3:299–308

Ross DJ (1975) Studies on a climosequence of soils in tussock grasslands-5. Invertase and amylase activities of topsoils and their relationships with other properties. NZ J Sci 18:511–518

Ross DJ (1976) Invertase and amylase activities in ryegrass and white clover plants and their relationships with activities in soils under pasture. Soil Biol Biochem 8:351–356

Ross DJ, Roberts HS (1970) Enzyme activities and oxygen uptakes of soils under pasture in temperature and rainfall sequences. J Soil Sci 21:368–381

Rotini OT (1935) La trasformazione enzimatica dell'urea nel terreno. Ann Labor Ric Ferm Spallanrani 3:143–154

Rubidge T (1977) The effect of moisture content and incubation temperature upon the potential cellulase activity of John Innes no. 1 soil (ISSN. 0020-6164). Int Biodeterior Bul 13:39–44

Schmidt G, Laskowski M Sr (1961) Phosphate ester cleavage (Survey). In: Boyer PD, Lardy H, Myrback K (eds) The enzymes, 2nd edn. Academic, New York, pp 3–35

Schussler A, Kluge M (2001) Geosiphon pyriforme, an endosymbiosis between fungus and cyanobacteria and its meaning as a model system for arbuscular mycorrhizal research. In: Esser K, Hock B (eds) The Mycota, 9. Springer, Heidelberg, pp 151–161

Shawale JG, Sadana J (1981) Purification, characterization and properties of β-glucosidase enzyme from *Sclerotium rolfsii*. Arch Biochem Biophys 207:185–196

Sinsabaugh RL, Antibus RK, Linkins AE (1991) An enzymic approach to the analysis of microbial activity during plant litter decomposition. Agric Ecosyst Environ 34:43–54

Sinsabaugh RL, Linkins AE (1989) Natural disturbance and the activity of Trichoderma viride cellulase complex. Soil Biol Biochem 21:835–839

Skujins J (1978) Soil enzymology and fertility index – a fallacy? History of abiotic soil enzyme research. In: Burns RG (ed) Soil enzymes. Academic, London, UK

Speir TW, Ross DJ (1978) Soil phosphatase and sulphatase. In: Burns RG (ed) Soil enzymes. Academic, London, UK, pp 197–250

Staddon WJ, Duchesne LC, Trevors JT (1998) Impact of clear-cutting and prescribed burning on microbial diversity and community structure in a Jack pine (*Pinus banksiana* Lamb.) clear-cut using BiOLOG gram-negative microplates. World J Microbiol Technol 14:119–123

Stevenson FJ (1986) Cycles of Soil-Carbon, Nitrogen, Phosphorus, Sulfur, Micronutrients; Wiley InterScience Publ, John Wiley & Sons, New York

Subba Rao NS (1997) Soil microbiology. IBH Pub, Oxford

Tabatabai MA (1977) Effect of trace elements on urease activity in soils. Soil Biol Biochem 9:9–13

Tabatabai MA (1982) Soil enzyme. In: Page AL (ed) Methods of soil analysis, part 2. American Society of Agronomy, Madison, WI, pp 903–948

Tabatabai MA (1994) Soil enzymes. In: Weaver RW, Angle JS, Bottomley PS (eds) Methods of soil analysis, part 2. Microbiological and biochemical properties, SSSA book series no. 5. Soil Science Society America, Madison, WI, pp 775–833

Thoma JA, Spradlin JE, Dygert S (1971) Plant and animal amylases. In: Boyer PD (ed) The enzymes. International Society of Soil-Science, Netherlands 5:115–189

Tisserant B, Gianinazzi-Pearson V, Gianinazzi S, Gollotte A (1993) In planta histochemical staining of fungal alkaline phosphatase activity for analysis of efficient arbuscular mycorrhizal infections. Mycol Res 97:245–250

Trevors JT (1984) Dehydrogenase activity in soil: a comparison between the INT and TTC assay. Soil Biol Biochem 16:673–674

Vincent PG, Sisler HD (1968) Mechanisms of antifungal action of 2,4,5,6-tetrachloroisopathalonitrile. Physiol Plant 21:1249–1264

Visscher PT, Vandenede FP, Schaub BEM, van Gemerden H (1992) Competition between anoxygenic phototropic bacteria and colorless sulfur bacteria in microbial mat. FEMS Microbiol Ecol 101:51–58

Wenzel WW, Pollak MA, Riedler C, Zischka RR, Blum WEH (1995) Influence of site conditions and heavy metals on enzyme activities of forest topsoil. In: Huang PM, Berthelin J, Bollag JM, McGill WN, Page AL (eds) Environmental impact of soil component interactions-metals, other inorganics, and microbial activities. CRC, Baton Rouge, LA, pp 211–225

White AR (1982) Visualization of cellulases and cellulose degradation. In: Brown RM Jr (ed) Cellulose and other natural polymer systems: biogenesis, structure, and degradation. Plenum, New York, pp 489–509

Wieland G, Neumann R, Backhaus H (2001) Variation of microbial communities in soil, rhizosphere, and rhizoplane in response to crop species, soil type, and crop development. Appl Environ Microbiol 67:5849–5854

Wilke BM (1991) Effect of single and successive additions of cadmium, nickel and zinc on carbon dioxide evolution and dehydrogenase activity in a sandy Luvisol. Biol Fertil Soils 11:34–37

Williams CH (1975) The chemical nature of sulphur compounds in soil. In: McLachlan KD (ed) Sulphur in Australasian agriculture. Sydney University Press, NSW, pp 21–30

Wright AL, Reddy KR (2001) Phosphorus loading effects on extracellular enzyme activity in Everglades wetland soil. Soil Sci Soc Am J 65:588–595

Yadav RS, Tarafdar JC (2001) Influence of organic and inorganic phosphorous supply on the maximum secretion of acid phosphatase by plants. Biol Fertil Soils 34:140–143

Yang Z, Liu S, Zheng D, Feng S (2006) Effects of cadium, zinc and lead on soil enzyme activities. J Environ Sci 18:1135–1141

Zantua MI, Bremner JM (1977) Stability of urease in soils. Soil Biol Biochem 9:135–140

Chapter 2
Role of Enzymes in Maintaining Soil Health

Shonkor Kumar Das and Ajit Varma

2.1 Introduction

Enzymes are the vital activators in life processes, likewise in the soil they are known to play a substantial role in maintaining soil health and its environment. The enzymatic activity in the soil is mainly of microbial origin, being derived from intracellular, cell-associated or free enzymes. A unique balance of chemical, physical, and biological (including microbial especially enzyme activities) components contribute to maintaining soil health. Evaluation of soil health therefore requires indicators of all these components. Healthy soils are essential for the integrity of terrestrial ecosystems to remain intact or to recover from disturbances, such as drought, climate change, pest infestation, pollution, and human exploitation including agriculture (Ellert et al. 1997). Deterioration of soil, and thereby soil health, is of concern for human, animal, and plant health because air, groundwater, and surface water consumed by humans, can be adversely affected by mismanaged and contaminated soil (Singer and Ewing 2000). As soil is the part of the terrestrial environment and supports all terrestrial life forms, protection of soil is therefore of high priority and a thorough understanding of soil enzymes activities is a critical factor in assuring that soil remains healthy. A better understanding of the role of this soil enzymes activity in maintaining the soil health will potentially provide a unique opportunity for an integrated biological assessment of soils due to their crucial role in several soil biological activities, their ease of measurement, and their rapid response to changes in soil management. Although there have been extensive studies on soil enzymes, little has been reported on their roles in maintaining soil health. Thus, it is authoritative to

S.K. Das (✉)
Department of Applied Chemistry and Biotechnology, Graduate School of Engineering, University of Fukui, 3-9-1, Bunkyo, Fukui 910-8507, Japan
e-mail: skdas76@yahoo.com

A. Varma
Amity Institute of Microbial Technology, Amity University, Sector 125, Noida 201303, Uttar Pradesh, India

G. Shukla and A. Varma (eds.), *Soil Enzymology*, Soil Biology 22,
DOI 10.1007/978-3-642-14225-3_2, © Springer-Verlag Berlin Heidelberg 2011

understand the roles of these enzymes' activity and their efficiency to maintain soil health for future betterment of soil research and soil biology.

2.2 Soil Enzymes

Soil enzymes are a group of enzymes whose usual inhabitants are the soil and are continuously playing an important role in maintaining soil ecology, physical and chemical properties, fertility, and soil health. These enzymes play key biochemical functions in the overall process of organic matter decomposition in the soil system (Sinsabaugh et al. 1991). They are important in catalyzing several vital reactions necessary for the life processes of micro-organisms in soils and the stabilization of soil structure, the decomposition of organic wastes, organic matter formation, and nutrient cycling, hence playing an important role in agriculture (Dick et al. 1994; Dick 1997).

All soils contain a group of enzymes that determine soil metabolic processes (McLaren 1975) which, in turn, depend on its physical, chemical, microbiological, and biochemical properties. The enzyme levels in soil systems vary in amounts primarily due to the fact that each soil type has different amounts of organic matter content, composition, and activity of its living organisms and intensity of biological processes. In practice, the biochemical reactions are brought about largely through the catalytic contribution of enzymes and variable substrates that serve as energy sources for microorganisms (Kiss et al. 1978). These enzymes may include amylase, arylsulphatases, β-glucosidase, cellulose, chitinase, dehydrogenase, phosphatase, protease, and urease released from plants (Miwa et al. 1937), animals (Kanfer et al. 1974), organic compounds, and microorganisms (James et al. 1991; Richmond 1991; Shawale and Sadana 1981) and soils (Gupta et al. 1993; Ganeshamurthy et al. 1995).

2.2.1 Kind of Soil Enzymes

2.2.1.1 Constitutive

Always present in nearly constant amounts in a cell (not affected by addition of any particular substrate – genes always expressed).

(Pyrophosphatase)

2.2.1.2 Inducible

Present only in trace amounts or not at all, but quickly increases in concentration when its substrate is present.

(Amidase)

Both types of enzymes are present in the soil.

2.2.2 *Origin and State of Soil Enzymes*

Although the general origins of soil enzymes are (a) microorganisms-living and dead, (b) plant roots and plant residues and, (c) soil animals; the state of soil enzymes in the soil is different as below

State-1: Role of clays

Most activity associated with clays
Increased resistance to proteolysis and microbial attack
Increases the temperature of inactivation

State-2: Role of organic matter

Humus material provides stability to soil nitrogen compounds
Enzymes attached to insoluble organic matrices exhibit pH and temperature changes
Inability to purify soil enzymes free of soil organic matters (bound to organic matter)

State-3: Role of clay–organic matter complexes

Lignin + bentonite (clay) protect enzymes against proteolitic attack, but not bentonite alone
Enzymes are bound to organic matter which is then bound to clay

2.2.3 *Importance of Soil Enzymes*

Release of nutrients into the soil by means of organic matter degradation
Identification of soils
Identification of microbial activity
Importance of soil enzymes as sensitive indicators of ecological change

2.2.4 *Application of Soil Enzymes*

Correlation with soil fertility
Correlation with microbial activity
Correlation with biochemical cycling of various elements in soil (C, N, S)
Degree of pollution (heavy metals, SO_4)
To assess the successional stages of an ecosystem
Forensic purposes
Rapid degradation of pesticides
Disease studies
Enzyme activity in soil fluctuates with environment.

2.3 Soil Health

2.3.1 Definition

The definition of soil health must be broad enough to encompass the many functions of soil, e.g., environmental filter, plant growth, and water regulation (Doran and Safley 1997). Definitions of air and water quality standards have existed for a long time, while a similar definition does not exist for soil. A definition of soil health based on this concept would encompass only a small fraction of the many roles soil play (Singh et al. 1999). Soil health is the net result of on-going conservation and degradation processes, depending highly on the biological component of the soil ecosystem, and influences plant health, environmental health, food safety, and quality (Halvorson et al. 1997; Parr et al. 1992).

Several definitions of soil health have been proposed during the last decades. Historically, the term soil quality described the status of soil as related to agricultural productivity or fertility (Singer et al. 1999). In the 1990s, it was proposed that soil quality was not limited to soil productivity but instead expanded to encompass interactions with the surrounding environment, including the implications for human and animal health. In this regard, several examples of definitions of soil quality have been suggested (Doran and Parkin 1994). In the mid-1990s, the term *soil health* was introduced. For example, a program to assess and monitor soil health in Canada used the terms quality and health synonymously to describe the ability of soil to support crop growth without becoming degraded or otherwise harming the environment (Acton and Gregorich 1995). Others broadened the definition of soil health to capture the ecological attributes of soil, and went beyond its capacity to simply produce particular crops. These attributes are chiefly associated with biodiversity, food web structure, and functional measures (Pankhurst et al. 1997).

Several numbers have been recognized for soil health which are as follows:

1. The continued capacity of soil to function as a vital living system, within the ecosystem and land-use boundaries, to sustain biological productivity, promote the quality of air and water environments, and maintain plant, animal, and human health (Doran and Safley 1997).
2. Soil health is an assessment of ability of a soil to meet its range of ecosystem functions as appropriate to its environment.
3. Soil health can also be defined as the continued capacity of a specific kind of soil to function as a vital living system, within natural or managed ecosystem boundaries, to sustain plant and animal productivity, to maintain or enhance the quality of air and water environments, and to support human health and habitation.

2.3.2 *Aspects of Soil Health*

The term soil health is used to assess the ability of a soil to

Sustain plant and animal productivity and diversity
Maintain or enhance water and air quality
Support human health and habitation

The underlying principle in the use of the term "soil health" is that soil is not just a growing medium; rather, it is a living, dynamic and ever-so-subtly changing environment. We can use the human health analogy and categorize a healthy soil as one

In a state of composite well-being in terms of biological, chemical, and physical properties

Not diseased or infirmed (i.e., not degraded, nor degrading), nor causing negative off-site impacts
With each of its qualities cooperatively functioning such that the soil reaches its full potential and resists degradation
Providing a full range of functions (especially nutrient, carbon, and water cycling), and in such a way that it maintains this capacity into the future.

2.3.3 *Interpretation of Soil Health*

Different soils will have different benchmarks of health depending on the "inherited" qualities, and on the geographic circumstance of the soil. The generic aspects defining a healthy soil can be considered as follows

"Productive" options are broad
Life diversity is broad
Absorbency, storing, recycling, and processing is high in relation to limits set by climate
Water runoff quality is of high standard
Low entropy
No damage to or loss of the fundamental components

This translates to

A comprehensive cover of vegetation
Carbon levels relatively close to the limits set by soil type and climate
Little leakage of nutrients from the ecosystem

Biological productivity relatively close to the limits set by the soil environment and climate
Only geological rates of erosion
No accumulation of contaminants
The ecosystem does not rely excessively on inputs of fossil energy

An unhealthy soil thus is the simple converse of the above.

2.3.4 *Pressures on Soil Health Towards Impacts*

The flow chart given below is a simple description of soil health factors and their impacts (Fig. 2.1).

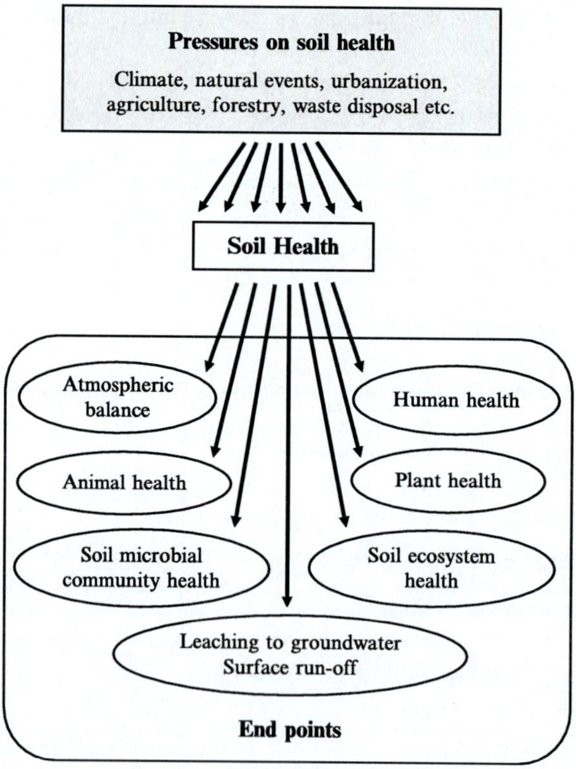

Fig. 2.1 Policy relevant end points of soil health monitoring

2.4 Indicators of Soil Health

2.4.1 Microorganism as Indicators of Soil Health

The biological activity in soil is largely concentrated in the topsoil, the depth of which may vary from a few to 30 cm. In the topsoil, the biological components occupy a tiny fraction (<0.5%) of the total soil volume and make up less than 10% of the total organic matter in the soil. These biological components consist mainly of soil organisms, especially microorganisms. Despite of their small volume in soil, microorganisms are key players in the cycling of nitrogen, sulfur, and phosphorus, and the decomposition of organic residues. Thereby they affect nutrient and carbon cycling on a global scale (Pankhurst et al. 1997). That is, the energy input into the soil ecosystems is derived from the microbial decomposition of dead plant and animal organic matter. The organic residues are, in this way, converted to biomass or mineralized to CO_2, H_2O, mineral nitrogen, phosphorus, and other nutrients (Bloem et al. 1997). Microorganisms are further associated with the transformation and degradation of waste materials and synthetic organic compounds (Torstensson et al. 1998). In addition to the effect on nutrient cycling, microorganisms also affect the physical properties of the soil. Production of extra-cellular polysaccharides and other cellular debris by microorganisms help in maintaining soil structure as well as soil health. Thereby, they also affect water holding capacity, infiltration rate, crusting, erodibility, and susceptibility to compaction (Elliott et al. 1996).

Microorganisms possess the ability to give an integrated measure of soil health, an aspect that cannot be obtained with physical/chemical measures and/or analyses of diversity of higher organisms. Microorganisms respond quickly to changes; hence they rapidly adapt to environmental conditions, and thus they can be used for soil health assessment, and changes in microbial populations and activities may therefore function as an excellent indicator of change in soil health (Kennedy and Papendick 1995; Pankhurst et al. 1995).

Microorganisms also respond quickly to environmental stress compared to higher organisms, as they have intimate relations with their surroundings due to their high surface to volume ratio. In some instances, changes in microbial populations or activity can precede detectable changes in the soil's physical and chemical properties, thereby providing an early sign of soil improvement or an early warning of soil degradation (Pankhurst et al. 1995). The impact of some chemicals on soil health is dependent on microbial activities. For example, the concentration of heavy metals in soil will not change over small time periods, but their bioavailability may. In this way, soil enzymes are acting as important indicators of soil.

2.4.2 Soil Enzymes as Indicators of Soil Health

Enzymes are the direct mediators for biological catabolism of soil organic and mineral components. Thus, these catalysts provide a meaningful assessment of

Table 2.1 Soil enzymes as indicators of soil health

Soil enzyme	Enzyme reaction	Indicator of microbial activity
Dehydrogenase	Electron transport system	C-cycling
β-glucosidase	Cellobiose hydrolysis	C-cycling
Cellulase	Cellulose hydrolysis	C-cycling
Phenol oxidase	Lignin hydrolysis	C-cycling
Urease	Urea hydrolysis	N-cycling
Amidase	N-mineralization	N-cycling
Phosphatase	Release of PO_4^-	P-cycling
Arylsulphatase	Release of SO_4^-	S-cycling
Soil enzymes	Hydrolysis	General organic matter degradative enzyme activities

reaction rates for important soil processes. Soil enzyme activities (1) are often closely related to soil organic matter, soil physical properties and microbial activity or biomass, (2) changes much sooner than other parameters, thus providing early indications of changes in soil health, and (3) involve simple procedures (Dick et al. 1996). In addition, soil enzyme activities can be used as measures of microbial activity, soil productivity, and inhibiting effects of pollutants (Tate 1995). Easy, well-documented assays are available for a large number of soil enzyme activities (Dick et al. 1996; Tabatabai 1994a, b). These include dehydrogenase, glucosidases, urease, amidases, phosphatases, arylsulphatase, cellulases, and phenol oxidases as shown in Table 2.1.

2.5 Potential Roles of Soil Enzymes in Maintaining Soil Health

A number of soil enzymes and their respective roles in maintaining soil health are stated below

2.5.1 *Amylase*

The starch hydrolyzing enzyme amylase (Ross 1976) is known to be constituted by α-amylase and β-amylase (King 1967; Thoma et al. 1971). The α-amylases are synthesized by plants, animals, and microorganisms, whereas, β-amylase is synthesized mainly by plants (Pazur 1965; Thoma et al. 1971). This enzyme is widely distributed in plants and soils so it plays a significant role in the breakdown of starch, which converts starch like substrates to glucose and/or oligosaccharides and β-amylase, which converts starch to maltose (Thoma et al. 1971). Studies have, however, indicated that the roles and activities of α-amylase and β-amylase enzymes may be influenced by different factors ranging from cultural practices, type of vegetation, environment and soil types (Pancholy and Rice 1973; Ross 1975). For example, plants may influence the amylase enzyme activities of soil by directly supplying enzymes from their residues or excreted compounds, or indirectly providing substrates for the synthetic activities of microorganisms. Greater understanding is required of the significance of these enzymes in the soil, and to enable proper management techniques to be devised to maximize the benefits that may be derived from such enzymes.

2.5.2 Arylsulphatases

This is due to the fact that certain proportions of sulphur in different soil profiles are bound into organic compounds and are indirectly available to plants. Arylsulphatases are typically widespread in nature (Dodgson et al. 1982) as well as in soils (Gupta et al. 1993; Ganeshamurthy et al. 1995). They are responsible for the hydrolysis of sulphate esters in the soil (Kertesz and Mirleau 2004) and are secreted by bacteria into the external environment as a response to sulphur limitation (McGill and Colle 1981). Its occurrence in different soil systems is often correlated with microbial biomass and rate of S immobilization (Klose and Tabatabai 1999; Vong et al. 2003). This enzyme has a role in the hydrolysis of aromatic sulphate esters (R–O–SO_3−) to phenols (R–OH) and sulfate, or sulfate sulfur (SO_4−2 or SO_4–S) (Tabatabai 1994a, b).

Soil is affected by various environmental factors (Burns 1982) such as heavy metal pollution (Tyler 1981); pH changes in the soil solution (Acosta-Martínez and Tabatabai 2000); organic matter content and its type (Sarathchandra and Perrott 1981); such as absorption to particles surfaces in soils, and the activity persistence of extra cellular arylsulfatases in the soil. Considering the importance of S in plant nutrition, a better understanding of the role(s) of arylsulfatases in S mobilization in agricultural soils is critical. So far, very little is known about specific microbial genera or species that play an important role in the soil organosulphur circle (Kertesz and Mirleau 2004) in which arylsulphatases is the key enzyme.

2.5.3 β-Glucosidase

Glucosidase is a common and predominant enzyme in soils (Eivazi and Tabatabai 1988; Tabatabai 1994a, b). It is named according to the type of bond that it hydrolyses. This enzyme plays an important role in soils because it is involved in catalyzing the hydrolysis and biodegradation of various β-glucosidase present in plant debris decomposing in the ecosystem (Ajwa and Tabatabai 1994; Martinez and Tabatabai 1997). Its final product is glucose, an important C energy source of life to microbes in the soil (Esen 1993). β-glucosidase is characteristically useful as a soil quality indicator, and may give a reflection of past biological activity, the capacity of soil to stabilize the soil organic matter, and can be used to detect management effect on soils (Bandick and Dick 1999; Ndiaye et al. 2000). This has greatly facilitated its adoption for soil quality testing (Bandick and Dick 1999). Some of the aglycons are known to be the precursors of the toxic substances, which cause soil sickness where plants are grown as monocrops (Patrick 1955; Borner 1958).

β-Glucosidase enzyme is very sensitive to changes in pH, and soil management practices (Acosta-Martínez and Tabatabai 2000; Madejón et al. 2001). Acosta-Martínez and Tabatabai 2000 reported β-glucosidase as sensitive to pH changes. This property can be used as a good biochemical indicator for measuring ecological changes resulting from soil acidification in situations involving activities of this enzyme. Consequently, more understanding of the β-glucosidase enzyme activities and factors influencing them in the ecosystem may contribute significantly to soil health studies.

2.5.4 Cellulases

Cellulose is the most abundant organic compound in the biosphere, comprising almost 50% of the biomass synthesized by photosynthetic fixation of CO_2 (Eriksson et al. 1990). Growth and survival of microorganisms important in most agricultural soils depends on the carbon source contained in the cellulose occurring in the soils (Deng and Tabatabai 1994). However, for carbon to be released as an energy source for use by the microorganisms, cellulose in plant debris has to be degraded into glucose, cellobiose and high molecular weight oligosaccharides by cellulases enzymes (White 1982). Cellulases are a group of enzymes that catalyze the degradation of cellulose, polysaccharides built up of β-1,4 linked glucose units (Deng and Tabatabai 1994). It has been reported that cellulases in soils are derived mainly from plant debris incorporated into the soil, and that a limited amount may also originate from fungi and bacteria in soils (Richmond 1991). Demonstrating the effects of increasing concentrations of fungicides on cellulases activities, Petker and Rai (1992) showed that there was a decreasing effect with fungicides captan, cosan, thiram, zinels, and sandolex. More recently, Arinze and Yubedee (2000) reported that fungicides benlate, calixin, and captan inhibited cellulase activity in *Fusarium monoliforme* isolates. Captatol inhibited cellulose activity in the sandy loam soil (Atlas et al. 1978), and chlorothalonil showed a clear reduction in cellulase activity under flooded or non-flooded conditions (Vincent and Sisler 1968). Studies have shown that activities of cellulases in agricultural soils are affected by several factors. These include temperature, soil pH, water and oxygen contents (abiotic conditions), the chemical structure of organic matter and its location in the soil profile horizon (Deng and Tabatabai 1994; Alf and Nannipieri 1995), quality of organic matter/plant debris and soil mineral elements (Sinsabaugh and Linkins 1989; Deng and Tabatabai 1994) and the trace elements from fungicides (Deng and Tabatabai 1994; Arinze and Yubedee 2000). Srinivasulu and Rangaswamy 2006 reported a significantly more stimulatory effect of cellulases in black soil than red soil. For instance, chitin in the presence of cellulose induces the synthesis of chitinase and other cell wall lytic enzymes which promote the release of the intramural β-glucosidase into the medium. All these findings suggest that activities of cellulases can be used to give preliminary indication of some of the physical chemical properties of soil, thus, easing agricultural soil management strategies. Since cellulases enzymes play an important role in global recycling of the most abundant polymer, cellulose in nature, it would be of critical importance to understand this enzyme better so that it may be used more regularly as a predictive tool in our soil fertility programs. More information on the role of this enzyme is needed since it is affected by different factors, which may jeopardize its involvement in the decomposition of cellulolytic materials in the soil for microbial use and improved soil health in agricultural ecosystems.

2.5.5 *Chitinase*

Chitinase or chitinolytic enzymes are key enzymes responsible for the degradation and hydrolysis of chitin (polyβ-1-4-(2-ncetamido-2-deoxy)-D-glucoside). They are also considered as the major structural component of many fungal cell walls that use the hyperparasitism mechanisms against pests/pathogen attack (Chet and Henis 1975; Chet 1987). These biological agents also reduce disease-producing agents by using other mechanisms such as antibiosis or competition mechanisms (Park 1960). This agriculturally important enzyme is produced or released by various organisms including plants and microorganisms (Deshpande 1986). Its presence in different forms in the ecosystem has demonstrated its effectiveness in the control of soil-borne diseases such as *Sclerotium rolfsii* and *Rhizoctonia solani* in beans and cotton, respectively (Ordentlich et al. 1988; Shapira et al. 1989). One of the mechanisms proposed involves lytic enzymes chitinase that cause the degradation of cell walls of pathogenic fungi (Ordentlich et al. 1988; Chet et al. 1990; Singh et al. 1999). As for its role in biological control of pests, moreover, due to environmental friendliness, there are so many avenues for the application of this enzyme for maintaining soil health and consequently, increase plant growth and final yields.

2.5.6 *Dehydrogenase*

The dehydrogenase enzyme activity is commonly used as an indicator of biological activity in soils (Burns 1978). This enzyme is considered to exist as an integral part of intact cells but does not accumulate extracellularly in the soil. Dehydrogenase enzyme is known to oxidize soil organic matter by transferring protons and electrons from substrates to acceptors. These processes are the part of respiration pathways of soil microorganisms and are closely related to the type of soil and soil air-water conditions (Kandeler 1996; Glinski and Stepniewski 1985). Since these processes are the part of respiration pathways of soil microorganisms, studies on the activities of dehydrogenase enzyme in the soil is very important as it may give indications of the potential of the soil to support biochemical processes which are essential for maintaining soil fertility as well as soil health. A study by Brzezinska et al. (1998) suggested that soil water content and temperature influence dehydrogenase activity indirectly by affecting the soil redox status. After flooding the soil, the oxygen present is rapidly exhausted so that a shift of the activity from aerobic to anaerobic microorganisms takes place. Such redox transformations are closely connected with respiration activity of soil microorganisms. They may serve as indicators of the microbiological redox systems in soils and can be considered a possible measure of microbial oxidative activity (Tabatabai 1982; Trevors 1984). For instance, lack of oxygen may trigger facultative anaerobes to initiate metabolic processes involving dehydrogenase activities and the use of Fe (III) forms as terminal electron acceptors (Bromfield 1954; Galstian and Awungian 1974), a process that may affect iron availability to plants in the ecosystem (Benckiser et al. 1984). Additionally, dehydrogenase enzyme is often used as a measure of any disruption caused by pesticides, trace elements or management practices to the soil (Reddy and Faza 1989; Wilke 1991; Frank and Malkomes 1993), as well as a direct measure of soil microbial activity (Trevors 1984; Garcia and Hernández 1997). It can also indicate the type and significance of pollution in soils. For example, dehydrogenase enzyme is high in soils polluted with pulp and paper mill effluents (McCarthy et al. 1994) but low in soils polluted with fly ash (Pitchel and Hayes 1990). Similarly, higher activities of dehydrogenases have been reported at low doses of pesticides, and lower activities of the enzyme at higher doses of pesticides (Baruah and Mishra 1986).

2.5.7 Phosphatases

In soil ecosystems, these enzymes are believed to play critical roles in P cycles (Speir and Ross 1978) as evidence shows that they are correlated to P stress and plant growth. Apart from being good indicators of soil fertility, phosphatase enzyme plays a key role in the soil system (Eivazi and Tabatabai 1977; Dick et al. 2000). For example, when there is a signal indicating P deficiency in the soil, acid phosphatase secretion from plant roots is increased to enhance the solubilization and remobilization of phosphate, thus influencing the ability of the plant to cope with P-stressed conditions (Karthikeyan et al. 2002; Mudge et al. 2002; Versaw and Harrison 2002). Understanding the dynamics of enzyme activities in these systems is crucial for predicting their interactions as their activities may, in turn, regulate nutrient uptake and plant growth, later on, where soil health is concerned.

2.5.8 Proteases

Proteases in the soil play a significant role in N mineralization (Ladd and Jackson 1982), an important process regulating the amount of plant available N and plant growth. This enzyme in the soil is generally associated with inorganic and organic colloids (Burns 1982; Nannipieri et al. 1996). The amount of this extra cellular enzyme activity may be indicative not only of the biological capacity of soil for the enzymatic conversion of the substrate, which is independent of the extent of microbial activity, but might also have an important role in the ecology of micro-organisms in the ecosystem (Burns 1982). There is a need to study the properties and factors affecting naturally occurring enzyme complexes such as those involving protease enzymes in the soil ecosystem as they may reveal some unknown role(s) in maintaining soil health and fertility.

2.5.9 *Urease*

Urease enzyme is responsible for the hydrolysis of urea fertilizers applied to the soil into NH_3 and CO_2 with the concomitant rise in soil pH (Andrews et al. 1989; Byrnes and Amberger 1989). This, in turn, results in a rapid N loss to the atmosphere through NH_3 volatilization (Simpson et al. 1984; Simpson and Freney 1988). Due to this role, urease activities in soils have received a lot of attention since it was first reported by Rotini (1935), a process considered vital in the regulation of N supply to plants after urea fertilization. Soil urease originates mainly from plants (Polacco 1977) and microorganisms found as both intra- and extra-cellular enzymes (Burns 1986; Mobley and Hausinger 1989). On the other hand, urease extracted from plants or microorganisms is rapidly degraded in soil by proteolytic enzymes (Pettit et al. 1976; Zantua and Bremner 1977). This suggests that a significant fraction of ureolytic activity in the soil is carried out by extracellular urease, which is stabilized by immobilization on organic and mineral soil colloids. Urease activity in soils is influenced by many factors. These include cropping history, organic matter content of the soil, soil depth, soil amendments, heavy metals, and environmental factors such as temperatures (Tabatabai 1977; Yang et al. 2006). For example, studies have shown that urease was very sensitive to toxic concentrations of heavy metals (Yang et al. 2006). Generally, urease activity increases with increasing temperature. It is suggested that higher temperatures increase the activity coefficient of this enzyme. Therefore, it is recommended that urea be applied at times of the day when temperatures are low. Since urease plays a vital role in the hydrolysis of urea fertilizer, it is important to uncover other unknown factors that may reduce the efficiency of this enzyme in the ecosystem.

2.6 Conclusion

It is very essential to understand the possible roles of soil enzymes in order to maintain soil health and its fertility management in ecosystems. These enzymes, usually found in the soil, may have significant effects on soil biology, environmental management, growth and nutrient uptake in plants growing in ecosystems. Their activities may, however, be influenced by unknown cultural management practices either in a major or minor amount. Studies focusing the discovery of new enzymes from microbial diversity in the soil might be the most suitable practices that may positively influence their activities for improved plant growth as well as rendering the friendly biological environments in order to sustain other living beings.

References

Acosta-Martínez V, Tabatabai MA (2000) Enzyme activities in a limed agricultural soil. Biol Fertil Soils 31:85–91

Acton DF, Gregorich EG (1995) Executive summary. In: Acton DF, Gregorich EG (eds) The health of our soils. Towards sustainable agriculture in Canada. Center for Land and Biological Resources, Research Branch Agriculture and Agri-food Canada, ON, Canada

Ajwa HA, Tabatabai MA (1994) Decomposition of different organic materials in soils. Biol Fertil Soils 18:175–182

Alf K, Nannipieri P (1995) Cellulase activity, methods in applied soil microbiology and biochemistry. Academic, London

Andrews RK, Blakeley RL, Zerner B (1989) Urease: a Ni (II) metalloenzyme. In: Lancaster JR (ed) The bioinorganic chemistry of nickel. VCH, New York, pp 141–166

Arinze AE, Yubedee AG (2000) Effect of fungicides on *Fusarium* grain rot and enzyme production in maize (*Zea mays* L.). Glob J Appl Sci 6:629–634

Atlas RM, Pramer D, Bartha R (1978) Assessment of pesticide effects on non-target soil microorganisms. Soil Biol Biochem 10:231–239

Bandick AK, Dick RP (1999) Field management effects on soil enzyme activities. Soil Biol Biochem 31:1471–1479

Baruah M, Mishra RR (1986) Effect of herbicides butachlor, 2,4-d and oxyfluorfen on enzyme activities and CO_2 evolution in submerged paddy field soil. Plant Soil 96:287–291

Benckiser G, Santiago S, Neue HU, Watanabe I, Ottow JCG (1984) Effect of fertilization and exudation, dehydrogenase activity, iron-reducing populations and Fe^{2+} formation in the rhizosphere of rice (*Oryza sativa* L.) in relation to iron toxicity. Plant Soil 79:305–316

Bloem J, de Ruiter P, Bouwman LA (1997) Food webs and nutrient cycling in agro-ecosystems. In: Van Elsas JD, Trevors JT, Wellington EMH (eds) Modern soil microbiology. Marcel Dekker, New York, pp 245–278

Borner H (1958) Untersuchungen uber den Abbau von Phlorizin im Boden. Ein Beitrag zum Problem der Bodenmudigkeit bei Obstgeholzen. Naturwiss 45:138–139

Bromfield SM (1954) Reduction of ferric compounds by soil bacteria. J Gen Microbiol 11:1–6

Brzezinska M, Stepniewska Z, Stepniewski W (1998) Soil oxygen status and dehydrogenase activity. Soil Biol Biochem 30:1783–1790

Burns RG (1978) Enzyme activity in soil: some theoretical and practical considerations. In: Bums RG (ed) Soil enzymes. Academic, London, pp 295–340

Burns RG (1982) Enzyme activity in soil: location and possible role in microbial ecology. Soil Biol Biochem 14:423–427

Burns RG (1986) Interaction of enzymes with soil mineral and organic colloids. In: Huang PM, Schnitzer M (eds) Interactions of soil minerals with natural organics and microbes. Soil Science Society of America, Madison, pp 429–452

Byrnes BH, Amberger A (1989) Fate of broadcast urea in a flooded soil when treated with *N*-(n-butyl) thiophospheric triamide, a urease inhibitor. Fertil Res 18:221–231

Chet I (1987) Trichoderma-application, mode of action, and potential as biocontrol agent of soil borne pathogenic fungi. In: Chet I (ed) Innovative approaches to plant disease control. Wiley, New York, pp 137–349

Chet I, Henis Y (1975) Sclerotial morphogenesis in fungi. Annu Rev Phytopathol 13:169–192

Chet I, Ordentlich A, Shapira R, Oppenheim A (1990) Mechanism of biocontrol of soil borne plant pathogen by rhizobacteria. Plant Soil 129:85–92

Deng SP, Tabatabai MA (1994) Cellulase activity of soils. Soil Biol Biochem 26:1347–1354

Deshpande MV (1986) Enzymatic degradation of chitin and its biological applications. J Sci Ind Res 45:273–281

Dick RP, Sandor JA, Eash NS (1994) Soil enzyme activities after 1500 years of terrace agriculture in the Colca Valley. Peru Agric Ecosyst Environ 50:123–131

Dick RP, Breakwell DP, Turco RF (1996) Soil enzyme activities and biodiversity measurements as integrative microbiological indicators. In: Doran JW, Jones AJ (eds) Methods of assessing soil quality. Soil Science Society of America, Madison, WI, pp 247–271

Dick RP (1997) Soil enzyme activities as integrative indicators of soil health. In: Pankhurst CE, Doube BM, Gupta VVSR (eds) Biological indicators of soil health. CABI, Wellingford, pp 121–156

Dick WA, Cheng L, Wang P (2000) Soil acid and alkaline phosphatase activity as pH adjustment indicators. Soil Biol Biochem 32:1915–1919

Dodgson KS, White G, Fitzgerald JW (1982) Sulphatase enzyme of microbial origin. Afr J Biotechnol Vol I. CRC, FL, pp 156–159

Doran JW, Parkin TB (1994) Defining and assessing soil quality. In: Doran JW, Coleman DC, Bezdicek DF, Stewart BA (eds) Defining soil quality for a sustainable environment. Soil Science Society of America, Madison, pp 3–21

Doran JW, Safley M (1997) Defining and assessing soil health and sustainable productivity. In: Pankhurst CE, Doube BM, Gupta VVSR (eds) Biological indicators of soil health. CABI, Wellingford, pp 1–28

Eivazi F, Tabatabai MA (1977) Phosphates in soils. Soil Biol Biochem 9:167–172

Eivazi F, Tabatabai MA (1988) Glucosidases and galactosidases in soils. Soil Biol Biochem 20:601–606

Ellert BH, Clapperton MJ, Anderson DW (1997) An ecosystem perspective of soil quality. In: Gregorich EG, Carter MR (eds) Soil quality for crop production and ecosystem health. Elsevier, Amsterdam, pp 115–141

Elliott LF, Lynch JM, Papendick RI (1996) The microbial component of soil quality. In: Stotzky G, Bollag JM (eds) Soil biochemistry. Marcel Dekker, New York, pp 1–21

Eriksson KEL, Blancbette RA, Ander P (1990) Biodegration of cellulose. In: Eriksson KEL, Blanchette RA, Ander P (eds) Microbial and enzymatic degradation of wood and wood components. Springer, New York, pp 89–180

Esen A (1993) β-glucosidases: overview. In: Esen A (ed) β-glucosidases and molecular biology. American Chemical Society, Washington, DC, pp 9–17

Frank T, Malkomes HP (1993) Influence of temperature on microbial activities and their reaction to the herbicide Goltix in different soils under laboratory conditions. Zentralblatt für Mikrobiol 148:403–412

Galstian AS, Awungian ZS (1974) Significance of the enzymes in oxidation of Fe and Mn oxides in soil (in Russian). Trans. 10th Intern. Congress Soil Sci III. Nauka Publishing House, Moscow, pp 130–135

Ganeshamurthy AM, Singh G, Singh NT (1995) Sulphur status and response of rice to sulphur on some soils of Andaman and Nicobar Islands. J Indian Soc Soil Sci 43:637–641

Garcia C, Hernández T (1997) Biological and biochemical indicators in derelict soils subject to erosion. Soil Biol Biochem 29:171–177

Glinski J, Stepniewski W (1985) Soil aeration and its role for plants. CRC, Boca Raton, FL

Gupta VVSR, Farrell RE, Germida JJ (1993) Activity of arylsuphatases in Saskatchewan soils. Can J Soil Sci 73:341–347

Halvorson JJ, Smith JL, Papendick RI (1997) Issues of scale for evaluating soil quality. J Soil Water Conserv 52:26–30
James ES, Russel LW, Mitrick A (1991) Phosphate stress response in hydroponically grown maize. Plant Soil 132:85–90
Kandeler E (1996) Nitrate. In: Schinner F, Öhlinger R, Kandeler E, Margesin R (eds) Methods in soil biology. Springer, Berlin, pp 408–410
Kanfer JN, Mumford RA, Raghavan SS, Byrd J (1974) Purification of β-glucosidase activities from bovine spleen affinity chromatography. Anal Biochem 60:200–205
Karthikeyan AS, Varadarajan DK, Mukatira UT, D'Urzo MP, Damaz B, Raghothama KG (2002) Regulated expression of *Arabidopsis* phosphate transporters. Plant Physiol 130:221–233
Kennedy AC, Papendick RI (1995) Microbial characteristics of soil quality. J soil water conserv. May–June:243–248
Kertesz MA, Mirleau P (2004) The role of soil microbes in plant sulphur nutrition. J Exp Bot 55:1939–1945
King NJ (1967) Glucoamylase of Coniophora cerebella. Biochem J 105:577–583
Kiss S, Dragan-Bularda M, Radulescu D (1978) Soil polysaccharidases: activity and agricultural importance. In: Burns RG (ed) Soil enzymes. Academic, London, pp 117–147
Klose S, Tabatabai MA (1999) Arylsulphatase activity of microbial biomass in soils. Soil Sci Soc Am J 63:569–574
Ladd JN, Jackson RB (1982) In: Stevenson FJ (ed) Nitrogen in agricultural soils. American Society of Agronomy, WI, pp 173–228
Madejón E, Burgos P, López R, Cabrera F (2001) Soil enzymatic response to addition of heavy metals with organic residues. Biol Fertil Soils 34:144–150
Martinez CE, Tabatabai MA (1997) Decomposition of biotechnology by-products in soils. J Environ Qual 26:625–632
McCarthy GW, Siddaramappa R, Reight RJ, Coddling EE, Gao G (1994) Evaluation of coal combustion by products as soil liming materials: their influence on soil pH and enzyme activities. Biol Fertil Soils 17:167–172
McGill WB, Colle CV (1981) Comparative aspects of cycling of organic C, N, S and P through soil organic matter. Geoderma 26:267–286
McLaren AD (1975) Soil as a system of humus and clay immobilised enzymes. Chem Scripta 8:97–99
Miwa T, Ceng CT, Fujisaki M, Toishi A (1937) Zur Frage der Spezifitat der Glykosidasen. I. Verhalted vonβ-d-glucosidases verschiedener Herkunft gegenuberdenβ-d-Glucosiden mit verschiedenen Aglykonen. Acta Phytochim (Tokyo) 10:155–170
Mobley HLT, Hausinger RP (1989) Microbial urease: significance, regulation and molecular characterization. Microbiol Rev 53:85–108
Mudge SR, Rae AL, Diatloff E, Smith FW (2002) Expression analysis suggests novel roles for members of Pht1 family of phosphate transporters in *Arabidopsis*. Plant J 31:341–353
Nannipieri P, Sequi P, Fusi P (1996) Humus and enzyme activity. In: Piccolo A (ed) Humic substances in terrestrial ecosystems. Elsevier, New York, pp 293–328
Ndiaye EL, Sandeno JM, McGrath D, Dick RP (2000) Integrative biological indicators for detecting change in soil quality. Am J Altern Agric 15:26–36
Ordentlich A, Elad Y, Chet I (1988) The role of chitinase of *Serratia marcescens* in biocontrol of *Sclerotium rolfsii*. Phytopathology 78:84–88
Pancholy SK, Rice EL (1973) Soil enzymes in relation to old field succession; amylase, cellulose, invertase, dehydrogenase and urease. Soil Sci Soc Am J 37:47–50
Pankhurst CE, Hawke BG, McDonald HJ, Kirkby CA, Buckerfield JC, Michelsen P, O'Brien KA, Gupta VVSR, Doube BM (1995) Evaluation of soil biological properties as potential bioindicators of soil health. Aust J Exp Agric 35:1015–1028
Pankhurst CE, Doube BM, Gupta VVSR (1997) Biological indicators of soil health: synthesis. In: Pankhurst CE, Doube BM, Gupta VVSR (eds) Biological indicators of soil health. CABI, Wallingford, Oxfordshire, pp 419–435

Park D (1960) Antagonism – the background of soil fungi. In: Parkinson D, Waid JS (eds) The ecology of soil fungi. Liverpool University Press, Liverpool, pp 148–159

Parr JF, Papendick RI, Hornick SB, Meyer RE (1992) Soil quality: attributes and relationship to alternative and sustainable agriculture. Am J Altern Agric 7:5–11

Patrick ZA (1955) The peach replant problem in Ontario. II. Toxic substances from microbial decomposition products of peach root residues. Can J Bot 33:461–486

Pazur JH (1965) Enzymes in the synthesis and hydrolysis of starch. In: Whistler R, Paschall EF (eds) Starch: chemistry and technology, vol 1, Fundamental aspects. Academic, New York, pp 133–175

Petker AS, Rai PK (1992) Effect of fungicides on activity, secretion of some extra cellular enzymes and growth of *Alternaria alternata*. Indian J Appl Pure Biol 7:57–59

Pettit NM, Smith ARJ, Freedman RB, Burns RG (1976) Soil urease: activity, stability and kinetic properties. Soil Biol Biochem 8:479–484

Pitchel JR, Hayes JM (1990) Influence of fly ash on soil microbial activity and populations. J Environ Qual 19:593–597

Polacco JC (1977) Is nickel a universal component of plant ureases? Plant Sci Lett 10:249–255

Reddy GB, Faza A (1989) Dehydrogenase activity in sludge amended soil. Soil Biol Biochem 21:327

Richmond PA (1991) Occurrence and functions of native cellulose. In: Haigler CH, Weimer PJ (eds) Biosynthesis and biodegradation of cellulose. Marcel Dekker, New York, pp 5–23

Ross DJ (1975) Studies on a climosequence of soils in tussock grasslands-5. Invertase and amylase activities of topsoils and their relationships with other properties. NZ J Sci 18:511–518

Ross DJ (1976) Invertase and amylase activities in ryegrass and white clover plants and their relationships with activities in soils under pasture. Soil Biol Biochem 8:351–356

Rotini OT (1935) La trasformazione enzimatica dell'urea nel terreno. Ann Labor Ric Ferm Spallanrani 3:143–154

Sarathchandra SU, Perrott KW (1981) Determination of phosphatase and arylsulphatase activity in soils. Soil Biol Biochem 13:543–545

Shapira R, Ordentlich A, Chet I, Oppenheim AB (1989) Control of plant diseases by chitinase expressed from cloned DNA in *Escherichia coli*. Phytopathology 79:1246–1249

Shawale JG, Sadana J (1981) Purification, characterization and properties of β-glucosidase enzyme from *Sclerotium rolfsii*. Arch Biochem Biophys 207:185–196

Simpson JR, Freney JR, Wetselaar R, Muirhead WA, Leuning R, Denmead OT (1984) Transformations and losses of urea nitrogen after application to flooded rice. Aust J Agric Res 35:189–200

Simpson JR, Freney JR (1988) Interacting processes in gaseous nitrogen loss from urea applied to flooded rice fields. In: Pushparajah E, Husin A, Bachik AT (eds) Proceedings of international symposium on urea technology and utilization. Malaysian Society of Soil Science, Kuala Lumpur, pp 281–290

Singh PP, Shin YC, Park CS, Chung YR (1999) Biological control of *Fusarium* wilt of cucumber by chitinolytic bacteria. Phytopathology 89:92–99

Singer MJ, Ewing S (2000) Soil quality. In: Sumner ME (ed) Handbook of soil science. CRC, Boca Raton, FL, pp 271–298

Sinsabaugh RL, Linkins AE (1989) Natural disturbance and the activity of *Trichoderma viride* cellulase complex. Soil Biol Biochem 21:835–839

Sinsabaugh RL, Antibus RK, Linkins AE (1991) An enzymic approach to the analysis of microbial activity during plant litter decomposition. Agric Ecosyst Environ 34:43–54

Speir TW, Ross DJ (1978) Soil phosphatase and sulphatase. In: Burns RG (ed) Soil enzymes. Academic, London, UK, pp 197–250

Srinivasulu M, Rangaswamy V (2006) Activities of invertase and cellulase as influenced by the application of tridemorph and captan to groundnut (*Arachis hypogaea*) soil. Afr J Biotechnol 5:175–180

Tabatabai MA (1977) Effect of trace elements on urease activity in soils. Soil Biol Biochem 9:9–13
Tabatabai MA (1982) Soil enzyme. In: Page AL (ed) Methods of soil analysis, Part 2. American Society of Agronomy, Madison, WI, pp 903–948
Tabatabai MA (1994a) Soil enzymes. In: Weaver RW, Angle JS, Bottomley PS (eds) Methods of soil analysis, part 2. Microbiological and biochemical properties. SSSA Book Series No. 5. Soil Science Society of America, Madison, WI, pp 775–833
Tabatabai MA (1994b) Soil enzymes. In: Mickelson SH (ed) Methods of soil analysis, Part 2. Microbiological and biochemical properties. Soil Science Society of America, Madison, WI, pp 775–833
Tate RL (1995) Soil microbiology. John Wiley, New York
Thoma JA, Spradlin JE, Dygert S (1971) Plant and animal amylases. In: Boyer PD (ed) The enzymes, 5th edn. Academic, New York, pp 115–189
Torstensson L, Pell M, Stenberg B (1998) Need of a strategy for evaluation of arable soil quality. Ambio 27:4–8
Trevors JT (1984) Dehydrogenase activity in soil: a comparison between the INT and TTC assay. Soil Biol Biochem 16:673–674
Tyler G (1981) Heavy metals in soil biology and biochemistry. In: Paul EA, Ladd JN (eds) Soil biochemistry, vol 5. Marcel Dekker, New York, pp 371–414
Versaw WK, Harrison MJ (2002) A chloroplast phosphate transporter, PHT2; 1, influences allocation of phosphate within the plant and phosphate-starvation responses. Plant Cell 14:1751–1766
Vincent PG, Sisler HD (1968) Mechanisms of antifungal action of 2,4,5,6-tetrachloroisopathalonitrile. Physiol Plant 21:1249–1264
Vong PC, Dedourge O, Lasserre-Joulin F, Guckert A (2003) Immobilized-S, microbial biomass-S and soil arylsulphatase activity in the rhizosphere soil of rape and barley as affected by labile substrate C and N additions. Soil Biol Biochem 35:1651–1661
White AR (1982) Visualization of cellulases and cellulose degradation. In: Brown RM (ed) Cellulose and other natural polymer systems: biogenesis, structure, and degradation. Plenum, New York, pp 489–509
Wilke BM (1991) Effect of single and successive additions of cadmium, nickel and zinc on carbon dioxide evolution and dehydrogenase activity in a sandy Luvisol. Biol Fertil Soils 11:34–37
Yang Z, Liu S, Zheng D, Feng S (2006) Effects of cadmium, zinc and lead on soil enzyme activities. J Environ Sci 18:1135–1141
Zantua MI, Bremner JM (1977) Stability of urease in soils. Soil Biol Biochem 9:135–140

Chapter 3
Agricultural and Ecological Significance of Soil Enzymes: Soil Carbon Sequestration and Nutrient Cycling

Wei Shi

3.1 Introduction

Organic matter decomposition is an important control of soil carbon sequestration and nutrient cycling. Due to sizable and complex nature of organic matter, decomposition often initiates with soil enzyme-catalyzed biochemical reactions. It is generally accepted that enzymatic depolymerization is a rate limiting step of decomposition. Therefore, research efforts are increasingly given to mechanisms that emphasize microbial enzyme production and soil enzyme activities that control carbon and nutrient dynamics.

Agricultural and ecological significance of soil enzymes have been progressively expanded since the first report on soil enzymes about a century ago. Soil enzymes, once used as descriptive parameters, have now been appreciated for their multifaceted functions in microbial activities, soil processes, and ecosystem responses to management and global environmental change (Sinsabaugh 1994; Waldrop et al. 2004; Fenner et al. 2005a; Lipson et al. 2005; Finzi et al. 2006). Decomposition models with the component of soil enzymes appear to be robust not only in addressing fundamental questions, such as carbon and nutrient limitations on microbial activity, but also in predicting the degree of decomposition in relation to organic carbon chemistry and nutrient availability and under specific environmental conditions (Schimel and Weintraub 2003; Moorhead and Sinsabaugh 2006). While numerous enzymes exist in a given soil, their relative importance in terms of specific soil function and/or ecosystem service may vary. For example, dehydrogenases are often used as the indicator of soil microbial activity and lignocellulases have been recognized as essential soil enzymes to control the decomposition of soil organic matter and plant litters (Sinsabaugh et al. 1992; Sinsabaugh 1994).

W. Shi
Department of Soil Science, North Carolina State University, Raleigh, NC 27695-7619, USA
e-mail: wei_shi@ncsu.edu

G. Shukla and A. Varma (eds.), *Soil Enzymology*, Soil Biology 22,
DOI 10.1007/978-3-642-14225-3_3, © Springer-Verlag Berlin Heidelberg 2011

This chapter aims to discuss soil enzymes that can be used to learn about soil carbon sequestration and nutrient cycling. Certainly, it is not my intention to provide an exhaustive list of soil enzymes that are responsible for organic matter decomposition and nutrient mineralization or detail impacts of environmental factors, such as temperature, water availability, and soil pH on the activity of soil enzymes. Instead, the focus of this chapter is on soil enzymes that have been recognized as the proximate control of decomposition, principles by which a microbial community generates extracellular enzymes, and implications that are relevant to ecosystem responses to management and global environmental change.

3.2 Soil Enzymes for Organic Matter Decomposition and Nutrient Cycling

3.2.1 Soil Enzymes of Primary Interest

Soil enzymes represent the enzymes that accumulate in soil and function extracellularly. Normally, a soil enzyme is associated with various abiotic and biotic soil constituents, such as soil aqueous phase, clay minerals, humic materials, cell debris, and viable but non-proliferating cells and, accordingly, its activity represents the sum of activities from these individual components (Ladd 1978; Burns 1982). While hundreds of extracellular enzymes may exist in a soil, only some has been quite often used to assess soil carbon and nutrient dynamics. Ligninase, cellulase, chitinase, and phosphatase are among extensively-exploited enzymes that catalyze organic matter decomposition and nutrient mineralization.

Ligninase is a group of oxidative enzymes that modify phenolic-containing organics including polyphenols, lignin, and humus. As a typical ligninase, for example, soil phenol oxidase acts upon phenolics of various complexities by catalyzing the release of oxygen radicals (Hammel 1997; Claus 2004). Through this modification, simple phenolics may be fully degraded and complicated ones are perhaps partially oxidized, thereby generating phenolic intermediates (Burke and Cairney 2002; Claus 2004; Toberman et al. 2008b). Phenol oxidase activity can be limited by low oxygen pressure, low temperature, and acidic pH (McLatchey and Reddy 1998; Criquet et al. 1999; Sinsabaugh et al. 2008). A number of microbial species from fungi, bacteria, and actinomycetes are able to produce phenol oxidase (Hammel 1997; Hullo et al. 2001; Burke and Cairney 2002; Endo et al. 2003; Fenner et al. 2005b).

Cellulase is a group of hydrolytic enzymes that catalyze the breakdown of glycosidic bonds in cellulose. Complete degradation of cellulose requires at least three enzymes: endo-β-1,4-glucanase, exo-β-1,4-glucanase, and β-glucosidase. Endoglucanase disrupts crystalline structure of cellulose via random cleavage of internal bonds; exoglucanase removes oligosaccharides, such as cellobiose and tetrasaccharides from the non-reducing ends of carbohydrate chains generated by endoglucanase; and glucosidase releases glucose from oligosaccharides (Alef and Nannipieri 1995). Several analytical methods have been developed for estimating the potential activity of soil endoglucanase, exoglucanase, glucosidase or their combination.

As the major component of exoskeleton of insects and cell walls of fungi, chitin represents one of dominant forms of organic nitrogen in soil (Paul and Clark 1996). Chitinase is an essential class of enzymes that hydrolyze glycosidic bonds in chitin and release acetyl-glucosamine. Three enzymes, chitinase, chitobiase, and *N*-acetyl-β-glucosaminidase act synergistically for the complete degradation of chitin. *N*-acetyl-β-glucosaminidase is often used as the indicator for the activity of soil chitinase.

Phosphatase characterizes a broad group of enzymes that hydrolyze phosphate ester and subsequently release phosphate. In general, phosphatase has low substrate specificity and thus can act upon a number of different structurally-related substrates (Alef et al. 1995). Extensively studied soil phosphatase includes acid phosphatase and alkaline phosphatase that are mainly found in acid and alkaline soils, respectively (Tabatabai 1994).

3.2.2 *Lignocellulolytic Enzymes as Predictors of Organic Matter Decomposition*

A relationship between soil enzyme activity and decomposition rate is highly expected given that extracellular enzyme-catalyzed depolymerization, is the rate limiting step of decomposition. Lignin- and cellulose-degrading enzymes appear to be favorably chosen for predicting decomposition, mainly because soil organic matter and plant materials are of carbohydrate- and phenolic-based structures. Sinsabaugh et al. (1992) developed a regression model to estimate the mass loss rate of birch sticks using five lignocellulose-degrading enzymes, including β-glucosidase, endocellulase, exocellulase, xylosidase, and phenol oxidase. Enzyme activities accounted fairly well for variations in mass loss rate across upland, riparian, and lotic systems, with r^2 ranging from 0.65 for endocellulase to 0.83 for β-glucosidase. Furthermore, the predictability of soil enzymes was improved by integrating activities of the five enzymes into a single and simple variable, referred to lignocellulase activity which explained 83% of mass loss variation. In a later study, an integrated cellulase activity was also used to estimate the decomposition rate of litters of various deciduous species (Sinsabaugh and Linkins 1993). Lignocellulolytic enzymes as predictors of organic matter decomposition appear to be supported by a number of nitrogen fertilization experiments, where changes in decomposition rate of plant litters are found to be positively associated with changes in cellulase and/or phenol oxidase activities (Carreiro et al. 2000; Saiya-Cork et al. 2002; DeForest et al. 2004; Waldrop et al. 2004; Sinsabaugh et al. 2005).

An oxidative enzyme-based "latch" mechanism has been proposed to elucidate soil organic carbon retention in peatlands (Freeman et al. 2001a). Under limited oxygen supply, low phenol oxidase activity slows down degradation of phenolic compounds in soil aqueous phase. Because of phenolic toxicity to hydrolytic enzymes (Pind et al. 1994; Freeman et al. 2001a), phenolic accumulation in soil solution can inhibit hydrolytic degradation of organic matter and therefore benefit soil carbon sequestration. Soil oxidative enzymes have also been considered as the proximate control for organic matter decomposition, in other ecosystems. Differences in soil carbon storage among various habitats appear to reflect differences in the activities of oxidative enzymes. Ecosystems in high and low latitudes both have a low rate of primary production, but their soil organic carbon contents differ considerably. In low latitude ecosystems where phenol oxidase and peroxidase are high, soil organic carbon is low. In contrast, great soil organic carbon exists in high altitude ecosystems where the activities of phenol oxidase and peroxidase are constrained by low pH, low temperature, and low oxygen availability (Sinsabaugh et al. 2008).

3.2.3 Enzymes as Indicators of Soil Nutrient Availability

Soil enzymes seem to be informative to capture soil nutrient supplying capacity. Positive correlations between the activity of soil enzymes and nutrient mineralization have been reported in agricultural soils. In long-term nirogen-fertilization cropping systems, for example, the activity of *N*-acetyl-glucosaminidase was tightly assoicated with nitrogen mineralization potential (Ekenler and Tabatabai 2002). Nitrogen minerlization from a broad range of soils in the North Center region of the United States was well correlated with the activities of *N*-acetyl-glucosaminidase and arylamidase (Ekenler and Tabatabai 2004; Dodor and Tabatai 2007).

Enzyme activity can also be used to indicate soil nutrient availability. A number of studies reveal that activities of soil phosphorus- and nitrogen-mineralization enzymes are negatively related to soil available phosphorous and nitrogen, respectively (Speir and Ross 1978; Sinsabaugh et al. 1993; Kang and Freeman 1999). In various cropping systems in the semi-arid region of Oregon, USA, for example, amidase and urease activitites have been found to be inversely related to application rates of synthetic nitrogen fertilizers (Dick 1992; Bandick and Dick 1999). Allison et al. (2008) also reported that nitrogen fertilization significantly suppressed the activities of protein- and chitin-degrading enzymes.

3.3 Microbial Enzyme Production as Affected by Organic Carbon and Nutrient Availability

3.3.1 Positive and Negative Feedback Mechanisms

During evolution, any type of microbes may have developed some cost-efficient strategy to explore resource. Microbes produce extracellular enzymes perhaps only when the enzymes can help to better use the resource and therefore lead to optimal microbial growth and metabolisms. This cost-efficient strategy predicts that microbial enzyme production can be low when the end products of enzymatic reactions, e.g., nutrients, are abundant or when the substrates of enzymatic reactions, e.g., complex organics, are limited (Allison and Vitousek 2005). Soil enzyme activities are expected to be suppressed by the addition of nutrients or induced by the addition of complex organics.

End product suppression on enzymes appears to be supported by numerous observations that the activity of a nutrient-mineralization enzyme is inversely related to the nutrient availability. Olander and Vitousek (2000) demonstrated that negative feedback was indeed the cause for the reduction of phosphatase and chitinase activities after long-term respective phosphorus and nitrogen fertilization. They argued that there were no direct and adverse effects of fertilization on enzyme activity because phosphatase and chitinase did not respond to short-term fertilization.

Inverse relationship between phosphorus availability and phosphatase activity seems to be more common than the one between nitrogen availability and nitrogen-mineralization enzymes. This is because organic phosphorus largely exists in the form of phosphate esters and is mineralized mainly through phosphatase catalysis. In contrast, organic nitrogen appears in various forms and can also be tightly associated with organic carbon. Not only do diverse nitrogen-mineralization enzymes catalyze mineral nitrogen release from complex organics, but also mineral nitrogen, can be produced as the byproduct of organic carbon decomposition. Therefore, a particular nitrogen-mineralization enzyme activity may not always have a negative relation with the available nitrogen (Allison and Vitousek 2005).

Above and below a threshold of nutrient availability, negative feedback may be irrelevant to microbial enzyme production. In the case of high nutrient availability, nutrient-mineralization enzyme activity can be low and thus fertilization will not generate additional suppressive effects (Olander and Vitousek 2000). When nutrient availability is low, the nutrient may greatly limit microbial growth and metabolisms. Thus, fertilization may stimulate microbial biomass and in turn microbial enzyme production. Iyyemperumal and Shi (2008) examined *N*-acetyl-glucosaminidase and protease activities in response to nitrogen fertilization rates. Soil microbial biomass increased with fertilization rates, and enzyme activities were, therefore, greater in fertilized soils than in the unfertilized controls.

Substrate induction on enzymes has been attested through substrate addition studies. For example, chitinase activity can be induced by chitin, phosphatase by organic phosphorus, β-glucosidase by cellobiose, and cellulase by cellulose (Smucker and Kim 1987; Chróst 1991; Shackle et al. 2000). However, positive induction may not always take place. Allison and Vitousek (2005) reported that cellulose addition had no any effect on β-glucosidase activity and also that collagen and cellulose phosphate addition did not increase the activities of respective glycine aminopeptidase and acid phosphatase. Lack of substrate induction on enzymes is thought to occur under some conditions, such as lack of a small amount of available nutrient that is required to activate inducible enzyme synthesis or enzyme production at low cost so that enzyme can be produced constantly rather than inducibly (Chróst 1991; Allison and Vitousek 2005).

3.3.2 *Microbial Resource Reallocation*

As a cost-efficient entity, the microbial community is expected to allocate resource proficiently for the syntheses of various carbon- and nutrient-acquiring enzymes (Sinsabaugh and Moorhead 1994). In response to a nutrient supply, microbial community likely shifts resource previously used for the synthesis of the nutrient-mineralization enzyme to the production of carbon- and other nutrient mineralization enzymes. Some supporting evidence includes increase in phosphatase activity due to improvement in nitrogen availability and increase in cellulose-degrading enzyme activity due to nitrogen and phosphorus fertilization (Dick et al. 1988; Zou et al. 1995; Allison and Vitousek 2005). A resource shift to the production of carbon-mineralization enzymes has been elegantly demonstrated in wetland soils with different available phosphorus and nitrogen (Fig. 3.1, Penton and Newman 2007).

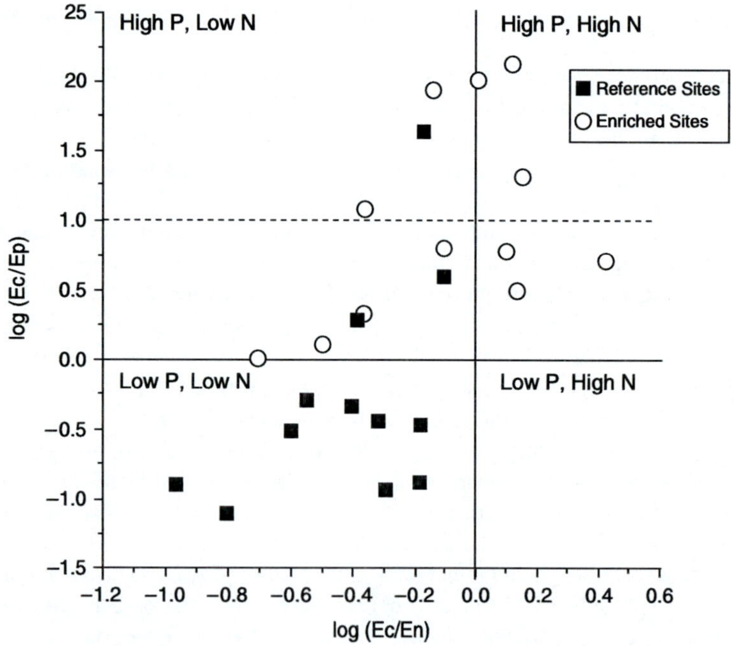

Fig. 3.1 Log plot of Ec/En vs. Ec/Ep for enriched and reference samples. Each point represents the mean from one sampling period. Ec/Ep values greater than 1 indicate high P availability. Adapted from Penton and Newman (2007)

In phosphorus and/or nitrogen rich soils, phosphorus- and nitrogen-mineralization enzyme activities were relatively low and instead carbon-mineralization enzyme activity increased.

It should be noted that resource partitioning among carbon-, nitrogen-, and phosphorus-mineralization enzymes is an integral feature of microbial community and thus perhaps follows some reaction stoichiometry (Sinsabaugh et al. 2008). Across various ecosystems, for example, some carbon-, nitrogen-, and phosphorus hydrolases are in close 1:1:1 relationship (Sinsabaugh et al. 2008). It is uncertain to what degree microbial resource reallocation for enzyme production can be modulated in terms of relative abundance of organic carbon and nutrient availability.

Interestingly, the magnitude of microbial resource allocation appears to be enzyme or nutrient specific. Allison and Vitousek (2005) found that enzymes involved in phosphorus mineralization were more responsive to nitrogen supply than nitrogen-mineralization enzymes to phosphorus availability. Penton and Newman (2007) also observed that carbon-mineralization enzymes were more responsive to phosphorus availability than to nitrogen availability. Effects of nitrogen and phosphorus availability on the production and activity of carbon-mineralization enzymes are intricate. An increase in nitrogen availability often stimulates the activity of cellulase. However, great nitrogen availability is found to reduce the activity of oxidative enzymes such as phenol oxidase and peroxidase.

3.3.3 *Microbial Nutrient Demand*

Microbial nitrogen mining has been proposed as a microbial process of obtaining nitrogen concealed within recalcitrant organic substances (Moorhead and Sinsabaugh 2006; Craine et al. 2007). Besides in the form of proteins, amino acids, and other labile compounds, nitrogen is tightly associated also with recalcitrant organic carbon. When nitrogen required for microbial growth and metabolisms cannot be met from labile nitrogen mineralization, microbes may "mine" nitrogen from recalcitrant organic matter and plant litters by producing lignin-degrading enzymes. Microbial nitrogen mining hypothesis seems to be in agreement with the common observation that ligninase activity is induced at low nitrogen availability but suppressed at high nitrogen availability (Kirk and Farrell 1987; Fog 1988). For the reason that phosphorous is normally unassociated with organic carbon, microbial phosphorous mining less likely occur (Craine et al. 2007). Microbial nitrogen mining helps to reconcile the suppressive effects of high available nitrogen on lignin-degrading enzymes and predicts that increase in available nitrogen may lead to reduction in decomposition rate.

3.4 Soil Enzyme Activity in Relation to the Quantity and Quality of Soil Organic Matter

As the major carbon and nutrient source, soil organic matter may have positive induction effects on microbial enzyme production. Furthermore, soil enzymes can be immobilized and thus accumulated on soil organic matter. Adsorption appears to provide good protection from decomposition, yet permit the retention of enzyme activity (Burns 1982). Accordingly, soil enzyme activities are expected to be positively related to soil organic matter. Sinsabaugh et al. (2008) examined the relationship between soil organic carbon content and the activity of seven soil enzymes using data from 40 terrestrial ecosystems. They observed that the activities of β-1,4-glucosidase, cellubiohydrolase, β-1,4-*N*-acetylglucosaminidase, and phosphatase were positively correlated with soil organic carbon content. However, the activity of phenol oxidase, peroxidase, and lucine aminopeptidase was independent of organic carbon content.

Soil organic matter is largely heterogeneous in terms of carbon and nutrient chemistry. Provided that substrate positive induction is an important mechanism for soil enzyme production, organic matter chemistry may be manifested in the relative abundance of various carbon and nutrient mineralization enzymes. There are some endeavors for investigating the relationship between organic matter chemistry and the activity of soil enzymes. For example, positive correlations between peptidase and protein and between β-1,4-*N*-acetyl glucosaminidase and polysaccharides were found in both light and heavy fraction of soil organic carbon (Grandy et al. 2007). In a long-term rye monoculture system, xylanase activity was correlated with the relative abundance of xylan and xylose (Leinweber et al. 2008). Shi et al. (2006) examined the relationship of soil enzyme activities with the chemical composition of soil organic matter in a turfgrass chronosequence. They found that integrated soil enzyme activity was unrelated to chemical compositions of organic matter characterized by Fourier transform infrared spectroscopy. No correlation between enzymes and organic matter is perhaps because soil organic matter reflects long-term decomposition processes whereas enzyme activity fluctuates with current conditions (Grandy et al. 2007).

3.5 Agricultural and Ecological Significance of Soil Enzymes in Soil Carbon Sequestration

3.5.1 Assessing Carbon Efflux in Peatlands Under Global Climate Change

Peatlands hold a vast amount of partially decayed vegetation matter due perhaps to phenol oxidase-associated low decomposition rate (Freeman et al. 2001a, 2004).

Peatland conditions that confine the activity of phenol oxidase, however, are projected to change with global climate change. Because of large organic carbon pool size, small shifts in decomposition rate may exert, through carbon efflux, enormous feedbacks to global environment.

Phenol oxidase has been considered as an "enzymatic latch" to hold organic carbon in peatlands (Freeman et al. 2001a). Oxygen that is required for phenol oxidase activity is often absent or limited in peatlands. Upon release of oxygen constraint by frequent droughts, phenol oxidase activity will increase, thereby accelerating the decomposition of soluble phenolics, the compounds inhibitory or toxic to hydrolytic enzymes. In turn, the activity of soil hydrolytic enzymes is improved. Lowering water table in peatlands has been found to significantly increase the activities of hydrolytic enzymes such as β-glucosidase, phosphatase, and sulphatase (Freeman et al. 1996). By eradicating phenol oxidase "latch", frequent droughts may generate sizeable carbon dioxide efflux into the atmosphere. Furthermore, reduction in organic carbon can impair peatland functions such as food preservation and water purification.

Peatlands also contribute greatly to aquatic carbon budget. Approximately, 20% of terrestrial dissolved organic carbon entering the ocean comes from peatlands (Fenner et al. 2007). Global warming can increase dissolved organic carbon export through its controls on phenol oxidase activity (Freeman et al. 2001b; Fenner et al. 2007). An increase of 10°C could lead to 36% increase in the activity of phenol oxidase (i.e., $Q_{10} = 1.36$) (Freeman et al. 2001b). In response, dissolved organic carbon increased by a Q_{10} value of 1.36 and soluble phenolics by a Q_{10} value of 1.72. Selective enrichment of soluble phenolics may further limit the decomposition of dissolved organic carbon through their inhibitory roles on hydrolytic enzymes, which can strengthen the loss of dissolved organic carbon from peatlands to aquatic systems.

Apparently, exports of dissolved organic carbon and soluble phenolics from peatlands are associated with phenol oxidase-degradation effects on peat matrix (Fenner et al. 2005a). Dissolved organic carbon discharge from peatlands can be quite large following a warm and dry period due to great fluctuation in phenol oxidase activity. After a short-term impeded drainage, for example, phenol oxidase activity can be improved and, as a result, soluble phenolics and dissolved organic carbon can increase (Toberman et al. 2008a). Carbon efflux from peatlands to aquatic systems may have significant impacts on regional redistribution of terrestrial carbon and exert significant controls over productivity and biogeochemical cycles in downstream ecosystems (Limpens et al. 2008).

3.5.2 *Predicting Soil Carbon Sequestration in Temperate Forests*

Lignocellulolytic enzymes have become the focal point to learn about the capacity of soil carbon sequestration in temperate forests under projected atmospheric nitrogen deposition. Because of opposite nitrogen effects on cellulose and lignin-degrading enzymes, magnitude and direction of nitrogen addition on decomposition appear to depend on the chemistry of organic carbon substances. Carreiro et al. (2000) examined the mass loss of plant litters in relation to relative abundance in lignin and cellulose. For dogwood litters with a low lignocellulose index (i.e., LCI), increase in nitrogen availability from an ambient level to 20 or 80 kg nitrogen ha^{-1} $year^{-1}$ stimulated the activities of both cellulase and phenol oxidase (Fig. 3.2). Cellulase activity alone could explain approximately 65% of variation in litter mass loss. For red oak litters with a high LCI value, increase in nitrogen availability stimulated cellulase activity but repressed phenol oxidase activity. Phenol oxidase alone accounted for about 51% of variation in decay coefficient.

Phenol oxidase associated with humified soil organic matter also responds to elevated nitrogen availability. In forests dominated with sugar maples and oaks, phenol oxidase activity in mineral soil was significantly reduced by elevated available nitrogen (Saiya-Cork et al. 2002; DeForest et al. 2004; Gallo et al. 2005), leading to increase in soil organic carbon storage. Over a 3-year period of nitrogen fertilization in temperate forests dominated with sugar maple-red oak and sugar maple-basswood, change in soil organic carbon content was found to be highly and negatively associated with change in soil phenol oxidase activity (Fig. 3.3), indicating the vital role of phenol oxidase in predicting soil carbon sequestration (Waldrop et al. 2004). Waldrop et al. (2004) proposed that ecosystems with highly lignified litters would accrue soil carbon and ecosystems with low lignin content would lose soil carbon in response to increase in soil available nitrogen. Over a three year period, soil carbon decreased by 20% in sugar maple-dominant ecosystem and increased by 10% in oak-dominant ecosystem.

Although stimulation on cellulolytic enzymes and suppression on oxidative enzymes have been described as the important mechanism dictating the decomposition of plant litters and soil organic matter under elevated nitrogen availability, positive, negative and no effects of nitrogen addition on both oxidative and hydrolytic enzymes have been reported. In general, it is more consistent for nitrogen stimulation effects on the activity of cellulolytic enzymes than for nitrogen suppressive effects on lignin-degrading enzymes (Carreiro et al. 2000; Saiya-Cork et al. 2002; Waldrop et al. 2004; Gallo et al. 2005; Sinsabaugh et al. 2005; Keeler et al. 2009). In addition, magnitude of change in soil enzyme activity in response to nitrogen availability seems to be related to the pool size of accumulative soil enzymes. Given that soil organic matter and clay minerals are two major components in stabilizing extracellular enzymes, enzyme response to an increase in available nitrogen is expected to be less in ecosystems with high soil organic matter and/or clay minerals. Zeglin et al. (2007) demonstrated that increase in nitrogen availability stimulated cellulolytic enzymes by about 40% in a soil with 0.5% of organic carbon, about 10% in a soil with 3% of organic carbon and almost 0% in a soil with 5% organic carbon. Enzyme-based prediction of soil carbon dynamics under atmospheric nitrogen deposition appears to be more significant and detectable in sandy-textured soils (Waldrop et al. 2004). In comparison with finer textured

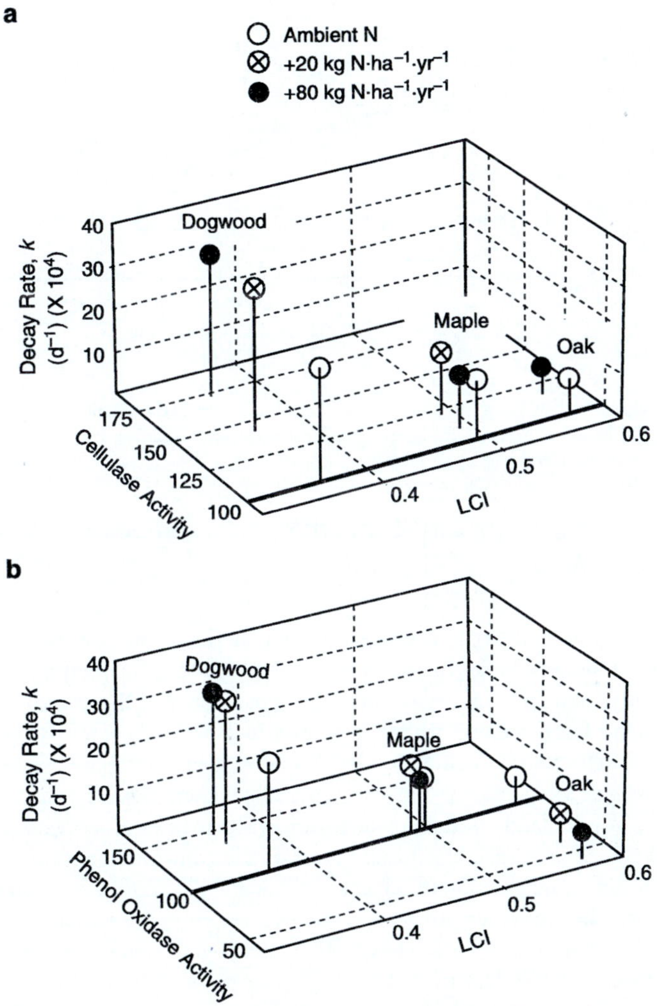

Fig. 3.2 Effects of two levels of N addition on the exponential decay rate coefficients (k) of dogwood, red maple, and red oak leaf litters and their relationships with litter LCI (lignocellulose index) and integrated enzyme activities relative to control levels (ambient N): (**a**) cellulase activity (combined activities of β-glucosidase and cellobiohydrolase); (**b**) activity of phenol oxidase, a ligninolytic enzyme. *Solid bold lines* represent the enzyme activities of ambient control levels, defined as 100%. Note especially the switch in direction of the N effect on phenol oxidase activity in the species with the highest LCI (i.e., lignin/[lignin + cellulose]). Adapted from Carreiro et al. (2000)

soil, a sandy soil has a low capacity of adsorption and thus protects soil organic matter poorly from decomposition. Therefore, a large proportion of organic matter can be subjected to enzyme-catalyzed degradation.

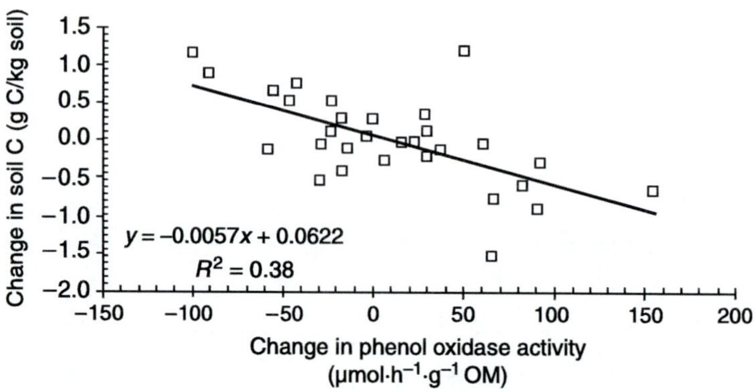

Fig. 3.3 Regression analyses of changes in phenol oxidase activity and changes in soil C following high levels of simulated N deposition (80 kg N ha^{-1} $year^{-1}$). *OM* stands for organic matter. Adapted from Waldrop et al. (2004)

3.5.3 Understanding Soil Carbon Storage Potential in Semi-arid Grasslands

Traditionally, low soil organic matter content in semi-arid grasslands has been attributed to low net primary production and thus low plant carbon input into soil. Until recently, oxidative enzymes have been considered as the culprit for limited accumulation of soil organic matter in semi-arid grasslands (Stursova and Sinsabaugh 2008). The activities of both phenol oxidase and peroxidase are found to be orders of magnitude greater in semi-arid grasslands than in other ecosystems such as forests and grasslands in temperate regions (Stursova et al. 2006; Zeglin et al. 2007). Microbial biomass in a semi-arid grassland in New Mexico was >40% lower than that in temperate grasslands in Kansas, USA and KwaZulu-Natal, South Africa and likewise the activities of soil hydrolytic enzymes including phosphatase, *N*-acetyl-glucosaminidase, cellobiohydrolase, and β-glucosidase were significantly low (Zeglin et al. 2007). However, the large pool size of oxidative enzymes seems to be unexpected from relatively small microbial biomass. Stursova and Sinsabaugh (2008) proposed that high activity of oxidative enzymes was due to selective stabilization under soil conditions in semi-arid grasslands and as a result, organic matter decomposition could remain at a rate equivalent to the net primary production.

Adapting to prolonged dryness and episodic precipitation, ecosystems in a semi-arid region may be more susceptible to global environmental change such as rainfall frequency, atmospheric nitrogen deposition, and elevated carbon dioxide. Similar to temperate forests, increase in soil nitrogen availability in semi-arid grasslands can stimulate the activity of soil hydrolytic enzymes such as cellobiohydrolase and β-glucosidase. However, soil nitrogen availability has little impact on the activity of oxidative enzymes (Henry et al. 2005; Stursova et al. 2006; Zeglin et al. 2007). No oxidative enzyme response appears to be a common phenomenon in grasslands where vegetations are of low lignin content (Zeglin et al. 2007). Henry et al. (2005) have provided enzyme-based insights on soil carbon dynamics in a semi-arid grassland under global environment change. While atmospheric nitrogen deposition accelerates decomposition via its positive effects on soil hydrolytic enzymes, increase in water availability can accelerate decomposition and thus soil carbon loss via improvement of oxidative enzyme activities.

3.5.4 Restoring Soil Organic Matter from the Perspective of Soil Enzymes

Decline in soil organic matter, which results often from intensive row-crop production, has been recognized as a critical issue limiting long-term land use and thus sustainable agricultural profits. To maintain or improve soil organic matter, diverse management practices such as no till farming, plant residue return, and crop rotations have been increasingly implemented in farming systems. Field-scale trials and observations have provided valuable information on soil organic matter dynamics, under various management practices. However, underlying mechanisms to elucidate management-associated organic matter decomposition appear to be overlooked. Recently, several studies have examined the response of soil lignocellulolytic enzymes to management practices (Henriksen and Breland 1999; Matocha et al. 2004; Zibliske and Bradford 2007).

Henriksen and Breland (1999) investigated the effects of soil nitrogen availability on carbon mineralization and soil enzyme activity in wheat straw amended soil in a 70-day incubation experiment. Increase in soil inorganic nitrogen concentration enhanced the activity of cellulolytic enzymes including exocellulase, endocellulase, and hemicellulase and consequently soil carbon mineralization. In wheat straw unamended soil, however, increase in available nitrogen suppressed the activity of endocellulase, hemicellulase, and exocellulase. Such negative effects were thought to result from negative nitrogen effects on humus degradation, thereby reducing the availability of carbohydrates structured within humus complex. In a year-long incubation study, Wang et al. (2004) also found that mineral nitrogen addition stimulated the decomposition during the first 100 days of plant materials of various biochemical compositions, including wheat straw and sugarcane litters. With progressive decomposition, however, negative nitrogen effects on recalcitrant fraction of plant materials appeared. At the end of incubation, mineral nitrogen addition generally had an overall negative effect on decomposition of plant materials. While available nitrogen that is about 1.2% of plant dry matter seems to be sufficient to the optimum decomposition of labile fraction of plan residues (Henriksen and Breland 1999), quantitative information between available nitrogen and its suppressive effect on recalcitrant fraction is limited. Information on phenol oxidase activity in relation to soil nitrogen availability may help to address long-term nitrogen effects on the decomposition of crop residues and soil organic matter.

Phenol oxidase appears to be manageable in agricultural soils through application of nitrogen fertilizers. Matocha et al. (2004) reported that nitrogen fertilization at 336 kg nitrogen ha^{-1} reduced soil phenol oxidase activity by 38% in no-till systems, but had no effect in the moldboard plow systems. In bermudagrass and tall fescue production systems, activity of soil phenol oxidase was reduced by ammonium nitrate fertilzation at >400 kg nitrogen ha^{-1} $year^{-1}$ and in contrast stimulated by swine lagoon effluent application (Iyyemperumal and Shi 2008). Inconsistent fertilization effects suggest that soil nitrogen availability may not be the sole factor regulating soil phenol oxidase activity. By examining soil phenol oxidase under various oxygen availabilities, Zibliske and Bradford (2007) found that the activity of phenol oxidase could be greatly suppressed under <10% oxygen.

In agricultural soil, phenol oxidase appears to act as an "enzymatic latch" to regulate soil organic carbon storage. With the incubation of agricultural soils amended with cowpea or sorghum residues under headspace oxygen concentrations ranging from 0.5 to 21%, the activities of phenol oxidase and peroxidase were found to be negatively related to soluble polyphenolics and dissolved soil organic carbon (Zibliske and Bradford 2007). It appears that reduction in phenol oxidase decreased the decomposition of soluble polyphenolics, which in turn inhibited the decomposition of dissolved organic carbon. These observations agree with those in peatlands and suggest that oxidative enzymes pool can be an important control to the accretion of soil organic carbon in agricultural soils. To develop a management tool from the concept that the activity of oxidative enzymes is an important control for soil carbon storage and nutrient mineralization, more research is needed to quantify the relations of soil oxidative enzymes with soil properties and management practices.

3.6 Conclusions

Soil enzymes have been acknowledged as important soil property to capture soil carbon and nutrient dynamics. Of numerous hydrolytic and oxidative soil enzymes, phenol oxidase appears to be the proximate control dictating the decomposition of soil organic matter. Environment that is conducive to the activity of soil phenol oxidase has been found to favor decomposition of soil organic matter and plant litters and, therefore, likely to impede soil carbon sequestration. To date, several ecosystem perturbation factors and management practices have been evaluated for their impacts on soil phenol oxidase activity. Other soil enzymes that are effective to forecast soil carbon and nutrient dynamics include cellulolytic enzymes and the enzymes for nitrogen and phosphorus mineralization.

As a cost-efficient entity, microbial community appears to possess several tactics for producing extracellular enzymes in relation to organic carbon chemistry and nutrient availability. End product suppression on enzymes represents one of microbial strategies and accounts for negative associations observed quite often between the activity of soil phosphatases and phosphorus availability and in some cases between the activity of N-acquiring enzymes and nitrogen availability. Nutrient induction for carbon- and other nutrient-mineralization enzymes reflects microbial resource reallocation in response to a nutrient supply. This strategy predicts that the activities of carbon- and phosphorous-mineralization enzymes will increase with nitrogen fertilization and the activities of carbon- and nitrogen-mineralization enzymes can be strengthened under the condition of abundant phosphate. Microbial nitrogen mining emerges from the observation that phenol oxidase activity is inhibited by an increase in available nitrogen. An ability of mining a nutrient concealed within recalcitrant organic substances seems to be another approach for microbial community to generate some carbon-mineralization enzymes.

There are a number of robust cases that soil enzyme activities can be predicted from microbial strategies for extracellular enzyme production and that changes in soil enzyme activities are able to manifest the direction and magnitude of organic carbon decomposition and nutrient mineralization. In forest ecosystems, for example, increase in cellulase activity and decrease in phenol oxidase activity under projected atmospheric nitrogen deposition seem to be in agreement with predictions based on microbial resource allocation and microbial nitrogen mining. Modifications in lignocellulolytic enzyme activities are, therefore, accountable for changes in decomposition rates and quantity of soil carbon retention. In peatlands, changes in phenol oxidase activity are tightly associated with carbon efflux to atmosphere and hydrosphere. However, there are some instances that neither microbial strategies for enzyme production nor relationships of soil enzyme activities with decomposition and mineralization are as expected. In order to reconcile different observations, some uncertainties need to be addressed, including thresholds for nutrient availability that determines whether nutrient suppression or induction on enzymes can take place, the relative abundance of soil organic matter and/or clay minerals that sustain a sufficient pool size of soil enzymes and therefore mask any effects associated with slight pool size change, and organic matter chemistry that determine the relative importance of cellulolytic enzymes and phenolics-degrading enzymes for predicting soil carbon decomposition.

Mechanisms that underline microbial enzyme production and the predictability of soil enzymes for soil carbon and nutrient dynamics have great implications in terms of global environmental change as well as management practices. To restore soil organic matter and optimize nutrient cycling in agricultural systems, we need to develop management practices that take the principles of soil enzyme activities into considerations.

References

Alef K, Nannipieri P (1995) Cellulase activity. In: Alef K, Nannipieri P (eds) Methods in applied soil microbiology and biochemistry. Academic, San Diego, CA, pp 345–349

Alef K, Nannipieri P, Trazar-Cepeda C (1995) Phosphatase activity. In: Alef K, Nannipieri P (eds) Methods in applied soil microbiology and biochemistry. Academic, San Diego, CA, pp 335–344

Allison SD, Vitousek PM (2005) Responses of extracellular enzymes to simple and complex nutrient inputs. Soil Biol Biochem 37:937–944

Allison SD, Czimczik CI, Treseder KK (2008) Microbial activity and soil respiration under nitrogen addition in Alaskan boreal forest. Global Change Biol 14:1156–1168
Bandick AK, Dick RP (1999) Filed management effects on soil enzyme activities. Soil Biol Biochem 31:1471–1479
Burke M, Cairney JWG (2002) Laccases and other polyphenol oxidases in ecto- and ericoid mycorrhizal fungi. Mycorrhiza 12:105–116
Burns RG (1982) Enzyme activity in soil: location and a possible role in microbial ecology. Soil Biol Biochem 14:423–427
Carreiro MM, Sinsabaugh RL, Repert DA, Parkhurst DF (2000) Microbial enzyme shifts explain litter decay responses to simulated nitrogen deposition. Ecology 81:2359–2365
Chróst RJ (1991) Environmental control of the synthesis and activity of aquatic microbial ectoenzymes. In: Chróst RJ (ed) Microbial enzymes in aquatic environments. Springer, New York, pp 29–59
Claus H (2004) Laccases: structure, reactions, distribution. Micron 35:93–96
Craine JM, Morrow C, Fierer N (2007) Microbial nitrogen limitation increases decomposition. Ecology 88:2105–2113
Criquet S, Tagger S, Vogt G, Iacazio G, Le Petit J (1999) Laccase activity of forest litter. Soil Biol Biochem 31:1239–1244
DeForest JL, Zak DR, Pregitzer KS, Burton AJ (2004) Atmospheric nitrate deposition and the microbial degradation of cellobiose and vanillin in a northern hardwood forest. Soil Biol Biochem 36:965–971
Dick RP (1992) A review: long-term effects of agricultural systems on soil biochemical and microbial parameters. Agr Ecosyst Environ 40:25–36
Dick RP, Rasmussen PE, Kerle EA (1988) Influence of long-term residue management on soil enzyme activities in relation to soil chemical properties of a wheat-fallow system. Biol Fertil Soils 6:159–164
Dodor DE, Tabatai MA (2007) Arylamidase activity as an index of nitrogen mineralization in soils. Commun Soil Sci Plan 38:2197–2207
Ekenler M, Tabatabai MA (2002) Beta-glucosaminidase activity of soils: effect of cropping systems and its relationship to nitrogen mineralization. Biol Fertil Soils 36:367–376
Ekenler M, Tabatabai MA (2004) Beta-glucosaminidase activity as an index of nitrogen mineralization in soils. Commun Soil Sci Plan 35:1081–1094
Endo K, Hayaski Y, Hibi T, Hosono K, Beppu T, Ueda K (2003) Enzymological characterization of EpoA, a laccase-like phenol oxidase produced by *Streptomyces griseus*. J Biochem 133:671–677
Fenner N, Freeman C, Reynold B (2005a) Observations of a seasonally shifting thermal optimum in peatland carbon-cycling processes: implications for the global carbon cycle and soil enzyme methodologies. Soil Biol Biochem 37:1814–1821
Fenner N, Freeman C, Reynolds B (2005b) Hydrological effects on the diversity of phenolic degrading bacteria in a peatland: implications for carbon cycling. Soil Biol Biochem 37:1277–1287
Fenner N, Freeman C, Lock MA, Harmens H, Reynolds B, Sparks T (2007) Interactions between elevated CO_2 and warming could amplify DOC exports from peatland catchments. Environ Sci Technol 41:3146–3152
Finzi AC, Sinsabaugh RL, Long TM, Osgood MP (2006) Microbial community responses to atmospheric carbon dioxide enrichment in a warm-temperate forest. Ecosystems 9:215–226
Fog K (1988) The effect of added nitrogen on the rate of decomposition of organic matter. Biol Rev 63:433–462
Freeman C, Liska G, Ostle NJ, Lock MA, Reynolds B, Hudson J (1996) Microbial activity and enzymic decomposition processes following peatland water table drawdown. Plant Soil 180:121–127
Freeman C, Ostle N, Kang H (2001a) An enzymic 'latch' on a global carbon storage. Nature 409:149

Freeman C, Evans CD, Monteith DT, Reynolds B, Fenner N (2001b) Export of organic carbon from peat soils. Nature 412:785

Freeman C, Olstle NJ, Fenner N, Kang H (2004) A regulatory role for phenol oxidase during decomposition in peatlands. Soil Biol Biochem 36:1663–1667

Gallo ME, Lauber CL, Cabaniss SE, Waldrop MP, Sinsabaugh RL, Zak DR (2005) Soil organic matter and litter chemistry response to experimental N deposition in northern temperate deciduous forest ecosystems. Global Change Biol 11:1514–1521

Grandy AS, Neff JC, Weintrau MN (2007) Carbon structure and enzyme activities in alpine and forest ecosystems. Soil Biol Biochem 39:2701–2711

Hammel KE (1997) Fungal degradation of lignin. In: Cadish G, Giller KE (eds) Driven by nature. CAB International, London, pp 33–45

Henriksen TM, Breland TA (1999) Nitrogen availability effects on carbon mineralization, fungal and bacterial growth, and enzyme activities during decomposition of wheat straw in soil. Soil Biol Biochem 31:1121–1134

Henry HA, Juarez JD, Field CB, Vitousek PM (2005) Interactive effects of elevated CO_2, N deposition and climate change on extracellular enzyme activity and soil density fractionation in a California annual grassland. Global Change Biol 11:1808–1815

Hullo MF, Moszer I, Danchin A, Martin-Verstraete I (2001) CoA of *Bacillus subtilis* is a copper-dependent laccase. J Bacteriol 183:5426–5430

Iyyemperumal K, Shi W (2008) Soil enzyme activities in two forage systems following application of different rates of swine lagoon effluent or ammonium nitrate. Appl Soil Ecol 38:128–136

Kang H, Freeman C (1999) Phosphatase and arylsulphatase activities in wetland soils: annual variation and controlling factors. Soil Biol Biochem 31:449–454

Keeler BL, Hobbie SE, Kellogg LE (2009) Effects of long-term nitrogen addition on microbial enzyme activity in eight forested and grassland sites: implications for litter and soil organic matter decomposition. Ecosystems 12:1–15

Kirk TK, Farrell RL (1987) Enzymatic combustion: the microbial degradation of lignin. Annu Rev Microbiol 41:465–505

Ladd JN (1978) Origin and range of enzymes in soil. In: Burns RG (ed) Soil enzymes. Academic, New York, pp 51–96

Leinweber P, Jandl G, Baum C, Eckhardt KU, Kandeler E (2008) Stability and composition of soil organic matter control respiration and soil enzyme activities. Soil Biol Biochem 40:1496–1505

Limpens J, Berendse F, Blodau C, Canadell JG, Freeman C, Holden J, Noulet N, Rydin H, Schaepman-Strub G (2008) Peatlands and the carbon cycle: from local processes to global implications – a synthesis. Biogeosciences 5:1475–1491

Lipson DA, Wilson RF, Oechel WC (2005) Effects of elevated atmospheric CO_2 on soil microbial biomass, activity and diversity in a charparral ecosystem. Appl Environ Microb 71:8573–8580

Matocha CJ, Haszler GR, Grove JH (2004) Nitrogen fertilization suppresses soil phenol oxidase enzyme activity in no-tillage systems. Soil Sci 169:708–714

McLatchey GP, Reddy KR (1998) Regulation of organic matter decomposition and nutrient release in a wetland soil. J Environ Qual 27:1268–1274

Moorhead DL, Sinsabaugh RL (2006) A theoretical model of litter decay and microbial interaction. Ecol Monogr 76:151–174

Olander LP, Vitousek PM (2000) Regulation of soil phosphatase and chitinase activity by N and P availability. Biogeochemistry 49:175–190

Paul EA, Clark FE (1996) Soil microbiology and biochemistry. Academic, San Diego, CA

Penton CR, Newman S (2007) Enzyme activity responses to nutrient loading in subtropical wetlands. Biogeochemistry 84:83–98

Pind A, Freeman C, Lock MA (1994) Enzymatic degradation of phenolic materials in peatlands – measurement of phenol oxidase activity. Plant Soil 159:227–231

Saiya-Cork KR, Sinsabaugh RL, Zak DR (2002) The effects of long-term nitrogen deposition on extracellular enzyme activity in an *Acer saccharum* forest soil. Soil Biol Biochem 34:1309–1315

Schimel JP, Weintraub MN (2003) The implications of exoenzyme activity on microbial carbon and nitrogen limitation in soil: a theoretical model. Soil Biol Biochem 35:549–563

Shackle VJ, Freeman C, Reynolds B (2000) Carbon supply and the regulation of enzyme activity in constructed wetlands. Soil Biol Biochem 32:1935–1940

Shi W, Dell E, Bowman D, Iyyemperumal K (2006) Soil enzyme activities and organic matter composition in a turfgrass chronosequence. Plant Soil 288:285–296

Sinsabaugh RL (1994) Enzymic analysis of microbial pattern and process. Biol Fertil Soils 187:69–74

Sinsabaugh RL, Linkins AE (1993) Statistical modeling of litter decomposition from integrated cellulase activity. Ecology 74:1594–1597

Sinsabaugh RL, Moorhead DL (1994) Resource allocation to extracellular enzyme production: a model for nitrogen and phosphorous control of litter decomposition. Soil Biol Biochem 26:1305–1311

Sinsabaugh RL, Antibus RK, Linkins AE, McClaugherty CA, Rayburn L, Repert D, Weiland T (1992) Wood decomposition over a 1st-order watershed – mass-loss as a function of ligno-cellulase activity. Soil Biol Biochem 24:743–749

Sinsabaugh RL, Antibus RK, Linkins AE, MaClaugherty CA, Rayburn L, Repert D, Weiland T (1993) Wood decomposition – nitrogen and phosphorus dynamics in relation to extracellular enzyme activity. Ecology 74:1586–1593

Sinsabaugh RL, Gallo ME, Lauber C, Waldrop MP, Zak DR (2005) Extracellular enzyme activities and soil organic matter dynamics for northern hardwood forests receiving simulated nitrogen deposition. Biogeochemistry 75:201–215

Sinsabaugh RL, Lauber CL, Weintraub MN, Ahmed B, Allison SD, Crenshaw C, Contosta AR, Causack D, Frey S, Gallo ME, Gartner TB, Hobbie SE, Holland K, Keeler BL, Powers JS, Stursova M, Takacs-Vesbach C, Wallenstein MD, Zak DR, Zeglin LH (2008) Stoichiometry of soil enzyme activity at global scale. Ecol Lett 11:1252–1264

Smucker RA, Kim CK (1987) Chitinase induction in an estuarine ecosystems. In: Llewellyn GC, O'Rear CE (eds) Biodegradation research. Plenum, New York, pp 347–355

Speir TW, Ross DJ (1978) Soil phosphatase and sulphatase. In: Burns RG (ed) Soil enzymes. Academic, New York, pp 197–250

Stursova M, Sinsabaugh RL (2008) Stabilization of oxidative enzymes in desert soil may limit organic matter accumulation. Soil Biol Biochem 40:550–553

Stursova M, Crenshaw CL, Sinsabaugh RL (2006) Microbial responses to long-term N deposition in a semiarid grassland. Microb Ecol 51:90–98

Tabatabai MA (1994) Soil enzymes. In: Weaver RW, Angle S, Bottomley P, Bezdicek D, Smith S, Tabatabai A, Wollum A (eds) Methods of soil analysis, part 2 – microbiological and biochemical properties. Soil Science Society of America Inc, Wisconsin, pp 775–833

Toberman H, Artz FC, RRE ECD, Fenner N (2008a) Impeded drainage stimulates extracellular phenol oxidase activity in riparian peat cores. Soil Use Manage 24:357–365

Toberman H, Evans CD, Freeman C, Fenner N, White M, Emmett BA, Artz RRE (2008b) Summer drought effects upon soil and litter extracellular phenol oxidase activity and soluble carbon release in an upland Calluna heathland. Soil Biol Biochem 40:1519–1532

Waldrop MP, Zak DR, Sinsabaugh RL, Gallo M, Lauber C (2004) Nitrogen deposition modifies soil carbon storage through changes in microbial enzymatic activity. Ecol Appl 14:1172–1177

Wang WJ, Baldock JA, Dalal RC, Moody PW (2004) Decomposition dynamics of plant materials in relation to nitrogen availability and biochemistry determined by NMR and wet-chemical analysis. Soil Biol Biochem 36:2045–2058

Zeglin LH, Stursova M, Sinsabaugh RL, Collins SL (2007) Microbial responses to nitrogen addition in three contrasting grassland ecosystems. Oecologia 154:349–359

Zibliske LM, Bradford JM (2007) Oxygen effects on carbon, polyphenols, and nitrogen mineralization potential in soil. Soil Sci Soc Am J 71:133–139

Zou X, Binkley D, Caldwell BA (1995) Effects of dinitrogen-fixing trees on phosphorus biogeochemical cycling in contrasting forests. Soil Sci Soc Am J 59:1452–1458

Chapter 4
Enzymes in Forest Soils

Petr Baldrian and Martina Štursová

4.1 Introduction

Forest soils typically accumulate dead organic material, known as plant litter, on the soil surface. When fresh, this material contains a variety of substrates, such as soluble saccharides, organic acids, amino acids or starch, as well as the plant cell wall-derived biopolymers, cellulose, hemicelluloses and lignin, which are utilized during the growth of soil decomposer microorganisms. The sequential transformation of plant litter in temperate forests leads to the formation of a distinct organic horizon derived from the transformation products of litter decomposition as well as the dead biomass of soil microorganisms. The organic horizon is known for its high content of humic substances. Compared to agricultural soils and grassland soils, forest soils contain more fungal biomass (van der Wal et al. 2006), and the mycorrhizal community is formed mainly by the ectomycorrhizal fungi associated with tree roots instead of the arbuscular mycorrhizal fungi, which dominate the grasslands.

Due to the high input of organic matter, enzymes participating in the decomposition of organic matter play an important role in forest soils. The enzymes relevant to decomposition studies are those related to the degradation of plant cell wall polymers (cellulose, hemicellulose, lignin, pectin) and microbial biomass (chitin and other polysaccharides). Besides these, enzyme systems involved in the acquisition of nitrogen, phosphorus, and sulfur are also of interest because these elements are indispensable for the growth of soil microbial biomass.

Plant litter decomposition is actually the most important process in shaping the formation of the soil matter gradient. This complex process involves numerous taxa of decomposer microorganisms and a wide array of enzymatic transformations of

P. Baldrian (✉) and M. Štursová
Laboratory of Environmental Microbiology, Institute of Microbiology of the ASCR, v.v.i, Vídeňská 1083, Praha 4 14220, Czech Republic
e-mail: baldrian@biomed.cas.cz

G. Shukla and A. Varma (eds.), *Soil Enzymology*, Soil Biology 22,
DOI 10.1007/978-3-642-14225-3_4, © Springer-Verlag Berlin Heidelberg 2011

individual molecules. It has been described as a successional process, where the initial chemical composition defines the microbial community which, in turn, secretes degradative enzymes. The produced enzymes change the chemical properties of the substrate. These changes further drive the decomposer community succession in the litter (Sinsabaugh et al. 2002a). In this chapter, factors affecting the activity of enzymes in forest soils with respect to their spatial distribution are discussed. The lignocellulose-degrading enzymes and their producers are the subject of another chapter in this book (Chap. 9).

4.2 Distribution of Enzymes in the Soil Profile

The stratification of organic compounds due to their sequential transformation by decomposer microorganisms leads to the formation of a sharp gradient of soil properties in the upper layers of forest soils. Along with the amount of total soil carbon and its readily utilizable forms, microbial biomass and activity (e.g., respiration) typically decrease with soil depth (Šnajdr et al. 2008b). This results in a sharp vertical gradient of activities of lignocellulose-degrading enzymes as well as of other enzymes (chitinases, phosphatases, peptidases, proteases, catalase, and invertase). This type of vertical distribution of enzyme activities was reported in several forest soils ranging from the Mediterranean evergreen forests and temperate forests to boreal ecosystems (Trasar-Cepeda et al. 2000; Prietzel 2001; Andersson et al. 2004; Wittmann et al. 2004; Zhang et al. 2004). Recent results on the vertical distribution of enzymes in a temperate *Quercus petraea* forest demonstrated that enzyme activities dramatically change with soil depth, even within individual soil horizons. Most apparently, the upper part of the organic horizon is significantly different from the lower part with respect to the organic carbon content, enzyme activities and microbial biomass (Šnajdr et al. 2008b; Fig. 4.1). Although the spatial differences of enzyme activities level-off in the deeper parts of soils, the activity of phosphatase, arylsulfatase, protease, and urease decrease with soil depth and soil organic matter down to depths greater than 1 m (Venkatesan and Senthurpandian 2006).

Microbial biomass analyses showed that soil is also stratified with respect to the relative abundance of individual microbial groups. The relative amount of fungal biomass tends to decrease with soil depth, while the relative amount of actinomycetes increases (Baldrian et al. 2008; Šnajdr et al. 2008b). This is reflected in the differences in the relative abundance of individual enzymes at different depths, since these partly reflect the abundance of their producers. Some studies have shown that enzymes targeting easily decomposable compounds (e.g., invertase or proteases) can be limited to the surface soil layers, while oxidases and peroxidases are significantly active at great depths (Zhang et al. 2005; Venkatesan and Senthurpandian 2006). While most enzymes show higher activity in the topsoil horizons, the highest arylsulfatase activity in a boreal pine forest was recorded in the deeper horizons where bacterial biomass dominated over fungal biomass (Wittmann et al. 2004).

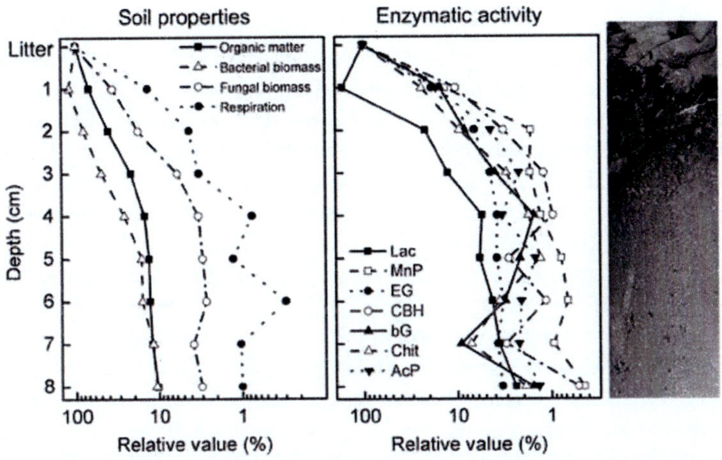

Fig. 4.1 The vertical gradient of soil organic matter content, microbial biomass, respiration and the activity of selected enzymes in a hardwood forest soil profile (*Quercus petraea*, Czech Republic, Europe) shows that enzyme activity decreases in the soil profile along with a decrease in soil organic matter content, soil microbial biomass and microbial activity. Bacterial biomass is based on the amount of bacteria-specific PLFA, and fungal biomass is based on ergosterol content. Depth was measured below the litter horizon. *Abbreviations*: *Lac* laccase, *MnP* Mn-peroxidase, *EG* Endo-1,4-β-glucanase, *CBH* cellobiohydrolase, *bG* β-glucosidase, *Chit* *N*-acetylglucosaminidase, *AcP* acid phosphatase. Data from Šnajdr et al. (2008b)

Litter quality and quantity is primarily responsible for the formation of soil property gradients with different slopes. In the same soil, the decrease of soil microbial biomass content and enzyme activity with depth was more pronounced under a hardwood tree (*Robinia pseudoacacia*) than in the same soil with a softwood *Pinus sylvestris*, the litter from which is more difficult to decompose (Landgraf et al. 2005).

4.3 Spatial Variability of Enzyme Activities in Soils

In contrast to uniform agricultural soils that are homogenized during land management, forest soils exhibit a significant level of horizontal variability in their properties. Relatively low variability is usually observed for physico-chemical variables, with coefficients of variation (CVs) of 5–10% for pH and approximately 25% for organic carbon, total carbon and total nitrogen in *Quercus* spp. and *Fagus* spp. soils (Morris 1999; Gömöryová 2004). Microbial biomass content is usually more variable: A CV of 44% was reported for fungal biomass content in a *Pinus sylvestris* forest humus layer and CVs of 36–48% were reported for both fungal and bacterial biomass in the upper 5 cm of a *Quercus rubra* forest soil (Morris 1999; Möttönen et al. 1999); CVs for individual PLFA molecules in the upper 5 cm of *Picea abies*/

Betula pubescens soil ranged from 16 to 37% (Saetre and Baath 2000). Soil enzyme activities are more widely varied and the recorded values often span several orders of magnitude (Prietzel 2001; Gömöryová 2004; Luis et al. 2005; Šnajdr et al. 2008b). For enzyme activities in *Quercus* spp. forest soils, the CVs were similar in the H and Ah horizons and ranged from 29 to 64% for phosphatase, β-glucosidase, endoglucanase, and other enzymes (Trasar-Cepeda et al. 2000; Šnajdr et al. 2008b).

The two most apparent factors that may affect the distribution of enzyme activities in soil are (1) the presence of their microbial producers and (2) the presence of substrates or inducers of enzyme expression. Enzymes produced constitutively by all groups of microorganisms, such as acid phosphatase (Criquet et al. 2004), can thus be expected to correlate with soil microbial biomass. Microbial carbon, respiration, and chitinase were shown to correlate in the H horizon of spruce and beech forest soil, but cellulase was not (Andersson et al. 2004). In studies with *Quercus ilex* litter, bacterial and fungal counts correlated with the activities of selected glycosyl hydrolases, but not with ligninolytic enzymes (Criquet et al. 2000, 2002). Fungi are often considered to be responsible for the production of several oxidative and hydrolytic enzymes in soils and the variation of fungal abundance is thus one of the important factors shaping the distribution of enzymes in soils (Moller et al. 1999; Caldwell 2005; Baldrian 2008). It was demonstrated in soil microcosms that the presence of saprotrophic basidiomycetes increased the activities of ligninolytic enzymes, polysaccharide hydrolases, chitinase, and acid phosphatase compared to non-colonized soil (Šnajdr et al. 2008a). Elevated activities of several enzymes were also recorded in soils under mycelia of fungi (Gramss 1997).

Soil quality is the second major factor that can potentially affect enzyme activity. In a mixed forest, bacterial biomass and community composition are regulated by the proximity of different tree species, presumably caused by the differences in litter quality, rhizosphere effects or the variation of soil moisture due to differential throughfall (Saetre 1999; Saetre and Baath 2000). In a *Fagus sylvatica* forest, catalase activity exhibited a patchy distribution with most of the activity found at tree crown perimeters (Gömöryová 2004). In contrast, phosphatase activity in rhizosphere soil of fir forests was found to be higher near tree stems as compared to the soil further from stems, which is probably due to the higher phosphate demand of tree root-colonized soil compared to bulk soil (Chen 2003).

In an attempt to locally detect enzyme activities in soils using immobilized substrates, it was demonstrated that the activity of several hydrolytic enzymes (e.g., phosphatase, aminopeptidase, chitinase, or β-glucosidase) in coniferous forest soils frequently co-localize with living biomass in the soil profile, such as roots, mycelial mats or ectomycorrhizal systems (Dong et al. 2007). In ectomycorrhizas, the expression of enzymes differs between the hartig net, fungal cords and fine mycelium. High phosphatase activity of fungal origin was found in fine *Suillus* hyphae and in mycorrhizal roots (Timonen and Sen 1998).

Cellulase and chitinase decreased with the decreasing C/N ratio in spruce litter and humus, but increased at the N-rich sites of beech litter microcosms (Andersson et al. 2004). An explanation for this may be that the decrease in the C/N ratio in litter corresponds with a decrease in accessible carbon sources in the soil during litter transformation, while extra nitrogen added to the litter increases its degradation. Humus under *Alnus* spp. with high nitrogen content exhibited higher microbial biomass and higher enzyme activities than humus under pine growing in the same type of soil (Niemi et al. 2007).

4.4 Factors Affecting Soil Enzyme Activities

4.4.1 Producers of Extracellular Enzymes

As mentioned above, fungi seem to be the only group of microorganisms that produce significant amounts of ligninolytic enzymes in soils and also seem to be the major producers of polysaccharide hydrolases (Moller et al. 1999; Baldrian 2008). This fact is underlined in forest soils by the high proportion of fungal biomass compared to other ecosystems. It was also demonstrated that fungi are responsible for the production of chitinase (*N*-acetylglucosaminidase) in soils, while the activity of arylsulfatase seems to be more reflective of the distribution of bacterial biomass (Moller et al. 1999; Baldrian et al. 2008). However, some enzymes, like phosphatase or β-glucosidase, are produced by a wide spectrum of soil organisms and their activity mainly corresponds to the amount of soil organic matter and microbial biomass (Amador et al. 1997).

4.4.2 Soil Chemistry and Nutrient Availability

The addition of low molecular mass organic compounds, such as monosaccharides or organic acids, affects the activity of several enzymes, such as acid phosphatase, phosphodiesterase, and urease, while other enzymes, such as protease, are unaffected. These additions also increase respiration (Renella et al. 2007). β-Glucosidase and phosphatase were also increased by the addition of carbon or nitrogen (Allison and Vitousek 2005).

Although the available phosphorus content in natural soil does not affect phosphatase activity (Venkatesan and Senthurpandian 2006), the lower demand for phosphorus after its addition typically leads to a decrease in phosphatase activity (Olander and Vitousek 2000). Similar results were also found for arylsulfatase in hardwood forest soil where the enzyme negatively correlated with SO_4^{2-} concentration and decreased by tens of percent after ammonium sulfate addition (Prietzel 2001).

The addition of cellulose increases both cellulolysis and total mineralization in soils, probably by the activation of cellulolytic enzymes. Although xylanase is often co-localized with cellulases in soils (Šnajdr et al. 2008b), it is not induced by the addition of cellulose (Scheu et al. 1993).

The effect of adding nitrogen to soils was frequently studied due to the fact that this process accompanies the anthropogenic pollution of the environment. Studies on lignocellulose degradation after nitrogen supplementation indicated that the effect depends greatly on the properties of the ecosystem. For more information, please see Chap. 9 of this book.

Chitinase activity was found to negatively correlate with soil nitrogen content and nitrogen addition decreased chitinase activity in soil with low nitrogen content (Olander and Vitousek 2000; Allison et al. 2008). This effect is probably a result of the fact that nitrogen addition decreases the relative amount of fungi in soils. There are, however, studies where neither fungal biomass content nor chitinase activity responded to simulated nitrate addition (Waldrop and Zak 2006) and the effect seems to be dependent on the properties of the ecosystem.

In general, the addition of nitrogen to hardwood forest soil has a more pronounced effect in the litter horizon than in the organic horizon. Urease, phosphatase, and several glycosidases increased by 7–56% in N-supplemented plots, while laccase and chitinase responded differently in soil and litter (Saiya-Cork et al. 2002).

The effect of elevated nitrogen input into soils is also apparent in cases where forests are invaded by trees with nitrogen-fixing symbionts, such as *Alnus* spp. or *Falcataria* spp. In both cases, a shift in the soil microbial community composition leading to the dominance of bacteria over fungi in high nitrogen conditions accompanied the changes in enzyme activities (Selmants et al. 2005; Allison et al. 2006).

The succession of vegetation and microbial communities is an important process, occurring in both natural and cultured ecosystems. The chemical changes accompanying the formation of a soil horizon are also reflected by the changes in microbial biomass and enzyme activity in top soils. Activities of extracellular enzymes and microbial biomass in topsoil tended to increase for at least 20 years during the spontaneous succession of coal mine deposits in temperate zones (Baldrian et al. 2008). In managed spruce forests, invertase and peroxidase activities increased until the site reached 23 years of age (Zhang et al. 2004). In both cases, the enzyme activities and microbial biomass decreased in the late stages of succession. In tropical soils that were reclaimed after bauxite mining, microbial biomass and enzyme activities increased much more rapidly after recovery and reached a plateau value after only 1 year of treatment; the increase of carbon and nitrogen content in the same soils was considerably slower and lasted for several years (Carneiro et al. 2008).

4.4.3 Seasonality

In biomes with substantial variation in temperature or moisture during the year, activities of extracellular enzymes in forest soils and litter show considerable seasonal variation. In previous studies, climatic factor such as temperature and soil moisture content have been identified as the main responsible factor for the observed seasonal differences in the case of laccase, polysaccharide hydrolases, phosphatase, urease, protease, and other enzymes (Dilly and Munch 1996; Criquet et al. 2000, 2002, 2004; Wittmann et al. 2004; Baldrian et al. 2008; Bastida et al. 2008). In hardwood forest with the litterfall season limited to autumn, the changes in the chemical composition of litter during its decomposition are another important cause of seasonal differences in enzyme activities (Dilly and Munch 1996; Fioretto et al. 2000).

In soils of hardwood forests developed during succession on mine deposits, seasonality was actually the most important factor in determining the variability of enzyme activities. Seasonality had a significant effect on glycosidases, arylsulfatase, chitinase, and especially phosphatase, accounting for 10–70% of the observed variability (Baldrian et al. 2008).

In temperate zones, higher enzyme activities are usually detected in the warm summer period as compared to the winter (Baum and Hrynkiewicz 2006; Niemi et al. 2007). Although the winter period is often regarded as a time of limited microbial activity in boreal forests where there are long periods of low soil temperature, as much as 7–32% of the enzyme-catalyzed annual turnover of organic compounds was estimated to occur during the 6 cold months of the year (Kahkonen et al. 2001; Wittmann et al. 2004). In ectomycorrhizal fungi, the activity of several hydrolases (particularly phosphatase and β-glucosidase) was higher in the winter when the supply of photosynthates from the host tree was limited as compared to the spring, resulting in a seasonal response of enzyme production based on nutrient availability (Mosca et al. 2007).

4.4.4 Soil Moisture Content

Soil moisture was previously reported to affect the microbial biomass, respiration in the soil and enzyme-catalyzed processes (Orchard and Cook 1983; Criquet et al. 2002). Most of the previous studies, however, focused on the effects of temporary drying and not on the effect of spatial variability. In dry periods, forest soils were reported to contain less microbial biomass and exhibited reduced enzyme activities (Ross 1987; Sardans and Penuelas 2005). In Mediterranean forests, the activities of laccase, β-glucosidase, acid phosphatase, urease, and protease were substantially reduced in the dry seasons (Criquet et al. 2002, 2004; Sardans and Penuelas 2005), and Mn-peroxidase in evergreen oak litter was only detected during humid periods (Criquet et al. 2000).

There is, however, also a significant level of spatial variation in soil moisture at any given time. The coefficients of variation of moisture in the upper part of different temperate forest soils were reported to be 10–60%, and the range of moisture content varied between 15 and 70%. The spatial differences in soil moisture content can partly to ascribed to the differences in throughfall (and the tree cover, since throughfall strongly correlates with the gap fraction) partly to the consumption by plant roots or translocation by fungal mycelia and partly to simple physical draining.

The spatial variation of soil moisture was found to affect the total microbial biomass (Stoyan et al. 2000; Gömöryová et al. 2006), although not necessarily the entire microbial community; in *Quercus rubra* forest soil, only bacterial and not fungal biomass was affected (Morris 1999). The spatial variability of soil moisture content also affects microbial processes. In different forest types, the moisture content positively correlated with the rate of N-transformation (Bengtson et al. 2005), phosphatase or catalase activity (Kramer and Green 2000; Gömöryová et al. 2006).

The effect of drying is very important since a relatively small moisture content reduction of 10% in Mediterranean evergreen forests caused 10–80% of the urease, protease and β-glucosidase activity. A higher reduction (21%) in moisture content caused by rainfall and runoff exclusion had an even more pronounced effect on the enzymes (Sardans and Penuelas 2005).

4.5 Litter Decomposition in Forest Ecosystems

Litter decomposition is the fastest process of soil organic matter decomposition due to the high initial amount of soluble and polymeric polysaccharides in the senescent plant material. Fresh leaf litter is typically composed of 3–7% soluble polysaccharides, 15–27% cellulose and 15–50% lignin. During litter transformation, soluble compounds including polysaccharides are removed first and lignin, which is degraded more slowly, usually accumulates (Osono and Takeda 2005). These changes in litter quality are reflected in the sequential production of individual enzymes participating in the decomposition of different litter components. Enzymes that act on soluble saccharides like invertase or α-glucosidase and β-glucosidase activity usually peak in the initial stages of decomposition (Sinsabaugh et al. 2002a; Fioretto et al. 2007; Papa et al. 2008). The activities of other cellulases usually increase more slowly and peak when the mass loss is at 40–80%. The activity of phenol oxidizing enzymes typically peak at late stages of decomposition when the available cellulose disappears (Sinsabaugh et al. 2002a). The higher laccase activity recorded during the late stages of decomposition on *Quercus ilex* and *Myrtus communis* litter was associated with an increase in the fungal biomass content of the litter (Fioretto et al. 2007; Papa et al. 2008).

The successive production of enzymes on decomposing litter reflects the successional changes of the decomposer community. With respect to fungi, it is generally accepted that the proportion of basidiomycetes that are able to produce ligninolytic enzymes increases with time, while ascomycetes dominate the earlier stages of decomposition, with many of them already being present in the senescent litter as endophytes (Aneja et al. 2006; Osono 2007). In beech litter degradation experiments, fungi colonizing fresh litter were mainly pectinase producers, while the middle phases of decomposition were dominated by cellulase-producing species and the late phases were characterized by cellulolytic and chitinolytic fungi (Kjoller and Struwe 1990, 2002).

Water soluble C content, cellobiohydrolase and laccase activities were found to correlate with mass loss in different litter types (Allison and Vitousek 2004). Linkins et al. (1990a, b) followed the decomposition of *Cornus florida*, *Quercus prinus*, and *Acer rubrum* leaves in a deciduous woodland and in vitro. Enzyme activities were affected by litter type and both the total mass loss and the loss of cellulose correlated with cellulase activities. However, clearly linking enzyme activity to litter decomposition is not always so simple. The decomposition of *Cornus florida* litter was accomplished with much less enzyme activity than that needed for the same mass loss in *Quercus prinus* or *Acer rubrum* litter (Carreiro et al. 2000).

Litter decomposition is a sensitive process which is affected by several external factors including forest management, nutrient supply and the presence and activity of microinvertebrates (Schlatte et al. 1998; Sinsabaugh et al. 2002b; Waldrop et al. 2003). Litter decomposition is also subject to seasonality. During the decomposition of different litter types in arid soils, the activities of lignocellulose-degrading enzymes and respiration were more affected by the changes of litter moisture content than by succession (Fioretto et al. 2007; Papa et al. 2008).

4.6 Conclusions

While there is sufficient information on the most important factors affecting enzymatic activity in forest soils, there are still many questions open for further research. The factors affecting the microscale spatial distribution of enzymes are only partly known and the data regarding the short-time dynamics of enzymatic processes (e.g., after the rewetting of soil) have not been addressed. Also, most of the data so far on the microbial community composition in a soil profile and the succession of biota on decomposing litter are based only on rough estimations of biomass and analyses of culturable microorganisms. However, to link the composition of the enzyme pool in litter with the active portion of the microbial population, it is necessary to simultaneously look for expressed genes and to analyze the total microbial populations using molecular methods. There is also a lack of knowledge about the biochemical properties of soil enzymes, since these enzymes have only scarcely been isolated (Baldrian 2006; Baldrian and Valášková 2008).

Acknowledgments Financial support from the Czech Science Foundation (526/08/0751) and from the Ministry of Education, Youth and Sports of the Czech Republic (OC08050) is gratefully acknowledged.

References

Allison SD, Vitousek PM (2004) Extracellular enzyme activities and carbon chemistry as drivers of tropical plant litter decomposition. Biotropica 36:285–296

Allison SD, Vitousek PM (2005) Responses of extracellular enzymes to simple and complex nutrient inputs. Soil Biol Biochem 37:937–944

Allison SD, Nielsen C, Hughes RF (2006) Elevated enzyme activities in soils under the invasive nitrogen-fixing tree *Falcataria moluccana*. Soil Biol Biochem 38:1537–1544

Allison SD, Czimczik CI, Treseder KK (2008) Microbial activity and soil respiration under nitrogen addition in Alaskan boreal forest. Global Change Biol 14:1156–1168

Amador JA, Glucksman AM, Lyons JB, Gorres JH (1997) Spatial distribution of soil phosphatase activity within a riparian forest. Soil Sci 162:808–825

Andersson M, Kjoller A, Struwe S (2004) Microbial enzyme activities in leaf litter, humus and mineral soil layers of European forests. Soil Biol Biochem 36:1527–1537

Aneja MK, Sharma S, Fleischmann F, Stich S, Heller W, Bahnweg G, Munch JC, Schloter M (2006) Microbial colonization of beech and spruce litter – influence of decomposition site and plant litter species on the diversity of microbial community. Microb Ecol 52:127–135

Baldrian P (2006) Fungal laccases – occurrence and properties. FEMS Microbiol Rev 30: 215–242

Baldrian P (2008) Enzymes of saprotrophic Basidiomycetes. In: Boddy L, Frankland JC, van West P (eds) Ecology of saprotrophic Basidiomycetes. Academic, Amsterdam, pp 19–41

Baldrian P, Valášková V (2008) Degradation of cellulose by basidiomycetous fungi. FEMS Microbiol Rev 32:501–521

Baldrian P, Trögl J, Frouz J, Šnajdr J, Valášková V, Merhautová V, Cajthaml T, Herinková J (2008) Enzyme activities and microbial biomass in topsoil layer during spontaneous succession in spoil heaps after brown coal mining. Soil Biol Biochem 40:2107–2115

Bastida F, Barbera GG, Garcia C, Hernandez T (2008) Influence of orientation, vegetation and season on soil microbial and biochemical characteristics under semiarid conditions. Appl Soil Ecol 38:62–70

Baum C, Hrynkiewicz K (2006) Clonal and seasonal shifts in communities of saprotrophic microfungi and soil enzyme activities in the mycorrhizosphere of *Salix* spp. J Plant Nutr Soil Sci 169:481–487

Bengtson P, Falkengren-Grerup U, Bengtsson G (2005) Relieving substrate limitation-soil moisture and temperature determine gross N transformation rates. Oikos 111:81–90

Caldwell BA (2005) Enzyme activities as a component of soil biodiversity: a review. Pedobiologia 49:637–644

Carneiro MAC, Siqueira JO, Moreira FMD, Soares ALL (2008) Soil organic carbon, total nitrogen, microbial biomass and activity in two rehabilitation chronosequences after bauxite mining. Revista Brasileira De Ciencia Do Solo 32:621–632

Carreiro MM, Sinsabaugh RL, Repert DA, Parkhurst DF (2000) Microbial enzyme shifts explain litter decay responses to simulated nitrogen deposition. Ecology 81:2359–2365

Chen HJ (2003) Phosphatase activity and P fractions in soils of an 18-year-old Chinese fir (*Cunninghamia lanceolata*) plantation. For Ecol Manage 178:301–310

Criquet S, Farnet AM, Tagger S, Le Petit J (2000) Annual variations of phenoloxidase activities in an evergreen oak litter: influence of certain biotic and abiotic factors. Soil Biol Biochem 32:1505–1513

Criquet S, Tagger S, Vogt G, Le Petit J (2002) Endoglucanase and beta-glycosidase activities in an evergreen oak litter: annual variation and regulating factors. Soil Biol Biochem 34: 1111–1120

Criquet S, Ferre E, Farnet AM, Le Petit J (2004) Annual dynamics of phosphatase activities in an evergreen oak litter: influence of biotic and abiotic factors. Soil Biol Biochem 36:1111–1118

Dilly O, Munch JC (1996) Microbial biomass content, basal respiration and enzyme activities during the course of decomposition of leaf litter in a black alder (*Alnus glutinosa* (L) Gaertn) forest. Soil Biol Biochem 28:1073–1081

Dong S, Brooks D, Jones MD, Grayston SJ (2007) A method for linking in situ activities of hydrolytic enzymes to associated organisms in forest soils. Soil Biol Biochem 39:2414–2419

Fioretto A, Papa S, Curcio E, Sorrentino G, Fuggi A (2000) Enzyme dynamics on decomposing leaf litter of *Cistus incanus* and *Myrtus communis* in a Mediterranean ecosystem. Soil Biol Biochem 32:1847–1855

Fioretto A, Papa S, Pellegrino A, Fuggi A (2007) Decomposition dynamics of *Myrtus communis* and *Quercus ilex* leaf litter: mass loss, microbial activity and quality change. Appl Soil Ecol 36:32–40

Gömöryová E (2004) Small-scale variation of microbial activities in a forest soil under a beech (*Fagus sylvatica* L.) stand. Pol J Ecol 52:311–321

Gömöryová E, Gregor J, Pichler V, Gömöry D (2006) Spatial patterns of soil microbial characteristics and soil moisture in a natural beech forest. Biologia 61:S329–S333

Gramss G (1997) Activity of oxidative enzymes in fungal mycelia from grassland and forest soils. J Basic Microbiol 37:407–423

Kahkonen MA, Wittmann C, Kurola J, Ilvesniemi H, Salkinoja-Salonen MS (2001) Microbial activity of boreal forest soil in a cold climate. Boreal Environ Res 6:19–28

Kjoller A, Struwe S (1990) Decomposition of beech litter: a comparison of fungi isolated on nutrient rich and nutrient poor media. Trans Mycol Soc Jpn 31:5–16

Kjoller AH, Struwe S (2002) Fungal communities, succession, enzymes, and decomposition. In: Burns RG, Dick RP (eds) Enzymes in the environment. Marcel Dekker, New York, Basel, pp 267–284

Kramer S, Green DM (2000) Acid and alkaline phosphatase dynamics and their relationship to soil microclimate in semiarid woodland. Soil Biol Biochem 32:179–188

Landgraf D, Wedig S, Klose S (2005) Medium- and short-term available organic matter, microbial biomass, and enzyme activities in soils under *Pinus sylvestris* L. and *Robinia pseudoacacia* L. in a sandy soil in NE Saxony, Germany. J Plant Nutr Soil Sci 168:193–201

Linkins AE, Sinsabaugh RL, McClaugherty CA, Melills JM (1990a) Cellulase activity on decomposing leaf litter in microcosms. Plant Soil 123:17–25

Linkins AE, Sinsabaugh RL, McClaugherty CA, Melillo JM (1990b) Comparison of cellulase activity on decomposing leaves in a hardwood forest and woodland stream. Soil Biol Biochem 22:423–425

Luis P, Kellner H, Zimdars B, Langer U, Martin F, Buscot F (2005) Patchiness and spatial distribution of laccase genes of ectomycorrhizal, saprotrophic, and unknown basidiomycetes in the upper horizons of a mixed forest cambisol. Microb Ecol 50:570–579

Moller J, Miller M, Kjoller A (1999) Fungal–bacterial interaction on beech leaves: influence on decomposition and dissolved organic carbon quality. Soil Biol Biochem 31:367–374

Morris SJ (1999) Spatial distribution of fungal and bacterial biomass in southern Ohio hardwood forest soils: fine scale variability and microscale patterns. Soil Biol Biochem 31:1375–1386

Mosca E, Montecchio L, Scattolin L, Garbaye J (2007) Enzymatic activities of three ectomycorrhizal types of *Quercus robur* L. in relation to tree decline and thinning. Soil Biol Biochem 39:2897–2904

Möttönen M, Järvinen E, Hokkanen TJ, Kuuluvainen T, Ohtonen R (1999) Spatial distribution of soil ergosterol in the organic layer of a mature Scots pine (*Pinus sylvestris* L.) forest. Soil Biol Biochem 31:503–516

Niemi RM, Vepsalainen M, Erkomaa K, Ilvesniemi H (2007) Microbial activity during summer in humus layers under *Pinus sylvestris* and *Alnus incana*. For Ecol Manage 242:314–323

Olander LP, Vitousek PM (2000) Regulation of soil phosphatase and chitinase activity by N and P availability. Biogeochemistry 49:175–190

Orchard VA, Cook FJ (1983) Relationship between soil respiration and soil moisture. Soil Biol Biochem 15:447–453

Osono T (2007) Ecology of ligninolytic fungi associated with leaf litter decomposition. Ecol Res 22:955–974

Osono T, Takeda H (2005) Decomposition of organic chemical components in relation to nitrogen dynamics in leaf litter of 14 tree species in a cool temperate forest. Ecol Res 20:41–49

Papa S, Pellegrino A, Fioretto A (2008) Microbial activity and quality changes during decomposition of *Quercus ilex* leaf litter in three Mediterranean woods. Appl Soil Ecol 40:401–410

Prietzel J (2001) Arylsulfatase activities in soils of the Black Forest/Germany – seasonal variation and effect of $(NH_4)_2SO_4$ fertilization. Soil Biol Biochem 33:1317–1328

Renella G, Landi L, Valori F, Nannipieri P (2007) Microbial and hydrolase activity after release of low molecular weight organic compounds by a model root surface in a clayey and a sandy soil. Appl Soil Ecol 36:124–129

Ross DJ (1987) Soil microbial biomass estimated by the fumigation-incubation procedure: seasonal fluctuations and influence of soil moisture content. Soil Biol Biochem 19:397–404

Saetre P (1999) Spatial patterns of ground vegetation, soil microbial biomass and activity in a mixed spruce-birch stand. Ecography 22:183–192

Saetre P, Baath E (2000) Spatial variation and patterns of soil microbial community structure in a mixed spruce-birch stand. Soil Biol Biochem 32:909–917

Saiya-Cork KR, Sinsabaugh RL, Zak DR (2002) The effects of long term nitrogen deposition on extracellular enzyme activity in an *Acer saccharum* forest soil. Soil Biol Biochem 34:1309–1315

Sardans J, Penuelas J (2005) Drought decreases soil enzyme activity in a Mediterranean *Quercus ilex* L. forest. Soil Biol Biochem 37:455–461

Scheu S, Wirth S, Eberhardt U (1993) Decomposition of C-14-labeled cellulose substrates in litter and soil from a beechwood on limestone. Microb Ecol 25:287–304

Schlatte G, Kampichler C, Kandeler E (1998) Do soil microarthropods influence microbial biomass and activity in spruce forest litter. Pedobiologia 42:205–214

Selmants PC, Hart SC, Boyle SI, Stark JM (2005) Red alder (*Alnus rubra*) alters community-level soil microbial function in conifer forests of the Pacific Northwest, USA. Soil Biol Biochem 37:1860–1868

Sinsabaugh RL, Carreiro MM, Alvarez S (2002a) Enzyme and microbial dynamics of litter decomposition. In: Burns RG, Dick RP (eds) Enzymes in the environment. Marcel Dekker, New York, Basel, pp 249–265

Sinsabaugh RL, Carreiro MM, Repert DA (2002b) Allocation of extracellular enzymatic activity in relation to litter composition, N deposition, and mass loss. Biogeochemistry 60:1–24

Šnajdr J, Valášková V, Merhautová V, Cajthaml T, Baldrian P (2008a) Activity and spatial distribution of lignocellulose-degrading enzymes during forest soil colonization by saprotrophic basidiomycetes. Enzyme Microb Technol 43:186–192

Šnajdr J, Valášková V, Merhautová V, Herinková J, Cajthaml T, Baldrian P (2008b) Spatial variability of enzyme activities and microbial biomass in the upper layers of *Quercus petraea* forest soil. Soil Biol Biochem 40:2068–2075

Stoyan H, De-Polli H, Bohm S, Robertson GP, Paul EA (2000) Spatial heterogeneity of soil respiration and related properties at the plant scale. Plant Soil 222:203–214

Timonen S, Sen R (1998) Heterogeneity of fungal and plant enzyme expression in intact Scots pine *Suillus bovinus* and *Paxillus involutus* mycorrhizospheres developed in natural forest humus. New Phytol 138:355–366

Trasar-Cepeda C, Leiros MC, Gil-Sotres F (2000) Biochemical properties of acid soils under climax vegetation (Atlantic oakwood) in an area of the European temperate-humid zone (Galicia, NW Spain): specific parameters. Soil Biol Biochem 32:747–755

van der Wal A, van Veen JA, Smant W, Boschker HTS, Bloem J, Kardol P, van der Putten WH, de Boer W (2006) Fungal biomass development in a chronosequence of land abandonment. Soil Biol Biochem 38:51–60

Venkatesan S, Senthurpandian VK (2006) Comparison of enzyme activity with depth under tea plantations and forested sites in south India. Geoderma 137:212–216

Waldrop MP, Zak DR (2006) Response of oxidative enzyme activities to nitrogen deposition affects soil concentrations of dissolved organic carbon. Ecosystems 9:921–933

Waldrop MP, McColl JG, Powers RF (2003) Effects of forest post harvest management practices on enzyme activities in decomposing litter. Soil Sci Soc Am J 67:1250–1256

Wittmann C, Kahkonen MA, Ilvesniemi H, Kurola J, Salkinoja-Salonen MS (2004) Areal activities and stratification of hydrolytic enzymes involved in the biochemical cycles of carbon, nitrogen, sulphur and phosphorus in podsolized boreal forest soils. Soil Biol Biochem 36:425–433

Zhang YM, Zhou GY, Wu N, Bao WK (2004) Soil enzyme activity changes in different-aged spruce forests of the eastern Qinghai-Tibetan plateau. Pedosphere 14:305–312

Zhang YM, Wu N, Zhou GY, Bao WK (2005) Changes in enzyme activities of spruce (*Picea balfouriana*) forest soil as related to burning in the eastern Qinghai-Tibetan Plateau. Appl Soil Ecol 30:215–225

Chapter 5
Extracellular Enzymes in Sensing Environmental Nutrients and Ecosystem Changes: Ligand Mediation in Organic Phosphorus Cycling

Thanh H. Dao

5.1 Introduction

Phosphorus (P) is a nonmetallic element that is essential to living organisms. Plant available P is growth limiting and organic P contributes to plant mineral nutrition only after dephosphorylation and release of inorganic phosphate, which is of particular importance in phosphate-deficient natural or cultivated ecosystems. Most of the organic P mineralizing activity occurs in the rhizosphere of plants where extracellular phosphohydrolases are released by plant roots (Tanaka et al. 1974; Helal 1990; Duff et al. 1994; Hunter and McManus 1999; Lopez-Gutierrez et al. 2004; Ma et al. 2009) and microorganisms (Greiner et al. 2000; Quan et al. 2003; Vohra and Satyanarayana 2003; Dao and Hoang 2008), allowing these organisms access to outlying P substrates that otherwise are not accessible (Li et al. 1991; Joner and Johansen 2000; Ezawa et al. 2005). However, multiple competing reactions are operating to regulate the solution-phase concentration and availability of these P-containing substrates and the released phosphate for plant uptake or dispersion in the environment (Fig. 5.1). Thus, to gain an accurate description of the transformations and cycling of organic P in soils and the environment, it is important to understand the interactions that occur between plant and their roots, microorganisms and fauna that populate terrestrial and aquatic environments, and the processes that are involved in the cycling of P-compounds in those biospheres. In this chapter, we will give particular attention to the ligand exchange process involved in the release of stabilized organic P, a forerunner process crucial to the action of phosphohydrolases.

The question of why these enzymes are extracellular has always been of intense interest. It is often wondered why plants and microorganisms externalize low quantities of such entities into their surrounding where the risks of sorption, deactivation,

T.H. Dao
US Department of Agriculture, Agriculture Research Service, Beltsville, MD, USA
e-mail: thanh.dao@ars.usda.gov

G. Shukla and A. Varma (eds.), *Soil Enzymology*, Soil Biology 22,
DOI 10.1007/978-3-642-14225-3_5, © Springer-Verlag Berlin Heidelberg 2011

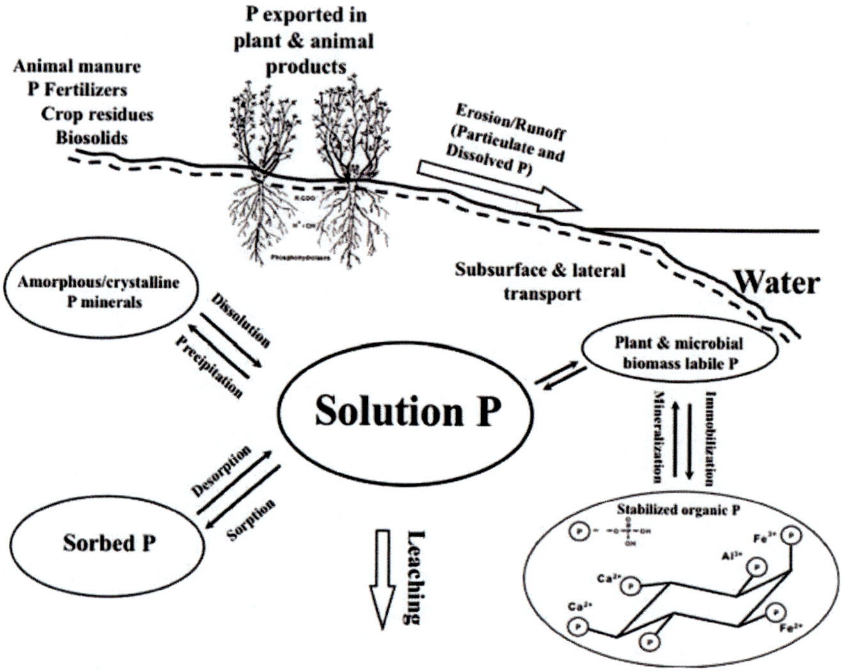

Fig. 5.1 Processes and reactions affecting phosphorus pools in soil

or degradation are relatively high in soil and aquatic environments (Burns and Dick 2002). Burns (1982) postulated that the localization of metabolic enzymes was partially determined by the size and the solubility of the target substrates, and the enzymes' activity regulated by whether they freely move in the soil or are in association with a mineral or organic surface of soil colloids. Although they may be short-lived, the extracellular enzymes are responsible for the hydrolysis of organic substrates that cannot pass through cell walls or membranes without extensive expenditures of cell's resources or that the substrate molecules are simply too large to do so.

Phosphohydrolases are synthesized to search out P-containing compounds and generate signals to which the producer's cells react by increasing expression and production of additional enzymes. Although enzymes' activity may be attenuated (Dick and Tabatabai 1987; Quiquampoix 1987; Quiquampoix et al. 2002; George et al. 2005), the apparent risk to externalizing enzymes is minimized by these surface interactions and associations (Burns 1982). A highly effective strategy of the cell, from a root of a plant, or an entire microbial community is to have a group of catabolic enzymes associated at or near the cell's surface, performing a sequence of processes in close proximity; an ideal biochemical microenvironment is thereby created, where complementary functions and metabolite concentrations are much higher than those of the surrounding to maximize the efficacy of individual metabolic processes. As plant and microorganisms have evolved multiple strategies to obtain the needed P, it is not surprising to observe the interweaving of physical and biochemical processes in the speciation and availability of organic P forms. Therefore, in our efforts to gain a better understanding of the processes of P sensing and cycling, a combination of ligand exchange and mineralization of organic P in an analytical tool for determining organic nutrient turnover or for detecting ecosystem changes appears to be a natural blend (Dao 2003, 2004, 2007; Bunemann 2008). From an environmental standpoint, the contribution of biologically mediated processes to the organic P release and movement must be fully understood to develop sustainable management practices for our agricultural production systems and mitigate eutrophic responses in sensitive aquatic ecosystems. In spite of major research and education efforts and public media scrutiny, agricultural P continues to be a significant pollutant of surface waters and estuaries around the globe. As soil fertility and nutrient accumulation rise, P-enriched soils remain a major contributor to P loading to runoff and surface and ground water as the result of land application of fertilizers and organic nutrients (Council for Agricultural Science and Technology CAST 1996, 2002; Zhang et al. 2006; Green et al. 2007; Dao et al. 2008). Algal blooms have often been related to increased P inputs from terrestrial ecosystems, often exceeding a critical soluble P concentration of 30 μg L^{-1} (Smith 1982; Jarvie et al. 2002; Mainstone and Parr 2002; Haukka et al. 2006; Hilton et al. 2006). Therefore, we will take another fundamental look at the biogeochemistry of P in the soil to understand the solubilization of organic P substrates, and the role of ligands in the enzyme-mediated mineralization and turnover of organic P in the supply of phosphate to meet plant and microbial nutritional needs.

5.2 Biogeochemistry of Phosphorus

Phosphorus is a decay product of Cl, S, and K (Lal et al. 1988). Radioactive decay series of ^{34}Ca, ^{33}K, ^{32}K, ^{31}F, ^{33}Ne, and ^{34}Ne isotopes have ^{31}P as a daughter product via beta and proton emissions. Phosphorus is always found combined with other elements, particularly oxygen. In a tetrahedral arrangement, four oxygen atoms bond to a P atom to form the phosphate (PO_4^{3-}) anion. Each oxygen atom carries a negative charge and can bond with hydrogen to yield, HPO_4^{2-}, $H_2\ PO_4^{-}$, and H_3PO_4, or phosphoric (V) acid, a major industrial product for the fertilizer and food industries. Phosphorus also can have an oxidation state of +3, phosphorus (III) which yields a large group of important industrial compounds and pesticides, including PH_3, PCl_3, P_2S_5, and phosphonates. The chemistry and metabolism of reduced organic P compounds have been extensively reviewed elsewhere (Devai et al. 1988; Ternan et al. 1998; Roels and Verstraete 2001; Quinn et al. 2007; White and Metcalfe 2007).

Phosphorus circulates through the environment in three natural cycles. In the inorganic cycle, phosphate-P in the earth crust is released to the soil, passes to lakes, streams and rivers to the seas and ocean floors. Under geologic pressure and temperature, P-containing sediments undergo transformations, resulting in metamorphic rocks that may be uplifted to form new landmasses to begin again the cycle, from land to sea floors. Phosphate rock is widely distributed across the globe in marine phosphorite deposits. Apatite, i.e., calcium phosphates, is the primary P compound in phosphate rock. Phosphorus in apatite minerals is only slightly soluble and of limited availability to plants and must be chemically processed to produce many agriculturally important forms of P fertilizers used in crop production, such as superphosphates, triple superphosphate, or ammonium phosphates when combined with ammonia.

Superimposed upon the inorganic P cycle, the other two cycles are about the moving of P through living organisms that are either aquatic or terrestrial dwellers. In these organisms, P is a required nutrient, playing a central role in numerous energetic metabolic processes. For example, P is a component of phospholipids, nucleic acids, the principal nucleotide cofactors required for energy transfer and catalysis in the cell, or in inositol phosphates involved in signal transduction and cellular regulation. Therefore, most P-containing compounds found in living systems are in the form of inorganic phosphate and organic phosphate mono- and diesters. The latter compounds are characterized by a carbon–oxygen–phosphorus bond, where P has a valence of +5. The biochemistry of P-containing compounds is considered relatively simple, consisting almost entirely of phosphate ester and phospho-anhydride formation and hydrolysis.

5.2.1 *Inorganic Phosphate*

Phosphate may undergo complex reactions with soil mineral and organic constituents of natural waters and soils. In dilute solutions, the following species are in thermodynamic equilibrium (Table 5.1).

Inorganic phosphate (i.e., orthophosphate) is a reactive chemical species, which is rapidly incorporated in various soil P pools, making most of the applied P unavailable for plant uptake. As a case study, Lindsay and Stephenson (1959) made a detailed study of the reactions affecting a granule of monocalcium phosphate monohydrate [$Ca(H_2PO_4)_2{\cdot}H_2O$], a common superphosphate fertilizer, when added to soil. Upon contact with the soil, the granule undergoes a series of complex reactions that includes dissolution and precipitation to transform it into more stable P minerals that render P even less available to plants. In sequence, a

Table 5.1 Dissociation constants for various phosphate species

	log *K*
$H_3PO_4^0 \rightleftharpoons H^+ + H_2PO_4^-$	−2.15
$H_2PO_4^- \rightleftharpoons H^+ + HPO_4^{2-}$	−7.20
$HPO_4^{2-} \rightleftharpoons H^+ + PO_4^{3-}$	−12.35
$H_2PO_4^- \rightleftharpoons 2H^+ + PO_4^{3-}$	−19.55
$2H_2PO_4^- \rightleftharpoons (H_2PO_4)_2^{2-}$	−0.35

Lindsay (1979)

supersaturated solution of Ca^{2+}, $H_2PO_4^-$, $CaHPO_4{\cdot}2H_2O$, and $CaHPO_4$ is formed at the water-fertilizer granule boundary. Super saturation induces the precipitation in the intermediary metastable solution of varying proportion of the two minerals, $CaHPO_4{\cdot}2H_2O$ and $CaHPO_4$ (Lindsay and Stephenson 1959). A sharp pH decrease also occurs as $H_3PO_4^0$ is formed. The highly acidic solution reacts with soil mineral surfaces, releasing Al^{3+}, Fe^{3+}, Fe^{2+}, Mn^{2+}, and other cations forming aquo-metal ions such as $Al(H_2O)_6^{3+}$ or $Fe(H_2O)_6^{3+}$; some of the acidity is neutralized in the process and initial reaction products may include aluminum and iron phosphates. Upon the addition of $Ca(H_2PO_4)_2{\cdot}H_2O$ to three soils, these researchers have identified some 32 aluminum ferric phosphate minerals of variable composition, (Al, Fe, X)-$PO_4{\cdot}nH_2O$ where X denotes the presence of cations other than Al and Fe (Lindsay et al. 1962).

In addition to reactions in the solution phase, Henmi and Huang (1985) demonstrated strong sorption of phosphate by edge surface AlO-functional groups in soils developed from volcanic ash and pumice. Components in Andisols, which are responsible for adsorption of P, are amorphous or poorly ordered hydrous aluminosilicates in the clay fraction, namely, allophanes and imogolite. Grinding or heating the clays at temperatures between 200° and 400°C increased P sorption. The increased capacity was attributed to the formation of the new P adsorption sites (AlOH–H_2O), as the result of breaking down Si–O–Al linkages located only on the edge surface of allophane and imogolite tubular structures in the clay fraction. In acid soils, phosphate strong sorption has been attributed to hydrous oxides of Fe and Al, and to (1:1) layer lattice clays (Juo and Fox 1977; Golberg and Sposito 1984b; Sanyal et al. 1993; Arai and Sparks 2002). In calcareous soils, the abundance of Ca and Mg results in the sorption and precipitation of pyrophosphates as insoluble compounds such as $Ca(NH_4)_2P_2O_7{\cdot}H_2O$ and $Mg(NH_4)_2P_2O_7{\cdot}4H_2O$, and in the process, reduces the amount of phosphate remaining in the soil solution for uptake by microorganisms or plant roots (Lindsay et al. 1962).

Phosphate adsorption on soils often has been described mathematically with the following generalized equation:

$$q(C) = \sum_{i=1}^{n} \frac{b_i K_i^{\beta_i} C^{\beta_i}}{\left(1 + K_i^{\beta_i} C^{\beta_i}\right)}$$

where $q(C)$ is the quantity of phosphate adsorbed per unit mass of soil solids, C is the solution- phase concentration of phosphate, and n, b_i, K_i, and β_i, are empirical parameters (Golberg and Sposito 1984). The equation takes the familiar form of the Langmuir adsorption model, $x/m = S_{\max}\ (KC/(1 + KC)$, when $n = 1$ and $\beta_i = 1$; and the equation reduces to the Freundlich equation, $S = KC^n$, when $n = 1$, $0 < \beta_i < 1$, and $KC << 1$.

Goldberg and Sposito (1984a) reported that the constant capacitance model developed by Schindler et al. (1976) and Stumm et al. (1976) provides more appropriately a quantitative description of phosphate adsorption by a variety of aluminum and iron hydrous oxides. The model describes phosphate adsorption based on a ligand exchange mechanism, and relates the charge density of a surface or plane of charge (σ_o) to the surface potential (Ψ_o), capacitance density C (farad m^{-2}), surface area, S (m^2 g^{-1}), the Faraday constant, F (coulomb mol^{-1}), and the suspension density (g L^{-1}). Thus, diverse reactions and transformations affect inorganic phosphates in the soil, illustrating the complexity of the chemical environment which plant roots and microorganisms face in their quest for P from their growth medium.

5.2.2 *Organic Phosphorus*

In terrestrial and aquatic ecosystems, microorganisms have evolved the ability to store a number of necessary metabolites in the form of polymers that are osmotically inert. Inorganic phosphate is one such metabolite that is stored within the cell as a high-molecular-weight inorganic polyphosphate; it is a linear polymer containing from a few to several hundred residues of phosphate that are linked by energy-rich phosphoanhydride bonds (Kulaev and Kulakovskaya 2000). In microbial cells, polyphosphate can confer the organism the increased resistance to stressful changes in its environment, regulation of enzyme activities, cation transport, as well as a source of inorganic phosphate to meet energy needs.

Another important group of stored phosphorylated organic compounds include inositol phosphate monoesters and phosphate diesters (i.e., phospholipids and nucleic acids). Plant reproductive organs and, in particular, seeds of wheat (*Triticum aestivum* L.), rice (*Oriza sativa* L.), corn (*Zea mays* L.), soybean [*Glycine max* (L.) Merr.], and mung bean [*Vigna radiata* (L.) R. Wilczek var. *radiata*] (Asada et al. 1969; Tanaka et al. 1974; Yamagata et al. 1980; Scott and Loewus 1986; Lott et al. 2000; Shi et al. 2005) store P as calcium, magnesium, and potassium salts of *myo*-inositol hexakis dihydrogenphosphate (hereafter, will be referred to as *myo*-inositol hexa*kis*phosphate). These compounds supply inositol and a range of phosphorylated substrates, inositol as C source, and cations to the growing embryo upon seed germination. The chemistry of inositol phosphate monoesters is found extensively in the recent literature due to their expansive roles in biological systems (e.g., cellular signal transduction; calcium regulation) and in plants and soil organisms (e.g., condensed P storage; antioxidant) (Cosgrove 1980; Stewart and Tiessen 1987; Voglmaier et al. 1996; Kulaev and Kulakovskaya 2000; Lemtiri-Chlieh et al. 2003; Turner et al. 2005, 2007).

The turnover of microbial tissues and plant residues at the soil surface contributes organic P compounds to the soil environment (Anderson 1980; Stewart and Tiessen 1987). Cultural practices such as the incorporation of green cover crops into the soil, and land application of animal manure also add plant-derived P to the soil P pools in an effort to build up soil fertility and thereby productivity (Biederbeck et al. 1993; Cavigelli and Thien 2003). *myo*-Inositol hexa*kis*phosphate and lower phosphomonoesters of inositol are the most abundant organic P compounds in soils, and may make up to 50% of total soil organic P (Wild and Oake 1966; Koopmans et al. 2003; Dao 2004). Of all the forms of organic P, the most prevalent forms found in soil are several steroisomers of inositol hexa*kis*phosphate as they are recalcitrant to microbial degradation. These phosphomonoesters have been identified using ^{31}P nuclear magnetic resonance spectroscopy (Cade-Menun et al. 2002; Koopmans et al. 2003; Turner et al. 2003; Jayasundera et al. 2005), high-performance liquid chromatography and mass spectrometry (Sandberg and Ahderinne 1986; Lehrfeld 1989, 1994; Cooper et al. 2007).

Information on the nature of organic P in manure is also increasingly available for different kinds of livestock and poultry and manure management practices (Peperzak et al. 1959; Gerritse and Eksteen 1978; He and Honeycutt 2001; Dao et al. 2006). As manure is composed of partly digested feed, direct spectroscopic measurements showed that the organic P fraction generally consists of phosphomonoesters of inositol, phospholipids, and nucleic acids derived from plant and microbial materials. In soils amended with animal and green manure from cover crops and plant residues, phytic acid (*myo*-inositol hexa*kis*phosphate) and other organic P are added to the near-surface soil P pool, in addition to inorganic phosphate (Dao 2004; Zhang et al. 2006). Although there have been many studies on the behavior of inorganic phosphate in soil as discussed in a previous section, desorption, speciation, and mineralization of organic P compounds is yet poorly understood. Mechanistic studies of the interactions between specific organic P compounds, metal hydrous oxides, and soil mineral and organic colloids as affected by solution properties such as pH, electrical conductivity, or other cation concentration are sparse.

5.3 Reactions Involving Organic Phosphorus in Soils

5.3.1 Chemical Processes

5.3.1.1 Adsorption–Precipitation

Although mathematically and mechanistically are very similar, differentiation between adsorption and precipitation is at the soil–liquid interface (Sposito 1986; Lutzenkirchen and Behra 1996). Many of the solute–surface interactions, namely, absorption, adsorption or enrichment at the interface of the solution and solid soil constituents, diffusion into the soil solids, and precipitation or the formation of a P solid phase at the interface have often been referred to as sorption when most if not all these mechanisms are operating to control P speciation and environmental behavior.

The mechanisms of of phosphate sorption on soils have been extensively studied; however, those of organic phosphate monoesters or organic P in general, have not received the same degree of scientific scrutiny. As they are biochemically reactive, interacting with soil components stabilize and shield phosphate monoesters from biodegradation, leading to their preferential accumulation in soils

(Anderson et al. 1974; Stewart and Tiessen 1987; Celi and Barberis 2007). The adsorption of phosphate monoesters on hydrous oxides has been modeled on inorganic phosphate adsorption on iron hydrous oxides (Ognalaga et al. 1994; Celi et al. 2001). Ognalaga et al. (1994) advanced a model of a binuclear surface complex for the adsorption of *myo*-inositol hexa*kis*phosphate on dried samples of goethite that was equilibrated with the inositol hexa*kis*phosphate (Fig. 5.2). In addition, whether the ring structure was *myo*-inositol or glucose, these researchers observed that the organic portion of the molecule did not appear to have any effect on the adsorption process, upon comparing the sorption of glucose-1-phosphate and *myo*-inositol hexa*kis*phosphate onto goethite.

This hypothesis was derived from the model of the adsorbed state of inorganic phosphate on ferrihydrite and other iron oxides (Hingston et al. 1974; Parfitt 1979; Parfitt et al. 1975) and soil constituents (Parfitt 1977). By ligand exchange, two surface hydroxyl ions (or water molecules) were replaced by a phosphate ion. Two of the oxygen atoms of the phosphate ion were coordinated to two different Fe^{3+} ions, resulting in a binuclear surface complex of the type Fe–O–P(O_2)–O–Fe (Hingston et al. 1974; Parfitt et al. 1975; Beek and Riemsdijk 1982). Molecular cross-sectional areas and surface coverage calculations and X-ray photoelectron spectroscopy further supported the binuclear model of phosphate adsorption (Martin and Smart 1987; Shin et al. 2004). In such an arrangement, the quantity

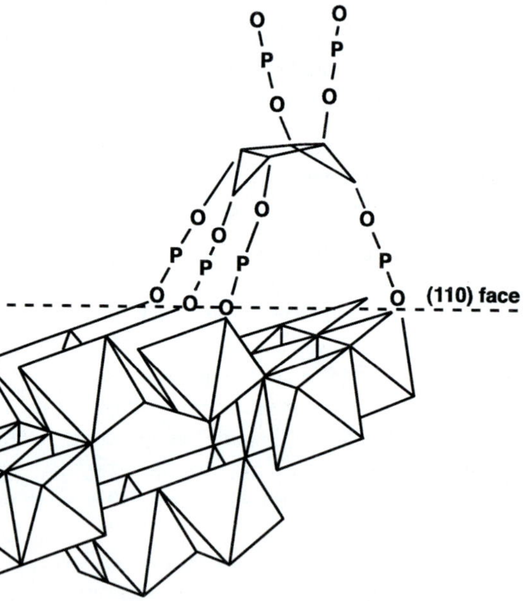

Fig. 5.2 Proposed adsorbed state of myo-inositol hexa*kis*phosphate on goethite. The boat conformation of the molecule was attributed to Cosgrove DJ 1966. The H– and OH– groups of the phosphate moieties are not shown (Adapted from Ognalaga et al. (1994))

of adsorbed phosphate per unit surface area averaged 2.5 μmol m^{-2} for a number of samples of natural and synthetic goethite (Golberg and Sposito 1984; Torrent et al. 1990). Variability in sorption capacities of iron and aluminum hydrous oxides evolve from differences in availability of singly coordinated –OH groups on the mineral surface (Parfitt 1989) (Fig. 5.3).

It is noted that, in the binuclear surface complex model for an organic phosphate monoester, *myo*-inositol hexa*kis*phosphate, the adsorbed molecule is forced into the boat conformation, deviating from its thermodynamically stable chair conformation. Furthermore, formation of monodentate complexes may well be an important mechanism of adsorption of phosphate monoesters in soil-water suspensions or colloids and sediments in aquatic ecosystems, as was the case for phosphate in the presence of water (Golberg and Sposito 1985). Shin et al. (2004) showed that phosphate adsorption occurred at least through a mixture of monodentate and binuclear bridges (Fig. 5.4), using a combination of infrared spectroscopy and X-ray photoelectron and diffraction spectroscopic techniques. In the author's opinion, one can also visualize binuclear and polynuclear complexes for the sorption of inositol phosphate monoesters, forming monodentate bridges with -FeO+ and -AlO+ of more than one surfaces of suspended colloids in soil suspensions and sediments.

5.3.1.2 Autohydrolysis and Metal-Mediated Hydrolysis

Other deactivation reactions involving organic P include autohydrolysis in presence of metal ions. *myo*-Inositol hexa*kis*phosphate may not be chemically hydrolyzed in vivo in the pH and temperature range of biological systems. However, its autohydrolysis may be triggered by the coordination of the metal ion with a phosphate group of *myo*-inositol hexa*kis*phosphate, promoting the hydrolysis of an adjacent phosphate group. Bullock et al. (1993) observed its spontaneous chemical hydrolysis at 95°C across a wide range of ambient pH. The reaction is significant in efforts to negate the binding and sequestration of essential trace minerals with the consumption of high-phytate food; boiling such food would destroy its binding and sequestering, for example dietary Fe to interfere with Fe absorption in humans on a high fiber diet (Sandberg et al. 1999). Specific instances of the autohydrolysis of *myo*-inositol hexa*kis*phosphate have also been observed at high temperature of 120°C, resulting in the formation of most of the phosphate esters of *myo*-inositol, with 39 chromatographically distinct *myo*-inositol mono-, bis-, tris-, tetra*kis*, penta*kis*-, and hexa*kis*phosphates in the presence of HCl (Hull et al. 1999). Metal-catalyzed dephosphorylation of nucleoside-5′ trisphosphates was observed in aqueous solutions by nucleophilic substitution with water in the presence of Cu^{2+}, Zn^{2+}, Ni^{2+}, Mn^{2+}, and Mg^{2+} (Sigel et al. 1984). Details on SN1 and SN2 nucleophilic substitution mechanisms are found in a review of the topic by Baldwin et al. (2005).

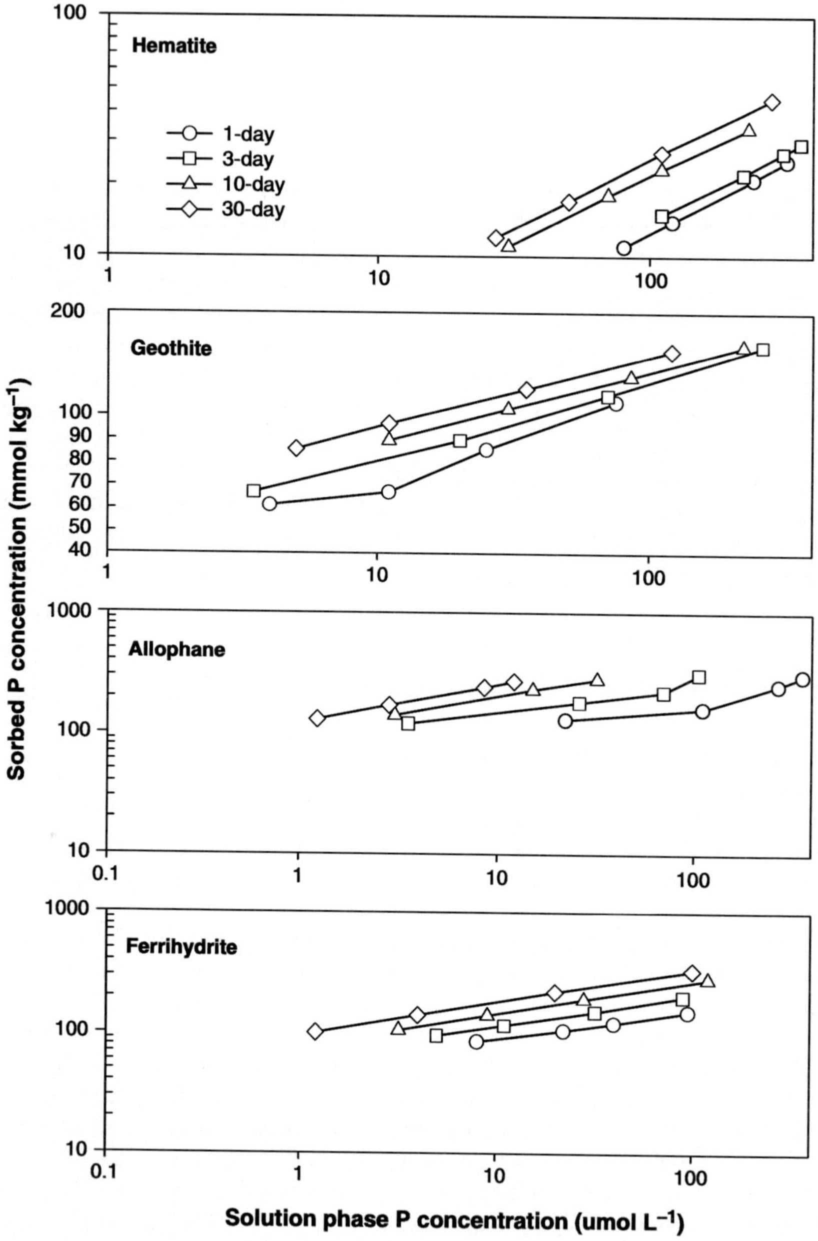

Fig. 5.3 Sorption of phosphate on aluminum and iron oxides (Parfitt 1989)

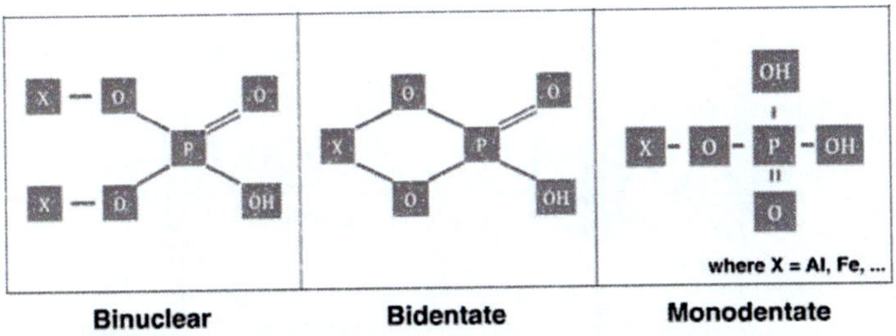

Fig. 5.4 Possible phosphate surface complexes (Golberg and Sposito 1985)

5.3.2 *Biological Processes Involved in Organic P Solubilization and Degradation*

As plants can only acquire P as inorganic phosphate, recycling of P has to occur through the turnover of biological tissues and the mineralization of organic P in its environment (Tadano and Sakai. 1991; Frossard et al. 2000; Hinsinger 2001). The process is mediated by phosphohydrolases induced and released by microbial systems, plant roots, or arbuscular mycorrhizal associations to mineralize organic P forms and release phosphate for plant assimilation (Harrison 2005). For example, wheat (*Triticum aestivum* L.) roots were observed to secrete 5–10 times greater concentrations of phosphohydrolases in the presence of organic P, when compared to a sterilized medium containing only inorganic P (Tarafdar and Claassen 2003). Roots of a wide range of plant species have shown similar ability to meet their P nutritional needs, e.g., white lupin (*Lupinus albus* L.) (Wasaki et al. 2003) and clover (Hunter and McManus 1999). It thus appears that mediated processes a plant uses to obtain P from its environment have an efficiency that by far exceeds that of abiotic processes to lead to the eventual release of phosphate for bringing the nutrient inside the cell via an active or passive process across its membrane (Ragothama 1999; Horst and Kahm 2004). Phosphate transporters in plant roots and cultured microbial cells are numerous and have been extensively studied (Ragothama 1999; Kulaev and Kulakovskaya 2000). Transporters are proteins that are an integral component of the cell membrane, capable of internalizing inorganic phosphate and other simple molecules; the transport is based on a chemiosmotic ion gradient and an example of a phosphate transporter in *Arabidopsis thaliana* has been described in Raghothama's review article (1999). Following is another basic rendition of the phosphate transporter interacting with organic anions in the rhizosphere by Horst and Kahm (2004) (Fig. 5.5).

5.3.2.1 Mineralization of Environmental Biological Tissues Containing Organic P

In native ecosystems, growing plants depend upon the mineralization of P-containing organic tissues and plant residues to meet their biotic needs, particularly in phosphate-deficient ecosystems. Suites of catabolic enzymes (proteases, glucosidases,

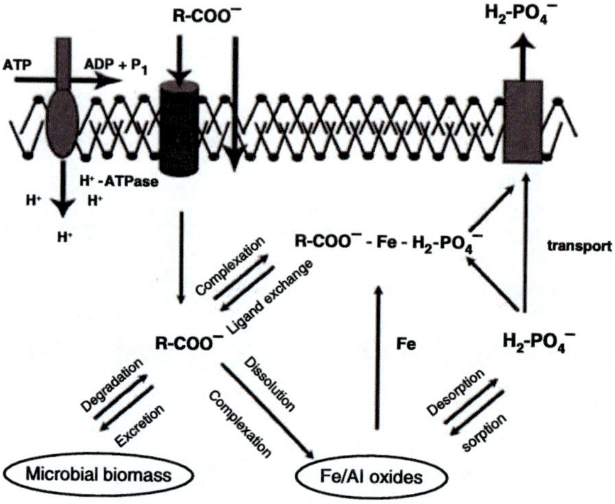

Fig. 5.5 Ligand-mediated phosphate release and transport across plant roots' membrane (Adapted from Horst and Kahm (2004))

oxidases, and phosphohydrolases) are induced and released to decompose organic matter and obtain the organic P contained within. Although much of the literature on the decomposition of plant residues and forest litter has been focused on the mineralization of C and N (Douglas et al. 1980; Christensen 1986; Bremer and van Kessel 1992; Dao 1998), significant interactions were observed between P and the mineralization of plant residue C and N (Kouno et al. 2002; Soon and Arshad 2002; Dao and Cavigelli 2003; Lupwayi et al. 2007). Mineralization of P was found to occur with substrates having C:P ratios near 30–50 though most plant residues have C:P rations exceeding 140:1.

5.3.2.2 Mycorrhizal Associations

To increase the efficiency of the scavenging for P-containing substrates, symbiotic association between a host plant and certain soil fungi, known as arbuscular mycorrhiza (AM) have been part of plant strategies to fulfill their P needs in natural and cultivated ecosystems. The arbuscular mycorrhizae (AM) hyphae can explore a large volume of soil scavenging for P substrates, inducing physiological changes that favor the establishment of P-mineralizing microorganisms in the hyphal micro-environment (Dighton 1983, 1991; Tarafdar and Marschner 1994; Singh and Kapoor 1998; Lopez-Gutierrez et al. 2004). Numerous studies have shown that mycorrhizae dissolve insoluble P minerals (Reid et al. 1984; Leyval and Reid 1991) through the excretion of H^+ and organic acids (Li et al. 1991; Lapeyrie et al. 1991; Neumann and Martinoia 2002), and hydrolysis of organic P by excreted phosphohydrolases. The contributions of AM to the cycling of P have been periodically reviewed (Bonfante-Fasolo 1984; Smith and Gianinazzi-Pearson 1988; Harrison 2005) and will not be further discussed in this chapter.

5.3.2.3 Extracellular Phosphohydrolases

Common to the above two mechanisms for obtaining P from their environment, microorganisms and plants are known to produce and release phosphohydrolases to the external environment, allowing the plant roots access to the outlying P-containing substrates otherwise inaccessible to the host plant (Li et al. 1991; Tadano and Sakai 1991; Jayachandran et al. 1992; Joner and Johansen 2000). By far, the biological dephosphorylation of inositol phosphates is the prevalent pathway of their breakdown, and is carried out by extracellular phosphohydrolases. Histidine acid- and alkaline phosphohydrolases are produced by a wide variety of organisms, including bacteria (i.e., *Bacillus subtilis*), yeasts (i.e., *Saccharomyces*), and fungi (i.e., *Aspergillus*) (Vohra and Satyanarayana 2003). Extensive reviews of organisms capable of utilizing inositol phosphates have been made (Greiner 2007). These enzymes are industrially produced as animal feed supplements (Vats et al. 2005). Many current developments include efforts to improve phosphohydrolases' activity at animal gastro-intestinal pH, their thermal stability and substrate specificity (Weaver et al. 2007).

From an environmental standpoint, the contribution of biologically-mediated processes to organic P release and movement must be fully understood to develop management practices to mitigate P loading to surface waters. Runoff from P-enriched cropping systems showed elevated P concentration (Green et al. 2007; Dao et al. 2008). While run off from P-enriched soils contains more P than the functionally defined molybdate-reactive P forms, information on the biological significance of the remaining unreactive fraction is limited. A study was conducted to characterize distributions of inorganic and enzyme-labile P forms in simulated runoff from soils under an orchardgrass-red clover and a no-till forage-type soybean-winter wheat systems showed significant relationships between water-extractable P (WEP) fluxes and soil bioactive P pools. Concentration and mass distributions of P forms in runoff over time were log-normally distributed. The strength of correlations between cumulative WEP mass loads and soil P pools was in the decreasing order of ligand-exchangeable organic P > ligand-exchangeable inorganic P > Mehlich-3 extractable P, suggesting that runoff P forms were directly associated with soil available P fractions that were partly derived from enzyme-mediated processes.

Enzymatic Dephosphorylation of Uncomplexed Organic Phosphate Monoesters

Stored organic P in seeds in the forms of calcium and magnesium salts of *myo*-inositol hexa*kis*phosphate is dephosphorylated to provide phosphate and inositol to the growing germ. Dephosphorylation pathways by phosphohydrolases from plants and microorganisms have common intermediate metabolites but they also do differ in stereochemistry and in end-products (Greiner et al. 2000, 2002; Quan et al 2003).

Greiner et al. (2000) reported that the dephosphorylation is stereochemically-specific. Stepwise dephosphorylation of *myo*-inositol hexa*kis*phosphate by *Escherichia coli* enzymes yields D/L-Ins(1,2,3,4,5) penta*kis*phosphate, D/L-Ins(2,3,4,5) tetra*kis*phosphate, D/L-Ins(2,4,5) trisphosphate or D/L-Ins(1,2,4) trisphosphate, D/L-Ins(1, 2)-, or (2, 5)-, or (4, 5) bisphosphate, to Ins(2) monophosphate, or Ins(5) monophosphate, where Ins = *myo*-inositol. Meanwhile, phosphohydrolases from wheat (*Triticum aestivum* L.) yield D-Ins(1,2,3,5,6) penta*kis*phosphate, D-Ins (1,2,5,6) tetra*kis*phosphate, D-Ins(1,2,6) trisphosphate, and D-Ins(1,2) bisphosphate, and Ins(2) monophosphate, whereas the phosphohydrolases LP11 and LP12 from lupin seeds yield D-Ins(1,2,4,5,6) penta*kis*phosphate, D-Ins(1,2,5,6) tetra*kis*phosphate, D-Ins(1,2,6) tri*s*phosphate, D-Ins(1,2) bisphosphate, and Ins(2) monophosphate as the final product. *Candida* phosphohydrolases were observed to yield in sequence: D/L-Ins(1,2,4,5,6) penta*kis*phosphate, D/L-Ins(1,2,3,5,6) penta*kis*phosphate, D/L-Ins(1,2,5,6) tetra*kis*phosphate, D/L-Ins(1,2,5) tri*s*phosphate, D/L-Ins (1, 2) bisphosphate, and D/L-Ins(2) monophosphate (Quan et al. 2003). The various steps of the dephosphorylation of *myo*-inositol hexa*kis*phosphate are shown as a flow chart to illustrate the sequence that varies in stereochemistry between microorganisms and plant species (Fig. 5.6).

In general, enzymes exhibiting an acidic pH optimum release from five to six phosphate groups of *myo*-inositol hexa*kis*phosphate and *myo*-inositol monophosphate and *myo*-inositol. The alkaline phosphohydrolases yield *myo*-inositol trisphosphate as the end product as these enzymes are unable to use *myo*-inositol phosphates with three or fewer phosphate groups as substrates (Barrientos et al. 1994).

Ligand Mediation of (or Chemically-Assisted) Enzymatic Dephosphorylation of Complexed Organic Phosphate Monoesters

The degradation of organic P-containing substances in the environment is often constrained by physical and biochemical factors affecting the availability of these substrates, more so than any attenuation in the activity of phosphohydrolases and extracellular enzymes in soil (Otani and Ae 1999; Dao 2003, 2004). *myo*-Inositol hexa*kis*phosphate has six phosphate moieties that provide twelve coordinate ligands for binding metal ions (Fig. 5.7). Metal interactions with the numerous phosphate ligands present on phosphate monoesters can lead to both intra- and intermolecular bonding resulting in the simultaneous formation of numerous monomeric and polymeric species. The phenomenon can lead to the co-precipitation of non-stoichiometric solid-phase mixtures, as the metal to ligand ratio increases. Polyvalent cation binding resulted in reductions in the availability of substrates, shielding organic P from hydrolysis by phosphate monoesterases (Dao 2003, 2007).

Phosphate monoesters, having differing number of phosphate moieties, exhibit differing abilities for binding metal ions (Sandberg and Ahderinne 1986;

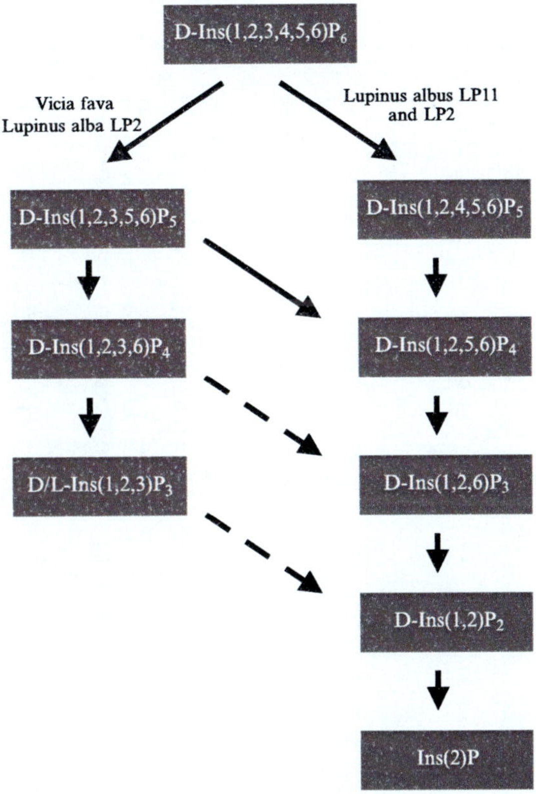

Fig. 5.6 Dephosphorylation pathways of *myo*-inositol hexa*kis*phosphate by phosphohydrolases in *Vicia fava* and *Lupinus albus*. Major pathway denoted by filled *solid arrows* (Adapted from Greiner et al. (2002))

Dao 2003), as well as differing affinity for charged surfaces of minerals and soil components (Fig. 5.8) (Anderson et al. 1974; McKercher and Anderson 1989). A consequence of these differences is reflected in their accessibility or susceptibility to enzymatic hydrolysis.

In the rhizosphere, plants and microorganisms modify the biochemistry of that environment to affect the solubility and release of complexed organic P and inorganic phosphate. Plant root exudates and siderophores have been long been implicated in the acquisition of P by plants grown in P-deficient soils (Ae et al. 1990; Plessner et al. 1993; Ragothama 1999; Hinsinger 2001; Jones et al. 2003). Excreted organic anions are a variety of products of the citric acid cycle, exchanging with bound phosphate and organic inositol hexa*kis*phosphate anions on soil exchange complex. For example, pigeon pea (*Cajanus cajan* [L.] Millsp.) releases piscidic acid (p-hydroxybenzyl tartaric acid) and its p-methoxybenzyl derivative in the rhizosphere, allowing the legume to utilize ironbound phosphate. Rape secretes malic and citric acids to release soluble P from a phosphate rock (Hoffland et al. 1989, 1992). Similar organic acids are released by roots of white lupin, when grown on a calcareous P deficient soil (Dinkelaker et al. 1989; Johnson et al. 1996;

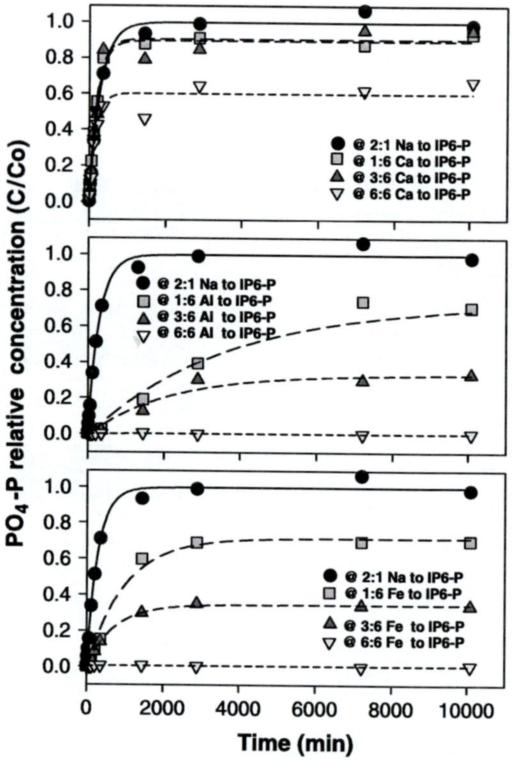

Fig. 5.7 Effect of counterions on the dephosphorylation of *myo*-Inositol hexa*kis*phosphate by *Aspergillus* phosphohydrolases (Dao 2003)

Veneklaas et al. 2003). In soil, commonly identified low molecular weight organic anions include formate, acetate, propionate, oxalate, and citrate, which are formed during microbial metabolism and decomposition of plant residues. Soil organic matter contributes aliphatic acids, phenols, phenolic acids, fulvic and humic substances (Freche et al. 1992; Haynes and Mokolobate 2001). These humic matter ligands are large and have multiple functional groups, binding cations and releasing soluble P to the soil solution. Animal manure and wastewater also contain large amounts of organic matter that likely generate complex ligands (Ma et al. 2001).

Owing to the ability of one ligand to participate in exchange reactions with another ligand, polydentate ligands of different size and charge characteristics such as CDTA (1,2-cyclohexane diamino-tetraacetate), DTPA (diethylene triamine-N, N,N′,N″,N″-pentaacetate), EDTA (ethylenediamine-N,N,N′,N′-tetraacetate); oxalic acid (ethanedioic acid), and phthalic acid (1,2-benzenedicarboxylic acid) were found to efficiently displace and exchange for myo-inositol hexakisphosphate and other organic phosphates (Dao, 2003, 2004). Adding ligands to a mixture of *myo*-inositol hexa*kis*phosphate and aluminum reversed the inhibitory effect of Al^{3+} on the dephosphorylation of the inositol phosphate. When Al^{3+} concentrations exceeded 0.75 mM, the extent of this phenomenon was in the following order:

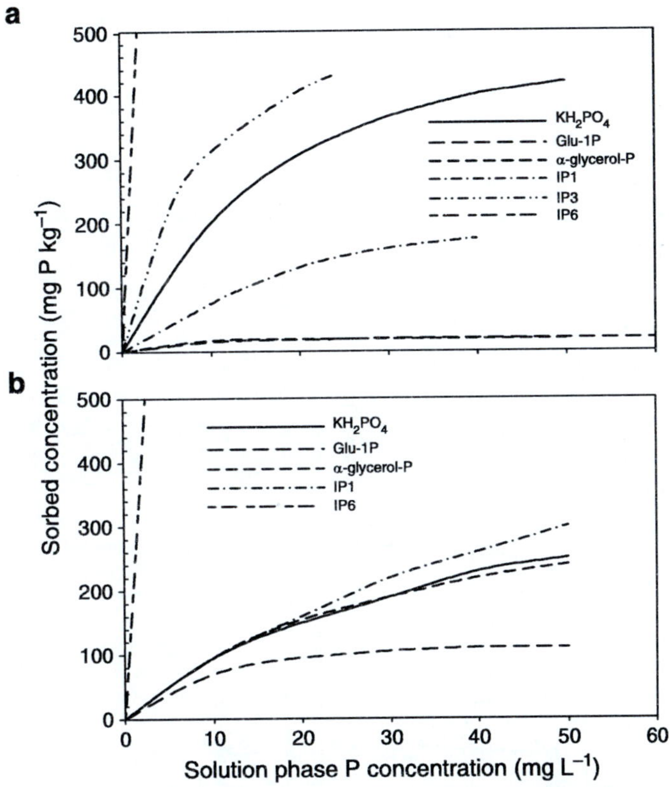

Fig. 5.8 Sorption of phosphate monoesters, with differing number of phosphate moieties on soil (**a**) and on peat moss (**b**) (Adapted from McKercher and Anderson (1989))

phthalate = oxalate < DPTA < EDTA = CDTA. The ligands CDTA and EDTA were able to reverse some, if not most of the inhibition of inositol phosphate hydrolysis at all Al^{3+} concentrations up to 1.5 mM at pH 4.5 and 6.0. The basis for this reversal of inhibition of dephosphorylation of complexed organic phosphates was found on the stoichiometry of electrical charge. Excess charge was needed to decouple and mobilize complexed *myo*-inositol hexa*kis*phosphate, and to attain the ligand to *myo*-inositol hexa*kis*phosphate charge concentration ratios between one and four equivalent molar concentration ratios had to be between 1.5 and 12-fold that of *myo*-inositol hexa*kis*phosphate, except in the case of phthalate (Fig. 5.9). However, there must be an upper limit to increasing charge concentration needed to overcome the inhibitory effect of polyvalent counterions, because molar concentrations eventually reach levels that would denature or precipitate the enzyme. The reader is referred to a previous review by the author for more detailed information of the chemistry of ligands and the stoichiometry needed to optimize the exchange process (Dao 2007).

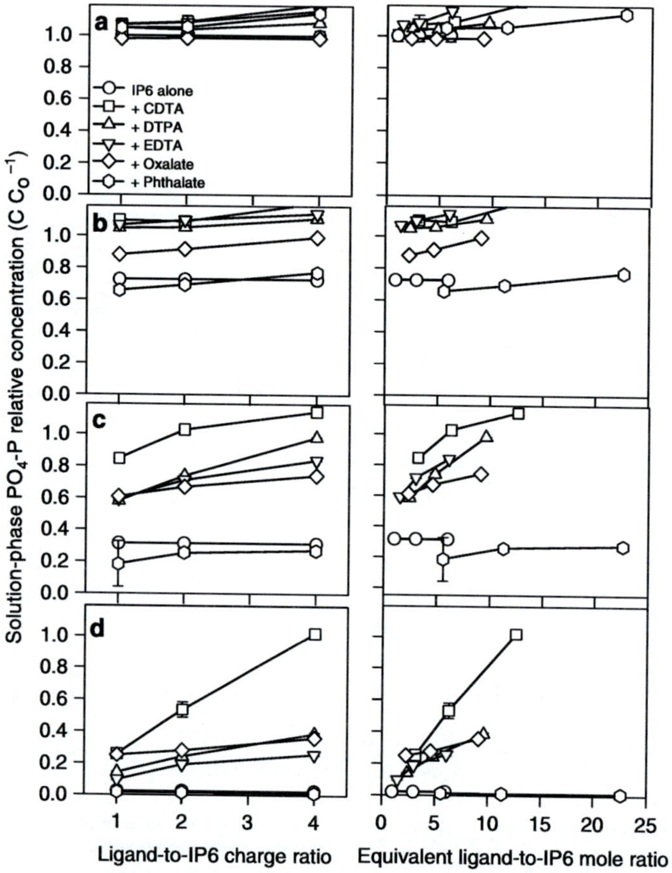

Fig. 5.9 Effects of selected organic ligands (LIGND), LIGND-to- *myo*-inositol hexa*kis*phosphate (IP6) charge ratios, and equivalent LIGND-to-IP6 mole ratios on the dephosphorylation of IP6 at four levels of Al^{3+} counterion, (**a** = 0; **b** = 0.25; **c** = 0.75, and **d** = 1.5 mM Al^{3+}), at suspension's pH 4.5 (Dao 2004)

5.4 An *In Vitro* Method to Emulate Ligand-Based Plant and Microbial Solubilization and Mineralization of Organic P Pools in Soil and Organic Bio-Nutrient Sources

Although soil enzyme diversity and activities have been used as indicators of ecosystem health (Burns 1982; Dick and Tabatabai 1987; Pankhurst et al. 1997; Dick 1992, 1994; Nannipieri et al. 2002), exogenous enzymes additions have also thought to be quantitative tools to detect changes in organic nutrients in varied environmental media and further elucidate soil nutrient pool sizes, pool turnover rates, and reinforce the role of enzymatic processes in organic nutrient cycling

(He and Honeycutt 2001; Turner et al. 2002; Dao 2003, 2004; Dao et al. 2006; Green et al. 2006, 2007; Bunemann 2008). In addition, extracellular phosphohydrolases are recently thought of as a means of detecting environmental P accumulation in soil and bioactive forms of P in animal manure and manure-amended soils. Making the concentration of specific enzymes or combinations of enzymes non-limiting allows quantitative determination of the size of substrate pools beyond chemical identification, as well as allows inference of rates of transformation and transport (Dao 2003, 2004; Dao et al. 2005; Vats et al. 2005; Green et al. 2007; Dao and Hoang 2008). Enzymatic methods have been applied to the identification of organic P-containing substrates in soil extracts, runoff, and surface water columns but have not been used as quantitative methods for soil P pools because of their low efficiency when compared to spectrometric techniques (Pant et al. 1994; He and Honeycutt 2001; Turner et al. 2002; Dao, 2007; Bunemann 2008). For example, Hayes et al. (2000) reported less than 2% of soil total P (6.2 mg kg^{-1}) was desorbed and hydrolyzed in a 50-mM citric acid extract and much less (0.3 mg kg^{-1}) in a water extract. He and Honeycutt (2001) reported a variety of organic P forms in manure and a soil, but achieved a low recovery under their experimental conditions. It must be recognized that the enzymatic intermediates and end products of organic P dephosphorylation are also highly reactive, with many biogeochemical sinks in soils contributing to the low recovery of desorbed and hydrolyzed P.

Improvements have recently been made to *in situ* enzymatic methods by enhancing substrate availability and therefore quantitative recovery of enzyme-labile P content of samples of manure, soils, or runoff (Dao 2003, 2004, 2007; He et al. 2006; Green et al. 2007; Dao and Zhang 2007). Dao (2003) first reported the approach of using polycarboxylate ligands in combination with extracellular enzymes to simultaneously extract, hydrolyze, and categorize P species, without separate steps of an extraction and isolation of the extract for enzymatic assay. The hydrolysis reaction is strongly driven to reach an equilibrium state toward the dephosphorylation of organic P. The rationale and mechanisms for enhanced efficiency have been extensively discussed (Dao 2004, 2007). The *in situ* experimental approach is distinctly different from other extraction and enzymatic methods (e.g., Hedley et al. 1982; Turner et al. 2002; He et al. 2006). In the two-step methods, the extraction of solid-phase P may not be complete or exhaustive and the extraction process is terminated upon separation and removal of the liquid phase for analysis of extracted P. In this ligand-based fractionation approach, bioavailable phosphate and phosphate monoesters, including inositol phosphates in manure (Dao 2003; Dao et al. 2006) and soils (Dao 2004; Dao et al. 2005; Rao and Dao 2008) can be differentiated into pools that reflect their potential solubilization. Fractions of manure P that are measured in the enzyme-hydrolysable P assay include (1) the water-extractable phosphate, (2) an EDTA-exchangeable phosphate pool that was not previously extracted by water alone, which has been found to be equivalent to the soil P pool extracted by the Mehlich-3 procedure (Dao et al. 2005; Rao and Dao 2008), (3) a water-extractable organic P pool that is hydrolysable by fungal phosphohydrolases, (4) an EDTA-exchangeable organic P pool that is hydrolysable by phosphohydrolases and includes inositol phosphates, and (5) a residual pool that is not extractable by water or dilute solutions of polycarboxylate ligands.

The ligand-based enzyme-hydrolysable P assay provides a relative scale of chemical and biological stability of P in manures. Phosphorus pools with greater stability, whether the substrates are inorganic phosphates, inositol phosphates, or other organic phosphates, may become biologically available over longer time scales in the soil. This was observed in a case study of the potential for dissolution and solubilization of immobilized P in soils treated with additives to reduce soluble P in soils. A freshly prepared iron hydroxide additive reduced water-extractable P in soil by 90% over a period of 16 weeks. In addition, a plant-available phosphate fraction (Mehlich-3 extractant) was also reduced in iron-treated soils, and both fractions remained unchanged up to 16 weeks. The ligand-based enzyme-hydrolysable P assay, on the other hand, revealed internal changes in P pools and showed that the effect of the iron additive was transitory. The inorganic EDTA extractable P fraction and the enzyme-hydrolysable P pool could be remobilized, reaching initial soil levels to nullify the P-immobilizing action of the iron additives by about the fourth week following soil treatment. The processes of solubilization and desorption of bioavailable P in unamended and $Fe(OH)_3$-amended soils were best described by single or sequential double exponential kinetic equations. This behavior would be consistent with the fact that these substances can range from metastable amorphous to semi crystalline physical states (Arai and Sparks 2002). More evidence of ligand mediation can be found in the enhanced dissolution of minerals (Wieland et al. 1988; Ullman et al. 1996).

5.5 Conclusions

The mineralization of organic matter and organic P compounds, in particular is performed by microorganisms with the ability to hydrolyze phosphorylated organic substrates with the aid of phosphomonoesterase and diesterase enzymes, along with enzymes of many other catabolic processes, occurring simultaneously or in specific sequences. For this reason, the knowledge of functional diversity of the community is as crucial as that of species diversity of that community, the relative contribution of each factor, and finally, the biogeochemical microenvironment they operate in. This chapter reviewed the complementary roles of extracellular enzymes and organic ligands in the solubilization and mineralization of dissolved and insoluble complexes of organic P in agricultural byproducts and in soils and water systems.

A high reactivity of organic P with soil constituents and/or suspended colloids resulted in low plant availability and the accumulation of such forms in many soils of the world. Ligand formation, ligand exchange, and extracellular phosphomonoesterase and diesterase enzymes mediate the turnover of organic P. The enzymatic dephosphorylation of uncomplexed organic phosphate monoesters, for example, *myo*-inositol hexa*kis*phosphate commonly found in soil may yield eventually 6 moles of phosphate and 1 mole of *myo*-inositol. The metabolites are stereochemically-specific, depending on whether the substrate is degraded by microbial systems or whether the substrate is metabolized by plant-derived phosphohydrolases. However, the enzymatic dephosphorylation of organic P-containing substances in the environment often is constrained by physical and biochemical factors affecting the solubility and availability of these substrates. Plant strategies to enhance the sensing and acquisition of P in their roots' environment have included a wide array of organic ligands and siderophores, and morphological modifications; and there is increasing evidence of complementary genetic adaptation of phosphate sensing and transporter systems, leading to major molecular efforts to research transgenic expression of high-affinity genes induced in response to phosphate deficiency in economically important plants and microbial systems.

In furthering our understanding of processes controlling organic P availability, an *in situ* ligand-based enzymatic method was described, which provides insight into the biological stability of environmental inositol phosphates. Selected polydentate ligands and fungal phosphohydrolases can differentiate various P pools that contribute to the solution-phase phosphate concentration in animal manures and manure-amended soils. A significant amount of intertwining of biotic and abiotic processes was observed in plant and microbes' arrays of strategies for sensing and obtaining P from their environment. What is clear is that another fundamental evaluation of the biogeochemistry of P in the soil appears necessary to improve our understanding of the solubilization of organic P substrates, and the role of ligands in the turnover of organic P. Although much progress have been made in the past two decades, knowledge gaps in the control mechanisms and the lack of practical effective solutions exist for enhancing the exchange of organic P and their accessibility by extracellular phosphohydrolases, novel, genetically modified, or otherwise. As P fertilizers and supplies are increasingly limited, the challenge to the agricultural communities is to mimic native plant strategies, but on the scale of industrial agricultural production to meet food and fiber needs on the global scale. Soils remain a large pool of stabilized organic P to manage sustainably. This soil P bank is a sizable pool, estimated to average >700 kg ha^{-1}. This pool is relatively stable, or at best, very slowly available. The limited finding the author made on the differential ligand strength between large polycarboxylate and small organic anions such as oxalate, and other aliphatic organic acids (i.e., malate, acetate, etc...) in decoupling metal–inositol hexa*kis*phosphate complexes (Dao 2004) apparently indicates a need for new approaches to further enhance the desorption and accessibility to stabilized organic P in soil and sediments. The challenge may include the improvement and translation of our knowledge of surface electrical properties of soil colloids into means and practices to alter the availability of sorbed organic P forms to extracellular enzymes' catabolic activity. In addition, soil is an extensive repository of a variety of municipal and industrial by-products, i.e., biosolids, coal-combustion ashes, foundry sands; animal manure and wastewater effluents are recycled in production agriculture in large quantities. It is highly desirable is highly desirable to continue research on the effect of soil management and extracellular enzymes of the major nutrient cycles of C and N, in addition to phosphohydrolases. It is imperative we improve our holistic understanding of nutrients' interaction on soil enzymes' induction and release in order to find solutions to constraints to their activity and the cycling of P to improve the use efficiency of this non-renewable resource.

References

Ae N, Ariahara J, Okada K, Yoshihara T, Johansen C (1990) Phosphorus uptake by pigeon pea and its role cropping systems of the Indian subcontinent. Science 248:477–480

Anderson G (1980) Assessing organic P in soils. In: Khasawneh FE, Sample EC, Kamprath EJ (eds) The role of phosphorus in agriculture. American Society Agronomy, Madison, WI, pp 411–431

Anderson G, Williams EG, Moir JO (1974) A comparison of sorption of inorganic phosphate and inositol hexaphosphate by six acid soils. J Soil Sci 25:51–62

Arai Y, Sparks DL (2002) ATR-FTIR spectroscopic investigation on phosphate adsorption mechanisms at the ferrihydrite–water interface. J Colloid Interface Sci 241:317–326

Asada K, Tanaka K, Kasai Z (1969) Formation of phytic acid in cereal grains. Ann NY Acad Sci 165:801–814

Baldwin DS, Howitt JA, Beattie JK (2005) Abiotic degradation of organic phosphorus compounds in the environment. In: Turner BL, Frossard E, Baldwin D (eds) Organic phosphorus in the environment. CABI Publishing, Oxfordshire, pp 75–88

Barrientos L, Scott JJ, Murthy PP (1994) Specificity of hydrolysis of phytic acid by alkaline phytase from lily pollen. Plant Physiol 106:489–1495

Beek J, van Riemsdijk WH (1982) Interaction of orthophosphate ions with soil. In: Bolt GH (ed) Soil chemistry. B. Physico-chemical models, 2nd edn. Elsevier, Amsterdam

Biederbeck VO, Bouman OT, Looman J, Slinkard AE, Bailey LD, Rice WA, Janzen HH (1993) Productivity of four annual legumes as green manure in dryland cropping systems. Agron J 85:1035–1043

Bonfante-Fasolo P (1984) Anatomy and morphology of VA mycorrhizae. In: Powell CL, Bagyaraj DJ (eds) VA mycorrhizae. CRC, Boca Raton, FL, pp 5–33

Bremer E, van Kessel C (1992) Plant-available nitrogen from lentil and wheat residues during a subsequent growing season. Soil Sci Soc Am J 56:1155–1160

Bullock JI, Duffin PA, Nolan KB (1993) In vitro hydrolysis of phytate at 95°C and the influence of metal ion on the rate. J Sci Food Agric 63:261–263

Bunemann EK (2008) Enzyme additions as a tool to assess the potential bioavailability of organically bound nutrients. Soil Biol Biochem 40:2116–2129

Burns RG (1982) Enzyme activity in soil: location and a possible role in microbial ecology. Soil Bio Biochem 14:423–427

Burns RG, Dick RP (2002) Enzymes in the environment: activity, ecology, and applications. Marcel-Dekker, New York, NY

Cade-Menun BJ, Liu CW, Nunlist R, McColl JG (2002) Soil and litter phosphorus-31 nuclear magnetic resonance spectroscopy: extractants, metals, and phosphorus relaxation times. J Environ Qual 31:457–465

Cavigelli MA, Thien SJ (2003) Phosphorus bioavailability following incorporation of green manure crops. Soil Sci Soc Am J 67:1186–1194

Celi L, Barberis E (2007) Abiotic reactions of inositol phosphates in soil. In: Turner BL et al (eds) Inositol phosphates: linking agriculture and the environment. CABI, Oxfordshire, UK, pp 207–220

Celi L, Presta M, Ajmore-Marsan F, Barberis E (2001) Effects of pH and electrolytes on inositol hexaphosphate interaction with goethite. Soil Sci Soc Am J 65:753–760

Christensen BT (1986) Barley straw decomposition under field conditions: effect of placement and initial nitrogen content on weight loss and nitrogen dynamics. Soil Biol Biochem 18:523–529

Cooper WT, Heerboth M, Salters VJM (2007) High-performance chromatographic separations of inositol phosphates and their detection by mass spectrometry. In: Turner BL et al (eds) Inositol phosphates: linking agriculture and the environment. CABI, Oxfordshire, UK, pp 23–40

Cosgrove DJ (1980) Inositol phosphates: their chemistry, biochemistry, and physiology. Elsevier, New York, NY

Council for Agricultural Science and Technology (2002) Animal diet modification to decrease the potential for N and P pollution Issue paper no 21. CAST, Ames, IA

Council for Agricultural Science and Technology (CAST) (1996) Integrated animal waste management. CAST, Ames, IA

Dao TH (1998) Tillage and crop residue effects on carbon dioxide evolution and carbon storage in a Paleustoll. Soil Sci Soc Am J 62:250–256

Dao TH (2003) Polyvalent cation effects on *myo*-inositol hexa*kis* dihydrogenphosphate enzymatic dephosphorylation in dairy wastewater. J Environ Qual 32:694–701

Dao TH (2004) Ligands and phytase hydrolysis of organic phosphorus in soils amended with dairy manure. Agron J 96:1188–1195

Dao TH (2007) Ligand effects on inositol phosphate solubility and bioavailability in animal manures. In: Turner BL, Richardson AE, Mullaney EJ (eds) Inositol phosphates: linking agriculture and environment. CABI, Oxfordshire, UK, pp 169–185

Dao TH, Cavigelli MA (2003) Mineralizable carbon, nitrogen, and water-extractable phosphorus release from stockpiled and composted manure, and manure-amended soils. Agron J 95:405–413

Dao TH, Hoang KQ (2008) Dephosphorylation and quantification of organic phosphorus in poultry litter by purified phytic-acid high affinity *Aspergillus* phosphohydrolases. Chemosphere 72:1782–1787

Dao TH, Zhang H (2007) Rapid composition and source screening of heterogeneous poultry litter by energy dispersive X-ray fluorescence spectrometry. Ann Environ Sci 1:69–79

Dao TH, Codling EE, Schwartz RC (2005) Time-dependent phosphorus extractability in calcium- and iron-treated high-phosphorus soils. Soil Sci 170:810–821

Dao TH, Lugo-Ospina A, Reeves JB, Zhang H (2006) Wastewater chemistry and fractionation of bioactive phosphorus in dairy manure. Comm Soil Sci Plant Anal 37:907–924

Dao TH, Guber AK, Sadeghi AM, Karns JS, van Kessel JS, Shelton DR, Pachepsky YA, McCarty G (2008) Loss of bioactive phosphorus and enteric bacteria in runoff from dairy manure applied to sod. Soil Sci 173:511–521

Devai I, Felfoldy L, Wittner I, Plosz S (1988) Detection of phosphine: new aspects of the phosphorus cycle in the hydrosphere. Nature 333:343–345

Dick RP (1992) A review: long-term effects of agricultural systems on soil biochemical and microbial parameters. Agric Ecosys Environ 40:25–36

Dick RP (1994) Influence of long-term tillage and crop rotation combinations on soil enzyme activities. Soil Sci Soc Am J 56:783–788

Dick WA, Tabatabai MA (1987) Kinetics and activities of phosphatase–clay complexes. Soil Sci 143:5–15

Dighton J (1983) Phosphatase production by mycorrhizal fungi. Plant Soil 71:455–462

Dighton J (1991) Acquisition of nutrients from organic resources by mycorrhizal autotrophic plants. Experientia 47:362–369

Dinkelaker B, Römheld V, Marschner H (1989) Citric acid and precipitation of calcium citrate in the rhizosphere of white lupin (*Lupinus albus* L.). Plant Cell Environ 12:285–292

Douglas CL Jr, Almaras RR, Rasmussen PE, Ramig RE, Roager RE Jr (1980) Wheat straw decomposition and placement effects on decomposition in dryland agriculture of the Pacific Northwest. Soil Sci Soc Am J 44:833–837

Duff SMG, Sarath G, Plaxton WC (1994) The role of acid phosphatases in plant phosphorus metabolism. Physiol Planta 90:791–800

Ezawa T, Hayatsu M, Saito M (2005) A new hypothesis on the strategy for acquisition of phosphorus in arbuscular mycorrhiza: up-regulation of secreted acid phosphatase gene in the host plant. Mol Plant-Microbe Interact 18:1046–1053

Freche M, Rouquet N, Koutsoukos P, Lacout JL (1992) Effects of humic compounds on the crystal growth of dicalcium phosphate dihydrate. Agrochimica 36:500–510
Frossard E, Condron LM, Oberson A, Sinaj S, Fardeau JC (2000) Processes governing phosphorus availability in temperate soils. J Environ Qual 29:15–23
George TS, Richardson AE, Simpson RJ (2005) Behavior of plant-derived extracellular phytase upon addition to soil. Soil Biol Biochem 37:977–988
Gerritse RG, Eksteen R (1978) Dissolved organic and inorganic phosphorus compounds in pig slurry: effect of drying. J Agric Sci 90:39–45
Goldberg S, Sposito G (1984a) A chemical model of phosphate adsorption by soils: I. Reference oxide minerals. Soil Sci Soc Am J 48:772–778
Golberg S, Sposito G (1984b) A chemical model of phosphate adsorption by soils. II. Noncalcareous soils. Soil Sci Soc Am J 48:779–783
Golberg S, Sposito G (1985) On the mechanism of specific phosphate adsorption by hydroxylated mineral surfaces: a review. Comm Soil Sci Plant Anal 16:801–821
Green VS, Dao TH, Stone G, Cavigelli MA (2006) Phosphorus fractions and dynamics among soil aggregate size classes of organic and conventional cropping systems. Soil Sci 171:874–885
Green VS, Dao TH, Stone G, Cavigelli MA, Baumhardt RL, Devine TE (2007) Bioactive phosphorus loss in simulated runoff from a phosphorus-enriched soil under two forage management systems. Soil Sci 172:721–732
Greiner R (2007) Phytate-degrading enzymes: regulation of synthesis in microorganisms and plants. In: Turner BL et al (eds) Inositol phosphates: linking agriculture and the environment. CABI, Oxfordshire, UK, pp 78–96
Greiner R, Carlsson NG, Alminger ML (2000) Stereospecificity of *myo*-inositol hexakisphosphate dephosphorylation by a phytate-degrading enzyme of *Escherichia coli*. J Biotechnol 84:53–62
Greiner R, Alminger ML, Carlsson NG, Muzquiz M, Burbano C, Cuadrado C, Pedrosa MM, Goyoaga C (2002) Pathway of dephosphorylation of *myo*-inositol hexa*kis*phosphate by phytases of legume seeds. J Agric Food Chem 50:6865–6870
Harrison MJ (2005) Signaling in the arbuscular mycorrhizal symbiosis. Annu Rev Microbiol 59:19–42
Haukka K, Kolmonen E, Hyder R, Hietala J, Vakkilainen K, Kairesalo T, Haario H, Sivonen K (2006) Effect of nutrient loading on bacterioplankton community composition in lake mesocosms. Microb Ecol 51:137–146
Hayes JE, Richardson AE, Simpson RJ (2000) Components of organic phosphorus in soil extracts that are hydrolyzed by phytase and acid phosphatase. Biol Fertil Soils 32:279–286
Haynes RJ, Mokolobate MS (2001) Amelioration of Al toxicity and P deficiency in acid soils by additions of organic residues: a critical review of the phenomenon and the mechanisms involved. Nutr Cycl Agroecosyst 59:47–63
He Z, Honeycutt CW (2001) Enzymatic characterization of organic phosphorus in animal manure. J Environ Qual 30:1685–1692
He Z, Dao TH, Honeycutt CW (2006) Insoluble iron-related inorganic and organic phosphates in animal manure and soil. Soil Sci 171:117–126
Hedley MJ, Stewart JWB, Chauhan BS (1982) Changes in inorganic and organic soil P fraction induced by cultivation practices and by laboratory incubations. Soil Sci Soc Am J 46:970–976
Helal HM (1990) Varietal differences in root phosphatase activity as related to the utilization of organic phosphates. Plant Soil 123:161–163
Henmi T, Huang PM (1985) Removal of phosphorus by poorly ordered clays as influenced by heating and grinding. Appl Clay Sci 1:133–144
Hilton J, O'Hare M, Bowes MJ, Jones JI (2006) How green is my river? A new paradigm of eutrophication in rivers. Sci Total Environ 365:66–83
Hingston FJ, Posner AM, Quirk JP (1974) Anion adsorption by goethite and gibbsite. II. Desorption of anions from hydrous oxide surfaces. J Soil Sci 25:16–26
Hinsinger P (2001) Bioavailability of soil inorganic phosphorus in the rhizosphere as affected by root-induced chemical changes: a review. Plant Soil 237:173–195

Hoffland E, Findenegg GR, Nelemans JA (1989) Solubilization of rock phosphate by rape. II. Local root exudation of organic acids as a response to P starvation. Plant Soil 113:161–165
Hoffland E, Boogaard RVD, Nelemans J, Findenegg G (1992) Biosynthesis and root exudation of citric and malic acids in phosphate-starved rape plants. New Phytol 122:675–680
Horst WJ, Kahm M (2004) Agronomic-based technologies toward more ecological use of phosphorus in agriculture. In: Valsami-Jones E (ed) Phosphorus in environmental technology: principles and applications. IWA, London, UK, pp 610–628
Hull SR, Gray JSS, Montgomery R (1999) Autohydrolysis of phytic acid. Anal Biochem 273:252–260
Hunter DA, McManus MT (1999) Comparison of acid phosphohydrolases in two genotypes of white clover with different responses to applied phosphate. J Plant Nutr 22:679–692
Jarvie HP, Neal C, Williams RJ, Neal M, Wickham HD, Hill LK, Wade AJ, Warwick A, White J (2002) Phosphorus sources, speciation and dynamics in the lowland eutrophic River Kennet, UK. Sci Total Environ 282–283:175–203
Jayachandran K, Schwab AP, Hetrick BD (1992) Mineralization of organic phosphorus by vesicular-arbuscular mycorrhizal fungi. Soil Biol Biochem 24:897–903
Jayasundera S, Schmidt W, Reeves JB, Dao TH (2005) Direct ^{31}P NMR spectroscopic measurement of phosphorous forms in dairy manures. J Food Agric Environ 3:328–333
Johnson JF, Allan DL, Vance CP (1996) Phosphorus deficiency in *Lupinus albus*. Altered lateral root development and enhanced expression of phosphoenolpyruvate carboxylase. Plant Physiol 112:31–41
Joner EJ, Johansen A (2000) Phosphatase activity of external hyphae of two arbuscular mycorrhizal fungi. Mycol Res 104:81–86
Jones DL, Dennis PG, Owens AG, van Hees PAW (2003) Organic acids behavior in soils – misconceptions and knowledge gaps. Plant Soil 248:31–41
Juo ASR, Fox RL (1977) Phosphate sorption characteristics of some benchmark soils of West Africa. Soil Sci 124:370–376
Koopmans GF, Chardon WJ, Dolfing J, Oenema O, van der Meer P, van Riemsdijk WH (2003) Wet chemical and phosphorus-31 nuclear magnetic resonance analysis of phosphorus speciation in a sandy soil receiving long-term fertilizer or animal manure applications. J Environ Qual 32:287–295
Kouno K, Wu J, Brookes PC (2002) Turnover of biomass C and P in soil following incorporation of glucose or ryegrass. Soil Biol Biochem 34:617–622
Kulaev I, Kulakovskaya T (2000) Polyphosphate and phosphate pump. Annu Rev Microbiol 54:709–734
Lal D, Chung Y, Platt T, Lee T (1988) Twin cosmogonic radiotracer studies of phosphorus recycling and chemical fluxes in the upper ocean. Limnol Oceanogr 33:1559–1567
Lapeyrie F, Ranger J, Vairelles D (1991) Phosphate solubilizing activity of ectomycorrhizal fungi in vitro. Can J Bot 69:342–346
Lehrfeld J (1989) High-performance liquid chromatography analysis of phytic acid on a pH-stable, macroporous polymer column. Cereal Chem 66:510–515
Lehrfeld J (1994) HPLC Separation and quantitation of phytic acid and some inositol phosphates in foods: problems and solutions. J Agric Food Chem 42:2726–2731
Lemtiri-Chlieh F, MacRobbie EA, Webb AA, Manison NF, Brownlee C, Skepper JN, Chen J, Prestwich GD, Brearley CA (2003) Inositol hexa*kis*phosphate mobilizes an endomembrane store of calcium in guard cells. Proc Natl Acad Sci 100:10091–10095
Leyval C, Reid CPP (1991) Utilization of microbial siderophores by mycorrhizal and non-mycorrhizal pine roots. New Phytol 119:93–98
Li XL, George E, Marschner H (1991) Extension of the phosphorus depletion zone in vesicular arbuscular mycorrhizal white clover in a calcareous soil. Plant Soil 36:41–48
Lindsay WL (1979) Chemical equilibria in soils. Wiley, New York, NY
Lindsay WL, Stephenson HF (1959) Nature of the reactions of monocalcium phosphate monohydrate in soils. II. Dissolution and precipitation reactions involving iron, aluminum, manganese, and calcium. Soil Sci Soc Am J 23:18–22

Lindsay WL, Frazier AW, Stephenson HF (1962) Identification of reaction products from phosphate fertilizers in soils. Soil Sci Soc Am J 26:446–452

Lopez-Gutierrez JC, Toro M, Lopez-Hernandez D (2004) Seasonality of organic phosphorus mineralization in the rhizosphere of the native savanna grass, *Trachypogon plumosus*. Soil Biol Biochem 36:1675–1684

Lott J, Ockenden I, Raboy V, Batten GD (2000) Phytic acid and phosphorus in crop seeds and fruits: a global estimate. Seed Sci Res 10:11–33

Lupwayi NZ, Clayton GW, O'Donovan JT, Harker KN, Turkington TK, Soon YK (2007) Phosphorus release during decomposition of crop residues under conventional and zero tillage. Soil Tillage Res 95:231–239

Lutzenkirchen J, Behra Ph (1996) On the surface precipitation model for cation sorption at the (Hydr)oxide water interface. Aquat Geochem 1:375–397

Ma H, Allen HE, Yin Y (2001) Characterization of isolated fractions of dissolved organic matter from natural waters and a wastewater effluent. Water Res 35:985–996

Ma X, Wright E, Ge Y, Bell J, Xi Y, Bouton JH, Wang Z (2009) Improving phosphorus acquisition of white clover (*Trifolium repens* L.) by transgenic expression of plant-derived phytase and acid phosphatase genes. Plant Sci 176:479–488

Mainstone CP, Parr W (2002) Phosphorus in rivers: ecology and management. Sci Total Environ 282–283:25–47

Martin RR, Smart RStC (1987) X-ray photoelectron studies of anion adsorption on goethite. Soil Sci Soc Am J 51:54–56

McKercher RB, Anderson G (1989) Organic phosphate sorption by neutral and basic soils. Commun Soil Sci Plant Anal 20:723–732

Nannipieri P, Kandeler E, Ruggiero P (2002) Enzyme activities and microbiological and biochemical processes in soil. In: Burn RG, Dick RP (eds) Enzymes in the environment. Marcel Dekker, New York, pp 1–33

Neumann G, Martinoia E (2002) Cluster roots – an underground adaptation for survival in extreme environments. Trends Plant Sci 7:162–167

Ognalaga M, Frossard E, Thomas F (1994) Glucose-1-phosphate and myo-inositol hexaphosphate adsorption mechanisms on goethite. Soil Sci Soc Am J 58:332–337

Otani T, Ae N (1999) Extraction of organic phosphorus in andosols by various methods. Soil Sci Plant Nutr 45:151–161

Pankhurst CE, Doube BM, Gupta VVSR (1997) Biological indicators of soil health: synthesis. In: Pankhurst CE, Doube BM, Gupta VVSR (eds) Biological indicators of soil health. CABI, Oxfordshire, UK, pp 419–435

Pant HK, Edwards AC, Vaughan D (1994) Extraction, molecular fractionation and enzyme degradation of organically associated phosphorus in soil solutions. Biol Fertil Soils 17:196–200

Parfitt RL (1977) Phosphate adsorption on an Oxisol. Soil Sci Soc Am J 41:1064–1067

Parfitt RL (1979) The nature of the phosphate-geothite complex formed with $Ca(H_2PO_4)_2$ at different surface coverage. Soil Sci Soc Am J 43:623–625

Parfitt RL (1989) Phosphate reactions with natural allophane, ferrihydrite, and goethite. J Soil Sci 40:359–369

Parfitt RL, Atkinson RJ, Smart RStC (1975) The mechanism of phosphate fixation by iron oxides. Soil Sci Soc Am J 39:837–841

Peperzak P, Caldwell AG, Hunziker RR, Black CA (1959) Phosphorus fractions in manures. Soil Sci 87:293–302

Plessner O, Klapatch T, Guerinot ML (1993) Siderophore utilization by *Bradyrhizobium japonicum*. Appl Environ Microbiol 59:1688–1690

Quan C, Fan S, Ohta Y (2003) Pathway of dephosphorylation of *myo*-inositol hexa*kis*phosphate by a novel phytase from *Candida krusei* WZ-001. J Biosci Bioeng 95:530–533

Quinn JP, Kulakova AN, Cooley NA, McGrath JW (2007) New ways to break an old bond: the bacterial carbon–phosphorus hydrolases and their role in biogeochemical phosphorus cycling. Environ Microbiol 9:2392–2400

Quiquampoix H (1987) A stepwise approach to the understanding of extracellular enzyme activity in soil II. Competitive effects on the adsorption of a beta-D-glucosidase in mixed mineral or organo-mineral systems. Biochimie 69:765–771

Quiquampoix H, Servagent-Noinville S, Baron M-H (2002) Enzyme adsorption on soil mineral surfaces and consequences for the catalytic activity. In: Burns RG, Dick RP (eds) Enzymes in the environment: activity ecology and applications. Marcel-Dekker, New York, NY, pp 285–306

Ragothama KG (1999) Phosphate acquisition. Annu Rev Plant Physiol Plant Mol Biol 50:665–693

Rao SC, Dao TH (2008) Relationships between phosphorus uptake in two grain legumes and soil bioactive P pools in fertilized and manure-amended soil. Agron J 100:1–6

Reid CPP, Crowley DE, Kim HJ, Powell PE, Szaniszlo PJ (1984) Utilization of iron by oat when supplied as ferrated synthetic chelate or as ferrated hydroxamate siderophore. J Plant Nutr 7:437–447

Roels J, Verstraete W (2001) Biological formation of volatile phosphorus compounds. Bioresour Technol 79:243–250

Sandberg AS, Ahderinne R (1986) High-performance liquid chromatographic method for determination of inositol tri-, tetra-, penta-, and hexaphosphates in foods and intestinal contents. J Food Sci 51:547–550

Sandberg A-S, Brune M, Carlsson NG, Hallberg L, Skoglund E, Rossander-Hulthén L (1999) Inositol phosphates with different numbers of phosphate groups influence iron absorption in humans. Am J Clin Nutr 70:240–246

Sanyal SK, De Datta SK, Chan PY (1993) Phosphate sorption-desorption behavior of some acidic soils of South and Southeast Asia. Soil Sci Soc Am J 57:937–945

Schindler PW, Furst B, Dick R, Wolf PU (1976) Ligand properties of surface silanol groups. I. Surface complex formation with Fe^{3+}, Cu^{2+}, Cd^{2+}, and Pb^{2+}. J Colloid Interface Sci 55: 469–475

Scott JJ, Loewus FA (1986) Phytate metabolism in plants. In: Graf E (ed) Phytic acid: chemistry and applications. Pilatus, Minneapolis, MN, pp 23–42

Shi J, Wang H, Hazebroek J, Ertl DS, Harp T (2005) The maize low-phytic acid 3 encodes a myo-inositol kinase that plays a role in phytic acid biosynthesis in developing seeds. Plant J 42:708–719

Shin E, Han JS, Min J, Min S-H, Park JK, Rowell RM (2004) Phosphate adsorption on aluminum-impregnated mesoporous silicates: surface structure and behavior of adsorbents. Environ Sci Technol 38:912–917

Sigel H, Hofstetter F, Martin RB, Milburn RM, Scheller-Krattiger V, Scheller KH (1984) General considerations on transphosphorylations: mechanism of the metal ion facilitated dephosphorylation of nucleoside 5′-triphosphates, including promotion of atp dephosphorylation by addition of adenosine 5′-monophosphate. J Am Chem Soc 106:7935–7946

Singh S, Kapoor KK (1998) Effects of inoculations of phosphate-solubilizing microorganisms and an arbuscular mycorrhizal fungus on mungbean grown under natural soil conditions. Mycorrhiza 7:249–253

Smith VH (1982) The nitrogen and phosphorus dependence of algal biomass in lakes: an empirical and theoretical analysis. Limnol Oceanogr 27:1101–1112

Smith SE, Gianinazzi-Pearson V (1988) Physiological interactions between symbionts in vesicular-arbuscular mycorrhizal plants. Annu Rev Plant Physiol Plant Mol Biol 39:221–244

Soon YK, Arshad MA (2002) Comparison of the decomposition and N and P mineralization of canola, pea, and wheat residues. Biol Fertil Soils 36:10–17

Sposito G (1986) Distinguishing adsorption from surface precipitation. In: Davis JA, Hayes KF (eds) Geochemical processes at mineral surfaces, vol 323, American chemical society symposium series. Washington, DC, pp 217–228

Stewart JWB, Tiessen H (1987) Dynamics of soil organic phosphorus. Biogeochem 4:41–60

Stumm WH, Hohl H, Dalang F (1976) Interaction of metal ions with hydrous oxide surfaces. Croat Chem Acta 48:491–504

Tadano T, Sakai H (1991) Secretion of acid phosphatases by roots of several crop species under phosphorus deficient conditions. Soil Sci Plant Nutr 37:129–140
Tanaka K, Yoshida T, Kasai Z (1974) Radioautographic demonstration of the accumulation site of phytic acid in rice and wheat grains. Plant Cell Physiol 15:147–151
Tarafdar JC, Claassen N (2003) Organic phosphorus utilization by wheat plants under sterile conditions. Biol Fertil Soils 39:25–29
Tarafdar JC, Marschner H (1994) Phosphatase activity in the rhizosphere and hyphosphere of VA mycorrhizal wheat supplied with inorganic and organic phosphorus. Soil Biol Biochem 26:381–395
Ternan NG, Mc Grath JW, Mc Mullan G, Quinn JP (1998) Review: organo-phosphonates: occurrence, synthesis and biodegradation by microorganisms. World J Microbiol Biotechnol 14:635–647
Torrent J, Barren V, Schwertmann U (1990) Phosphate adsorption and desorption by goethites differing in crystal morphology. Soil Sci Soc Am J 54:1007–1012
Turner BL, McKelvie ID, Haygarth PM (2002) Characterization of water-extractable soil organic phosphorus by phosphatase hydrolysis. Soil Biol Biochem 34:27–35
Turner BL, Mahieu N, Condron LM (2003) Phosphorus-31 nuclear magnetic resonance spectral assignments of phosphorus compounds in soil NaOH–EDTA extracts. Soil Sci Soc Am J 67:497–510
Turner BL, Frossard E, Baldwin DS (2005) Organic phosphorus in the environment. CABI, Oxfordshire, UK
Turner BL, Richardson AE, Mullaney EJ (2007) Inositol phosphates: linking agriculture and environment. CABI, Oxfordshire, UK
Ullman WJ, Kirchman DL, Welch SA, Vandevivere P (1996) Laboratory evidence for microbially mediated silicate mineral dissolution in nature. Chem Geol 132:11–17
Vats P, Bhattacharyya MS, Banerjee UC (2005) Use of phytases (*myo*-inositol hexa*kis*phosphate phosphohydrolases) for combating environmental pollution: a biological approach. Crit Rev Environ Sci Technol 35:469–486
Veneklaas EJ, Stevens J, Cawthray GR, Turner S, Grigg AM, Lambers H (2003) Chick pea and white lupin rhizosphere carboxylates vary with soil properties and enhance phosphorus uptake. Plant Soil 248:187–197
Voglmaier SM, Bembenek ME, Kaplin AI, Dorman G, Olszewski JD, Prestwich GD, Snyder SH (1996) Purified inositol hexakisphosphate kinase is an ATP synthase: diphosphoinositol penta*kis*phosphate as a high-energy phosphate donor. Proc Natl Acad Sci 93:4305–4310
Vohra A, Satyanarayana T (2003) Phytases: microbial sources, production, purification, and potential biotechnological applications. Crit Rev Biotechnol 23:29–60
Wasaki J, Yamamura T, Shinano T, Osaki M (2003) Secreted acid phosphatase is expressed in cluster roots of lupin in response to phosphorus deficiency. Plant Soil 248:129–136
Weaver JD, Mullaney EJ, Lei XG (2007) Altering the substrate specificity site of *Aspergillus niger* PhyB shifts the pH optimum to pH 3.2. Appl Microbiol Biotechnol 76:117–122
White AK, Metcalfe WW (2007) Reduced phosphorus compounds. Annu Rev Microbiol 61:379–400
Wieland E, Wehrli B, Stumm W (1988) The coordination chemistry of weathering. III. A generalization on the dissolution rates of minerals. Geochim Cosmochim Acta 52:1969–1981
Wild A, Oake OL (1966) Organic phosphate compounds in calcium chloride extracts of soils: identification and availability to plants. J Soil Sci 17:356–371
Yamagata H, Tanaka K, Kasai Z (1980) Purification and characterization of acid phosphatase in aleurone particles of rice grains. Plant Cell 21:1449–1460
Zhang H, Dao TH, Basta NT, Dayton EA, Daniel TC (2006) Remediation techniques for manure nutrient loaded soils. In: Rice JM, Cadwell DF, Humenik FJ (eds) National Center for Manure and Animal Waste Management White Papers. American Society Agriculture Biological Engineers. Pub. 913C0306. St. Joseph, MI, pp 483–503

Chapter 6
Importance of Extracellular Enzymes for Biogeochemical Processes in Temporary River Sediments during Fluctuating Dry–Wet Conditions

Annamaria Zoppini and Jürgen Marxsen

6.1 Introduction

The special hydrological circumstances that characterize temporary rivers (Fig. 6.1) make them particularly sensitive to environmental alterations. Temporary waters are found throughout the world and include intermittent streams and ponds, episodic rain puddles and seasonal limestone lakes. These natural bodies of water experience a recurrent dry phase of varying duration and spatial extent (Uys and O'Keeffe 1997). Temporary waters are widespread in semi-arid regions worldwide where they play an important role as water sources (e.g., water supply, irrigation and hydroelectric power generation). Climate change represents an emerging problem for these ecosystems by increasing the frequency and duration of drought periods, with potentially important effects on fluvial biogeochemical processes.

From the examination of new findings from the past 6 years of research the Intergovernmental Panel on Climate Change Working Group I (IPCC WGI 2007) concluded that warming of the climate system is unequivocal. An ensemble of 12 climate models (Milly et al. 2005) projects changes in stream flow with 10–30% decreases in runoff in southern Europe by the year 2050. This trend has already been experienced over the last century not only in the Mediterranean basins (WFD/EUWI 2006), but also in streams and rivers in temperate European regions. Here decreasing runoffs and extended periods of desiccation have been observed, especially in headwaters (Wilby et al. 2006; Sutherland et al. 2008).

A. Zoppini (✉)
Istituto di Ricerca Sulle Acque, Consiglio Nazionale delle Ricerche, Area della Ricerca di Roma 1, via Salaria Km 29.300, CP10, 00015 Monterotondo, Roma, Italy
e-mail: zoppini@irsa.cnr.it

J. Marxsen
Limnologische Fluss-Station des Max-Planck-Instituts für Limnologie, 36110 Schlitz, Germany
Institut für Allgemeine und Spezielle Zoologie, Tierökologie, Justus-Liebig-Universität Gießen, Heinrich-Buff-Ring 26–32, 35392 Gießen, Germany

G. Shukla and A. Varma (eds.), *Soil Enzymology*, Soil Biology 22,
DOI 10.1007/978-3-642-14225-3_6,

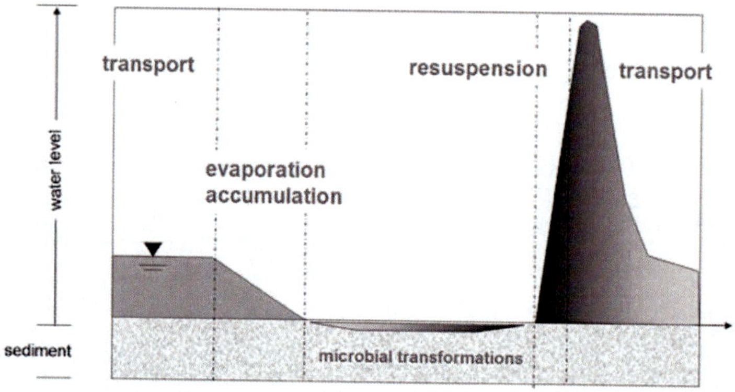

Fig. 6.1 Changes in hydrological conditions in temporary rivers (after Kirkby and Froebrich (2006), modified)

Temporary rivers are the dominant surface water bodies in the Mediterranean region although they are rarely monitored. Until recently dry reaches have been described as "biologically inactive" (Stanley et al. 1997) and there is still little information available on the biogeochemical processes. Tzoraky et al. (2007) observed that in semi-arid catchments the channel bed processes continue even at a low level (40%) of sediment moisture. As a result, the first water flush entering into contact with sediments after a drought period could be extremely modified in its chemistry compared to the base flow, causing in turn drastic modifications in the chemistry of receiving water-bodies (e.g., lakes, rivers, and coastal waters) (Fig. 6.1).

Aquatic sediments act as a sink and source of nutrients. Microbial degradation and transformation of the organic matter deposited in a river channel bed are key processes with regard to the carbon cycling in the lotic food web, which links sedimentary organic matter to the upper level of the community, including carnivores (Fischer and Pusch 2001; Findlay et al. 2003; Mulholland 2003; Marxsen 2006). Via the microbial food chain, complex organic substrates are solubilized by extracellular enzymes in a series of steps from particulate organic matter to high molecular weight dissolved organic carbon and low molecular weight substrates (Chrost 1991). A strong coupling between bacterial activities and deposition of organic material in sediment has been found for both marine and freshwater systems (Sander and Kalff 1993; Goedkoop et al. 1997; Wobus et al. 2003). Hence the efficiency of extracellular enzymes represents a critical step and is able to influence the incorporation of organic carbon into bacterial cells and the consequent transfer to the food chain (Marxsen and Fiebig 1993).

Drying and rewetting are well known climatic factors influencing the physiological status of microbial biomass in soil ecosystems, with significant effects on mineralization (e.g., Raubuch et al. 2002; Griffiths et al. 2003; Mikha et al. 2005). Although benthic microbial processes have been recognized as having an important role in carbon and nutrient flux, few studies have documented the response to dry–wet cycles in temporary waters. The exact role of microbes in mediating these processes is still largely unresolved because methodological constraints make it difficult to determine whether the effects are biologically or physically driven. In addition, in situ measurements are very difficult because of high temporal variability and spatial heterogeneity. This issue has been recently addressed by the European Commission through its funding of Research Projects on temporary rivers (TempQsim and Mirage) with the aim to provide advanced tools to significantly improve the efficiency of integrated water management in Mediterranean semiarid river catchments. This action has encouraged experimental studies on the role of the benthic microbial community in carbon, nutrient and energy flux during drastic changes in water availability (Tzoraky et al. 2007; Amalfitano et al. 2008; Fazi et al. 2008; Zoppini et al. 2010).

6.2 The Response of Benthic Microbial Communities to Drought: A Matter of Survival

Drought has been cited as the most serious of natural disasters as regards loss of life and its impact on agricultural production and economics (Wilhite 2000), although it is a natural feature of aquatic ecosystems in most regions of the world (McMahon and Finlayson 2003).

Temporary waters result in sediment desiccation during an extended period of the year with the consequence of exposing the benthic organisms living there to the air. Biota that inhabits these ecosystems must be morphologically, physiologically and behaviorally adaptable to survive such conditions until the first flood arrives.

Desiccation has been reported to alter the chemistry (e.g., De Groot and Van Wijck 1993) and mineralogy (e.g., Baldwin 1996) of the sediment or soil and kills up to three-quarters of the microbes (e.g., Qiu and McComb 1995). In semiarid regions of the planet drought events have a major effect on benthic community functions, including a delay in litter decomposition together with a decrease in invertebrate density (Pinna and Basset 2004; Fonnesu et al. 2004; Larned et al. 2007).

Tolerance mechanisms to desiccation are poorly understood despite the fact that numerous prokaryotic and eukaryotic organisms are capable of surviving more or less complete dehydration. Drying-rewetting imposes physiological constraints that few genera of bacteria, called anhydrobiotes, can tolerate (Potts 1994; Billi and Potts 2002). Bacteria can cope with this problem through strategies such as forming polymers and spores with the ability to resist physical blows. The production and storage of intracellular solutes, like amino acids and low molecular weight carbohydrates, for acclimatizing to low water conditions (Halverson et al. 2000), is especially important for Gram-negative bacteria, which are unable to form spores and are more susceptible than Gram-positives to disruption by osmotic stress because of their less stable cell walls (Fierer et al. 2003; Schimel et al. 2007).

Examples of spore-forming bacteria are quite widespread among Gram-positive bacteria and they colonize various habitats, including the aquatic environment. In their dormant state, spores have no detectable metabolism although recent findings indicate that dormant bacterial spores, belonging to the genus *Bacillus*, can significantly influence the distribution of heavy metals in the sedimentary marine environment by enzymatic catalysis (Francis and Tebo 2002; Dick et al. 2006). Moreover over the long term (millions of years) catalase was found in freeze-dried permafrost samples (Gilichinsky et al. 1992), while in deep-sea sediments enzyme activity was detected even in a 124,000 year-old sapropel layer (Coolen and Overmann 2000). Overall, these studies suggest that the commonly held view that bacteria in the dormant state are inactive should be revised.

Lowland river–floodplain systems are characterized by a high degree of variability where inundation marks the shift from a terrestrial ecosystem to an aquatic one. In these systems partial drying of wet (previously inundated) sediments results in an increased sediment affinity for phosphorus and produces a zone where there is nitrification coupled with denitrification causing a reduction of nutrient availability (Baldwin and Mitchell 2000). A complete desiccation of sediments may lead to the death of bacteria (and subsequent mineralization of nutrients), a decrease in the affinity of P for iron minerals, a decrease in microbial activity and a cessation of all anaerobic bacterial processes (e.g., denitrification).

For semi-permanent stream sediments Rees et al. (2006) showed that changes in microbial community structure were preserved even one month after rewetting, with a significant difference from the pre-drought and drought microbial communities.

The functional properties of microbial communities were investigated in the sediment of ephemeral rivers (Larned et al. 2007). Non-specific esterase activity, used as an assay for total enzyme activity, was negatively related to the dry period length. The esterase activity decreased by a negative exponential model in the dry periods of 1–7 days accompanied by decreasing respiration rates. Moreover a prolonged period of drought (up to 417-days) further affected esterase activity, reducing the rates by an order of magnitude to close to the detection limit.

Extracellular enzymes may be stable in dry soils or aquatic biofilms for weeks. For phosphatases Perez-Mateos et al. (1991) observed that 65% of indigenous enzymes remained active after 50 days of soil storage at 22°C, while Romaní and Sabater (1997) found that extracellular enzyme activity recovered immediately in stromatolitic riverine communities when rewetted after a summer drought.

Benthic cyanobacterial mats from marshes of northern Belize, an oligotrophic environment, are exposed to extreme conditions in terms of hydrology, nutrient availability and salinity (Sirová et al. 2006). In the cyanobacterial mats alkaline phosphatase exhibited the highest extracellular enzyme activity, followed by leucine-aminopeptidase, arylsulphatase, and β-glucosidase. During the period of drought dry mats retained the same level of potential phosphatase activity after 10 weeks of desiccation (350–450 μmol MUF g^{-1} ash free dry weight min^{-1}). In this environment the preservation of enzyme activity may be favored by the levels of polysaccharide-rich extracellular polymeric substances (EPS) that characterise benthic cyanobacterial mats. Most of the phosphatase activity, visualized using artificial ELF®97 phosphate, appeared free and located in the EPS matrix throughout the mat, with a decoupling from its source in both space and time.

Many studies on the effect of droughts on benthic microorganisms encounter difficulties in disentangling the relative effects of the spatial and temporal extents of low flows and of these from the river bed drying pattern. The alternative approach to gathering information is to simulate a drought event in the laboratory.

Amalfitano et al. (2008) collected wet sediments from four temporary European rivers and let them dry under controlled conditions until complete desiccation. Despite the different origin of the microbial communities the responses of structural and functional parameters to drought were very similar (Fig. 6.2). Bacterial carbon production exponentially decreased, nearly ceasing in dry conditions, followed by a slower decrease in bacterial abundance, with an overall reduction of 74%. By the end of the experiment, live cells (14% of the initial value) were depressed in their main metabolic functions. Hence a conspicuous number of live cells were still abundant at the end of the experiment, but mostly depressed in their metabolic activity. As a result, the significant decrease in per-cell production resulted in a very substantial increase in the community turnover time. Community composition shifted, with an increase in *Alpha-* and *Betaproteobacteria* when sediment was dried. In the same experiment extracellular enzyme activities, measured fluorometrically (Wobus et al. 2003), involved in P, N, and C cycling were weakly or not at all affected by the progressive decreasing of water availability (Zoppini 2007). Aminopeptidase activity was affected by drying and displayed a slow decreasing trend in all tested rivers although it preserved between 40 and 60% of the initial rates in dry sediments (Fig. 6.3). Overall the capacity of alkaline phosphatase to hydrolyze phosphorilated organic matter did not change significantly during the progressive loss of moisture: only one in four tested sediments (the river Krathis) showed a negative trend with desiccation but still preserved 80% of its initial capacity during dry conditions. Similarly the hydrolyzing capacity of polymers like polysaccharides and lipids (β-glucosidase and lipase activities) was preserved under drought conditions without showing any significant trend. This enzyme activity survival capacity recalls mechanisms found in desiccation-tolerant bacterial cells capable of accumulating proteins, of which many are able to remain stable (Billi and Potts 2002), or the capacity of polysaccharides-rich extracellular polymeric substances (EPS) to favor enzyme preservation (Sirová et al. 2006).

To sum up the data, microbial communities are significantly affected by water stress conditions and there are changes in microbial community structure with a drastic reduction of cell abundance, vitality and metabolic activity. Hydrolytic enzymes constitute an exception to this trend. They are probably preserved even if the cells in which they originated become non-viable. We postulate that the preservation of these enzymes represents an important mechanism for the fast recovery of surviving microbial cells after drought. If the hydrolysis of organic matter continues in the dry sediment we can infer that the pool of organic compounds contained in the first water flush could be enriched with labile nutrients that accumulate in the sediments as bacteria are died or temporarily

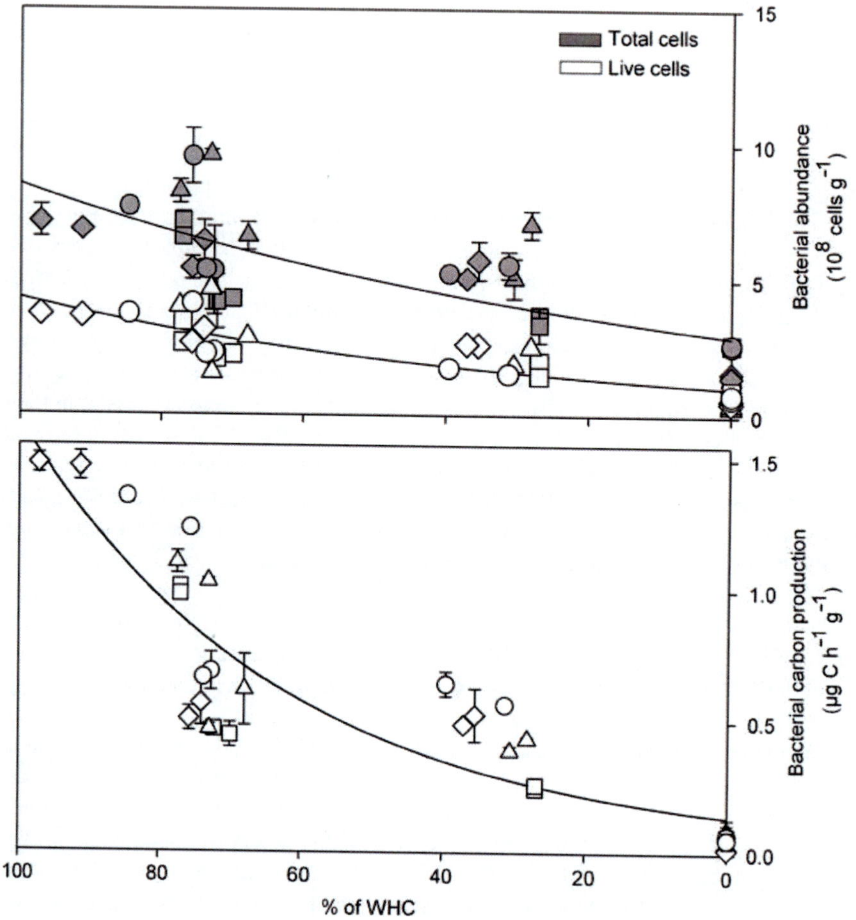

Fig. 6.2 Total and live (Live/Dead® BacLight™ dye technique) bacterial cell abundance (*top panel*) and carbon production (^{3}H-Leucine inc.) (*bottom panel*) versus sediment water content, expressed as a percentage of sediment water hold capacity (WHC). Note the *x*-axis reverse scale: 100% identifies wet sediment at the beginning of the experiment, 0% corresponds to the ending dry sediment. All data are normalized to grams of dry sediment. Error bars indicate standard deviations of three independent measurements. Regression curves are shown originating from temporary rivers (total cell abundance: $r^2 = 0.69$, $P < 0.05$; live cell abundance $r^2 = 0.76$, $P < 0.05$; ^{3}H-leucine inc. $r^2 = 0.79$, $P < 0.05$). (*square* Tagliamento, Italy; *triangle* Krathis, Greece; *diamond* Mulargia, Italy; *circle* Pardiela, Portugal). (after Amalfitano et al. (2008), modified)

inactive. However, in the case of a complete stop of extracellular enzyme activity during extreme dryness, enzymes are able to become active immediately upon rewetting (Marxsen et al. 2010), thus resulting in immediate delivery of these compounds.

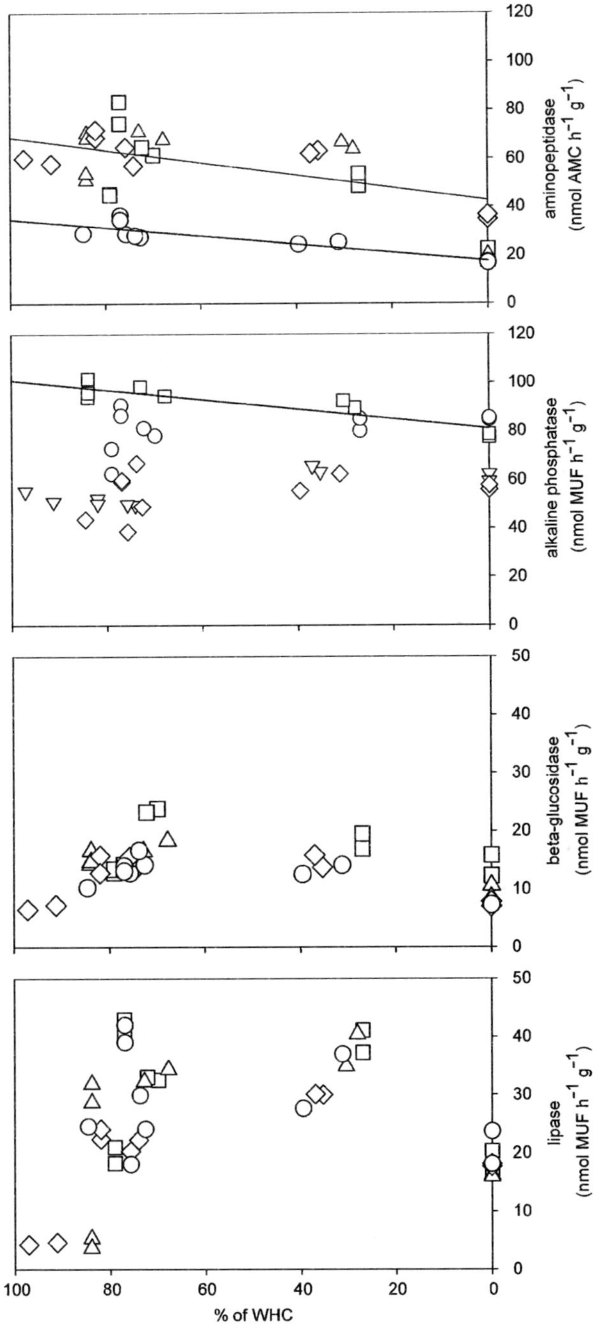

Fig. 6.3 Extracellular enzyme activities versus sediment water content expressed as percentage of sediment water hold capacity (WHC) (see Fig. 3 for symbols). All data are normalized to gram of

6.3 Benthic Microbial Community Awakening After Flooding

The intense run-off and flushing associated with the heavy storms typically occur shortly after the end of a dry period. These events can severely impact microbial distribution as river discharges can increase by several orders of magnitude with respect to their regular flow (Holmes et al. 1998).

Inundation of previously dry river reaches can trigger ecological "hot moments" during which biogeochemical reactions or biological processes begin to accelerate after long quiescent periods (McClain et al. 2003). Inundation can activate microbial and algal cells, delivers chemical substrates to reaction sites, and stimulates enzyme-controlled nutrient transformations and organic matter mineralization (Baldwin and Mitchell 2000; Burns and Ryder 2001; Belnap et al. 2005; Romanì et al. 2006).

Several findings suggest that biota response to inundation depends on the duration of the preceding dry phase (Baldwin and Mitchell 2000; Larned et al. 2007). After 17 days of inundation esterase enzyme activity in sediments that had been dry for 256–523 days was comparable to the potential activity in non-inundated sediments from a similar dry-period range (Larned et al. 2007). After the same dry-period range and 17 days of inundation there was no detectable relationship between sediment respiration rate and dry period duration, although rates tended to be higher for inundated sediments than for non-inundated sediments.

Strong temporal peaks in enzyme activity were observed in sediments within 7 days following inundation (Burns and Ryder 2001). The short response time of α- and β-glucosidase activities after 24 h flooding suggests that even short pulses in high flows may stimulate bacterial activity as dissolved organic carbon loads also peak at this time. However, a longer wetting time may be needed to drive hydrolysis of the proteins, fatty acids and longer chain polysaccharides. This indicates a rapid use of available carbon by microbial communities. A general decline in enzyme activity rates was found in the 21 days following the first week from inundation, which was thought to be the result of substrate limitation within the flooded cores or inhibition by hydrolysis end products.

There have been laboratory-scale experiments conducted on dry sediment simulating an inundation event in order to demonstrate the role of sediments as a source of microbial populations, and related activities, in "first-flush" water (Fazi et al. 2008).

Within 9 h after inundation of dry sediments, benthic bacteria colonized the overlaying water and approximately 20% of total cells exhibited DNA de novo synthesis (bromodeoxyuridine-positive). The primary microbial colonizers of the overlaying water – as determined by 16S rRNA gene sequence analysis – were related to at least six different phylogenetic lineages of *Bacilli*, a group with many

---

Fig. 6.3 (continued) dry sediment. Data reported are from duplicated microcosms and single values are means of four measurements. Regression curves are shown (aminopeptidase $r^2 = 0.78$ Pardiela river and $r^2 = 0.52$ in the rest of rivers, $P < 0.05$; alkaline phosphatase $r^2 = 0.80$, $P < 0.05$) (Zoppini 2007) spore-forming members (Onyenwoke et al. 2004), and *Alphaproteobacteria* (*Brevundimonas* spp. and *Caulobacter* spp.). The microbial awakening was accompanied by C production rates similar to those measured in highly productive eutrophic systems with a prevalence of biomass synthesis (^{3}H-leucine incorporation) over cell division (^{3}H-thymidine incorporation) (Fig. 6.4). Significant extracellular enzyme activities were also observed: within 28 h after inundation aminopeptidase activity reached 70% of the peak value, which was after 72 h (431 nmol MCA L^{-1} h^{-1}), while alkaline phosphatase activity reached 83% (99 nmol MUF l^{-1} h^{-1}). The analysis of the aminopeptidase to alkaline phosphatase activity ratio showed the different role played by these enzymes in the metabolic awakening of bacterial cells. The lower hydrolysis rate of organic phosphorus compared to proteins, described by the increasing AMA:APA ratio, indicated a slower P-remobilization compared to N-remobilization. Aminopeptidase and alkaline phosphatase activities were also significantly correlated to bacterial carbon and cell production ($P < 0.01$; $n = 12$). This confirmed the key role played by extracellular enzymes in making available organic compounds for both the synthesis of intracellular proteins and the production of new biomass.

A more advanced tool for studying metabolic properties in streambed sediments is represented by the perfused core technique (Marxsen and Fiebig 1993). The application of this approach enables the acquisition of new information on the time and mode of recovery of benthic microbial communities after flooding without major disturbances of the sediment core (Marxsen et al. 2010). The response of two river sediments, originating from semi-arid (Mulargia, Italy) and temperate (Breitenbach, Germany) climatic regions, were analyzed. Both sediments were similar in microbial community composition, determined via CARD-FISH, in that they were dominated by *Betaproteobacteria* (39–45%) and *Alphaproteobacteria* (27–31%). However, the Mulargia sediment contained a higher percentage of Gram-positive bacteria (24%) than the Breitenbach one (9%). After rewetting bacterial cell abundances did not change significantly from their initial values, which were similar to those observed in the wet sediments (Mulargia 10×10^8 cells mL^{-1} sediment and Breitenbach 30×10^8 cells mL^{-1} sediment) (Fig. 6.5).

The functional awakening of the microbial community was marked by rapid bacterial carbon production, which reached the maxima rates within 48 h in both sediments (Fig. 6.5). This trend was also accompanied by a rapid increase in extracellular enzyme activities (Fig. 6.6). A few hours after flooding aminopeptidase activity reached about half the level measured in non-desiccated sediments. This activity increased further until the end of the experiment, when values reached those measured in unaffected sediments. Alkaline phosphatase was reactivated within a few hours although in both sediments it underwent a progressive decrease. These experimental findings are in accordance with previous observations. Alkaline phosphatase is involved in phosphorous remobilization and its activity is proportional to phosphorous demand. It has been estimated that 30–60% of the microbial biomass carbon contained in the soil may be released during an individual rewetting event along with water soluble phosphorous deriving, for example, from the

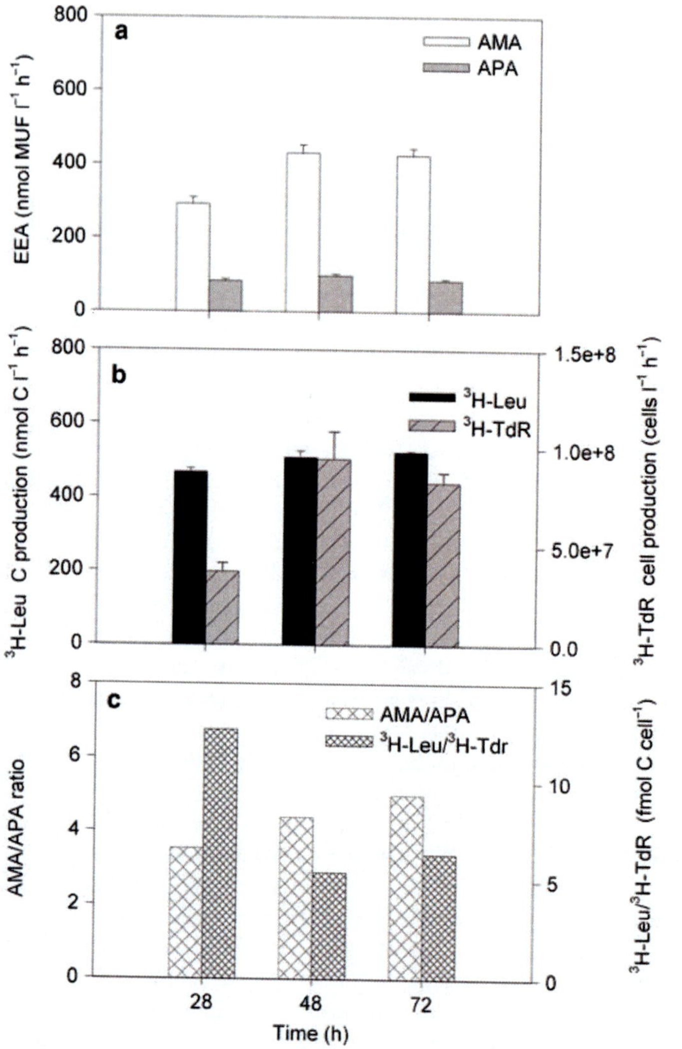

Fig. 6.4 (**a**) Extracellular enzyme activities (AMA, aminopeptidase activity; APA, alkaline phosphatase activity). (**b**) Bacterial carbon production (^{3}H-leucine inc.) and cell production (^{3}H-thymidine inc.). (**c**) The AMA to APA ratios and ^{3}H-Leu to ^{3}H-TdR ratios. Data are means and standard deviation of duplicate measurements from three independent microcosms (after Fazi et al. (2008), modified)

microbial cell rupture caused by osmotic shock (Kieft et al. 1987; Halverson et al. 2000; Baldwin and Mitchell 2000; Austin et al. 2004).

The hydrolysing activity related to polysaccharides (β-glucosidase) and lipids (lipase) also showed a fast recovery with differences among sediments (Fig. 6.6). The high activity levels measured at the beginning of rewetting in both

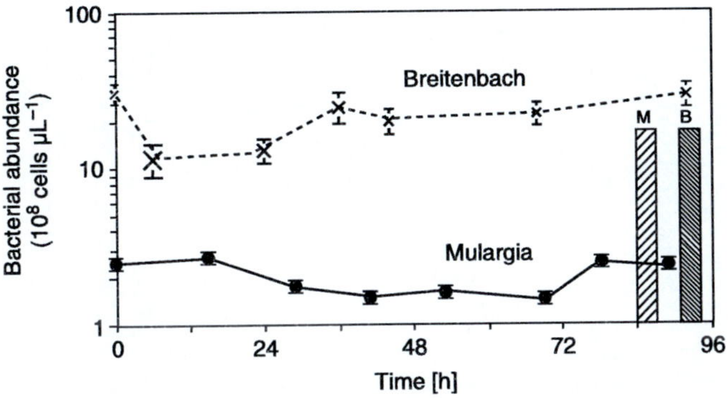

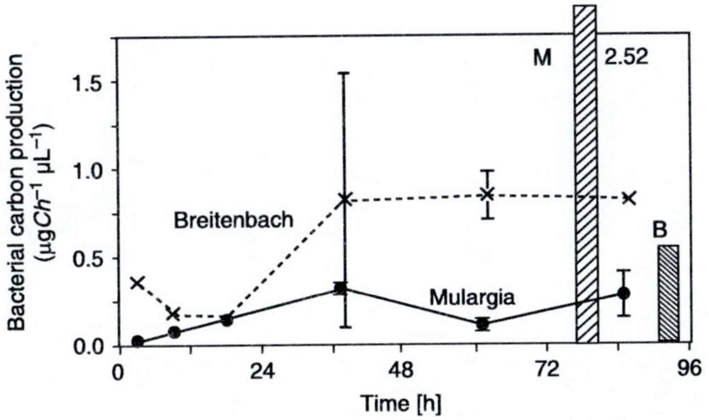

Fig. 6.5 (**a**) Bacterial abundance and (**b**) bacterial carbon production (BCP, ^{14}C-leucine incorporation) after rewetting of desiccated streambed sediments. Average values and standard deviations are shown (if not visible, sd is hidden behind symbols). The columns give data from non-desiccated sediments (*B* Breitenbach, *M* Mulargia) (after Marxsen et al. (2010), modified)

environments, especially those for enzymes involved in polymer degradation (β-glucosidase, peptidase and lipase), can be taken as an indication of outlasting of extracellular enzymes during drought, confirming previous findings on the preservation of enzymes during drying (Sirová et al. 2006; Fazi et al. 2008; Zoppini 2007). Their activities can play an important role in fuelling the bacterial metabolism as they supply organic substrates rich in energy (glucose and fatty acids) and amino acids ready to use for synthesising new biomass.

6.4 Conclusions

Overall our understanding of the effect of dry–wet conditions on the ecology of temporary rivers is still limited. From the information available on the functional properties of microbial communities we can infer that their role in carbon, nutrient and energy flux in water stress conditions is important.

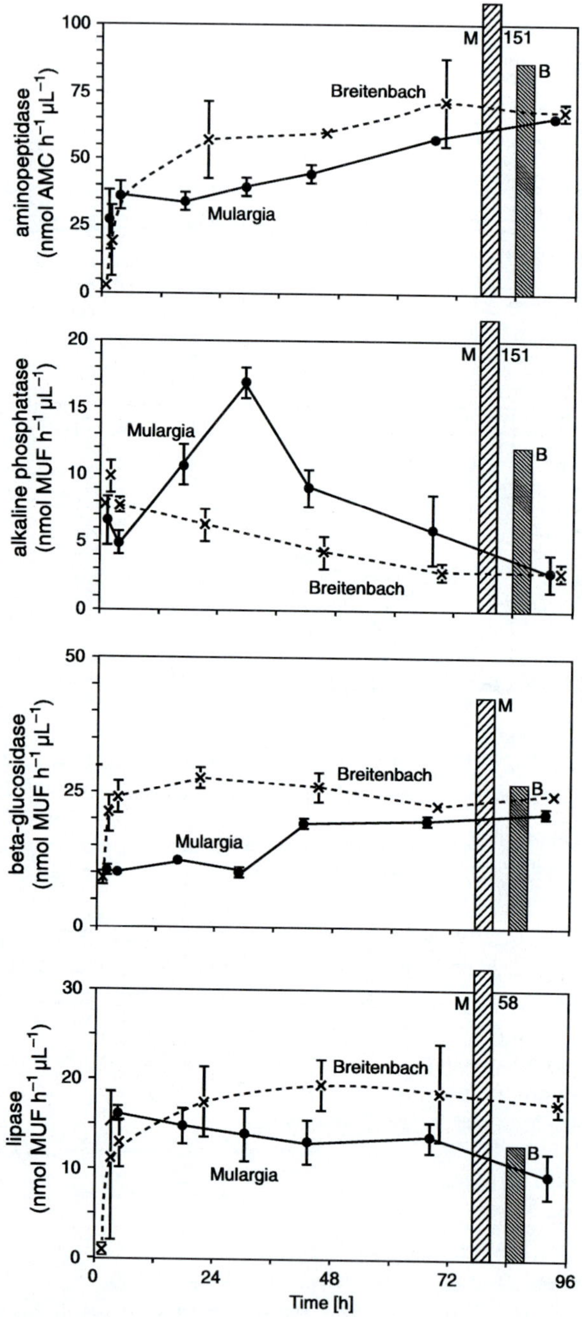

Fig. 6.6 Extracellular enzyme activities after rewetting of desiccated streambed sediments. Average values and standard deviations are shown (if not visible, sd is hidden behind symbols). The columns give data from non-desiccated sediments (*B* Breitenbach, *M* Mulargia) (after Marxsen et al. (2010), modified)

A common feature of rewetted sediments (and soil) is a large flush of mineral phosphorous and nitrogen in the initial phase. Microbial communities can significantly contribute to this via different mechanisms. The preservation of enzymes during periods of drought, even in prohibitive conditions for the rest of the microbial metabolism, determines an excess of labile organic matter and nutrients ready to be taken up later by bacteria at water arrival or to be delivered immediately upon rewetting. In this scenario the first flood delivers labile low molecular organic matter and nutrients to the receiving water bodies with the potential to accelerate microbial processes and affect water quality (i.e., the onset of anoxia).

The fast resumption of enzyme activity favors the rapid re-establishment of sediment functions due to aquatic microbial communities. It appears that drying-rewetting plays a key role in a number of metabolic processes in temporary rivers and the length of drought periods is important in affecting biogeochemical cycling.

Increased periods of drought, due to climate change, could thus represent a bottleneck for bacterial communities living in temporary waters and they will need to adopt their metabolic strategies to survive them.

References

Amalfitano S, Fazi S, Zoppini A, Barra Caracciolo A, Grenni P, Puddu A (2008) Responses of benthic bacteria to experimental drying in sediments from Mediterranean temporary rivers. Microb Ecol 55:270–279

Austin AT, Yahdjian L, Stark JM, Belnap J, Porporato A, Norton U, Ravetta DA, Schaeffer SM (2004) Water pulses and biogeochemical cycles in arid and semiarid ecosystems. Oecologia 141:221–235

Baldwin DS (1996) Effects of exposure to air and subsequent drying on the phosphate sorption characteristics of sediments from a eutrophic reservoir. Limnol Oceanogr 41:1725–1732

Baldwin DS, Mitchell AM (2000) The effects of drying and re-flooding on sediment and soil nutrient dynamics of lowland river–floodplain systems: a synthesis. Regulated Rivers: Res Manag 16:457–467

Belnap J, Welter JR, Grimm NB, Barger N, Ludwig JA (2005) Linkages between microbial and hydrologic processes in arid and semiarid watersheds. Ecology 86:298–307

Billi D, Potts M (2002) Life and death of dried prokaryotes. Res Microbiol 153:7–12

Burns A, Ryder DS (2001) Response of bacterial extracellular enzymes to inundation of floodplain sediments. Freshw Biol 46:1299–1307

Chrost RJ (1991) Environmental control of the synthesis and activity of aquatic microbial ectoenzymes. In: Chrost RJ (ed) Microbial enzymes in aquatic environments. Springer, New York, pp 29–59

Coolen MJL, Overmann J (2000) Functional exoenzymes as indicators of metabolically active bacteria in 124,000-year-old sapropel layers of the eastern Mediterranean sea. Appl Environ Microbiol 66:2589–2598

De Groot C, Van Wijck C (1993) The impact of desiccation of a freshwater marsh (Garcines Nord, Camargue, France) on sediment–water–vegetation interactions. Part one: the sediment chemistry. Hydrobiologia 252:83–94

Dick GJ, Lee YE, Tebo BM (2006) Manganese(II)-oxidizing bacillus spores in Guaymas Basin hydrothermal sediments and plumes. Appl Environ Microbiol 72:3184–3190

Fazi S, Amalfitano S, Piccini C, Zoppini A, Puddu A, Pernthaler J (2008) Colonization of overlaying water by bacteria from dry river sediments. Environ Microbiol 10:2760–2772

Fierer N, Schimel JP, Holden PA (2003) Influence of drying-rewetting frequency on soil bacterial community structure. Microb Ecol 45:63–71

Findlay SEG, Sinsabaugh RL, Sobczak WV, Hoostal M (2003) Metabolic and structural response of hyporheic microbial communities to variations in supply of dissolved organic matter. Limnol Oceanogr 48:1608–1617

Fischer H, Pusch M (2001) Comparison of bacterial production in sediments, epiphyton and the pelagic zone of a lowland river. Freshw Biol 46:1335–1348

Fonnesu A, Pinna M, Basset A (2004) Spatial and temporal variations of Detritus breakdown rates in the river Flumendosa Basin (Sardinia, Italy). Int Rev Hydrobiol 89:443–452

Francis CA, Tebo BM (2002) Enzymatic manganese(II) oxidation by metabolically dormant spores of diverse Bacillus species. Appl Environ Microbiol 68:874–880

Gilichinsky DA, Vorobyova EA, Erokhina LG, Fyodorov-Dayvdov DG, Chaikovskaya NR (1992) Long-term preservation of microbial ecosystems in permafrost. Adv Space Res 12:255–263

Goedkoop W, Gullberg KR, Johnson RK, Ahlgren I (1997) Microbial response of a freshwater benthic community to a imulated diatom sedimentation event: interactive ejects of benthic fauna. Microb Ecol 34:131–143

Griffiths RI, Whiteley AS, O'Donnell AG, Bailey MJ (2003) Physiological and community responses of established grassland bacterial populations to water stress. Appl Environ Microbiol 69:6961–6968

Halverson LJ, Jones TM, Firestone MK (2000) Release of intracellular solutes by four soil bacteria exposed to dilution stress. Soil Sci Soc Am J 64:1630–1637

Holmes RM, Fisher SG, Grimm NB, Harper BJ (1998) The impact of flash floods on microbial distribution and biogeochemistry in the parafluvial zone of a desert stream. Freshw Biol 40:641–654

IPCC (2007) Climate change 2007: the physical science basis. Contribution of working group I to the fourth assessment report of the intergovernmental panel on climate change. Cambridge University Press, Cambridge, 996

Kieft T, Soroker E, Firestone M (1987) Microbial biomass response to a rapid increase in water potential when dry soil is wetted. Soil Biol Biochem 19:119–126

Kirkby M, Froebrich J (2006) Introduction. In: Froebrich J, Bauer M (eds) Critical issues in the water quality dynamics of temporary waters – evaluation and recommendations from the TempQsim Project. Booklet – enduser summary. Institute for water quality and waste management, University of Hannover, Hannover, pp 8–11

Larned ST, Datry T, Robinson C (2007) Invertebrate and microbial responses to inundation in an ephemeral river reach in New Zealand: effects of preceding dry periods. Aquat Sci 69:554–567

Marxsen J (2006) Bacterial production in the carbon flow of a central European stream, the Breitenbach. Freshw Biol 51:1838–1861

Marxsen J, Fiebig DM (1993) Use of perfused cores for evaluating extracellular enzyme activity in stream-bed sediments. FEMS Microbiol Ecol 13:1–11

Marxsen J, Zoppini A, Wilczek S (2010) Microbial communities in streambed sediments recovering from desiccation. FEMS Microb Ecol 71:374–386

McClain ME, Boyer EW, Dent CL, Gergel SE, Grimm NB, Groffman PM, Hart SC, Harvey J, Johnston CA, Mayorga E, McDowell WH, Pinay G (2003) Biogeochemical hot spots and hot moments at the interface of terrestrial and aquatic ecosystems. Ecosystems 6:301–312

McMahon TA, Finlayson BL (2003) Droughts and anti-droughts: the low-flow hydrology of Australian rivers. Freshw Biol 48:1147–1160

Mikha MM, Riceb CW, Millikenc GA (2005) Carbon and nitrogen mineralization as affected by drying and wetting cycles. Soil Biol Biochem 37:339–347

Milly PCD, Dunne KA, Vecchia AV (2005) Global pattern of trends in streamflow and water availability in a changing climate. Nature 438:347–350

Mulholland PJ (2003) Large-scale patterns in dissolved organic carbon concentration, flux and sources. In: Findlay SE, Sinsabaugh RL (eds) Aquatic ecosystems, Interactivity of dissolved organic matter. Academic, Elsevier, San Diego, pp 139–159

Onyenwoke RU, Brill JA, Farahi K, Wiegel J (2004) Sporulation genes in members of the low G+C Gram-type- positive phylogenetic branch (Firmicutes). Arch Microbiol 182:182–192
Perez-Mateos M, Busto MD, Rad JC (1991) Stability and properties of alkaline phosphate immobilized by a rendzina soil. J Sci Food Agr 55:229–240
Pinna M, Basset A (2004) Summer drought disturbance on plant detritus decomposition processes in three River Tirso (Sardinia, Italy) sub-basins. Hydrobiologia 522:311–319
Potts M (1994) Desiccation tolerance of prokaryotes. Microbial Rev 58:755–805
Qiu S, McComb AJ (1995) Planktonic and microbial contributions to phosphorus release from fresh and air-dried sediments. Mar Freshw Res 46:1039–1045
Raubuch M, Dyckmans J, Joergensen RG, Kreutzfeldt M (2002) Relation between respiration, ATP content, and adenilate energy charge (AEC) after incubation at different temperature and after drying and rewetting. J Plant Nutr Soil Sci 165:435–440
Rees GN, Watson GO, Baldwin DS, Mitchell AM (2006) Variability in sediment microbial communities in a semipermanent stream: impact of drought. J North Am Benthol Soc 25:370–378
Romanì AM, Sabater S (1997) Metabolism recovery of a stromatolitic biofilm after drought in a Mediterranean stream. Archiv für Hydrobiologie 140:262–271
Romanì AM, Vazquez E, Butturini A (2006) Microbial availability and size fractionation of dissolved organic carbon after drought in an intermittent stream: biogeochemical link across the stream–riparian interface. Microb Ecol 52:501–512
Sander BC, Kalff J (1993) Factors controlling bacterial production in marine and freshwater sediments. Microb Ecol 26:79–99
Schimel J, Balser TC, Wallenstein M (2007) Microbial stress–response physiology and its implications for ecosystem function. Ecology 8:1386–1394
Sirová D, Vrba J, Rejmánková E (2006) Extracellular enzyme activities in benthic cyanobacterial mats: comparison between nutrient-enriched and control sites in marshes of northern Belize. Aquat Microb Ecol 44:11–20
Stanley EH, Fisher SG, Grimm NB (1997) Ecosystem expansion and contraction in streams. Bioscience 47:427–435
Sutherland WJ, Bailey MJ, Bainbridge IP et al (2008) Future novel threats and opportunities facing UK biodiversity identified by horizon scanning. J Appl Ecol 45:821–833
Tzoraky O, Nikolaidis NP, Amaxidis Y, Skoulikidis NTH (2007) In-stream biogeochemical processes of a temporary river. Environ Sci Technol 41:1225–1231
Uys CM, O'Keeffe J (1997) Simple words and fuzzy zones: early directions for temporary river research in South Africa. Environ Manage 21:517–531
WFD/EUWI (2006) Mediterranean joint process WFD/EUWI, Water Scarcity Drafting Group, Tool Box (Best practices) on water scarcity. Draft Version 9, to be modified, 13 Feb 2006
Wilby RL, Whitehead PG, Wade AJ, Butterfield D, Davis RJ, Watts G (2006) Integrated modelling of climate change impacts on water resources and quality in a lowland catchment: River Kennet, UK. J Hydrol 330:204–220
Wilhite DA (2000) Drought as a natural hazard. Concept and definition. In: Wilhite DA (ed) Drought: a global assessment, vol 1. Routledge, London, New York, pp 3–18
Wobus A, Bleul C, Maassen S, Scheerer C, Schuppler M, Jacobs E, Röske I (2003) Microbial diversity and functional characterization of sediments from reservoirs of different trophic state. FEMS Microbiol Ecol 46:331–347
Zoppini A (2007) Extracellular enzyme activity in temporary river sediments under drying and rewetting conditions. In: Dick RP (ed) Proceedings of the third international conference, enzymes in the environment: ecology, activity and applications. Viterbo (Italy), 15–19 July 2007, p 29
Zoppini A, Amalfitano S, Fazi S, Puddu A (2010) Dynamics of a benthic microbial community in a riverine environment subject to hydrological fluctuations (Mulargia River, Italy). Hydrobiologia (in press) DOI: 10.1007/s10750-010-0199-6

Chapter 7
Soil Enzymes as Indication of Soil Quality

Ayten Karaca, Sema Camci Cetin, Oguz Can Turgay, and Ridvan Kizilkaya

7.1 Introduction

Soil quality is the phenomenon that has been developed to evaluate the factors effecting soil functionality. It is mainly concerned with sustainable use of soil resources in terms of enhanced agricultural productivity, environmental quality and human health. The reasons for these focusing are rapid growth of the world population, which is demanding an increasing emphasis on sustainable agricultural soil management, and as second ongoing degradation of soil resources, which is also closely associated with the loss of soil quality by climate changes, wildfires, erosion, salinization and agricultural–industrial pollution problems.

Early definitions of "soil quality" were actually formalized efforts that had been restricted to analysis of various soil characteristics and evaluation of soil conditions but, in course of time, it has been noticed that analysis of soil properties alone could not be adequate to measure soil quality, as long as the criteria evaluated are associated with the role or function of the soil. Therefore, the term of soil quality has been modified several times related to the ongoing development beyond soil management and the interactions with other ecological systems. For example, in 1980s, soil quality was defined as "the sustained capability of a soil to accept, store and recycle water, nutrients and energy" (Anderson and Gregorich 1984). On the other hand, agricultural activities diffused through much larger ecological systems by time and it has been understood that soil serves many functions not only within or between agricultural lands but also beyond environmental ecosystems.

A. Karaca (✉) and O.C. Turgay
Faculty of Agriculture, Department of Soil Science, Ankara University, Diskapi, Ankara, Turkey
e-mail: akaraca@agri.ankara.edu.tr

S.C. Cetin
Agricultural Faculty, Soil Science Department, Gaziosmanpasa University, Tokat, Turkey

R. Kizilkaya
Agricultural Faculty, Soil Science Department, Ondokuz Mayis University, Samsun, Turkey

G. Shukla and A. Varma (eds.), *Soil Enzymology*, Soil Biology 22,
DOI 10.1007/978-3-642-14225-3_7, © Springer-Verlag Berlin Heidelberg 2011

This development required a revision in the definition of soil quality and a more detailed definition has been stated as "Soil quality is the capacity of a specific kind of soil to function, within natural or managed ecosystem boundaries, to sustain plant and animal productivity, maintain or enhance water and air quality, and support human health and habitation" by Soil Science Society of America (1995).

Today, the common framework to evaluate soil quality is generally based on a group of facts mainly including soil functions, processes, attributes, indicators and methodology. In this stepwise evaluation, soil function is the answer of the question of "what the soil does" and soil processes can be defined as the characteristic features of a soil use at anytime. Soil quality attributes describe the measurable soil characteristics that affect the ability of the soil to achieve various functions. In many cases our ability to assess soil quality is complicated because it is not usually possible to directly measure the rate of specific soil processes due to their simultaneity, diverse and conflicting natures and interactions in time, space, and intensity. Therefore soil quality indicators that are associated with specific soil processes can be used as an indirect and useful measure of soil quality changes. In the literature, there have been many attempts to discover possible links between soil quality indicators and specific soil functionality (i.e., plant production) (Doran and Parkin 1994; Gregorich et al. 1994; Larson and Pierce 1994). These works revealed that soil quality indicators could represent either a single or a variety of soil attributes, be easily measured, have some sensitivity to soil management in a wide range, and have also a relatively low sampling error. Supplying available methodology with the convenience of duplication and high accuracy and speed would be also effective factors on the selection of soil quality indicators.

Basic soil quality indicators is expected to integrate the combined effects of the soil's physical, chemical and biological properties and processes, be applicable in many various management and climate conditions, accessible to wide group of users with different expertise and be included by existing soil databases.

Soil physical quality is simply an organization of mineral particles, pores and water. The indications of the relationships between these components are usually stated as topsoil depth, bulk density, porosity, aggregate stability, texture, crusting, and compaction and primarily reflect the conditions of root growth, seedling emergence, infiltration, or movement of water within the soil profile (Dexter 2004). The poor soil physical quality implies poor water infiltration, run-off of water from the surface, hard-setting, poor aeration, and poor rootability, while opposite or absence of these conditions is attributed to good soil physical quality. Since the changes in soil physical conditions occur simultaneously, several authors have mentioned that no single parameter of soil physical conditions has been considered as a measure of soil physical quality (Dexter and Czyz 2000), and a range of soil physical properties should be integrated to obtain an overall assessment of soil physical quality (Canarache 1990).

The soil quality in terms of soil chemical aspects is related to three main functions in soil, (a) recycling of soil organic matter, (b) storage and gradual release of nutrients, and (c) buffering of potentially toxic elements (Warkentin 1995). These functions actually regulate nutrient flow and the transport of toxic elements through soil solution where they are taken up by the plants. Recycling of organic materials depends on the content of mineralisable organic matter and microbial biomass responsible for the continuous supply of nutrients to the soil solution. Function (b) above is controlled by the maintenance of optimum nutrient concentration in the soil solution whereas function (c) needs lower concentrations of toxic elements in the soil solution.

The attributes and indicators of the soil's physical and chemical processes summarized above could be often considered to be adequate to evaluate man-induced changes in soil quality, but on the other hand, they usually represent slowly changing features of soil (i.e., soil structure, organic matter pool and nutrient balance) and since many of the changes in soil physical and physico-chemical conditions take place over the long term, a soil quality evaluation based on these features would require combined analysis of a large group of soil physical and chemical parameters (Gil-Sotres et al. 2005).

On the other hand, over the last two decades, there has been an increasing scientific focus on soil biological-biochemical aspects, which have been reported to be more sensitive to the slight modifications pertaining to soil management practices and land use changes (Klein et al. 1985; Powlson et al. 1987; Nannipieri et al. 1990; Ellert and Gregorich 1995; Yakovchenko et al. 1996) and thus may provide more dynamic indications of soil quality changes. Among the indigenous soil biological components, soil microorganisms have a key role in a number of important biochemical processes such as soil microbial activities, responsible for the cycles of bio-elements (C, N, P, and S) and energy transfer in soil ecosystem.

Basic source of soil microbial activities is soil organic matter and depending on the land use and other soil characteristics, microorganisms are in a continuous labor to govern soil organic matter and, in most cases like stress conditions caused by adverse anthropogenic effects, this can be rapidly reflected either to the microbial diversity level or to biologically active soil organic matter components i.e., microbial biomass, enzymes and other ephemeral organic compounds i.e., proteins and carbohydrates.

Recent developments in the studies of soil microbial diversity have provided an innovative perspective to soil scientists and enabled them to have an insight into the structure and function of soil microbial communities which used to be unknown and hence called as "black box" until 2 decades ago. In relation to the concept of soil quality, the soil microbial community changes can be stressed in two distinct ways (1) verification of the composition and distribution of different functional groups of soil microorganisms and (2) analysis of the dynamics of specific organisms or communities under changing soil conditions (Visser and Parkinson 1992). In other words, a change in entire or specific soil microbial communities could be reckoned as a sensitive indication of a change in soil quality conditions. However, these sensitive tools are usually based on the use of laboring molecular techniques (analysis of intra- and extra-cellular markers i.e., DNA, RNA and cell-wall fatty acids) and generally requires microbiological expertise, costly reagents and well-equipped laboratory facilities. This actually may help to explain increasing attention on soil quality studies focusing on microbial and biochemical soil quality indicators and taking general (microbial activity) and specific (hydrolytic enzymes) biochemical parameters into consideration.

As a useful tool of soil biochemical quality, soil enzyme activities have been often suggested as sensitive indicators of soil ecological quality because (1) they measure principal microbial reactions involving nutrient cycles in soil, (2) they may easily respond to changes in soil by natural or anthropogenic factors, and (3) they can be easily measured (Gianfreda and Bollag 1996; Calderon et al. 2000; Drijber et al. 2000; Nannipieri et al. 2002).

Nevertheless, up until now, the validation of soil biochemical properties including soil enzymes as a universal soil quality criterion has still been under discussion due to the contradictory results obtained from the studies using these properties as an individual parameter or an index component. The reasons for questionable findings on biochemical approaches in the literature have been ascribed to the factors such as different methodological perspectives, the lack of standard analysis and reference values, inadequate databases for high quality soils, high degree of variability between biochemical properties and seasonal or edaphic influences (Gil-Sotres et al. 2005). These drawbacks can certainly be overcome by increasing scientific efforts to understand the behaviors of different soil characteristics and how they function and relate to each other in both disturbed and undisturbed soils.

With these in the mind, we will concentrate on the assessment of soil biochemical quality in specific regard to soil enzyme activities. The purposes of this chapter is to emphasis soil enzymes as a soil component, their ecological distribution and behaviours under different environments and finally their potential as a soil quality indicator reflecting ecosystem disturbance.

7.2 Soil Enzymes: Classification, Sources, States, Affecting Factors and Activation or Inhibition

7.2.1 Enzyme Classification

Enzymes classification in biological systems is based on reaction type and classified as follows: (1) oxidoreductases: oxidation-reduction, (2) transferases: transfer of functional groups, (3) hydrolases: hydrolysis, (4) lyases: elimination groups for forming double-bond, and (5) izomerases: izomerisation and (6) ligases: forming bond with ATP hydrolysis (Voet and Voet 1995). Soil enzymes mainly belong to hydrolases and the rest of them belong to other classes such as oxidoreductases, transferases, and lyases (Dick and Tabatabai 1992).

Location of enzymes is considered another classification of soil enzymes: endocellular or intracellular (in organism) and extracellular (out of organism) (Kunito et al. 2001). Arylsulphatase is an extracellular enzyme whereas dehydrogenase is an intracellular enzyme. Commonly studied soil enzymes can be summarized as: (1) phosphatases (EC 3.1.3.1): it hydroles compounds of oganic phosphorus and transforms them into different forms of inorganic phosphorus, (2) urease (EC 3.5.1.5): acting hydrolysis of urea to carbon dioxide and ammonia (3) dehydrogenase: oxidation of organic matter (intracellular), (4) β-glucosidase (EC 3.2.1.21): it catalyzes the hydrolysis of β-D-glucopyranoside and is one of the three or more enzymes involved in the saccharification of cellulose, (5) arylsulphatase (EC 3.1.6.1): involving in the hydrolsis of arylsulphate esters by fission of the oxygen–sulphur (O–S) bond, (6) catalase: involving in microbial oxidoreductase metabolism (intracellular) and may be related to the metabolic activity of aerobic organisms. (7) dehydrogenase: these enzymes are found in all living organisms and take part in many reactions involved in energy transfer in microbial metabolic reactions.

7.2.2 *Sources of Soil Enzymes*

Sources of soil enzymes are showed in Fig.7.1. Enzymes are originated from mostly microorganisms and as well as plants and animals (Bandick and Dick 1999).

7.2.3 *States*

State of enzymes has been used to define the phenomenon that enzymes exist in the soil (Dick and Tabatabai 1992). When an enzyme enters into a soil system, enzyme

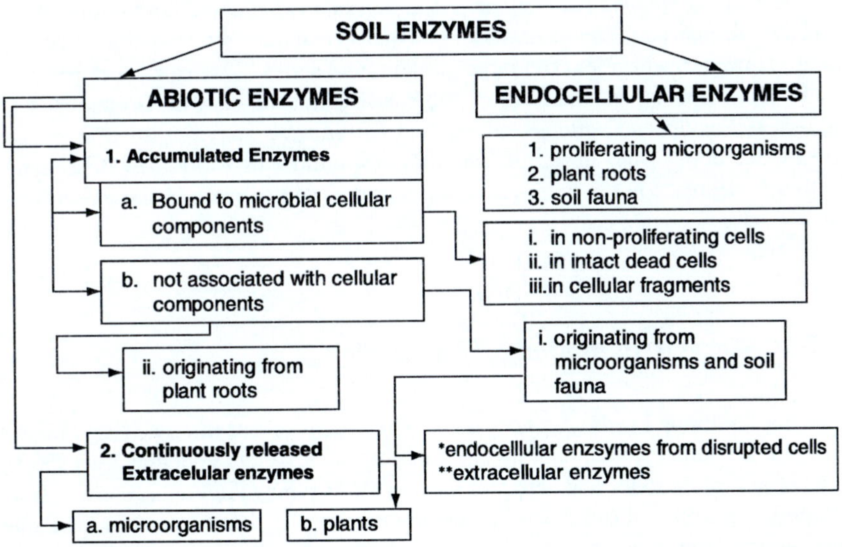

Fig. 7.1 Sources of enzymes in soils (Adapted from Dick and Tabatabai (1992) according to Skujins (1978))

can be found in different states in the soil. These states are: adsorption, microencapsulation, cross-linking, copolymer formation, entrapment, ion-exchange, adsorption and cross-linking and covalent attachment. These mechanisms are considered as the protective effect of soil on extracellular enzyme activity. Adsorption of enzyme by clay minerals is a protective stabilization from microbial attack (Esminger and Gieseking 1942). When free enzymes enter into a soil system, either it can complex with humic colloids or can be stabilized on clay surfaces and organic matter (Boyd and Mortland 1990). Soil enzymes are stabilized by soil organic matter rather than by inorganic components (Dick and Tabatabai 1992). Organic matter protects urease against microbial attack and other processes (Conrad 1940) and enzyme activities are significantly correlated with the organic matter content of soils. In addition, free enzymes can complex with humic acid and these complexes are much more stable to heat and enzymatic degradation than the free enzymes (Serban and Nissenbaum 1986).

7.2.4 Affecting Factors

Enzymatic reactions mostly depend on pH, ionic state, temperature and the presence or absence of inhibitor and activator (Burns 1978; Tabatabai 1982, 1994). In general, every enzyme has an optimum pH value and lower the activity up and below this optimum. Temperature influences enzyme activity in different ways than chemical reactions. Actually, with increasing of temperature every 10°C, chemical reaction rate is doubled but enzymatic reaction rate is increased until optimum pH and then decreased (Tabatabai 1994). Enzymatic reactions are temperature-dependent. However, enzymes are large protein molecules that excess temperature (>50°C) causes of denaturating of protein structure of enzyme (Campbell and Smith 1993). Special cations are required for enzyme activities such as Ni for urease for acting role in active site (Mathews and van Holde 1995). Enzyme activities decrease in the presence of the inhibitor and increase in the presence of an activator.

7.2.5 Enzyme Inhibition and Activation

Enzyme inhibitor can be defined as reducing agent of enzyme activity whereas enzyme activators are stimulating agent of enzyme activity (Voet and Voet 1995). The effects of inhibitor and activator on enzyme are shown in Fig. 7.2. The effects of both agents are mostly on Km. parameter. Km. is defined as “the substrate concentration at which the reaction rate is half maximal” (Stryer 1995). As seen in the figure, Km. values increased in the presence of inhibitor and decreased in the presence of the activator and V.

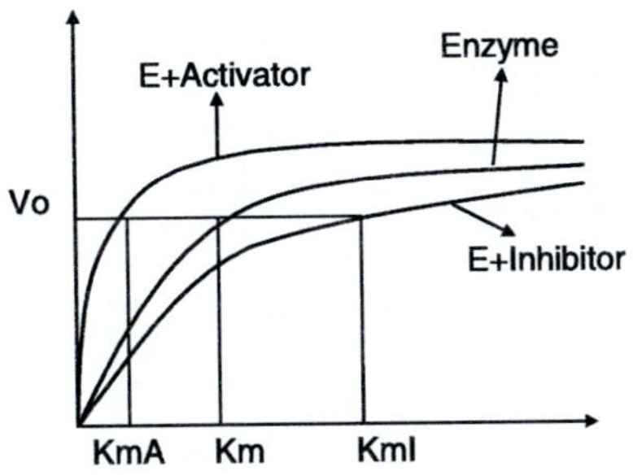

Fig. 7.2 The effects of inhibitor and activator on enzyme activity (Voet and Voet 1995)

7.3 Indication of Soil Enzymes

Soil enzymes are important in ecosystem functioning with their catalytic feature on nutrient cycling (Bandick and Dick 1999). Soil enzymes have a crucial role in C (β-glucosidase and β-galactosidase), N (urease), P (phosphatases), and S (sulphatase) cycle. Dick and Tabatabai (1992) explain the importance of soil enzymes as "soil enzymes are useful in describing and making predictions about an ecosystem's function, quality and the interactions among subsystems". Bandick and Dick (1999) reported that soil enzymes are used as soil quality indicators due to "their relationship to soil biology, ease of measurement and rapid response to changes in soil management".

Using soil enzymes as soil quality indicators can be classified into three distinct areas: (1) pollution indicators, (2) ecosystems perturbations indicators and (3) agricultural practice indicators. Soil is a vital resource not only an environmental filter providing the quality of both water and atmosphere but also being a very complicated medium pursuing sustainability of ecosystems (Trasar-Cepeda et al. 2000). Depending on environmental degradation, soil quality also changes; however, soil quality standards' has been established later than air and water quality standards (Dick 1997). Soil quality is measured by all soil physical, chemical and biological properties. Microorganisms and biochemical activities have a crucial role in soil ecosystems. Soil enzymes are significant in soil functioning because of the following features: (1) having a role in decomposition of organic inputs, (2) transformation of soil organic matter, (3) releasing nutrients from the unavailable to the available form to plants, (4) participating in N_2 fixation, (5) detoxification of xenobiotics (unnatural compounds such as pesticides, industrial wastes, etc.), and (6) participating in nitrification and denitrification processes (Dick 1997).

The effects of different type of pollutants on soil enzyme activities are summarized by Dick (1997). Table 7.1 exhibits the sum of his review paper. Generally, heavy metals inhibit soil enzyme activities; pesticides behave differently depending on the type of pesticide either by inhibition/stimulation or have little effect or even no effect. Industrial amendments or contaminants (fly ash, pulp and paper mill effluent) influence positively or negatively soil enzyme activities depending on the

Table 7.1 Pollution effects on soil enzyme aactivities in literature (summarized from review of Dick (1997), references therein)

Pollution type	Studied enzyme	Influence	Recommended indicator	References
Heavy metals (Hg, Ag, Cr, Cd)	SP, LG, CL, LA, BG	Inhibition		Frankenberger and Tabatabai (1981, 1991a, b), Eivazi and Tabatabai (1990), Deng and Tabatabai (1995)
Heavy metals	ARS AC, UR, IN	High inhibition Low inhibition	ARS	Al-Khafaji and Tabatabai (1979), Bardgett et al. (1994), Yeates et al. (1994)
Pesticides	Enzyme activities	Little or no effect		Burns (1978), Ladd (1985)
Insecticides, fungicides, herbicides	DH	Stimulation		Tu (1981)
Pesticides	PH, UR	Slight increase or little effect		Davies and Greaves (1981), Baruah and Mishra (1986), Tu (1992, 1993)
Glycophosate, paraquat, Carboryl Atrazine	IN UR, PH UR, IN PH UR, PH, IN	Stimulation No effect Inhibition No effect No effect (except very high concentration)		Gianfreda et al. (1993).
Paraquat, glyphosate Purified Complex with montmorillonite	UR	No effect Stimulation		Gianfreda et al. (1994)
Pesticides (accidental spill) Alachlor (10,000 mg kg^{-1}) alone or Mixture atrazine and metochlor	DH ES	Inhibition Inhibition (with herbicide mixture)		Dzantor and Felsot (1991)
Herbicide (imaxethapyr)	DH PR, FDA	Inhibition Stimulation		Perrucci and Scarponi (1994)
Pesticides atrazine	PH, IN, BG, UR	Inhibition		Voets et al. (1974)
Herbicide 2,4-D salt formulation 2,4-D isoctyl ester formulation	DH, UR DH, UR	Little effect Inhibition		Rai (1992)
Industrial amendments or contaminants Fly ash	UR, ARS DH	Stimulation Inhibition		McCarty et al. (1994)
Fly ash	PH, SP, DH, IN CT	Inhibition Stimulation		Pichtel and Hayes (1990)
Pulp and paper mill effluent	IN, DH	Stimulation	IN	Kannan and Oblisami (1990)
Hydrocarbons Jet fuel (50 and 135 mg g^{-1})	FDA	Inhibition		Song and Bartha (1990)

(*continued*)

Table 7.1 (continued)

Pollution type	Studied enzyme	Influence	Recommended indicator	References
Atmospheric pollution	Soil enzymes	Inhibition		Bitton and Boylan (1985)
Acid precipitation (lab. conditions)	DH, PH, UR	No effect		Bitton et al. (1985)
Acid precipitation (field conditions)				
Industrial air pollutants (N and S gases)	CL	Inhibition		Ohtonen et al. (1994)
Traffic pollution (heavy metals)	CL, AML	Inhibition		Joshi et al. (1993)

SP sulphatase, *LG* L-glutaminase, *CL* cellulase, *LA* L-aspariginase, *BG* β-glucosidase, *ARS* arylsulfatase, *AC* acid phosphatase, *UR* urease, *IN* invertase, *DH* dehydrogenase, *PH* phosphatase, *ES* esterase, *PR* protease, *FDA* 3,6-diacetylfluorescein hydrolysis, *CT* catalase, *AML* amylase

type of enzyme. Hydrocarbons, acid precipitation, industrial air pollution (N and S gases) and traffic pollution (heavy metals) adversely affect soil enzyme activities.

Soil enzymes are used as pollution indicators by many researchers due to agricultural practices and organic pollution (Gianfreda et al. 2005), irrigation by polluted water (Zhang et al. 2008; Filip et al. 1999; Barton et al. 2000), sewage sludge and municipal waste application (Fernandes et al. 2005; Kizilkaya and Hepşen 2004; Kizilkaya and Bayrakli 2005; Bastida et al. 2008; Antolin et al. 2005), pesticides application (Gundi et al. 2005; Cai et al. 2007; Fragoeiro and Magan 2008; Yu et al. 2006; Kucharski and Wyszkawska 2008), heavy metals (Kizilkaya et al. 2004; Bhattacharya et al. 2008; Hinojosa et al. 2008; Kunito et al. 2001).

Indicators are very important for resource managers in order to understand ecological change (Dale et al. 2008). Dale and Beyeler (2001) summarized the criteria for ecological indicators: (1) easy to measure, (2) sensitive on system stresses, (3) respond to stress, (4) anticipation of change in the ecological system, (5) predict changes, (6) being integrative, (7) ability to respond to natural disturbances, anthropogenic stresses and changes over time, (8) variable with response, (9) having attention of measured parameters spatial and temporal change (Dale et al. 2008).

Soil enzymes are used as the ecosystem's perturbations indicators by many scientists related to changing land use (Sicardi et al. 2004; Acosta-Martinez et al. 2003, 2007; Trasar-Cepeda et al. 2008a; Gil-Sotres et al. 2005; Lin et al. 2004; devegetation and revegetation (Bastida et al. 2006; Izquierdo et al. 2005), forest fires (Staddon et al. 1998; Fioretto et al. 2005; Zhang et al. 2005; Boerner et al. 2008; Camci Cetin et al. 2009), and changing climatic conditions (Sardans and Penuelas 2005; Sowerby et al. 2005). Many authors suggested soil enzyme activities were sensitive, reliable and early indicators for discriminating land use, reflecting soil restoration and fire stress (Sicardi et al. 2004; Izquierdo et al. 2005; Caravaca et al. 2003; Fioretto et al. 2005; Camci Cetin et al. 2009).

Farming practices may cause modification in soil environment and providing good conditions for plant growth such as irrigation, tillage, applying fertilizers and pesticides, addition of amendments, applying of plant residues, and using crop rotations. All agricultural practices can affect nutrient turnover and microbial activity by influencing soil biological properties (Curci et al. 1997). Soil enzymes play a crucial role in catalytic reactions involved in organic matter decomposition and nutrient cycling (Ajwa et al. 1999). Many studies showed that agricultural practices (crop rotation, mulching, tillage, application of fertilizers, and pesticides) might have different effects on both soil enzymes and microbial activities (Ladd 1985; Dick et al. 1987; Tabatabai 1994). Enzyme activity was found to be the most strongly depressed soil property under intensive agronomic use compared with other biochemical parameters (Saviozzi et al. 2001).

Soil enzymes are used as agricultural practices indicators by many researchers due to irrigation (Zhang and Wang 2006), application of fertilizers (Kandeler et al. 1999; Yang et al. 2008; Melero et al. 2007; Bell et al. 2006; Saha et al. 2008a; Marinari et al. 2000; Hu and Cao 2007; Chang et al. 2007; Acosta-Martinez et al. 1999; Patra et al. 2006), application or amendments (Saha et al. 2008b; Martens et al. 1992; Rajashekhararao and Siddaramappa 2008; Leon et al. 2006), different management and farming systems (FlieBbach et al. 2007; Melero et al. 2008; Gajda and Martyniuk 2005; Monokrousos et al. 2006; Sarapatka 2002; Benitez et al. 2006; Landgraf and Klose 2002; Gajda and Martyniuk 2005; Kremer and Li 2003; Bandick and Dick 1999), crop rotation (Benintende et al. 2008; Dodor and Tabatabai 2005, 2003), and tillage (Mina et al. 2008; Ekenler and Tabatabai 2004, 2003; Acosta-Martinez and Tabatabai 2001; Madejon et al. 2007; Curci et al. 1997; Deng and Tabatabai 1997).

Trasar-Cepeda et al. (2000) indicated that soil enzymes as indicators of soil pollution were limited due to their ability for reflecting soil degradation caused by pollution. Doran and Parkin (1994) and Elliott (1997) suggested that pollution indicators should have the following features: (1) pollutant sensitivity, (2) reflecting ability of different levels of pollution, (3) response either increases or decreases, (4) sensitivity of different pollutants, (5) discriminating between pollutant effect and prior degradation of the polluted soil and (6) differentiating all pollutants based on different degrees of soil degradation they cause (Trasar-Cepeda et al. 2000 and references therein). For that reason, many authors (Beck 1984; Stefanic 1994; Perucci 1992; Sinsabaugh 1994; Stefanic 1994; and Yakovchenko et al. 1996) suggested that using soil enzyme activities as indicators of soil contamination would be acceptable using soil enzymes and other biochemical properties in combination to develop more complex expressions for minimizing limitation of soil enzyme as a pollution indicator (Trasar-Cepeda et al. 2000 and references therein).

7.3.1 *Soil Enzymes as Pollution Indicators*

Intensive agricultural practices and organic pollution also change soil enzyme activities. Gianfreda et al. (2005) found that arylsulfatase, β-glucosidase, phosphatase, urease, dehydrogenase, and fluorescein diacetate hydrolase were lower or had no activity in non-cultivated soils (heavily or moderately polluted by organic contaminants) when compared to agricultural soils.

Irrigation systems influence soil systems depending on irrigation water quality. Long-term sewage irrigation resulted in an increase of heavy metal quantity in soil and caused a positive impact on the microbial community and soil quality (Zhang et al. 2008). Similarly, soil enzyme activities (β-glucosidase, β-acetyl-glucosaminidase, proteinase, and phosphatase) and microbial biomass were greater in wastewater irrigated soils than in wastewater unirrigated soils (Filip et al. 1999). Denitrifiers and denitrification enzyme activity were higher in wastewater-irrigated soils than wastewater-unirrigated soils, although both the types of soils had the same initial denitrifiers and denitrification enzyme activity in forest soil (Barton et al. 2000).

Sewage sludge application in agricultural areas is preferred due to containing valuable nutrients and improving soil fertility (Antolin et al. 2005). Addition of sewage sludge increased basal respiration, microbial biomass, metabolic quotient and enzyme activities and their values were positively correlated with increased sewage sludge doses (Fernandes et al. 2005). Kizilkaya and Hepşen (2004) showed sewage sludge amended soil have higher enzyme activity such as urease, phosphomonoesterase, arylsulphatase than unamended soil. They found that amended sewage sludge might have quickly decomposed resulting in a higher enzyme activity in soil, and sewage sludge treatment had stimulated microbial production of enzyme activity or had more of these enzymes accessible to substrate.

Effects of adding different doses (0, 100, 200, and 300 t/ha dry weight) and C/N ratios (3:1, 6:1 and 9:1) of the sewage sludge on activities of β-glucosidase, alkaline phosphatase, arylsulphatase and urease in a clay loam soil at 25°C and 60% water holding capacity were studied by Kizilkaya and Bayrakli (2005). Nitrogen was added in the form of $(NH_4)_2SO_4$ solution to the sludge to obtain different C/N ratios. Rapid and significant increase in the soil enzymatic activity has been noted at different doses and C/N ratios of the sewage sludge amendments as compared to unamended ones. Enzyme activities varied with differences in incubation period. Soils with the highest C/N ratio and sludge dose had the highest β-glucosidase activity. Alkaline phosphatase and aryl sulphatase showed an increment in their activity during the first 30 days of incubation followed by a pronounced decrease compared to unamended soil. Urease activity, however, showed an increase within 15 days, and thereafter activity declined. The highest activities of urease, alkaline phosphatase and arylsulphatase were observed in soil amended with a low C/N ratio and the highest dose of sludge.

Another type of waste usage in agricultural soils is solid municipal waste application. Soil enzyme activities (β-glucosidase, urease, alkaline phosphatase, and o-diphenol oxidase; humus-associated enzymes) increased with the dose of application of solid municipal waste at different levels (low, medium, high, and very high) but only a certain level, after this level either sustain the same level or decreased (Bastida et al. 2008).

Both natural events and anthropogenic activities continuously influence soil quality (Puglisi et al. 2006). Application of sewage sludge promotes soil biological activity (Saviozzi et al. 1999). Basal respiration, microbial biomass and enzyme activities (urease, BAA-protease, phosphatase, and β-glucosidase) under barley cultivation, were improved by repeated sewage sludge application (Antolin et al. 2005).

Pesticides application in agricultural system affects soil enzyme activities in different ways depending on type, concentration, application duration, number of pesticides and other amendments in soils. Also, different soil enzymes response differently either increase or decrease. Insecticides (monocrotophos, quinalphos, and cypermethrin in single or combination) significantly increased dehyrdogenase activity but decreased in the highest concentration (Gundi et al. 2005). Dehydrogenase activity was decreased because of high residual acetochlor (Cai et al. 2007). However, application of organic fertilizer in soils treated by acetochlor resulted in stimulation of dehydrogenase activity in soil. Addition of pesticides (simazine trifuralin, and dieldrin; inoculation of soil microcosms with *Trametes versicolor* and *Phanerochaeto chrysosporium*) in soil decreased dehyrdogenase activity in most treatments (Fragoeiro and Magan 2008). Acid phosphatase, alkaline phosphotase, urease, catalase, and invertase activities inhibited only first and second pesticide application (chlorotholanil) and especially urease, catalase and acid phosphatase were affected (Yu et al. 2006). Kucharski and Wyszkawska (2008) found that urease and dehydrogenase were the least tolerant and alkaline phosphatase was the most tolerant to the effect of herbicide (Apyros 75 WG).

The effect of heavy metals was evaluated many authors on soil enzyme activities. Studies concerning the effects of heavy metals in soils were shown in Table 7.2. Heavy metals are known to cause long-term toxic effects within ecosystems and can have a negative influence on soil enzymatic processes (Kizilkaya et al. 2004). They can also affect microbial proliferation and enzyme activities by masking catalytically active groups, altering protein conformation or competing with other metals involved in the formation of enzyme–substrate complexes (Eivazi and Tabatabai 1990). Fluorescein diacetate, β-glucosidase, urease, phosphatase, and arylsulphatase were negatively correlated with soils metals (Cd, Cr, Cu, and Pb) in

Table 7.2 Studies concerning the effects of heavy metals in soils (summarized from the study of Puglisi et al. (2006), references therein)

Treatment or application	Enzyme	References
Copper contamination mine and arable land	BG, PH, UR	Leiros et al. (1999)
Metal contaminated mine	ARS, PH, UR, DH, XY, PR	Majer et al. (2002)
Bioremediation monitoring	BG, PH, UR, DH, PR	Pascual et al. (2000)
Metal-contaminated grassland	BG, PH	Kuperman and Carriero (1997)
Cadmium and sludge	PH, DH, AOA	Dar (1996)
Tanning and landfill effluent, hydrocarbon contaminated	ARS, BG, PH, UR, DH	Trasar-Cepeda et al. (2000)
Bioremediation monitoring	UR, DH, CT, LP	Margesin et al. (2000)
Heavy metal contamination	ARS, BG, PH, UR, DH	Hinojosa et al. (2004)

BG β-glucosidase, *ARS* arylsulfatase, *UR* urease, *DH* dehydrogenase, *PH* phosphatase, *PR* protease, *CT* catalase, *XY* xylanase, *AOA* arginine ammonification activity, *LP* lipase long-term irrigation of sewage treatments (Bhattacharya et al. 2008). The highest enzyme activities were found in non-polluted soils compared with polluted but restored and polluted but un-restored (Hinojosa et al. 2004). Acid phosphatase, alkaline phosphatase, arylsulfatase, cellulose, dehydrogenase, protease, urease, and invertase (except β-D-glucosidase) increased with addition of sewage sludge or compost; however, the ratio of enzyme activities to microbial biomass showed a decreasing trend with these applications (Kunito et al. 2001).

7.3.2 Soil Enzymes as Ecosystems Perturbations Indicators

Changing land use from one type to another type generally affects the soil ecosystems. Land use conversion from natural grazed pastures to commercial *Eucalyptus grandis* plantations, showed that no significant effect was found on the number of cellulolytic aerobes, P-solubilizers and *Azotobacter* ssp. communities, whereas significant effect on soil respiration, the C-mineralization coefficient, dehydrogenase, fluorescein diacetate hydrolysis and acid and alkaline phosphatase activities (Scardi et al. 2004). Scardi et al. (2004) reported that land use and management practices alter the total amount and composition of soil organic matter (Reaves 1997) and significantly change the enzyme activities (Dick 1997).

Natural systems and distributed systems are different from their biological activities. β-glucosidase and acid phosphatase, and arylsulfatase (except α-glucosidase and β-glucosaminidase) based on land use showed the following orders, respectively: pasture > forest > agriculture and forest = pasture > agriculture (Acosta-Martinez et al. 2007).

Land use type and the selected enzyme kind respond in different ways. Land use changed the level of organic matter and enzyme activities (acid phosphomonoesterase, β-glucosidase, phosphodiesterase, arylsulfatase, urease, protease, hydrolyzing bezoyl-argininamidase, invertase and carboxymethyl-cellulase) based on the type of land use (agricultural and forest) and type of enzyme (Trasar-Cepeda et al. 2008a). Although biomass-C, urease and β-glucosidase were used as possible indicators of land use on soil (Gil-Sotres et al. 2005), land use did not always modify biochemical properties in the same way and no biochemical property can be used as an universal indicator for discriminating land use (Trasar-Cepeda et al. 2008a). Dehydrogenase activity was lower in plastic-greenhouse vegetable cultivation compared with rice-wheat rotation; however, lower activities (urease, invertase, and phosphatase) were found in open-field vegetable cultivation soils than rice-wheat rotation soils (Lin et al. 2004). Excessive use of N and P fertilizers can be harmful to long-term soil quality due to acidification, nutrient enrichment and influencing microbial activity especially influenced by land use patterns.

Natural systems changed to agricultural systems and not only vegetation but also soil biological properties altered in soil ecosystems. The higher enzyme activities (β-glucosidase, β-glucosaminidase, arylamidase, alkaline, and acid phosphatase, phosphodiesterase, and arylsulfatase) were found in the conservation reserve program, native grassland and rotation with other crops (wheat or sorghum) compared with continuous cotton. The results showed that crop rotation provided higher enzyme activities in soils (Acosta-Martinez et al. 2003).

Devegatation and revegetation influence soil quality compared with undistributed soils. Dehydrogenase and protease activities were lower in devegetated soil (devegetation of *Pinus halepensis* and natural shrubs) than undistributed soil (Bastida et al. 2006). Elimination of vegetation caused a long-term negative influence on the biochemical state and microbial activity of soil. Soil quality has not been recovered even after 15 years due to deforestation. Protease and β-glucosidase activities were higher in revegetation with *Casuarina equisetifolia* than with *Anacardium occidentale*; however, urease, protease, acid phosphatase, and β-glucosidase activities were significantly greater in revegetated soils (soil restoration: after revegetation of a mining area) than in the bare soil 4 years after planting (Izquierdo et al. 2005).

Forest fires are considered as natural disturbances in forest ecosystem (Gonzalez-Perez et al. 2004; Zhang et al. 2005) and cause the most dramatic changes in forest ecosystems (Shakesby et al. 2007). Due to the low volatilization temperature of N, most of the nitrogen found in biomass and soil is lost to the atmosphere when forest fires occur. After a forest fire, many complex influences occur in the forest ecosystems. These influences can be summarized into two main groups: (1) reduction of biomass and (2) alteration of below-ground quality–quantity and functionality (Zhang et al. 2005; Neary et al. 1999). Many studies have showed the effect of forest fire on soil enzyme activities (Zhang et al. 2005; Ajwa et al. 1999; Boerner and Brinkman 2003; Boerner et al. 2000, 2005; Saa et al. 1993, 1998; Senthilkumar et al. 1997). Hart et al. (2005) reported that soil biological properties are more sensitive than the other soil properties to soil heating because of their low fatal temperature (<100°C) (DeBano et al. 1998).

Only some enzyme activities discriminate the fire effect on forest ecosystems as bio-indicators. Among soil enzymes (cellulose, xylanase, invertase, trehalase, and protease), only invertase activity decreased in burned plots throughout the study period (Fioretto et al. 2005) There was no differences between unburned (*Quercus*-dominated forests) and burned sites based on soil enzyme activities(acid phosphatase, α-glucosidase, phenol oxidase, chitinase and L-glutaminase) sampled in early spring and late spring/early summer; however, there was a clear separation of burned and unburned soils due to L-glutaminase activity in late summer/early autumn (Boerner et al. 2008). Acid phosphatase activity could be used for evaluating fire effects on soils compared with alkaline phosphatase and arylsulfatase activities (Staddon et al. 1998).

Different soil enzymes were investigated by different researchers for discriminating fire stress on soil quality. Some activities were decreased and others were increased. Invertase and proteinase activities were declined by burning but acid phosphatase, polyphenoloxidase, and peroxidase activities were increased (Zhang et al. 2005). Phenol oxidase activity was decreased by the fire effect relative to control (Boerner et al. 2008). Urease activity declined by the fire effect and this negative effect on urease activity was time-dependent and recovered after 12 years of burning (Camci Cetin et al. 2009).

Climatic conditions influence all living things as well as soil quality. The reduction of 10% of soil moisture decreased urease (10–67%), protease (15–66%) and β-glucosidase (10–80%) activities while decreasing of 21% of soil moisture declined urease (42–60%), protease (35–54%), β-glucosidase (35–83%) and acid phosphatase (31–40%) activities and no significant influences were found on alkaline phosphatase activity (Sardans and Penuelas 2005). N-cycling enzymes (protease and urease) were the most influenced by drought. The effect of drought on soil enzyme activities (glucosidase, sulfatase, phosphatase, and leucine amino peptidase) was higher and more pronounced in the Northern European sites than in the Southern European due to having moisture limited feature (Sowerby et al. (2005). The effect of temperature enhanced soil enzyme activities in all sites (Denmark, UK, The Netherlands and Spain); however, Northern Europe soils were more sensitive to changes in rainfall regimes than more moisture limited Southern European soils.

7.3.3 Soil Enzymes as Agricultural Practices Indicators

One of the agricultural practices is irrigation that provides adequate moisture level in soil for plant growth. Zhang and Wang (2006) investigated the impact of subsurface irrigation (−10, −16, −25, −40, and −63 KPa) on phosphatase, urease, and catalase activities under tomato cultivation in a greenhouse experiment. Phosphatase and catalase activities increased in more frequent irrigation (−10 and −16 KPa) and urease activities decreased under irrigation.

Fertilization of soils are conducted in soils by using different fertilizers such as mineral, manure, green manure, compost, and vermicompost. Kandeler et al. (1999) showed that farmyard manure enhanced microbial biomass, urease, deaminase, and alkaline phosphatase activity in soils compared with other treatments (mineral fertilizers) under crop rotations. The stage of plant growth should be the cause of concern on enzyme activities in soils in terms of evaluating the impact of fertilities. Similarly, soil enzyme activities (phosphatase, catalase, invertase, and urease) under different fertilizers (no fertilization, organic manure, organic manure + N, organic manure + NP, organic manure + NK, and organic manure + NPK) were lower in the early growth stages of cucumber, but enhanced in the late stages (Yang et al. 2008). Also, the type of mineral fertilizers used influences soil enzyme activity depending on the soil enzyme involved in which there is, nutrient cycling (N, P, C, and S). Soil enzyme activities were inhibited by N fertilizer while they were increased by P and K fertilizers. Decrease of urease activity could be explained by activation of nitrification and denitrification causing suppression in urease production (Aon et al. 2001).

Soil quality demonstrates agroecosystem sustainability and productivity (Melero et al. 2007). Using organic manures rather than chemical fertilizers in crop production is assumed to be a more sustainable type of agriculture (Chang et al. 2007). Excessive use of mineral fertilizer can decline soil organic matter as well as soil fertility (Melero et al. 2007). Thus, application of organic fertilizers becomes favorable for environmentalists, agriculturists and consumers (Chang et al. 2007). Compost fertilized plots (vegetal compost and animal compost) showed increase in enzyme activities (dehydrogenase, protease, glucosidase, and alkaline phosphatase) under dryland agriculture system (Melero et al. 2007). Bell et al. (2006) found increased phosphates activities following manure application likely due to enhanced P mineralization. Dehyrogenase, cellulose, protease, urease, acid phosphatase, and alkaline phosphatase activities (except urease) increased with application of manure; however, acid and alkaline phosphatase activities were adversely affected by chemical fertilizer treatment (Saha et al. 2008a). Dick (1997) reported that phosphatase inhibition was related to the level of PO_4 (high level) in soil solution (Chunderova and Zubets 1969) and phosphate may inhibit phosphatase synthesis and orthophosphate was known as a competitive inhibitor of acid and alkaline phosphatase activity (Juma and Tabatabai 1978). Animal manure enhances soil enzyme activities than mineral fertilizers and soil microbial and biochemical activities (Deng et al. 2006).

Organic fertilizers can be used in agricultural systems especially in organic farming. Compost application is important for establishing and maintaining soil organic matter to a certain level in organic farming (Chang et al. 2007). Organic treatments (sludge vermicompost and stabilized dairy manure) enhanced soil biological activity (microbial biomass and acid phosphatase, dehydrogenase, and protease BAA) due to an enrichment of soil organic matter (Marinari et al. 2000). Hu and Cao (2007) showed that alkaline phosphatase and urease activities were significantly greater in compost systems than in chemical fertilizers and control systems. Chang et al. (2007) found that soil enzyme activities (dehyrogenase cellulose, β-glucosidase, protease, urease, arylsulfatase and acid and alkaline phosphatase) and as well as other microbial properties increased significantly in the compost-treated soils compared with chemical-fertilized soils; however, no significant elevation was observed in studied enzyme activities after the compost dose of 540 kg N ha^{-1} $year^{-1}$.

The combination of mulch and fertilizers stimulate enzyme activities compared with fertilizers alone. β-glucosidase and dehydrogenase activities increased with increasing of leaf and N fertilization rate (Acosta-Martinez et al. 1999). Combined used of green manure and nitrification inhibitor increased content of mineral-N (NH_4^+ and NO_3^-); however, dehydrogenase and nitrate reductase activities decreased with combined application (Patra et al. 2006).

Using organic manures and crop residues is recommended in sustainable soil management due to providing maximum benefits and high environmental quality (Rajashekhararao and Siddaramappa 2008). Dehyrogenase activity was higher in composted cattle manure (44–200%) and vermicompost (22–108%) than in control (Saha et al. 2008b) Saha et al. (2008b) concluded that: (1) organic amendments applications enhanced organic matter contents and microbial biomass and this provided better potential for higher enzyme production and greater enzyme activities, (2) additions of organic amendments showed different responses on soil enzyme activities depending on organic amendment types, and (3) addition of organic amendments (either cattle manure compost or vermicompost) improved soil quality, increasing soil organic matter content and stimulating biological and biochemical properties.

Different kinds of organic amendments originated from animal or vegetal sources can be used in agricultural systems. Organic amendment residues (poultry manure, sewage sludge, barley straw and green alfalfa) applications provided higher enzyme activities (phosphomonoesterases, arylsulfatase, *N*-acetyl-β-glucosaminidase, β-glucosidase, β-glactosidase, invertase, dehyrogenase, amidase, and urease) in the first year of experiment; however, following subsequent additions did not maintain high enzyme activities (Martens et al. 1992). The most effective amendment increasing soil enzyme activities (except urease) was the straw amendment.

Organic amendments influence soil microbial and biochemical properties in different ways depending on the nutrient content. Application of higher rates of organic amendments (rice residues and tree litters: high C content) was favorable soil quality parameters (microbial biomass C, microbial quotient, urease and acid phosphatase activities) (Rajashekhararao and Siddaramappa 2008). Although they evaluated the other soil health parameters (extraction yield of humus and composition of humus) microbial quotient was the most sensitive indicator for reflecting the decline in soil quality. The authors reported that microbial biomass, microbial quotient and soil enzyme activities are used for measuring soil quality as parameters (Doran and Zeis 2000). Addition of different C sources enhanced urease activity at different levels of elevation depending on N levels (Rajashekhararao and Siddaramappa 2008). Higher level of N stimulated urease activity in different C sources.

Organic amendments can be used for suppression of plant diseases. Root rot severity was strongly adversely correlated with total C, arylsulfatase and β-glucosidase activities (Leon et al. 2006). β-Glucosidase was not accepted as an useful indicator of disease suppression because it varied over time. Arylsulfatase was the best indicator for reflecting disease suppression. They indicated that applying organic amendments to soils can cause disease suppression by enhancing antagonist microorganisms (Cook 1990) and microbial biomass and activity can be related to microbial competition with pathogens.

Conventional agriculture can cause to decrease soil organic matter and reduction in soil fertility (Melero et al. 2008). In terms of this view, sustainable agricultural management practices are crucial in the production of food and human nutrition (Niemi et al. 2008). Conventional cropping systems is required to use mineral fertilizers and pesticides for high crop yield; whereas organic cropping systems is preferred to use plant residues, green manure, farmyard manure and plant rotation for adequate crop yield and high soil quality (Niemi et al. 2008). Inorganic systems, plant production depends primarily on nutrient cycling in soils that are controlled by microbes and their activities and soil enzymes (Monokrousos et al. 2006).

Different farming systems may change soil parameters especially soil microorganisms and soil enzymes.

FlieBbach et al. (2007) investigated the impact of different farming systems [organic: bio-dynamic (farmyard manure compost + slurry), and bio-organic (farmyard manure + slurry); conventional: stacked farmyard manure + slurry, and mineral: mineral fertilizers; and unfertilized] on dehyrogenase activity on soil microbial biomass under 7 year crop rotation. Dehyrogenase activity was lower 39–42% in conventional and 62% in mineral farming systems compared with organic (bio-dynamic) farming system. Transition of farming system (conventional through organic) affects soil enzyme activities. Protease and phosphatase activities increased in organically fertilized soils (Melero et al. 2008). Similarly, arylsulfatase, phosphomonoesterase and esterase activities were greater in organic managed systems than in conventional systems and application of peat amendments increased phosphmonoesterase, phosphodiesterase, leucine aminopeptidase, chitinase, cellobiosidase, α-glucosidase and esterase activities, but decreased arylsulfatase and alanine aminopeptidase (Niemi et al. 2008). Sarapatka (2002) evaluated the effect of different farming system (conventional + no animal manure: CN, organic + no animal manure: OA, organic + no animal manure + standard cultivation: ONS, and organic + no animal manure + minimum soil management: ONM) on acid and alkaline phosphatase activities. The order of acid and alkaline phosphatase activities was: ONM > OA > CN > ONS. The author suggested that soil phosphatase activity was directly related to organic matter content and was affected by farming activities.

Monoculture cultivation causes different response on soil enzyme activities as well as conventional crop management compared with organic farming system. Dehyrogenase, acid and alkaline phosphatase activities and microbial biomass C and N levels were lower under monoculture system than under organic and conventional-short rotation systems (Gajda and Martyniuk 2005).

Changing farming system to organic farming, especially transition period influences soil biochemical properties. After conversion to farming system from conventional to organic, L-asparaginase, L-glutaminase, urease, acid and alkaline phosphatase were greater under organic management system than conventionally managed asparagus fields (Monokrousos et al. 2006). The experiments were designed as organic farming lasted 2, 3, 5, and 6 years. Acid phosphatase activity increased from the newest to oldest organic managed areas.

Many authors evaluated the impact of different management systems on soil quality parameters. Higher enzyme activities (dehydrogenase, *o*-diphenol oxidase, β-glucosidase, and phosphatase) were found under organic management system compared with conventional and integrated systems (Benitez et al. 2006). The authors suggested that using synthetic chemicals caused the reduction of biochemical activity in soils.

Different management systems can affect soil quality the way of soil perturbation. Greater enzyme activities (β-glucosidase and L-asparaginase) were found under succession fallow compared to agriculture and forestry systems, and this elevation could be explained by the lack of tillage that led to higher microbial and biochemical activities under successive fallow system (Landgraf and Klose 2002) Similarly, Gajda and Martyniuk (2005) found substantial disturbances in microbial activity of monoculture soil and higher enzyme activities (dehydrogenase and phosphatases) and microbial biomass C and N contents in organic and conventional systems. Declined populations (biomass and respiration rate) and activity of microorganisms in the monoculture soil might cause increasing deleterious microorganisms or decreasing nutrient availability to crops. Fluorescein diacetate hydrolase, dehydrogenase and phosphatase activities were the highest in the native-prairie ecosystem, organic farming, integrating cropping (no-tillage and crop rotation), crop rotation systems than compared with other systems (conventional + monoculture and conventional + high agrichemical input) and greater enzyme activities were found in higher weed-suppressive activity, organic matter content and water-stable soil aggregates (Kremer and Li 2003). Bandick and Dick (1999) showed that higher enzyme activities were found in the continuous grass and continuous pasture than in cultivated fields and the highest activity was observed in cover crops plots.

Due to monoculture systems decreased soil microbial activities,sustainable agricultural management requires crop rotation. Microbiological and biochemical properties (microbial biomass C and N, metabolic quotient, urease and fluorescein diacetate hydrolysis) were sensitive variables to determine soil rotations' effects and they could be used as soil quality indicators (Benintende et al. 2008). Also, microbial parameters were more effective management indicators in soil quality than biochemical parameters and all of them should be used as bioindicators when considering the impact of management on soil quality. Trasar-Cepeda et al. (2008b) showed that intensive agricultural system (with crop rotation) has led to low hydrolytic enzyme activities (acid phosphatase, β-glucosidase, phosphodiesterase, arylsulfatase, urease, BAA-protease, invertase, and carboxymethylcellulase) and organic matter content compared to climax soils (oak forest soil).

The influence of crop rotation [NERC site: continuous soybean (SSSS), continuous corn (CCCC), corn-soybean (CSCS), corn-corn-oat-alfalfa (CCOM); and CWRC site: CCCC, CSCSC, CCOM, and COMM] and fertilization (0 and 180 kg N ha^{-1}) on soil enzyme activities (α- and β-glucosidases and α- and β-galactosidases) were studied by Dodor and Tabatabai (2005). The results showed that : (1) glycosidases activities were significantly influenced by crop rotation but not by nitrogen application at the two sites; (2) greater activities were found in meadow or oat plots and the lowest in continuous corn or soybean plots; and (3) monocropping suppressed glycosidases activities whereas multicropping stimulated them. Since-different cropping systems change soil environment, a soil enzyme should be chosen that is directly involved in soil organic matter cycle because a more accurate way is the measurement of changes than measurement of C and N (Dodor and Tabatabai 2005). Multicropping systems enhanced amidohydrolases activities: however, monocropping systems reduced amidohydrolases activities and the mineralization of organic N in soil was provided by microorganisms through enzyme activities and amidohydrolases activities can be used to anticipate N mineralization and monitor soil quality (Dodor and Tabatabai 2003). Microbiological properties are accepted as more sensitive and more responsive indicators to agronomic treatments relative to soil physical and chemical properties as soil quality indicators (Biederbeck et al. 2005).

Tillage application may change soil quality through altering soil physico-chemical, hydrological (Rahman et al. 2008), microbiological and biochemical properties and thus influences soil microbial communities and the production of soil enzymes (Acosta-Martinez and Tabatabai 2001). Tillage affects soil nutrient levels and its availability (Etena et al. 1999), distribution of organic matter in the soil profile (Kandeler et al. 1999), soil water and oxygen content (Curci et al. 1997), and soil fertility. Tillage especially influence soil organic matter by exposing more soil organic matter to microbial attack (Mina et al. 2008) and finally rapid loss of soil organic matter (Madejon et al. 2007). Losing soil organic matter causes decline of soil biological activity and crop productivity and increase soil erosion (Madejon et al. 2007). To sum, tillage causes a great perturbation in soil environment.

Many researchers conducted the impact of tillage on soil quality parameters as well as soil enzymes. Dehydrogenase activity increased under continuous zero-tillage practice and alkaline phosphatase and protease activities were higher in the zero-tillage system over conventional practice; however, cellulose activity was greater in conventional practice compared with other management (Mina et al. 2008) Ekenler and Tabatabai (2004) found that L-glutaminase was the most sensitive N cycling enzyme followed by L-asparaginase, amidase, arylamidase, urease and L-aspartase for discriminating the effect of liming and tillage amendments (no-till, ridge-till, and chisel-plow). Also, the most sensitive enzymes reflecting soil management practices were β-glucosidase and β-glucosaminidase whose activities could be accepted early and reliable indicators of changes in soil properties affected by liming and tillage systems (Ekenler and Tabatabai 2003).

The combined effect of tillage and residue placements alters soil enzyme activities. Arylamidase activity was highly influenced by tillage and residue placements and the greatest arylamidase activity was found in treatments of chisel/mulch, moldboard plow/mulch, and no-till/double mulch whereas the lowest activity was observed in treatments of moldboard plow/normal and no-till/bare (Acosta-Martinez and Tabatabai 2001). Dehydrogenase, alkaline phosphatase, β-glucosidase, and urease were higher under conservation tillage than under traditional tillage (Madejon et al. 2007).

Soil enzyme activities are accepted early and are more reliable indicators than soil physico-chemical properties under different tillage systems. Curci et al. (1997) evaluated the influence of conventional tillage systems (shallowing plowing: 20 cm, deep plowing: 40 cm and scarification: 50 cm) at different depths (0–20, 20–40, 40–50, and 50–70 cm) on soil enzyme activities (acid phosphatase, alkaline phosphatase, phosphodiesterase, pyrophosphatase, arylsulfatase, dehydrogenase, α- and β-glucosidase, α- and β-glalactosidase, urease and nitrate reductase). The results showed that: (1) glycosidase, galactosidase, nitrate reductase, and dehydrogenase activities were influenced by tillage systems, (2) their activities were greater in shallow plowing and scarification than deep plowing plots in the upper layer (0–20 cm) of soils, and (3) no significant differences were found in soil physico-chemical properties under different tillage systems.

No-till systems provide better enzyme activities in soils. Deng and Tabatabai (1997) showed that acid phosphatase, alkaline phosphatase, phosphodiesterase, inorganic pyrophosphatase and arylsulfatase were significantly higher in no-till/double mulch than in other treatments (no-till/bare, no-till/normal, chisel/normal, chisel/mulch, moldboard/normal, and moldboard/mulch).

7.4 Conclusion

Soil enzymes are important soil components that are closely associated with physicochemical and biological characteristics of soil. However, human activities, agricultural practices and environmental pollution severely influence their existence and activities in soil. Intracellular enzymes are more susceptible to anthropogenic activities. However, extracellular enzymes held in clay and organic components have no linkage with the microorganisms by which it had been previously synthesized and thus their responses are usually slow and permanent while those of intracellular enzymes are more sensitive to the changes in soil environment. On the other hand, it is likely that extracellular enzymes would provide reliable results in the evaluation of long-term impacts of human activities on soil.

As a conclusion, depending on their origin, soil enzymes are powerful tools applied in the assessment of short- or long-term changes in soil. Majority of the anthropogenic or environmental factors affecting soil quality can be addressed to the changes in the soil's physical and chemical aspects, but at the same time they provoke the changes in soil enzyme pool as an indication of soil biological quality, meaning that overall soil quality parameters are closely related to each other. For example, agricultural practices such as organic matter applications, irrigation, fertilization and tillage change soil aggregation processes, nutrient recycling and also soil biological activities governing soil enzyme production. The aggregation of soil mineral particles is one of the major soil processes that is known to develop subsequently the microbiological–biochemical activities. Both aggregation and disaggregation processes are actually long-term events, but on the contrary, changes in soil microbiological–biochemical conditions are fast and occurs in a short span of time.

Soil enzymes that substantially originated from soil microorganisms may therefore be more preferable compared to soil physical and chemical quality parameters. For example, the index values of structure stability, erosion rate and erodibility are common soil criterion applied for the determination of soil erosion sensitivity, but in most cases these parameters are time-consuming and economically not reasonable. On the other hand, extracellular soil enzymes can be considered as an indication of soil erodibility (Kizilkaya et al. 2003 and they also enable researchers efficiently, to monitor the impacts of a wide group of agricultural management practices and the differences between remediated and severely degraded soil conditions. Due to these advantages over soil physical and chemical quality attributes, soil enzyme activity is considered to be an integral index of soil health but it should be kept in mind that an accurate evaluation of soil health needs to be complemented by physical, chemical and other biological parameters. Although many scientific communities have recognized the potential value of soil enzyme activities as a proper indication of soil quality, a number of problems i.e., Differences in methodological perspectives and the lack of standard methods can complicate the interpretation of soil enzyme activities.

The variability in soil enzyme values is also related to the fact that natural changes in climatical conditions and soil characteristics occurs simultaneously with the changes resulted from agricultural activities or environmental pollution. This actually makes more difficult to answer the question of which factor would virtually operate soil biochemical processes and to compare soil enzyme data obtained from different regions or climate zones. Nevertheless, soil enzyme activity is a fundamental indication of most changes in soil condition and hence soil quality, but it seems necessary to develop site-specific evaluation criteria for more accurate use of soil enzyme activities as a soil quality parameter.

References

Acosta-Martinez V, Tabatabai MA (2001) Tillage and residue management effects on arylamidase activity in soils. Biol Fertil Soils 34:21–24

Acosta-Martinez V, Reicher Z, Bischoff M, Turco RF (1999) The role of tree leaf mulch and nitrogen fertilizer on turfgrass soil quality. Biol Fertil Soils 29:55–61

Acosta-Martinez V, Klose S, Zobeck TM (2003) Enzyme activities in semiarid soils under conservation reserve program, native grassland and cropland. J Plant Nutr Soil Sci 166:699–707

Acosta-Martinez V, Cruz L, Sotomayer-Ramirez D, Perez-Alegria L (2007) Enzyme activities as affected by soil properties and land use in a tropical watershed. Appl Soil Ecol 35:35–45

Ajwa HA, Dell CJ, Rice CW (1999) Changes in enzyme activities and microbial biomass of tallgrass prairie soil as related to burning and nitrogen fertilization. Soil Biol Biochem 31:769–777

Al-Khafaji AA, Tabatabai MA (1979) Effects of trace elements on arylsulfatase activity in soils. Soil Science 127:129–133

Anderson DW, Gregorich EG (1984) Effect of soil erosion on soil quality and productivity. In: Soil erosion and degradation. Proceedings of 2nd ann. western provincial conf. rationalisation of water and soil research and management, Saskatoon, Saskatchewan, pp 105–113

Antolin M, Pascual I, Garcia C, Polo A, Sanchez-Diaz M (2005) Growth, yield and solute content of barley in soils treated with sewage sludge under semiarid Mediterranean conditions. Field Crop Res 94:224–237

Aon MA, Gabella MN, Sarena DE, Colaneri AC, Franco MG, Burgos JL, Cortassa S (2001) İ. Spatio-temporal patterns of soil microbial and enzymatic activities in an agricultural soil. Appl Soil Ecol 18:255–270

Bandick AK, Dick RP (1999) Field management effects on soil enzyme activities. Soil Biol Biochem 31:1471–1479

Bardgett RD, Speir TW, Ross DJ, Yeates GW, Kettles HA (1994) Impact of pasture contamination by copper, chromium and arsenic timber preservative on soil microbial properties and nematodes. Biology and Fertility of Soils 18:71–79

Barton L, Schipper LA, Smith CT, Mclay CDA (2000) Denitrification enzyme activity is limited by soil aeration in a wastewater-irrigated forest soil. Biol Fertil Soils 32:385–389

Baruah M, Mishra RR (1986) Effect of herbicides butachlor, 2,4-D and oxyfluorfen on enzyme activities and CO_2 evolution in submerged paddy field soil. Plant and Soil 96:287–291

Bastida F, Moreno JL, Hernandez T, Garcia C (2006) Microbiological activity in a 15 years after its devegetation. Soil Biol Biochem 38:2503–2507

Bastida F, Kandeler E, Hernandez T, Garcia C (2008) Long-term effect of municipial solid waste amendment on microbial abundance and humus-associated enzyme activities under semiarid conditions. Microb Ecol 55:651–661

Beck (1984) Methods and application of soil microbiological analysis at the Landensanstalt für Bodenkultur und Pflanzenbau (LBB) in Munich for the determination of some aspects of soil fertility. In: Nemes MP, Kiss S, Papacostea P, Stefanic Ç, Rusan M (eds) Research concerning a Biological Index of soil fertility. Fifth symposium on soil biology. Romanian National Society of Soil Science; Bucharest, pp 13–20

Bell JM, Robinson CA, Schwartz RC (2006) Changes in soil properties and enzymatic activities following manure applications to a rangeland. Rangeland Ecol Manage 59:314–320

Benintende SM, Benintende MC, Sterren MA, De Battista JJ (2008) Soil microbiological indicators of soil quality in four rice rotations systems. Ecol Indic 8:704–708

Benitez E, Nogales R, Compos M, Ruano F (2006) Biochemical variability of olive-orchard soils under different mamagement systems. Appl Soil Ecol 32:221–231

Bhattacharya P, Tripathy S, Chakrabarti K, Chakraborty A, Banik P (2008) Fractionation and bioavailability of metals and their impacts on microbial properties in sewage irrigated soil. Chemosphere 72:543–550

Biederbeck VO, Zenther RP, Campbell CA (2005) Soil microbial populations and activities as influenced by legume gren fallow in a semiarid climate. Soil Biol Biochem 37:1775–1784

Bitton G, Boylan RR (1985) Effect of acid precipitation on soil microbial activity: I. Soil core studies. Journal of Environmental Quality 14:66–69

Bitton G, Volk BG, Graetz DA, Bossart JM, Boylan JM, Byers GE (1985) Effect of acid precipitation on soil microbial activity: II. Field studies. Journal of Environmental Quality 14:69–71

Boerner REJ, Brinkman JA (2003) Fire frequency and soil enzyme activity in Southern Ohio oak-hickory forests. Appl Soil Ecol 23:137–146

Boerner REJ, Decker KLM, Sutherland EK (2000) Prescribed burning effects on soil enzyme activity in a Southern Ohio, harwood forest: a landscape-scale analysis. Soil Biol Biochem 32:899–908

Boerner REJ, Brinkman JA, Smith A (2005) Seasonal variations in enzyme activity and organic carbon in soil of a burned and unburned hardwood forest. Soil Biol Biochem 37:1419–1426

Boerner REJ, Giai C, Huang J, Miesel JR (2008) Initial effects of fire and mechanical thinning on soil enzyme activity and nitrogen transformations in eight North American forest ecosystems. Soil Biol Biochem 40:3076–3085

Boyd SA, Mortland MM (1990) Enzyme interactions with clays and clay-organic matter complexes. In: Bollag JM, Stotzky G (eds) Soil biochemistry. Marcel Dekker, New York, pp 1–28

Burns RG (1978) Enzyme activity in soil: some theoretical and practical considerations. In: Burns RG (ed) Soil enzymes. Academic, New York, pp 295–340

Cai X, Sheng G, Liu W (2007) Degradation and detoxification of acetochlor in soils treated by organic and thiosulfate amendments. Chemosphere 66:286–292

Calderon JF, Jackson LE, Scow KM, Rolston DE (2000) Microbial responses to simulated tillage in cultivated and uncultivated soils. Soil Biol Biochem 32:1547–1559

Camci Cetin S, Ekinci H, Kavdir Y, Yuksel O (2009) Using soil urease enzyme activity as soil quality for reflecting fire influence in forest ecosystem. Fresen Environ Bull (communicated)

Campbell PN, Smith AD (1993) Biochemistry illustrated: an illustrated summary of the subject for medical and other students of biochemistry. Logman Singapore Publishers, Singapore, pp 355–378

Canarache A (1990) Fizica Solurilor Agricole. Ceres, Bucarest, p 268

Caravaca F, Alguacil M, Figueroa D, Barea JM, Roldan A (2003) Re-establishment of Retana sphaerocarpa as a target species for reclamation of soil physical and biological properties in a semiarid Mediterranean land. For Ecol Manage 182:49–58

Chang EH, Chung RS, Tsai YH (2007) Effect of different application rates of organic fertilizer on soil enzyme activity and microbial population. Soil Sci Plant Nutr 53:132–140

Chunderova AI, Zubets T (1969) Phosphatase activity in derropodzolic soils. Pochovedeniye 11:47–53

Conrad JP (1940) The nature of the catalyst causing the hydrolysis of urea in soils. Soil Sci 54:367–380

Cook RJ (1990) Twenty-five years of progress towards biological control. In: Hornby D (ed) Biological control of soil-borne pathogens. CABI, Wallingford, UK

Curci M, Pizzigallo MDR, Crecchio C, Minninni R, Ruggiero P (1997) Effects of conventional tillage on biochemical properties of soils. Biol Fertil Soils 25:1–6

Dale VH, Beyeler SC (2001) Challenges in the development and use of ecological indicators. Ecol Indic 1:3–10

Dale VH, Peacock AD, Garten CT, Sobek F, Walfe AK (2008) Selecting indicators of soil, microbial and plant conditions to understand ecological changes in Georgia pine forests. Ecol Indic 8:818–827

Dar GH (1996) Effects of sewage-sludge on soil microbial biomass and enzyme activities. Bioresource Technology 56:141–145

Davies HA, Greaves MP (1981) Effects of some herbicides on soil enzyme activities. Weed Research 21:205–209

DeBano LF, Neary DG, Ffolliott PF (1998) Fire's effects on ecosystems. Wiley, New York, USA

Deng SP, Tabatabai MA (1995) Cellulase activity of soils: effect of trace elements. Soil Biology and Biochemistry 27:977–979

Deng SP, Tabatabai MA (1997) Effect of tillage and residue management on enzyme activities in soils: III. Phosphatases and arylsulfatase. Biol Fertil Soils 24:141–146

Deng SP, Parham JA, Hattey JA, Babu D (2006) Animal manure and anhydrous ammonia amendment alter microbial carbon use efficiency, microbial biomass and activities of dehydrogenase and amidohydrolases in semiarid agroecosystems. Appl Soil Ecol 33:258–268

Dexter AR (2004) Soil physical quality. Part I. Theory, effects of soil texture, density, and organic matter, and effects on root growth. Geoderma 120:201–214

Dexter AR, Czyz EA (2000) Soil physical quality and the effects of management practices. In: Wilson MJ, Maliszewska-Kordybach B (eds) Soil quality, sustainable agriculture and environmental security in Central and Eastern Europe. NATO Science Series 2, Environmental Security, Vol 69. Kluwer Academic Publishers, Dordrecht, pp 153–165

Dick RP (1997) Soil enzyme activities as integrative indicators of soil health. In: Pankhurst CE, Doube BM, Gupta VVSR (eds) Biological indicators of soil health. CABI, Wallingford, Oxfordshire, pp 121–156

Dick WA, Tabatabai MA (1992) Significance and potential uses of soil enzymes. In: Meeting FB Jr (ed) Soil microbial ecology. Marcel Dekker, New York, Basel, Hong Kong, pp 95–127

Dick RP, Rasmussen PE, Kerle EA (1987) Kinetic parameters of enzyme activities as influenced by organic residues and N fertilizer management. In: Agronomy Abstracts, 79th Annual Meeting ,Atlanta, GA. American Society of Agronomy, Madison, WI

Dodor DE, Tabatabai MA (2003) Amidohydrolases in soils as affected by cropping systems. Appl Soil Ecol 24:73–90

Dodor DE, Tabatabai MA (2005) Glycosidases in soils as affected by cropping systems. J Plant Nutr Soil Sci 168:749–758

Doran JW, Parkin TB (1994) Defining and assessing soil quality. In: JW Doran et al (eds) Defining soil quality for a sustainable environment. Soil Science Society of America. Special Publication No. 15, American Society of Agronomy, Madison, WI, pp 3–21

Doran JW, Zeis MR (2000) Soil health and sustainability: managing the biotic component of soil quality. Appl Soil Ecol 15:3–11

Drijber RA, Doran JW, Parkhurst AM, Lyon DJ (2000) Changes in soil microbial community structure with tillage under long-term wheat-fallow management. Soil Biol Biochem 32:1419–1430

Dzantor EK, Felsot AS (1991) Microbial responses to large concentrations of herbicides in soil. Environmental Toxicology and Chemistry 10:649–655

Eivazi F, Tabatabai MA (1990) Factors affecting glucosidase and galactosidase activities in soil. Soil Biol Biochem 22:891–897

Ekenler M, Tabatabai MA (2003) Effects of liming and tillage systems on microbial biomass and glycosidases in soils. Biol Fertil Soils 39:51–61

Ekenler M, Tabatabai MA (2004) Arylamidase and amidohydrolases in soils as affected by liming and tillage systems. Soil Tillage Res 77:157–168

Ellert BH, Gregorich EG (1995) Management-induced changes in the actively cycling fractions of soil organic matter. In: Carbon forms and functions in forest Soils. Soil Science Society of America, Madison, WI, pp 119–138

Elliott ET (1997) Rationale, for developping bioindicators of soil health. In: Pankhurst CE, Doube BM, Gupta VVSR (eds) Biological indicators of soil health. CABI, Wellington, pp 49–78

Esminger LE, Gieseking SE (1942) Resistance of clay-adsorbed proteins to proteolytic hydrolysis. Soil Sci 53:205–209

Etena A, Hakansson I, Zagal E, Bucas S (1999) Effect of tillage depth on organic carbon content and physical properties in five Swedish soils. Soil Till Res 52:129–139

Fernandes SAP, Bettiol W, Cerri CC (2005) Effect of sewage sludge on microbial biomass, basal respiration, metabolic quotient and soil enzymatic activity. Appl Soil Ecol 30:65–77

Filip Z, Karazawa S, Berthelin J (1999) Characterization of effects of a long-term wastewater irrigation on soil quality by microbiological and biochemical parameters. J Plant Nutr Soil Sci 162:409–413

Fioretto A, Papa S, Pellegrino A (2005) Effects of fire on soil respiration, ATP content and enzyme activities in Mediterranean maquis. Appl Veg Sci 8:13–20

FlieBbach A, Oberholzer H, Gurnst L, Mader P (2007) Soil organic matter and biological soil quality indicators after 21 years of organic and conventional farming. Agric Ecosyst Environ 118:273–284

Fragoeiro S, Magan N (2008) Inpact of trameters versicolor and *Phanerochaete chrysosporium* on differential breakdown of pesticide mixtures in soil microcosms at two water potentials and associated respiration and enzyme activity. Int Biodeterior Biodegradation 62:376–383

Frankenberger WT Jr, Tabatabai MA (1981) Amidase activity in soils: IV. Effects of trace elements and pesticides. Soil Science Society of America Journal 45:1120–1124

Frankenberger WT Jr, Tabatabai MA (1991a) Factors affecting L-asparaginase activity in soils. Biology and Fertility of Soils 11:1–5

Frankenberger WT Jr, Tabatabai MA (1991b) Factors affecting L-glutaminase activity in soils. Soil Biology and Biochemistry 23:875–879

Gajda A, Martyniuk S (2005) Microbial biomass C and N and activity of enzymes in soil under winter wheat grown in different crop management systems. Pol J Environ Stud 14:159–163

Gianfreda L, Bollag JM (1996) Influence of natural and anthropogenic factors on enzyme activity in soil. In: Stotzky G, Bollag JM (eds) Soil biochemistry. Marcel Dekker, New York, pp 123–194

Gianfreda L, Rao MA, Piotrowska A, Palumbo G, Colombo C (2005) Soil enzyme activities as affected by anthropogenic alterations: intensive agricultural practices and organic pollution. Sci Total Environ 341:265–279

Gianfreda L, Sannino F, Filazzola MT, Violante A (1993) Influence of pesticides on the activity and kinetics of invertase, urease and acid phosphatase enzymes. Pesticide Science 39:237–244

Gianfreda L, Sannino F, Ortega N, Nannipieri P (1994) Activity of free and immobilized urease in soil: effects of pesticides. Soil Biology and Biochemistry 26:777–784

Gil-Sotres F, Trasar-Cepeda C, Leiros MC, Seoane S (2005) Different approaches to evaluating soil quality using biochemical properties. Soil Biol Biochem 37:877–887

Gonzalez-Perez JA, Gonzalez-Villa FJ, Gonzola A, Knicker H (2004) The effect of fire on soil organic matter-a review. Environ Int 30:855–870

Gregorich EG, Carter MR, Angers DA, Monreal CM, Ellert BH (1994) Towards a minimum data set to assess soil organic matter quality in agricultural soils. Can J Soil Sci 74:367–385

Gundi VAKB, Narashima G, Reddy BR (2005) Interaction effects of insecticides on microbial populations and dehydrogenase activity in a black clay soil. J Environ Sci Health B 40:269–283

Hart SC, DeLuca TH, Newman GG, MacKenzie MD, Boyle SJ (2005) Post fire vegetative dynamics as drivers of microbial community structure and function in forest soils. For Ecol Manage 220:166–185

Hinojosa MB, Garcia-Ruiz R, Vinegla B, Carreira JA (2004) Microbiological rates and enzyme activities as indicators of functionality in soils affected by the Aznalcollar toxic spill. Soil Biol Biochem 36:1637–1644

Hinojosa MB, Carreira JA, Rodriguez-Moroto JM, Garcia-Ruiz R (2008) Effects of pyrite sludge pollution on soil enzyme activities: ecological dose-response model. Sci Total Environ 396:89–99

Hu C, Cao Z (2007) Size and activity of soil microbial biomass and soil enzyme activity and microbial population. Soil Sci Plant Nutr 53:132–140

Izquierdo I, Caravaca F, Alguacil MM, Hernandez G, Roldan A (2005) Use of microbiological indicators for evaluating success in soil restoration after revegetation of a mining area under subtropical conditions. Appl Soil Ecol 30:3–10

Joshi SR, Sharma GD, Mishra RR (1993) Microbial enzyme activities related to litter decomposition near a highway in a sub-tropical forest of North east India. Soil Biology and Biochemistry 25:1763–1770

Juma NG, Tabatabai MA (1978) Distribution of phosphomonoesterases in soils. Soil Sci 126:101–108

Kanan K, Oblisami G (1990) Influence of paper mill effluent irrigation on soil enzyme activities. Soil Biology and Biochemistry 22:923–926

Kandeler E, Tscherko D, Spiegel H (1999) Long-term monitoring of microbial biomass, N mineralisation and enzyme activities of a Chernozem under different tillage management. Biol Fertil Soils 28:343–351

Kizilkaya R, Bayrakli B (2005) Effects of N-enriched sewage sludge on soil enzyme activities. Appl Soil Ecol 30:192–202

Kízílkaya R, Hepşen FŞ (2004) Effect of biosolid amendment on enzyme activities in earthworm (*Lumbricus terrestris*) casts. J Plant Nutr Soil Sci 167:202–208

Kizilkaya R, Aşkin T, Bayrakli B, Sağlam M (2004) Microbial characteristics of soils contaminated with heavy metals. Eur J Soil Biol 40:95–102

Kízílkaya R, Askin T, Ozdemir N (2003) Use of enzyme activities as a soil erodibility indicator. Indian J Agric Sci 73:446–449

Klein DA, Sorensen DL, Redente EF (1985) Soil enzymes: a predictor of reclamation potential and progress. In: Tate RL, Klein DA (eds) Soil reclamation processes. Microbiological analyses and applications. Marcel Dekker, New York, pp 273–340

Kremer RJ, Li J (2003) Developing weed-suppressive soils through improved soil quality management. Soil Tillage Res 72:193–202

Kucharski J, Wyszkawska J (2008) Biological properties of soil contaminated with the herbicide Apyros 75 WG. J Elementol 13:357–371

Kunito T, Saeki K, Goto S, Hayashi H, Oyaizu H, Matsumoto S (2001) Copper and zinc fractions affecting microorganisms in long-term sludge-amended soils. Bioresour Technol 79:135–146

Kuperman RG, Carriero MM (1997) Soil heavy metal concentrations microbial biomass and enzyme activities in a contaminated grassland system. Soil Biology and Biochemistry 29:179–190

Ladd JN (1985) Soil enzymes. In: Vaughan D, Malcolm RE (eds) Soil organic matter and biological activity. Nijhoff, Dordrecht, Boston, pp 175–221

Landgraf D, Klose S (2002) Mobile and readily available C and N fractions and their relationship to microbial biomass and selected enzyme activities in a sandy soil under different management systems. J Plant Nutr Soil Sci 165:9–16

Larson WE, Pierce FJ (1994) The dynamics of soil quality as a measure of sustainable management. In: JW Doran et al (eds) Defining soil quality for a sustainable environment. Soil Science Society of America. Spec. Pub. No. 15, American Society of Agronomy, Madison, WI, pp 37–51

Leiros MC, Trasar-Cepeda C, Garcia-Fernandez F, Gil-Sotres F (1999) Defining the validity of a biochemical index of soil quality. Biology and Fertility of Soils 30:140–146

Leon MCC, Stone A, Dick RP (2006) Organic soil amendments: impacts on snap bean common root rot (*Aphanomyces euteiches*) and soil quality. Appl Soil Ecol 31:199–210

Lin XG, Yin R, Zhang HY, Huang JF, Chen RR, Cao ZH (2004) Changes of soil microbiological properties caused by land use changing from rice-wheat rotation to vegetable cultivation. Environ Geochem Health 26:119–128

Madejon E, Moreno F, Murillo JM, Pelagrin F (2007) Soil biochemical response to long-term conservation tillage under semiarid Mediterranean conditions. Soil Tillage Res 94:346–352

Majer BJ, Tscherko D, Paschke A, Wennrich R, Kundi M, Kandeler E, Knasmüller S (2002) Effects of heavy metal contamination of soils on micronucleus induction in Tradescantia and on microbial enzyme activities: a comparative investigation. Mutation Research 515:11–124

Margesin R, Zimmebauer A, Schinner F (2000) Monitoring of soil bioremediation by soil biological activities. Chemosphere 40:339–346

Marinari S, Masciandaro G, Ceccanti B, Grego S (2000) Influence of organic and mineral fertilizers on soil biological and physical properties. Bioresour Technol 72:9–17

Martens DA, Johanson JB, Frankenberger WT (1992) Production and persistence of soil enzymes with repeated addition of organic residues. Soil Sci 153:53–61

Mathews CK, van Holde KE (1995) Biochemistry, 2nd edn. The Benjamin/Cummings Publishing Company Inc, New York, pp 360–414

McCarty GW, Siddaramappa R, Wright RJ, Codling EE, Gao G (1994) Evaluation of cool combustion by products as soil liming materials: their influence on soil pH and enzyme activities. Biology and Fertility of Soils 17:167–172

Melero S, Madejon E, Ruiz JC, Herencia JF (2007) Chemical and biochemical properties of a clay soil under dryland agriculture system as affected by organic fertilization. Eur J Agron 26:327–334

Melero S, Madejon E, Herencia JF, Ruiz JC (2008) Effect of implementing organic farming on chemical and biochemical properties of an irrigated loam soil. Argon J 100:136–144

Mina BL, Saha S, Kumar N, Srivastva AK, Gupta HS (2008) Changes in soil nutrient content and enzymatic activity under conventional and zero-tillage practices in an Indian sandy clay loam soil. Nutr Cycl Agroecosyst 82:273–281

Monokrousos N, Papatheodorou EM, Diamantopoulos JD, Stomou GP (2006) Soil quality variables in organically and conventionally cultivated field sites. Soil Biol Biochem 38:1282–1289

Nannipieri P, Ceccanti B, Grego S (1990) Ecological significance of biological activity in soil. In: Bollag JM, Stotzky G (eds) Soil biochemistry, Vol 6. Marcel Dekker, New York, pp 293–355

Nannipieri P, Kandeler E, Ruggiero P (2002) Enzyme activities and microbiological and biochemical processes in soil. In: Burns RG, Dick RP (eds) Enzymes in the environment. Activity, ecology and applications. Marcel Dekker, New York, pp 1–33

Neary DG, Klopatek CC, DeBano LF, Ffolliott PF (1999) Fire effects on belowground sustainability: a review and synthesis. For Ecol Manage 122:51–71

Niemi RM, Vepsalainen M, Wallenius K, Erkomaa K, Kukkonen S, Palojarvi A, Vestberg M (2008) Conventional versus organic cropping and peat amendment: impacts on soil microbiota and their activities. Eur J Soil Biol 44:419–428

Ohtonen R, Lahdesmaki P, Markkola AM (1994) Celulase activity in forest humus along an industrial pollution gradient in Oulu, Northern Finland. Soil Biology and Biochemistry 26:97–101

Pascual JA, Garcia G, Hernandez T, Moreno JL, Ros M (2000) Soil microbial activity as a biomarker of degradation and remediationprocesses. Soil Biology and Biochemistry 32:1877–1883

Patra AK, Chhonkar PK, Khan MA (2006) Effect of gren manure Aesbania sesban and nitrification inhibitor encapsulated calcium carbide (ECC) on soil mineral-N, enzyme activity and nitrifying organisms in a rice-wheat cropping system. Eur J Soil Biol 42:173–180

Perrucci P, Scarponi L (1994) Effects of the herbicide imazethapyr on soil microbial biomass and various soil enzyme activities. Biol Fertil Soils 17:237–240

Perucci P (1992) Enzyme activity and microbial biomass in a field soil amended with municipal refuse. Biol Fertil Soils 1:111–115

Pichtel JR, Hayes JM (1990) Influence of fly ash on soil microbial activity and populations. Journal of Environmental Quality 19:593–597

Powlson DS, Brookes PC, Christensen BT (1987) Measurement of soil microbial biomass provides an early indication of changes in total soil organic matter due to straw incorporation. Soil Biol Biochem 19:159–164

Puglisi E, Del Re AAM, Rao MA, Gianfreda L (2006) Development and validation of numerical indexes integrating enzyme activities of soils. Soil Biol Biochem 38:1673–1681

Rahman MH, Okuba A, Sugiyama S, Mayland HF (2008) Physical, chemical and microbiological properties of an Andisol as related to land use and tillage practice. Soil Tillage Res 101:10–19

Rai JPN (1992) Effects on long-term 2,4-D application on microbial populations and biochemical processes in cultivated soil. Biology and Fertility of Soils 13:187–191

Rajashekhararao BK, Siddaramappa R (2008) Evaluation of soil quality parameters in a tropical paddy soil amended with rice residues and tree litters. Eur J Soil Biol. doi:10.1016/j.ejsobi.2008.04.002

Reaves DW (1997) The role of soil organic matter in maintaining soil quality in continuous cropping systems. Soil Till Res 43:131–167

Saa A, Trasar-Cepeda MC, Gill-Sotres F, Carballas T (1993) Changes in soil phosphorus and acid phosphatase activity immediately following forest fires. Soil Biol Biochem 25: 1223–1230

Saa A, Trasar-Cepeda MC, Carballas T (1998) soil P status and phosphomonoesterase activity of recently burnt and unburnt soil following laboratory incubation. Soil Biol Biochem 30:419–428

Saha S, Prakash V, Kundu S, Kumar N, Mina BL (2008a) Soil enzymatic activity as affected by long term application of farm yard manure and mineral fertilizer under a rainfed soybean-wheat system in N–W Himalaya. Eur J Soil Biol 44:309–315

Saha S, Mina BL, Gopinath KA, Kundu S, Gupta HS (2008b) Organic amendments affect biochemical properties of a subtemperate soil of the Indian Himalayas. Nutr Cycl Agroecosyst 80:233–242

Sarapatka B (2002) Phosphatase activity of eutric cambisols (uppland Sweden) in relation to soil properties and farming systems. Acta Agriculturae Biochemica 33:18–24

Sardans J, Penuelas J (2005) Drought decreases soil enzyme activity in a Mediterranean *Quercus ilex* L. forest. Soil Biol Biochem 37:455–461

Saviozzi A, Biasci A, Riffaldi F, Levi-Minzi R (1999) Long-term effects of farmyard manure and sewage sludge or some soil biochemical characteristics. Biol Fertil Soils 30:100–106

Saviozzi A, Levi-Minzi R, Cardelli R, Riffaldi R (2001) A comparison of soil quality in adjacent cultivated, forest and native grassland soils. Plant and Soil 233:251–259

Senthilkumar K, Marian S, Vdaiyan K (1997) The effect of burning on soil enzyme activities natural grassland in southern India. Ecol Res 12:21–25

Serban A, Nissenbaum A (1986) Humic acid association with peroxidase and catalase. Soil Biol Biochem 18:41–44

Shakesby RA, Wallbrink PJ, Doerr SH, English PM, Chafer CJ, Humphreys GS, Blake WH, Tomkins KM (2007) Distinctiveness of wildfire effects on soil erosion in southeast Australian eucalypt forests assessed in a global context. For Ecol Manage 238:347–364

Sicardi M, Garcia-Prechac F, Frioni L (2004) Soil microbial indicators sensitive to land use conversion from pastures to commercial *Eucalyptus grandis* (Hill ex Maiden) plantations in Uruguay. Appl Soil Ecol 27:125–133

Sinsabaugh RL (1994) Enzymic analysis of microbial pattern and processes. Biol Fertil Soils 17:69–74

Skujins JJ (1978) History of abiotic soil enzyme research. In Soil Enzymes. R.G. Burns (ed). Academic Pres. New York, pp 1–49

Soil Science Society of America (1995) Statement on soil quality. Agronomy News, June.

Song HG, Bartha R (1990) Effects of jet fuel spills on the microbial community of soil. Applied and Environmental Microbiology 56:646–651

Sowerby A, Emmett B, Beier C, Tietema A, Penuelas J, Estiarte M, Meeteren MJMV, Hughes S, Freeman C (2005) Microbial community changes in healthland soil communities along a geographical gradient: interaction with climate change manipulations. Soil Biol Biochem 37:1805–1813

Staddon WJ, Duchesne LC, Trevors JT (1998) Acid phosphatase, alkaline phosphatase and arylsulfatase activities in soils from a jack pine (*Pinus banksiana* Lamb.) ecosystem after clear-cutting prescribed burning and scarification. Biol Fertil Soils 27:1–4

Stefanic G (1994) Biological definition, quantifying method and agricultural interpretation of soil fertility. Romanian Agric Res 2:107–116

Stryer L (1995) Biochemistry, 4th edn. WH Freeman and Company, New York, pp 181–206

Tabatabai MA (1982) Soil enzymes. In: Page AL, Miller RH, Keeney DR (eds) Methods of soil analysis. Part 2. Agronomy 9, 2nd edn. American Society of Agronomy, Madison, WI, pp 903–947

Tabatabai MA (1994) Soil enzymes. In: Page AL, Miller RH, Keeney DR (eds) Methods of soil analysis. American Society of Agronomy, Madison, WI, pp 775–833

Trasar-Cepeda C, Leiros MC, Seoane S, Gil-Sotres F (2000) Limitations of soil enzymes as indicators of soil pollutions. Soil Biol Biochem 32:1867–1875

Trasar-Cepeda C, Leiros MC, Gil-Sotres F (2008a) Hydrolytic enzyme activities in agricultural and forest soils. Some implications for their use as indicators of soil quality. Soil Biol Biochem 40:2146–2155

Trasar-Cepeda C, Leiros MC, Seoane S, Gil-Sotres F (2008b) Biochemical properties of soils under crop rotation. Appl Soil Ecol 39:133

Tu CM (1981) Effects of some pesticides on enzyme activities in an organic soil. Bulletin of Environmental Contamination 27:109–114

Tu CM (1993) Influence of ten herbicides on activities of microorganisms and enzymes in soils. Bulletin of Environmental Contamination and Toxicology 49:120–128

Visser S, Parkinson D (1992) Soil biological criteria as indicators of soil quality: soil microorganisms. Am J Altern Agric 7:33–37

Voet D, Voet JG (1995) Biochemistry. Introduction to enzymes. Chapter 12, 2nd edn. Wiley, New York, pp 332–344

Voets JP, Meerschman P, Verstraete W (1974) Soil microbiological and biochemical effects of long-term atrazine applications. Soil Biology and Biochemistry 6:149–152

Warkentin BP (1995) The changing concept of soil quality. J Soil Water Conserv 50:226–228

Yakovchenko VI, Sikora LJ, Rauffman DD (1996) A biologically based indicator of soil quality. Biol Fertil Soils 21:245–251

Yang L, Li T, Li F, Leucoff JH, Cohen S (2008) Fertilization regulates soil enzymatic activity and fertility dynamics in a cucumber field. Sci Hortic 116:21–26

Yeates GW, Orchard VA, Speir TW, Hunt JL, Hermans MCC (1994) Reduction in soil biological activity following pasture contamination by copper, chromium, arsenic timber preservative. Biology and Fertility of Soils 18:200–208

Yu YL, Shan M, Fang H, Wang X, Chu XQ (2006) Responses of soil microorganisms and enzymes to repeated applications of chlorothalonil. J Agric Food Chem 54:10070–10075

Zhang YL, Wang YS (2006) Soil enzyme activities with greenhouse subsurface irrigation. Pedosphere 16:512–518

Zhang YM, Wu N, Zhou GY, Bao WK (2005) Changes in enzyme activities of spruce (*Picea balfouriana*) forest soil as related to burning in the eastern Qinghai-Tibetan plateau. Appl Soil Ecol 30:215–225

Zhang YL, Dai JL, Wang RQ, Zhang J (2008) Effects of long-term sewage irrigation on agricultural soil microbial structural and functional characterizations in Shandong, China. Eur J Soil Biol 44:84–91

Chapter 8
Enzyme Activities in the Rhizosphere of Plants

Dilfuza Egamberdieva, Giancarlo Renella, Stephan Wirth, and Rafiq Islam

8.1 Introduction

The rhizosphere was first defined by Hiltner in 1904 as the soil volume surrounding the root surface. Generally, the rhizosphere has a thickness of 1–2 mm, but functionally, the rhizosphere can be defined as the soil portion physically and chemically influenced by growth and activity of the root. Therefore, the dimension of the rhizosphere can vary depending on the mass and architecture of the plant roots. The rhizosphere is a unique hot spot in soil from the viewpoint of microbial ecology, as soil micro-organisms are considerably stimulated by the activity of the roots (Jones et al. 2004; Hinsinger et al. 2006). Microbial communities are part of a complex food web that uses considerable amount of carbon fixed by the plant and released into the rhizosphere (i.e., rhizodeposits). In fact, increased microbial activity in the rhizosphere is sustained by nutrients secreted by plant roots in the form of soluble exudates, such as carbohydrates, aminoacids, low molecular weight organic acids and other photosynthates (Nannipieri et al. 2003). Generally, the rhizosphere soil contains more soluble sugars but less insoluble material than the surrounding soil and there is less nitrogen but more polyphenols (Somers et al. 2004; Raaijmakers et al. 2009). However, through the exudation of a wide variety of compounds, the plant roots may not only regulate positively or negatively the

D. Egamberdieva (✉)
Faculty of Biology and Soil Sciences, The National University of Uzbekistan, Tashkent 100174, Uzbekistan
e-mail: dilfuza_egamberdiyeva@yahoo.com

G. Renella
Plant Nutrition and Soil Science, University of Florence, Florence, Italy

S. Wirth
Leibniz-Centre for Agricultural Landscape Research, Institute of Landscape Matter Dynamics, Müncheberg, Germany

R. Islam
Soil and Water Resources, Ohio State University South Centers, Piketon, OH, USA

G. Shukla and A. Varma (eds.), *Soil Enzymology*, Soil Biology 22,
DOI 10.1007/978-3-642-14225-3_8, © Springer-Verlag Berlin Heidelberg 2011

microbial activity in the rhizosphere but also regulate complex ecological interactions between plant and herbivores, plant beneficial microrganisms, and change the chemical and physical properties of the soil, and inhibit the growth of competing plant species (Nardi et al. 2000).

Since plant nutrient uptake process occur through the rhizosphere, the activity of rhizosphere microbial community is of great importance for plant growth (Chanway 2002). The overall enzyme activity of the rhizosphere as well as bulk soil can depend on enzymes localized in root cells, root remains, microbial cells, microbial cell debris, microfaunal cells and the related cell debris, free extracellular enzymes or enzymes adsorbed onto or occluded into the soil colloids (Nannipieri et al. 2003). The study of different hydrolase enzyme activities in the rhizosphere soil and their changes is important since they indicate the potential of a soil to carry out specific biochemical reactions, and these hydrolytic enzymes are important in maintaining soil fertility and plant productivity (Burns 1982). Because plant nutrient uptake occurs through the rhizosphere, the activity of rhizosphere microbial community is of great importance for plant growth (Chanway 2002).

Soil enzymes are involved in the catalysis of a large number of reactions necessary for life processes of microorganisms in soils, decomposition of organic residues, cycling of nutrients, and formation of organic matter and soil structure (Dick et al. 1996; Drijber et al. 2000; Colombo et al. 2002; Nannipieri et al. 2002). These enzymes include amylase, arylsulphatases, β-glucosidase, cellulase, chitinase, dehydrogenase, phosphatase, protease, urease and others, derived from plant, animal, or microbial origins (Dick and Tabatabai 1984; James et al. 1991; Gupta et al. 1994; Nannipieri et al. 2003). These enzymes can be accumulated, stabilized, and/or decomposed in the soil (Dick 1997).

Plant roots have been considered a source of extracellular enzymes in soil. In early studies Estermann and McLaren (1961) found that root caps disruption of barley (*Hordeum vulgare*) possessed phosphatase activity. Juma and Tabatabai (1988) reported that corn (*Zea mays*) and soybean (*Glycine max*) roots contain acid phosphatase, but no detectable alkaline phosphatase activity.

Enzyme activity in the rhizosphere can be of intracellular origin, released after microbial cell disuption or root cell sloughing (Hawes et al. 2003), and may be associated with soil colloids and cell debris, or enzymes may be actively secreted by plant roots or root associated microrganisms. Although the former enzyme location are supposed to contribute less to the whole rhizosphere enzyme activity being mineralized by microbial communities, it should be underlined that the methods today available for measuring the soil enzyme activity do not allow to discriminate the contribution of the different enzyme locations to the whole soil or rhizosphere enzyme activity (Nannipieri et al. 2003).

Generally, low molecular organic compounds are passively released due to the existing steep intracellular/extracellular gradient whilst larger and more complex compounds or polymers are actively secreted through exocytosis (Neumann and Römheld 2007). It is important to take into account that the root exudation rate is higher in the root apical zone and decreases in the senescent root districts and that the root exudation profile is different in the various parts of the root.

Rhizosphere microrganisms release extracellular enzymes for the initial degradation of high molecular polymers such as cellulose, chitin and lignin, proteins, leading to their mineralization to mineral N, P, and S (Burns 1982; Nannipieri et al. 1996). Badalucco et al. (1996) reported that in wheat rhizosphere various hydrolyse activites were higher in the rhizosphere than in the bulk soil with an increasing gradient towards the rhizoplane. Tarafdar and Jungk (1987) reported increasing of mineral P, decreasing organic P and progressive P depletion moving from bulk soil to the rhizoplane which was correlated to increasing phosphatase activity.

For example, to our knowledge very few studies on the effects of the root mucilages on the enzyme stabilization and on the substrate diffusion have been conducted. Concerning the importance of enzyme diffusion, an interesting experiment was presented by Pilar et al (2009), showing that resorcinol-immobilized phosphatases increased the organic P fraction, enhanced the seed germination and had a positive effect on plant biomass and plant P content. Donegan and Seidler (1999) reported reduced dehydrogenase and alkaline phosphatase activities in the rhizosphere of transgenic alfalfa regardless of association with recombinant nitrogenfixing soil *Sinorhizobium meliloti*, as compared to rhizosphere soil sampled from parental alfalfa. George et al (2005) reported that phytase released by transgenic *Trifolium subterraneum* L. was inactivated by its interaction with soil colloids, or by proteolysis.

Some information on the contribution of plant roots and microrganisms to the whole rhizosphere enzyme activity is provided by experiments based on transgenic plants, or based on rhizobacteria inocula. Richardson and Brinson (2001) reported that *Arabidopsis thaliana* containing a phytase gene (*phy*A) from *Aspergillus niger* absorbed P from various organic P sources, while George et al. (2005) reported that *T. subterraneum* L expressing a phytase gene (*phy*A) from *Aspergillus niger* released phytase. Lung et al. (2005) reported that *Nicotiana tabacum* carrying a phytase gene from *Bacillus subtilis* (168*phy*A), secreted extracellular phytase and used phytate as the sole P source.

Concerning the contribution of rhizobacteria, information has been gained by studies involving the inoculation of genetically modified microrganisms. Naseby and Lynch (1997, 1998) were among the first to evaluate the effects of genetically modified rhizobatceria on the enzyme activities in the rhizosphere, and reported that inoculation of different *Pseudomonas fluorescens* strains had negative, neutral or positive effects on rhizosphere hydrolase activity. These findings may be attributable to the influence of other microbial groups, the availability of nutrients in the rhizosphere, plant species composition and its nutritional status, and also by the type of genetic modification (Naseby and Lynch 2002). Viterbo et al. (2002) reported the production of chitobiosidase, endochitinase, endo-β-1-3-glucanase and *N*-acetylglucosaminidase by the soil borne fungus *Trichoderma harzianum* Th008, soybean (*Glycine max* cv. Williams 82) roots, and with the exception of endochitinase. Furthermore, they found that *T. Harzianum* Th008 was the source of the endochitinase in the rhizosphere, whilst *N*-acetylglucosaminidase was mainly produced by the soybean roots. Idriss et al. (2002) reported that incoluation of *Bacillus amyloliquefaciens* FZB45 stimulated growth of maize growth under

P limitating conditions in the presence of phytate, whereas a phytase-negative mutant strain (FZB45/M2) did not stimulate plant growth. However, in this case the experimental set-up did not exclude the contribution of plant phosphatases.

Increasing production of phenol oxidase activities in sorghum root exudates of plants exposed to increased phenanthrene concentrations reported by Muratova et al. (2009) indicate an active role of enzymes released by root exudates, thus providing a positive plant response in the rhizosphere degradation of PAHs and derivatives in plants growing in contaminated soils. However, the importance of the production of intermediates in the degradation of aromatic compounds mediated by oxo-reductase enzyme activties illustrates once again the importance of the synergistic contribution of both plant and microbe derived enzymes in the rhizosphere (Sipilä et al. 2008).

8.2 Effect of Root Exudates on Rhizosphere Soil Enzyme Activities

In the rhizosphere, up to 50% of the photosynthates are moved into roots, about 1% is actively released as root exudates but 10% is the total loss as root debris (Uren 2007). Root exudates represent a carbon-rich substrate for the rhizosphere microorganisms. Typical soluble root exudates are organic acids such as citrate, malate, fumarate, oxalate and acetate and carbohydrates such as glucose, xylose, fructose, maltose, sucrose, galactose, and ribose (Lugtenberg and Bloemberg 2004). Roots also release inorganic compound such as CO_2, inorganic ions, protons and anions as a consequence of the root metabolic activity. Soluble organic root exudates are generally readily available to the rhizosphere and rhizoplane microorganisms and may diffuse at a longer distance from the rhizoplane than high-molecular weight compounds,which must be firstly hydrolysed in smaller compounds before they can be taken up by microbial cells (Nannipieri 2007). Root exudates can be a source of easily degradable N-compounds, such as amino acids and small peptides, able to induce protease synthesis (Garcia-Gil et al. 2004).

Because of the high complexity of the rhizosphere environment, the mineralization of the different root exudates and their stimulatory effects on microbial activity have been approached, studying the effects of single low molecular weight organic molecules in simple systems mimicking the rhizosphere environment (Kozdrój and van Elsas 2000; Badalucco and Kuikman 2001; Falchini et al. 2003; Baudoin et al. 2003).

It is well established that N transformations in the rhizosphere soil are related to C dynamics and release of available C from roots (Qian et al. 1997). Gross N mineralization and immobilization rates of the rhizosphere soil are higher than those of the bulk soil, this is due to higher microbial activity in the rhizosphere than in the bulk soil. Landi et al. (2006) studied the effects of various model root exudates on N immobilization rates and reported that glucose was more effective than oxalic acid presumably because the first stimulated a larger proportion of soil micro organisms (Anderson and Domsch 1978), whereas the latter is decomposed by specialized microorganisms (Messini and Favilli 1990). In fact, fewer changes in the bacterial community induced during glucose decomposition compared to oxalic acid in model rhizosphere systems have been reported by Falchini et al. (2003), Baudoin et al. (2003), and Landi et al. (2006). In a follow up study, Renella et al (2007) reported that the effects of root exudates on N mineralization also depended on the soil chemical properties. Concerning the nitrogen dynamics in the rhizosphere, generally higher diversity of functional genes such as amoA and nifH genes have been detected in the rhizosphere than in bulk soil (Briones et al. 2003; Cocking 2003).

Using the model rhizosphere system described by Badalucco and Kuikman (2001), Falchini et al. (2003) showed that selected root exudates were mineralized to different extents and had different stimulatory effects on microbial growth and on hydrolase activities, mostly localized in the rhizosphere zone. These studies have confirmed that root exudation is the main factor controlling microbial activity and community structure in the rhizosphere.

The complex role of flavonoids released by white lupin roots in P mobilization from soil under P-deficiency conditions was studied by Tomasi et al. (2008) who showed that such polyphenolic compounds mobilize insoluble P, but also selectively inhibit the soil phosphohydrolase activity and reduce the microbial activity, possibly preventing the microbial P immobilization in the rhizosphere.

8.3 The Role of Lytic Enzymes and Plant Growth Regulators

Rhizosphere bacteria produce and release lytic enzymes that can hydrolyze a wide variety of polymeric compounds, including chitin, proteins, cellulose, and hemicellulose (Burns 1982; Lugtenberg and Bloemberg 2004; Pal and McSpadden Gardener 2006; Nannipieri 2007). The production of lytic enzymes by rhizosphere microorganisms can result in the suppression of plant pathogenic fungi directly. The rhizosphere bacterial species such *Pseudomonas*, *Bacillus, Enterobacter, Alcaligenes*, and *Pantoea* produced lytic enzymes such lipase, protease, collagenase, amylase, pectinase, cellulase, protease, lecithinase and were active against soil fungi (Lugtenberg et al. 2001; Egamberdiyeva and Hoflich 2003a, b, 2004; Egamberdieva et al. 2008). *Pseudomonas* is known to produce lytic enzymes and some isolates have been shown to be effective at suppressing fungal plant pathogens in tomatoes (Lugtenberg et al. 2001). Dunne et al. (2000) showed that overproduction of extracellular protease in the mutant strains of *Stenotrophomonas maltophilia* W81 resulted in improved biocontrol of *Pythium ultimum*. Nielsen and Sorensen (1997) demonstrated that isolates of *P. fluorescens* antagonistic to *R. solani* and *P. ultimum*, produced lytic enzymes.

Plant growth promoting regulators such as auxin, or gibberellins are present in root exudates and thus enter the rhizosphere. Auxins are a class of plant hormones: the most common and well characterised is indol-3-acetic acid (IAA), which is known to stimulate both rapid (e. g., increases in cell elongation) and long term (e. g., cell division and differentiation) responses in plants (Cleland 1990; Hagen 1990). The auxin level was usually higher in the rhizosphere than in the free bulk soil, probably as a consequence of an increased microbial population or of accelerated metabolism owing to the presence of root exudates (Frankenberger and Arshad 1995; Somers et al. 2004).

Microorganisms in the rhizosphere of various crops appear to have a greater potential to synthesize and release auxins as secondary metabolits because of the rich supply of substances and it is an important factor in soil fertility (Patten and Glick 2002; Frankenberger and Arshad 1995). According to several reports 86% of the bacterial isolates from the rhizosphere of various plants produced phytohormones such auxins, gibberelins, kinetin-like substances, but also different hydrolytic enzymes such protease, lipase, pectinase, and amylase (Hagen 1990; Frankenberger and Arshad 1995; Nielsen and Sorensen 1999; Lugtenberg et al. 2001). Lebuhn and Hartmann (1993) suggested the presence of higher auxin content in rhiosphere soils because of root colonization with *Azospirilla* and *Rhizobium* capable of excreting auxin without the addition of tryptophan (Patten and Glick 2002). Plant growth promoting rhizobacteria (PGPR) can affect plant growth directly by the synthesis of phytohormones, nitrogen fixation for plant use, improvement of nutrient uptake, solubilization of inorganic phosphate, and mineralization of organic phosphate (Dobbelaere et al. 2003). These physiological changes are linked to increases in enzyme activity.

The exploitation of enzymes in the rhizosphere, weather or not extracellular, that can rapidly convert precursors to biologically active molecules, holds potential for increasing crop production (Zahir et al. 2001). In our study we observed that indole-3-acetic acid (IAA) and gibberellic acid (GA) effect positively on alkaline phosphomonoestrase, phosphodiestrase, protease and urease activities in the model root surface (MRS) of a simplified rhizosphere system.

Acid and alkaline phosphatase activities in wheat rhizosphere were strongly correlated with the depletion of organic P (Tarafdar and Jungk 1987). The plant growth regulators did not have an effect on acidic phosphomonoesterase activity in the rhizosphere (MRS) (Fig. 8.1).

Protease activity is involved in the hydrolysis of N compounds to NH_4, using low-molecular-weight protein substrates and microorganisms is responsible for breaking down urea into ammonium (Tabatabai 1982). Urease enzyme is responsible for the hydrolysis of urea fertiliser applied to the soil into NH_3 and CO_2 with the concomitant rise in soil pH (Byrnes and Amberger 1989). Other authors reported that urease released by plant root and rhizosphere micro organisms as intra- and extra-cellular enzymes (Burns 1986; Mobley and Hausinger 1989). Their activities in rhizosphere soil were strongly stimulated by IAA and GA (Fig 8.2).

In early studies, Doyle and Stotzky (1993) found no difference in enzyme activities (arylsulfatase, phosphatases, and dehydrogenase) when an *Escherichia coli* strain was introduced into soil. Mawdsley and Burns (1994) reported that inoculating wheat seedling with *Flavobacterium* spp. increased activity of

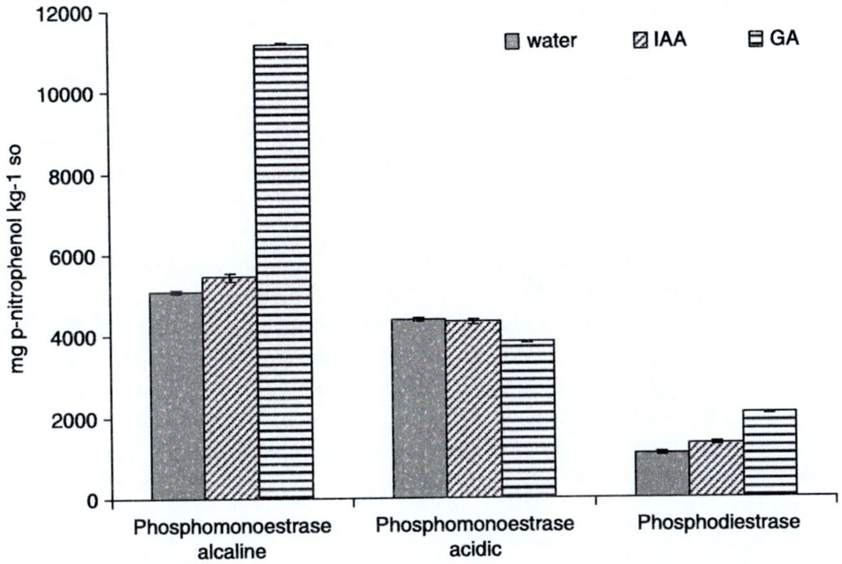

Fig. 8.1 The effect of indol-3-acetic acid (IAA), and gibberellic acid (GA) on phosphomonoestrases and phosphodiesterase activities in rhizosphere soil (mg *p*-nitrophenol kg^{-1} soil h^{-1})

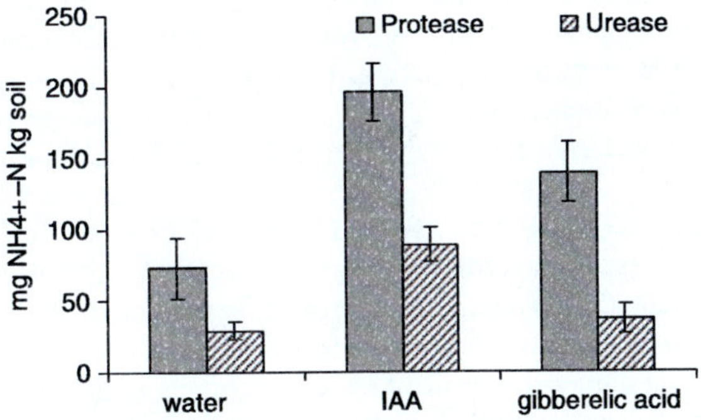

Fig. 8.2 The effect of indol-3-acetic acid (IAA), and gibberellic acid (GA) on protease and urease activities in rhizosphere soil (mg NH_4^+-N g^{-1} soil h^{-1})

OL-galactosidase, I-galactosidase, a-glucosidase, and I-glucosidase. This strain is able to produce plant growth stimulating substances such as auxin. Naseby and Lynch (1997) observed that bacterial inoculant *P. fluorescens* increased urease activity and decreased alkaline phosphomonoesterase and phosphodiesterase activity in the rhizosphere soil of wheat. This decrease explained by Naseby and

Lynch (1997) that the inoculant inhabited niches of other microbes in the community would respond in a more dynamic fashion to the change in conditions. The first possibility is that the inoculant is competitively excluding certain microbial populations (De Leij et al. 1993). Other possible effects of the inoculant could be metabolic activity of the strain, that might have directly affected the indigenous microbial community (Keel et al. 1992) or modified root secretions (Mozafar et al. 1992).

In earlier studies of plant growth regulators, the activities of rhizosphere bacteria including nitrogen fixation (Lindberg et al. 1985), production of cytokinin, auxin (Lebuhn et al. 1997; Timmusk et al. 1999), or hydrolytic enzymes such protease, lipase, pectinase, amylase (Nielsen and Sorensen, 1997) increased the N, P, and K uptake of plant components (Hoflich et al. 1997; Egamberdiyeva and Hoflich 2004). Tarafdar and Jungk (1987) observed that total P and organic P contents decreased in the rhizosphere soil of wheat, whereas the inorganic P content increased in the vicinity of the rhizoplane. Such an increase was correlated with the increased acid and alkaline phosphatase activities and fungal and bacterial population in the rhizosphere soil.

8.4 Effect of Soil Management and Heavy Metals on Rhizosphere Soil Enzyme Activities

Soil enzyme activities are considered to hold the potential to discriminate between soil management treatments, probably because they are closely related to microbial biomass, which is frequently reported to be sensitive indicator to such treatments (Dick 1997). Measurement of soil hydrolases provides an early indication of changes in soil fertility, since they are related to the mineralization of such important nutrients elements as N, P, C (Ceccanti and Garcia 1994). The crop management practices, and crop varieties affect the amount of acid phosphatase exuded by plant roots (Izaguirre-Mayoral et al. 2002; Wright and Reddy 2001). Yadav and Tarafdar (2001) reported that legumes secrete more phosphatase activity than cereal, and chickpea roots secrete more than maize (Li et al. 2004), which could be due to a higher requirement of P by legumes in the symbiotic nitrogen fixation process as compared to cereals. Other studies (Kai et al. 2002; Li et al. 2002) reported that P deficiency in the soil increased acid phosphotase secretion from plant root, which enhance the solubilisation and remobilisation of phosphate. It influences positively the ability of the plant to cope with P-stressed conditions. Urease enzyme also affected by soil type, temperature, organic matter content, soil management practices, crops and heavy metals (Yang et al. 2006).

Gianfreda and Ruggiero (2006) reported that the enzyme activities in soil are affected by the presence and nature of the plant cover. Soils with a long history of continuous corn monoculture, without proper amendment with organic matter, showed low organic matter contents and low dehydrogenase, invertase,

Table 8.1 Acid, alcalic phosphomonoestrases, phosphodiestrase, galactosidase, glucosidase, urease, protease and FDA activities in non saline (NS), mid saline (MS) and strong saline (SS) soils

Treatments	(mg *p*-nitrophenol kg^{-1} soil × h^{-1})					(mg NH_4^+–N kg^{-1} soil × h^{-1})		
	Acid phosphatase	Alkaline phosphatase	Phosphodiestrase	Galactosidase	Glucosidase	Urease	Protease	FDA
EC								
NS (1.3 dS m^{-1})	821.3	2,811.5	914.8	442.7	256.9	7.9	23.9	23.9
MS (5.6 dS m^{-1})	456.1	2,367.8	761.2	482.0	200.1	6.5	22.2	12.0
SS (7.1 dS m^{-1})	952.2	1,884.0	796.2	491.6	127.6	5.8	17.0	6.1

arylsulphatase, and β-glucosidase as compared with continuous corn-fertilized soils. A long-term intensive monoculture usually supplies lower amounts and diversity of organic matter than crop rotation, thus suppressing microbial activities ad consequently decreasing enzymatic ones (Klose and Tabatabai 2000). Soil enzyme activities are inhibited in trace element contaminated soils (Tyler et al. 1989) mainly due to direct interactions between trace elements with enzyme molecules, or substrates of enzyme–substrate complexes.

In the rhizosphere the effects of trace elements on the enzyme activity are even more complex than in bulk soil due to the larger concentrations of trace element ligands in rhizodeposits, acidification and root trace element uptake. In spite of the relevance of the topic, there are relatively few studies on the hydrolase activity in the rhizospehere of trace element contaminated soils. Renella et al. (2005) reported a reduced hydrolase actvity in reposnse to the release of model root exudates by a model rhizosphere system in Cd contaminated soils. Possible causes of lower enzyme production in the rhizosphere of trace element contaminated soils could be both microbial metabolic stress (Renella et al. 2006) and lower mineralization of low molecular weight organic acids complexed with trace elements by soil microbial communities. To our knowledge, no information is available on the changes in the release of enzymes by plants exposed to trace element contaminated soils. More studies are surely needed to better understad the effects of trace element pollution on the hydrolase activity in the rhizosphere.

8.5 Impact of Salinity on Rhizosphere Enzyme Activities

Salinity is a major concern for irrigated agriculture in arid and semi-arid regions of the world (Shirokova et al. 2000; Egamberdiyeva et al. 2007). Eivazi and Tabatabai (1988), García et al. (1994), and Batra and Manna (1997) showed that the activities of different soil enzymes were seriously reduced in saline soils affecting the capacity of the soil to recycle nutrients and to release them for their use by plants. Frankenberger and Bingham (1982) also reported that soil enzymes such dehydrogenase, phosphatase, sulfatase, amylase, and β-glucosidase activity was severely inhibited in salinized soils and their variation in soils seemed to be related to the physico-chemical microbial properties of soils (Nannipieri et al. 1990; Zahir et al. 2001).

Among the variables assessed, enzymatic activities (ureases, proteases, phosphatases, and glucosidases) may be sensitive indicators to detect changes occurring in soils under field conditions. The decline in enzymes activity with increasing salinity appeared to be associated with change in osmotic potential of the soil due to higher salt concentrations and specific ion toxicity (Zahir et al. 2001). Matsuguchi and Sakai (1995) investigated the effect of soil salinity with intensive cultivation on microbial communities in the soil-root system, and they found a negative effect of salt stress on microbial populations. Frankenberger and Bingham (1982) have pointed out the inhibition of different enzymatic activities caused by soil salinity. Dash and Panda (2001) established that NaCl salt stress induced changes in the growth and enzyme activities in blackgram (*Phaseolus mungo*) seeds, this negative effect was even greater on hydrolases (phosphatase and β-glucosidase).

In spite of the large information on the effects of soil secondary salinization on soil enzyme activity, very few studies have focussed on the effects of salinization on the enzyme activity in the rhizosphere. In Table 8.1 we report rhizosphere soil enzyme activities in cotton grown on various levels of saline soil. We observed that salinity inhibited urease, protease, alkalic phosphomonoestrases and phosphodiestrase activity of rhizosphere soil. Non-saline soil showed highest alkaline phosphomonoestrase, phosphodiestrase, glucosidase, protease, urease, and fluorescein diacetate (FDA) (Table 8.1). However, alkaline phosphatase was higher in all soils than other hydrolase enzymes. Generally, alkaline phosphatase predominated in soils with neutral or slightly alkaline pH (Tripathi et al. 2007). Since, higher plants are devoid of alkaline phosphatase, the alkaline phosphatase of soils seems to be derived totally from microorganisms (Dick and Tabatabai 1983; Juma and Tabatabai 1988). Gianfreda and Bollag (1994) observed a direct inhibitory effect of soil organic constituents on the activities of acid phosphatase.

Acid phosphomonoestrase, phosphodiestrase, and galactosidase were not affected by an increase of soil salinity, and did not correlate with soil electrical conductivity, organic matter content, Cl, and Na (Table 8.1). Other enzymes such alkaline phosphomonoestrases, proteases, glucosidases, ureases, and fluorescein diacetate hydrolase were inhibited in saline soils, in relation to the soil organic C and microbial biomass (Table 8.1). Carrasco et al. (2006) found that the changes in microbiological activity along the spatial vegetation gradient were also revealed by the variations in protease-BAA, phosphatase, urease and b-glucosidase activities.

Urease and protease activities appear to be more sensitive to salinity than phosphatases. García et al. (1994) reported that the reduction in enzyme activities can be due to lower microbial biomass that releasing less enzymes but also by the fact that in semi-arid soils the enzyme activity is mainly extracellular, stable and form complexes with the organic and mineral colloids. In saline soils, salt tolerant bacteria produce enzymes, whose activity has a greater salt requirement than that of corresponding enzymes produced by non- salt tolerant bacteria (Zahran 1997).

This decrease in enzymatic activity with soil salinity can be explained by the fact that in the semi arid soils many of the enzymes are extracellular, stable and form complexes with the organic and mineral colloids (García et al. 1994; Garcia and Hernandez 1996). The increase of conductivity disperses the clays and the stable enzymes remain unprotected and therefore, more susceptible to denaturation (Frankenberger and Bingham 1982; Pathak and Rao 1998). Garcia and Hernandez (1996) in their work reported that salinity negatively affects biological and biochemical fertility of the soils and is more pronounced with NaCl than Na_2SO_4 which can be attributed to the toxic effect of a particular ion in saline soils on microbial growth. Probably toxicity of chlorides was greater than the toxicity by sulphates with those enzymes and microbial biomass being the most sensitive. In addition, high salt concentrations tend to reduce the solubility and denature enzyme proteins through disruption of the tertiary protein structure which is essential for enzymatic activity (Frankenberger and Bingham 1982;

Zahran 1997). Reboreda and Caçador (2008) reported that extracellular enzyme activity in salt marshes sediments colonised by *Spartina maritima* were positively correlated with root biomass indicating the role of halophyte roots on nutrient cycling and microbial functioning sediments.

8.6 Conclusions

Research on the enzyme activity in the rhizosphere suffer major methodological limitations due to the complexity of the rhizosphere in terms of sampling, and due to the susceptibility of the rhizosphere to artifacts during rhizosphere sampling. For such reasons, enzyme activity in the rhizosphere has been studied by means of artificial systems mimicking the release of selected root exudates under controlled conditions. Although this approach is useful to describe specific effects, we all are aware that they are very far from the real plant soil microbe situation.

A possible approach is the direct visualization of the rhizosphere environment. The use of soil thin sections, treated for ultracytochemical localization of enzymes in rhizobacteria and root extracellular polysaccharides and cell wall by transmission electron microscopy were proposed by R.C. Foster (see Ladd et al. 1996), but these approaches were not sufficiently developed and can not be applied for enzymes associated to electron-dense particles and minerals.

Concerning the relationship between microbial diversity and enzyme activity in the rhizosphere, in spite of the increasing knowledge on the microbial community composition, no comparisons with the enzyme activity are normally carried out. This kind of comparitve work may be of great interest from theoretical and practical aspects. For example, it may clarify whether microbial inocula are capable of increasing specific enzyme activities for various purposes, such as reducing the crop fertilization, or enhancing the phytodegradation of organic pollutants.

Finally, a transition from studies determining only the enzyme activity to studies aiming at identifyng and quantifying the enzyme molecules by the proteomic approach is in our opinion, a better approach for understanding of the source and relevance of hydrolase activity in the rhizosphere. For example, Wen et al. (2007) showed that enzymes released in the pea root secretome may play a role in whole protection mechanisms in the presence of pathogen fungi. Identification of these enzymes may be useful to improve the crop protection strategy. De la Peña et al. (2008) studied the role of proteins secreted by secreted roots of *Medicago sativa* and *Arabidopsis thaliana* in the presence of symbiotic (*Sinorhizobium meliloti*) and pathogenic (*Pseudomonas syringae*) bacteria, and found that the interaction between the two plants increased the secretion of various proteins, prevalently hydrolases, peptidases, as compared to control plants. Implementation of rhizosphere proteomics for disentangling the complexity of the enzyme secretion by plants and microrganisms in the plant-microbial chemical communication in the rhizosphere may likely improve our knowledge on the enzyme activity in the rhizosphere more than other technical improvements. It is considered a key issue for elucidating the importance and the location of the enzyme activity in the rhizosphere in the next few years from both applicative and basic research viewpoints.

References

Anderson JPE, Domsch KH (1978) A physiological method for the quantitative measurement of microbial biomass in soils. Soil Biol Biochem 10:215–221

Badalucco L, Grego S, Dell'Orco S, Nannipieri P (1996) Effect of liming on some chemical, biochemical, and microbiological properties of acid soils under spruce (*Picea abies* L.). Biol Fertil Soils 14:76–83

Badalucco L, Kuikman PJ (2001) Mineralisation and immobilisation in the rhizosphere. In: Pinton A, Varanini Z, Nannipieri P (eds) The rhizosphere. Biochemistry and organic substances at the soil–plant interface. Marcel Dekker, New York, pp 159–196

Batra L, Manna MC (1997) Dehydrogenase activity and microbial biomass carbon in salt affected soils of semi arid regions. Arid Soil Res Rehabil 3:293–303

Baudoin E, Benizri E, Guckert A (2003) Impact of artificial root exudates on the bacterial community structure in bulk soil and maize rhizosphere. Soil Biol Biochem 35:1183–1192

Briones AM, Okabe S, Umemiya Y, Ramsing NB, Reichardt W, Okuyama H (2003) Ammonia-oxidizing bacteria on root biofilms and their possible contribution to N use efficiency of different rice cultivars. Plant Soil 250:335–348

Burns RG (1982) Enzyme activity in soil: location and a possible role in microbial ecology. Soil Biol Biochem 14:423–427

Burns RG (1986) Interaction of enzymes with soil mineral and organic colloids. In: Huang PM, Schnitzer M (eds) Interactions of soil minerals with natural organics and microbes. Soil Science Society of America, Madison, WI, pp 429–452

Byrnes BH, Amberger A (1989) Fate of broadcast urea in a flooded soil when treated with N-(n-butyl) thiophospheric triamide, a urease inhibitor. Fertil Res 18:221–231

Carrasco L, Caravaca F, Alvarez-Rogel J, Roldán A (2006) Microbial processes in the rhizosphere soil of a heavy metals-contaminated Mediterranean salt marsh: a facilitating role of AM fungi. Chemosphere 64:104–111

Ceccanti B, Garcia C (1994) Coupled chemical and biochemical methodologies to characterize a composting process and the humic substances. In: Senesi N, Miano T (eds) Humic Substances in the global environment and its implication on human health. Elsevier, New York, NY, pp 1279–1285

Chanway CP (2002) Plant growth promotion by *Bacillus* and relatives. In: Berkeley R, Heyndrickx M, Logan N, De Vos P (eds) *B subtilis* for biocontrol in variety of plants. Blackwell, Malden, MA, pp 219–235

Cleland RE (1990) Auxin and cell elongation. In: Davies PJ (ed) Plant hormones and their role in plant growth and development. Kluwer Academic, Dordrecht, The Netherlands, pp 132–148

Cocking EC (2003) Endophytic colonisation of plant roots by nitrogen-fixing bacteria. Plant Soil 252:169–175

Colombo C, Palumbo G, Sannino F, Gianfreda L (2002) Chemical and biochemical indicators of managed agricultural soils. In: 17th World congress of soil science, Bangkok. Thailand, 1740-2, pp 1–9

De-la-Peña C, Lei Z, Watson BS, Sumner LW, Vivanco JM (2008) Root-microbe communication through protein secretion. Journal of Biological Chemistry 283:25247–25255

De Leij F, Whips JM, Lynch HJ (1993) The use of colony development for the characterization of bacterial communities in soil and on roots. Microb Ecol 27:81–97

Dash M, Panda SK (2001) Salt stress induced changes in growth and enzyme activities in germinating *Phaseolus muingo* seeds. Biol Plant 44:587–589

Dick WA, Tabatabai MA (1983) Activation of soil pyrophos-phatase by metal ions. Soil Biol Biochem 15:359–363

Dick WA, Tabatabai MA (1984) Kinetic parameters of phosphatase in soils and organic waste materials. Soil Sci 137:7–15

Dick RP, Breakwell DP, Turco RF (1996) Soil enzyme activities and biodiversity measurements as integrative microbiological indicators. In: Doran JW, Jones AJ (eds) Methods for assessing soil quality. Special Publication No 49. Soil Science Society America, Madison, WI, USA, pp 247–271

Dick RP (1997) Soil enzyme activities as integrative indicators of soil health. In: Pankhurst CE, Doube BM, Gupta VVSR (eds) Biological indicators of soil health. CAB International, Wellingford, pp 121–156

Donegan KK, Seidler RJ (1999) Effects of transgenic plants on soil and plant microorganisms. Recent Res Devel Microbiol 3:415–424

Drijber RA, Doran JW, Parkhurst AM, Lyon DJ (2000) Changes in soil microbial community structure with tillage under long-term wheat–fallow management. Soil Biol Biochem 32:1419–1430

Dunne C, Monne-Loccoz Y, de Bruijn FJ, O'Gara F (2000) Overproduction of an inducible extracellular serine protease improves biological control of *Pythium ultimum* by *Stenotrophomonas maltophilia* strain W81. Microbiology 146:2069–2078

Dobbelaere S, Vanderleyden J, Okon Y (2003) Plant growth-promoting effects of diazotrophs in the rhizosphere. Crit Rev Plant Sci 22:107–149

Doyle JD, Stotzky G (1993) Methods for the detection of changes in the microbial ecology of soil caused by the introduction of microorganisms. Microb Releases 2:63–72

Egamberdiyeva D, Hoflich G (2003a) Influence of growth-promoting bacteria on the growth of wheat in different soils and temperatures. Soil Biol Biochem 35:973–978

Egamberdieva D, Kamilova F, Validov SH, Gafurova L, Kucharova Z, Lugtenberg B (2008) High incidence of plant growth-stimulating bacteria associated with the rhizosphere of wheat grown on salinated soil in Uzbekistan. Environ Microbiol 10:1–9

Egamberdiyeva D, Gafurova L, Islam KR (2007) Salinity effects on irrigated soil chemical and biological properties in the Syr Darya Basin of Uzbekistan. In: Lal R, Sulaimanov M, Stewart B, Hansen D, Doraiswamy P (eds) Climate change and terrestrial C sequestration in Central Asia Taylor-Francis, New York, pp 147–162

Egamberdiyeva D, Hoflich G (2003b) The effect of associative bacteria from different climates on plant growth of pea at different soils and temperatures. Arch Agron Soil Sci 49:203–213

Egamberdiyeva D, Hoflich G (2004) Importance of plant growth promoting bacteria on growth and nutrient uptake of cotton and pea in semi-arid region Uzbekistan. J Arid Environ 56:293–30

Eivazi F, Tabatabai MA (1988) Glucosidases and galactosidases in soils. Soil Biol Biochem 20:601–606

Estermann EF, McLaren AD (1961) Contribution of rhizoplane organisms to the total capacity of plants to utilize organic nutrients. Plant Soil 15:243–260

Falchini L, Naumova N, Kuikman PJ, Bloem J, Nannipieri P (2003) CO_2 evolution and denaturing gradient gel electrophoresis profiles of bacterial communities in soil following addition of low molecular weight substrates to simulate root exudation. Soil Biol Biochem 36:775–782

Frankenberger JWT, Arshad M (1995) Microbial synthesis of auxins. In: Frankenberger WT, Arshad M (eds) Phytohormones in soils. Marcel Dekker, New York, pp 35–71

Frankenberger WTJ, Bingham FT (1982) Influence of salinity on soil enzyme activities. Soil Sci Soc Am J 46:1173–1177

Garcia-Gil JC, Plaza C, Senesi N, Brunetti G (2004) Effects of sewage sludge amendment on humic acids and microbiological properties of a semiarid Mediterranean soil. Biol Fertil Soils 39:320–328

García C, Hernández T, Costa F, Ceccanti B (1994) Biochemical parameters in soils regenerated by addition of organic wastes. Waste Manage Res 12:457–466

Garcia C, Hernandez T (1996) Influence of salinity on the biological and biochemical activity of a calciorthird soil. Plant Soil 178:225–263

George TS, Simpson RJ, Hadobas PA, Richardson AE (2005) Expression of a fungal phytase gene in *Nicotiana tabacum* improves phosphorus nutrition of plants grown in amended soils. Plant Biotechnol J 3:129–140

Gianfreda L, Ruggiero P (2006) Enzyme activities in soil. In: Nannipieri P, Smalla K (eds) Nucleic acids and proteins in soil. Springer, Heidelberg, Germany, pp 257–311

Gianfreda L, Bollag J (1994) Effect of soils on behaviour of immobilized enzymes. Soil Sci Soc Am J 58:1672–1681

Gupta VSR, Roper MM, Kinkegaard JA, Angus JF (1994) Changes in microbial biomass and organic matter levels during the first year of modified tillage and stubble management practices on red earth. Aust J Soil Res 32:1339–1354

Hagen G (1990) The control of gene expression by auxin. In: Davies PJ (ed) Plant hormones and their role in plant growth and development. Kluwer Academic, The Netherlands, pp 149–163

Hawes MC, Bengough G, Cassab G, Ponce G (2003) Root caps and rhizosphere. J Plant Growth Regul 21:352–367

Hinsinger P, Plassard C, Jaillard B (2006) The rhizosphere: a new frontier in soil biogeochemistry. J Geochem Explor 88:210–213

Hoflich G, Tappe E, Kuhn G, Wiehe W (1997) Einfluß associativer rhizospharenbakterien auf die nährstoffaufnahme und den ertrag von mais. Arch Acker Pfl Boden 41:323–333

Idriss EE, Makarewicz O, Farouk A, Rosner K, Greiner R, Bochow H (2002) Extracellular phytase activity of *Bacillus amyloliquefaciens* FZB45 contributes to its plant-growth-promoting effect. Microbiology 148:2097–2109

Izaguirre-Mayoral ML, Flores S, Carballo O (2002) Determination of acid phosphatases and dehydrogenase activities in the rhizosphere of nodulated legume species native to two contrasting savannah sites in Venezuela. Biol Fertil Soils 35:470–472

James ES, Russel LW, Mitrick A (1991) Phosphate stress response in hydroponically grown maize. Plant Soil 132:85–90

Jones DL, Hodge A, Kuzyakov Y (2004) Plant and mycorrhizal regulation of rhizodeposition. New Phytol 163:459–480

Juma NG, Tabatabai MA (1988) Hydrolysis of organic phosphates by corn and soybean roots. Plant Soil 107:31–38

Kai M, Takazumi K, Adachi H, Wasaki J, Shinano T, Osaki M (2002) Cloning and characterization of four phosphate transporter cDNAs in tobacco. Plant Sci 163:837–846

Keel C, Schnider U, Maurhofer M, Voisard C, Laville J, Burger U, Wirthner P, Hass D, Defago G (1992) Suppression of root diseases by *Pseudomonas fluorescens* CHAO: importance of the bacterial metabolite 2,4-diacetylphloroglucinol. Mol Plant-Microbe Interact 5:4–13

Klose S, Tabatabai MA (2000) Urease activity of microbial biomass in soil as affected by cropping system. Biol Fertil Soils 31:191–9

Kozdrój J, van Elsas JD (2000) Response of the bacterial community to root exudates in soil polluted with heavy metals assessed by molecular and cultural approaches. Soil Biol Biochem 32:1405–1417

Ladd JN, Foster RC, Nannipieri P, Oades JM (1996) Soil structure and biological activity. In: Stotzky G, Bollag JM (eds), Soil Biochemistry vol 9. Marcel Dekker, New York, pp 23–78

Landi L, Valore F, Asher J, Renella G, Falchini L, Nannipieri P (2006) Root exudates effects on the bacterial communities, CO_2 evolution, nitrogen transformations and ATP content of rhizosphere and bulk soils. Soil Biol Biochem 38:509–516

Lebuhn M, Hartmann A (1993) Method for the determination of indole-3-acetic acid and related compounds of L-tryptophan catabolism in soils. J Chromatogr 629:255–266

Lebuhn M, Heulin T, Hartmann A (1997) Production of auxin and other indolic and phenolic compounds by *Paenobacillus polymyxa* strains isolated from different proximity to plant roots. FEMS Microbiol Ecol 22:325–334

Li Y, Guohua M, Fanjun C, Jianhua Z, Fusuo Z (2004) Rhizosphere effect and root growth of two maize (*Zea mays* L.) genotypes with contrasting P efficiency at low P availability. Plant Sci 167:217–223

Li D, Zhu H, Liu K, Liu X, Leggewie G, Udvardi M, Wang D (2002) Purple acid phosphatases of *Arabidopsis thaliana*.. J Biol Chem 277:27772–27781

Lindberg T, Granhall U, Tomenius H (1985) Infectivity and acetylene reduction of diazotrophic rhizosphere bacteria in wheat (*Triticum aestivum*) seedlings under gnotobiotic conditions. Biol Fertil Soils 1:123–129

Lung SC, Chan WL, Yip W, Wang L, Yeung EC, Lim BL (2005) Secretion of beta-propeller phytase from tobacco and *Arabidopsis* roots enhances phosphorus utilization. Plant Sci 169:341–349

Lugtenberg BJJ, Dekkers L, Bloemberg GV (2001) Molecular determinants of rhizosphere colonization by Pseudomonas. Annu Rev Phytopathol 39:461–490

Lugtenberg BJJ, Bloemberg GV (2004) Life in the rhizosphere. In: Ramos JL (ed) Pseudomonas, vol 1. Kluwer Academic/Plenum Publishers, New York, NY, pp 403–430

Matsuguchi T, Sakai M (1995) Influence of soil salinity on the populations and composition of fluorescent pseudomonads in plant rhizosphere. Soil Sci Plant Nutr 41:497–504

Mawdsley JL, Burns RG (1994) Inoculation of plants with *Flavobacterium* P25 results in altered rhizosphere enzyme activities. Soil Biol Biochem 26:871–882

Messini A, Favilli F (1990) Calcium oxalate decomposing microorganisms a microbial group of the rhizosphere of forest plants. Ann Microbiol Enzimol 40:93–102

Muratova A, Pozdnyakova N, Golubev S, Wittenmayer L, Makarov O, Merbach W, Turkovskaya O (2009) Oxidoreductase activity of Sorghum root exudates in a phenanthrene-contaminated environment. Chemosphere 74:1031–1036

Mobley HL, Hausinger RP (1989) Microbial ureases: significance, regulation, and molecular characterization. Microbiol Rev 53:85–108

Mozafar A, Duss F, Oertli JJ (1992) Effect of *Pseudomonas fluorescens* on the root exudates of 2 tomato mutants differently sensitive to Fe chlorosis. Plant Soil 144:167–176

Nannipieri P, Ceccanti B, Grego S (1990) Ecological significance of the biological activity in soil. In: Bollag JM, Stotzky G (eds) Soil biochemistry, vol 6. Martin Dekker, New York, pp 293–366

Nannipieri P, Sequi P, Fusi P (1996) Humus and enzyme activity. In: Piccolo A (ed) Humic substances in terrestrial ecosystems. Elsevier, New York, pp 293–328

Nannipieri P, Kandeler E, Ruggiero P (2002) Enzyme activities and microbiological and biochemical processes in soil. In: Burns RG, Dick RP (eds) Enzymes in the environment. Marcel Dekker, New York, USA, pp 1–33

Nannipieri P, Ascher J, Ceccherini MT, Landi L, Pietramellara G, Renella G (2003) Microbial diversity and soil functions. Eur J Soil Sci 54:655–670

Nardi S, Concheri G, Pizzeghello D, Sturaro A, Rella R, Parvoli G (2000) Soil organic matter mobilization by root exudates. Chemosphere 5:653–658

Nannipieri P (2007) Functions of microbial communities and their importance in soil. CAB reviews: perspective in agriculture, veterinary science, nutrition and natural resources. doi:10.1079/PAVSNNR20072050

Naseby DC, Lynch JM (2002) Enzymes and microorganisms in the rhizosphere. In: Burns RG, Dick RP (eds) Enzymes in the environment, activity, ecology and applications. Marcel Dekker, New York, pp 109–123

Naseby DC, Lynch JM (1997) Rhizosphere soil enzymes as indicators of perturbations caused by enzyme substrate addition and inoculation of a genetically modified strain of *Pseudomonas fluorescens* on wheat seed. Soil Biol Biochem 29:1353–1362

Naseby DC, Lynch JM (1998) Impact of wild-type and genetically modified *Pseudomonas fluorescens* on soil enzyme activities and microbial population structure in the rhizosphere of pea. Mol Ecol 7:617–625

Neumann G, Römheld V (2007) The release of root exudates as affected by the plant physiological status. In: Pinton R, Varanini Z, Nannipieri P (eds) The rhizosphere. biochemistry and organic substances at the soil-plant interface. Taylor and Francis Group, Boca Raton, FL, pp 23–72

Nielsen P, Sorensen J (1997) Multi-target and medium-independent fungal antagonism by hydrolytic enzymes in *Paenibacillus polymyxa* and *Bacillus pumillus* strains from barley rhizosphere. FEMS Microbiol Ecol 22:183–192

Nielson MN, Sorensen J (1999) Chitinolytic activity of Pseudomonas fluorescens isolates from barley and sugar beet rhizosphere. FEMS Microbiol Ecol 30:217–227

Pal KK, McSpadden Gardener B (2006) Biological control of plant pathogens. The Plant Health Instructor. doi:10.1094/PHI-A-2006-1117-02

Pathak H, Rao DLN (1998) Carbon and nitrogen mineralization from added organic matter in saline and alkali soils. Soil Biol Biochem 30:695–702

Patten CL, Glick BR (2002) Role of *Pseudomonas putida* indole acetic acid in development of the host plant root system. Appl Environ Microbiol 68:3795–3801

Pilar MC, Ortega N, Perez-Mateos M, Busto MD (2009) Alkaline phosphatase–polyresorcinol complex: characterization and application to seed coating. J Agric Food Chem 57:1967–1974

Qian JH, Doran JW, Walters DT (1997) Maize plant contributions to root zone available carbon and microbial transformations of nitrogen. Soil Biol Biochem 29:1451–1462

Raaijmakers JM, Paulitz TC, Steinberg C, Alabouvette C, Moënne-Loccoz Y (2009) The rhizosphere: a playground and battlefield for soilborne pathogens and beneficial microorganisms. Plant Soil 321:341–361

Reboreda R, Caçador I (2008) Enzymatic activity in the rhizosphere of *Spartina maritima*: potential contribution for phytoremediation of metals. Mar Environ Res 65:77–84

Renella G, Egamberdiyeva D, Landi L, Mench M, Nannipieri P (2006) Microbial activity and hydrolase activities during decomposition of root exudates released by an artificial root surface in Cd-contaminated soils. Soil Biol Biochem 38:702–708

Renella G, Landi L, Valori F, Nannipieri P (2007) Microbial and hydrolase activity after release of low molecular weight organic compounds by a model root surface in a clayey and a sandy soil. Appl Soil Ecol 36:124–129

Renella G, Mench M, Landi L, Nannipieri P (2005) Microbial activity and hydrolase synthesis in long-term Cd-contaminated soils. Soil Biol Biochem 37:133–139

Richardson JL, Brinson MM (2001) Wetlandsoils and the hydrogeomorphic classification of wetlands. In: Richardson JL, Vepraskas MJ (eds) Wetlands soils: genesis, hydrology, landscapes, and classification. CRC Press LLC, Boca Raton, FL, pp 209–227

Shirokova Y, Forkutsa I, Sharafutdinova N (2000) Use of electrical conductivity instead of soluble slats for soil salinity monitoring in Central Asia. Irrig Drain Syst 14:199–205

Sipilä TP, Keskinen AK, Akerman ML, Fortelius C, Haahtela K, Yrjälä K (2008) High aromatic ring-cleavage diversity in birch rhizosphere: PAH treatment-specific changes of I.E.3 group extradiol dioxygenases and 16S rRNA bacterial communities in soil. ISME J 9:968–981

Somers E, Vanderleyden J, Srinivasan M (2004) Rhizosphere bacterial signalling: a love parade beneath our feet. Crit Rev Microbiol 30:205–23

Tabatabai MA (1982) Soil enzymes. In: Page AL, Miller RH, Keeney DR (eds) Methods of soil analysis. Part 2. American Society of Agronomy, Soil Science Society of America, Madison, WI, pp 903–947

Tarafdar JC, Jungk A (1987) Phosphatase activity in the rhizosphere and its relation to the depletion of soil organic phosphorus. Biol Fertil Soils 3:199–204

Tyler G, Balsberg Pahlsson AM, Bengtsson G, Baath E, Tranvik M (1989) Heavy-metal ecology of terrestrial plants, micro-organisms invertebrates. Water Air Soil Pollut 47:189–215

Timmusk E, Niccander B, Granhall U, Tillberg E (1999) Cytokinin production by *Bacillus polymyxa*.. Soil Biol Biochem 31:1847–1852

Tomasi N, Weisskopf L, Renella G, Landi L, Pinton R, Varanini Z, Nannipieri P, Torrent J, Martinoia E, Cesco S (2008) Flavonoids of white lupin roots participate in phosphorus mobilization from soil. Soil Biol Biochem 40:1971–1974

Tripathi RD, Srivastava S, Mishra S, Singh N, Tuli R, Gupta DK, Maathuis JM (2007) Arsenic hazards: strategies for tolerance and remediation by plants. Trends Biotechnol 25:158–165

Uren NC (2007) Types, amounts, and possible functions of compounds released into the rhizosphere by soil-grown plants. In: Pinton R et al (eds) The rhizosphere, 2nd edn. CRC, Boca Raton, FL, pp 1–22
Viterbo A, Ramot O, Chemin L, Chet I (2002) Significance of lytic enzymes from *Trichoderma* spp. in the biocontrol of fungal plant pathogens. Antonie Leeuwenhoek 81:549–556
Wen F, Van Etten HD, Tsaprailis G, Hawes MC (2007) Extracellular proteins in pea root tip and border cell exudates. Plant Physiololgy 143:773–783
Wright AL, Reddy KR (2001) Phosphorus loading effects on extracellular enzyme activity in everglades wetland soil. Soil Sci Soc Am J 65:588–595
Yadav RS, Tarafdar JC (2001) Influence of organic and inorganic phosphorous supply on the maximum secretion of acid phosphatise by plants. Biol Fertil Soils 34:140–143
Yang Z, Liu S, Zheng D, Feng S (2006) Effects of cadmium, zinc and lead on soil enzyme activities. J Environ Sci 18:1135–1141
Zahir ZA, Malik MAR, Arshad M (2001) Soil enzymes research: a review. Online J Biol Sci 1:299–307
Zahran HH (1997) Diversity, adaptation and activity of the bacterial flora in saline environments. Biol Fertil Soils 25:211–223

Chapter 9
Lignocellulose-Degrading Enzymes in Soils

Petr Baldrian and Jaroslav Šnajdr

9.1 Introduction

Terrestrial soils contain the largest pool of organic carbon in the biosphere (cca 1,800 Pg). Mineralization of this organic matter by heterotrophic microorganisms thus significantly affects the global carbon cycle. The primary effectors of soil organic matter decomposition are extracellular enzymes that deconstruct plant and microbial cell wall polymers and deliver soluble substrates for microbial assimilation (Burns and Dick 2002).

Since the polymers contained within or derived from plant biomass form by far the largest pool of soil carbon and represent the most important input of organic material into soils, the decomposition of lignocellulose attracts considerable attention. However, in contrast to the physiology and ecology of lignocellulose degradation by wood-associated fungi, which is well characterized, most of the reports on lignocellulose decomposing enzymes in the soil are limited to the measurement of enzymatic activity; only a few attempts have been made to study the physiology of lignocellulose-decomposing microorganisms in the soil (Baldrian 2008a). Enzymes in the soil environment are most often studied as the effectors of transformative processes, rather than as products of specific microorganisms (Burns and Dick 2002; Sinsabaugh et al. 2008).

Lignocellulose is composed primarily of the polysaccharidic polymers cellulose and hemicelluloses, and the polyphenolic polymer lignin. During transformation in soils, humic substances (humin, humic, and fulvic acids) are formed from both lignocellulose and structural components of microbial decomposers. It is important to note that, while polysaccharides are sources of both carbon and energy-acquisition by soil microorganisms, the degradation of lignin, and likely humic substances as

P. Baldrian (✉) and J. Šnajdr
Laboratory of Environmental Microbiology, Institute of Microbiology of the ASCR, v.v.i, Vídeňská 1083, 14220 Praha 4, Czech Republic
e-mail: baldrian@biomed.cas.cz

G. Shukla and A. Varma (eds.), *Soil Enzymology*, Soil Biology 22,
DOI 10.1007/978-3-642-14225-3_9, © Springer-Verlag Berlin Heidelberg 2011

well, does not provide enough energy to maintain decomposition, and thus does not play the primary nutritional role. The degradation of lignin, cellulose and humic acids has been the topic of several recent reviews (Hatakka 2001; Kästner and Hofrichter 2001; Lynd et al. 2002; Baldrian 2008b; Baldrian and Valášková 2008). Some recent papers also *discussed* the topic of the relative contribution of soil fungi and bacteria to the degradation of lignocellulose and other organic compounds in soils (de Boer et al. 2005; Ekschmitt et al. 2008). In this chapter, we summarise the data on individual enzymes catalyzing lignocellulose decomposition, their ecology in different habitats and soil factors that affect their production and activity. The enzymes in specific, usually highly spatially heterogeneous forest soils as well as their role in litter decomposition are treated in more detail in Chap. 4.

9.2 Degradation of Cellulose and Hemicelluloses

Cellulose is the main polymeric component of litter and is the most abundant polysaccharide on Earth. The chemical composition is simple: it consists of D-glucose residues linked by β-1,4-glycosidic bonds to form linear polymeric chains of over 10,000 glucose residues. Cellulose contains both highly crystalline regions where individual chains are linked to each other, and less-ordered amorphous regions. The degradation of crystalline regions is much slower than that of the amorphous ones and some microorganisms are able to attack only amorphous cellulose (Baldrian and Valášková 2008).

A typical efficient system for cellulose decomposition includes endo-type hydrolases (endo-1,4-β-glucanases, EC 3.2.1.4), exo-type hydrolases (cellobiohydrolases (CBH) EC 3.2.1.4) and 1,4-β-glucosidases (EC 3.2.1.21); the activities of these enzymes are synergistic. Typical cellulolytic systems of saprotrophic cellulose-degrading fungi (e.g., *Trichoderma* or saprotrophic basidiomycetes) consist of multiple enzymes representing all of these three groups. Cellobiohydrolases are produced with specificity for either the reducing or nonreducing ends of cellulose polymer (Lynd et al. 2002; Baldrian and Valášková 2008). The cellulolytic systems of bacteria are different from those of fungi; they are often formed by a complex of enzymes associated to form so-called "cellulosomes" and are frequently associated with bacterial cell walls (Lynd et al. 2002).

Hemicelluloses are low molecular mass linear or branched polymers, usually containing several different sugar units and substituted side chains. Xylans, consisting of xylose units, and glucomannans, consisting of glucose and mannose units, are the main hemicelluloses of angiosperm and conifer trees, respectively, while other lignocellulosic materials may additionally contain considerable amounts of arabinogalactans and galactans (Baldrian 2008b). The polymers of hemicelluloses are typically branched and contain neutral and/or acidic side groups that render hemicelluloses noncrystalline or poorly crystalline. Enzymatic decomposition of hemicelluloses requires a complex set of different enzymes, reflecting the structural complexity of the substrate. Hemicellulose hydrolysis proceeds through the concerted action of endo-type enzymes, side-group cleaving enzymes and exotype enzymes as well as enzymes that cleave the side groups (e.g., acetyl esterases). Cleavage ultimately results in the liberation of monomeric sugars and acetic acid.

The hemicellulases most typically assayed in soils are endo-1,4-β-xylanase (EC 3.2.1.8) and 1,4-β-xylosidase (EC 3.2.1.37), but several other enzymes are known to be produced by saprotrophic soil fungi and bacteria, including endomannanases, β-mannosidases, galactosidases, arabinosidases, and acetyl esterases, as well as debranching enzymes (Soponsathien 1998; Steffen et al. 2007a; Valášková et al. 2007). The decomposition of hemicellulose is not limited by its physical structure, but rather by the diversity of its chemical composition and intramolecular bonding. Many cellulases and hemicellulases have recently been demonstrated to have broad substrate specificities and thus, it is not always simple to link a specific enzyme with a target substrate (Baldrian and Valášková 2008).

In addition to enzymatic hydrolysis, polysaccharides in wood have also been demonstrated to be degraded by nonenzymic radical-producing systems based on cellobiose dehydrogenase, quinone cycling or small glycopeptides. Their role in the soil is probably more limited than in wood, but cellobiose dehydrogenase has been observed in saprotrophic, parasitic and mycorrhizal fungi (Baldrian and Valášková 2008).

9.3 Enzymes Degrading Lignin and Humic Substances

Lignin is a branched polymer of substituted phenylpropane units joined by carbon-carbon and ether linkages. Due to the diversity of chemical bonds and the complexity of its three-dimensional structure, it is the most recalcitrant component of lignocellulose. Ligninolytic systems consist of oxidases, peroxidases and hydrogen peroxide-producing enzymes. Ligninolytic oxidase – laccase – oxidizes its substrates using molecular oxygen, while the peroxidases need a supply of extracellular hydrogen peroxide, which is formed by the oxidation of different organic compounds.

Lignin peroxidase (EC 1.11.1.14) and manganese peroxidase (MnP; EC 1.11.1.13) are able to cleave the lignin polymer and perform lignin mineralization (Hatakka 2001; Hofrichter 2002). Laccase (phenoloxidase, polyphenol oxidase, EC 1.10.3.2) can oxidise phenolic compounds, including lignin and its derivatives. However, although it might be involved in some lignin transformation pathways, the enzyme alone cannot cleave or mineralize lignin or humic compounds (Leonowicz et al. 2001; Baldrian 2006).

Laccase is the most frequently measured oxidative enzyme in soils, and a target of several past studies (Luis et al. 2004; Blackwood et al. 2007). However, laccase is an enzyme with multiple roles, spanning from interspecific interactions over defence against the toxicity of phenols or heavy metals, to morphogenesis (Baldrian 2004, 2006). Due to its inability to transform lignin, however, the ecological role of laccase in C turnover seems to be frequently largely overestimated. Mn-peroxidase activity has not often been addressed, but it has been observed in forest soils (Criquet et al. 2000; Šnajdr et al. 2008b) and found to be produced by litter-decomposing fungi growing on litter or in the presence of humic substances (Steffen et al. 2002; 2007b). Lignin peroxidase was not found in soil, except in soils amended with white-rot species of wood-associated basidiomycetes (Baldrian 2008a). In addition to the above enzymes, activity of peroxidases in soils has been reported by several authors, e.g., Sinsabaugh et al. (2008). Although it was intended to use these data for quantification of lignin transformation, the identity and real substrate of these enzymes is difficult to assess. Any measurement taken is likely to be a composite of the activity of several enzymes, including plant root-derived enzymes that might be able to transform phenols, but not necessarily polyphenols. While the role of ligninolytic peroxidases in wood is linked to the need of their fungal producers to penetrate wood masses, the role of these enzymes in soils, where the lignocellulose compounds of litter are more accessible, is unclear.

As mentioned above, ligninolytic enzymes also require hydrogen peroxide for activity. Enzymes producing hydrogen peroxide have not often been found in soil, but one example, aryl alcohol oxidase, was detected in cultures of litter-decomposing basidiomycetes (Steffen et al. 2000).

Since humic compounds are primarily composed of lignin residues, ligninolytic enzymes are probably the most important in the degradation of soil humic substances (Hofrichter 2002; Valášková et al. 2007). The producers of the most important enzyme, Mn-peroxidase, the litter-decomposing basidiomycetes, are thus thought to play a major role in the transformation of these compounds (Steffen et al. 2002). However, due to the heterogeneous nature of humic substances, other enzymes including horseradish peroxidase, β-glucosidase and Mn^{3+} or H_2O_2 are able to cleave or decolorize them, as well as the radical-producing systems involved in polysaccharide degradation (Gramss et al. 1999; Baldrian and Valášková 2008).

9.4 Lignocellulolytic Systems of Soil Microorganisms

The ability to degrade cellulose and hemicelluloses is limited to certain groups of soil bacteria. Moreover, there is a distinct difference in cellulolytic strategy between the aerobes (*Acidothermus*, *Bacillus*, *Erwinia*, *Micromonospora*, *Pseudomonas*, *Rhodotermus*, and *Streptomyces*) and the anaerobes (*Acetivibrio*, *Clostridium*, *Eubacterium*, *Fibrobacter*, *Ruminococcus*, *Spirochaeta*, and *Thermotoga*). With relatively few exceptions, anaerobes degrade cellulose primarily *via* complex cellulase systems exemplified by the well-characterized polycellulosome organelles of the thermophilic bacterium *Clostridium thermocellum*. Aerobic cellulolytic and hemicellulolytic bacteria (e.g., the *Actinomycetes*) employ a strategy similar to that of fungi: enzymes are released into the environment and do not form complexes, although they do act synergistically. Adherence to substrate is probably not essential (McCarthy 1987; Lynd et al. 2002). The ability of bacteria to decompose lignin is limited to some actinomycete species which, moreover, cause only minor lignin and humic acid mineralization (McCarthy 1987; Hatakka 2001). The identity of extracellular enzymes involved in this process was, moreover, not yet clarified, with the exception of a cell surface-associated peroxidase that was isolated from *Streptomycetes* with the ability to degrade soil humic acids (Dari et al. 1995). Laccases are also produced by bacteria (e.g., *Bacillus* spp.), but the enzyme is usually cell wall- or spore-associated, and implicated in cell wall or spore coat formation (Claus and Filip 1997).

Among fungi, the degradation of hemicelluloses is typically found in saprotrophic species (Steffen et al. 2007a; Valášková et al. 2007; Baldrian 2008b). Degradation of cellulose can be performed by many fungal species from Zygomycota, Ascomycota, and Basidiomycota, although not all species contain the whole set of cellulolytic enzymes; β-glucosidase is the most commonly produced enzyme. In contrast to polysaccharide degradation, production of ligninolytic enzymes is limited to certain groups of fungi. Laccase is produced by Ascomycetes and Basidiomycetes (both saprotrophic and mycorrhizal species) and some lichens (Baldrian 2006; Zavarzina and Zavarzin 2006). The production of Mn-peroxidase is limited to saprotrophic cord-forming basidiomycetes (Hofrichter 2002), and lignin peroxidase production seems to be limited to genera inhabiting wood and not soil (Morgenstern et al. 2008).

Since the saprotrophic fungi seem to be rare in deeper soil horizons (O'Brien et al. 2005; Lindahl et al. 2007), the saprotrophic abilities of ectomycorrhizal (ECM) basidiomycetes that dominate deeper layers of forest soils have been frequently studied. The saprotrophic abilities of ECM fungi to degrade plant litter and to produce β-glucosidase, β-xylosidase, and cellobiohydrolase were found to be limited compared to saprotrophic species (Colpaert and van Laere 1996; Colpaert and van Tichelen 1996). Mycorrhizal fungi also have limited ability to decompose lignin model compounds. Although the ability of ericoid mycorrhizal fungi is greater than that of ECM fungi, it is still far lower than in saprotrophic fungi (Bending and Read 1997). Recently published analyses of whole genomes of mycorrhizal versus saprotrophic fungi show that the gene pool of ECM fungi is limited: they do not contain ligninolytic peroxidases, cellobiohydrolase and have fewer genes encoding other polysaccharide hydrolases than saprotrophs (Martin and Selosse 2008; Nagendran et al. 2009). Saprotrophic basidiomycetes thus seem to be the most efficient soil lignocellulose degraders, although the link between enzyme production and lignocellulose mineralization is not always clear (Fig. 9.1).

Fungi seem to dominate over bacteria in lignocellulose degradation, based on their ability to decompose lignin and the fact that most soil cellulolytic activity is of fungal origin (Kjoller and Struwe 2002). Soil colonised by saprotrophic fungi exhibits significantly higher activity of laccase and MnP than soil with no apparent fungal colonisation (Gramss 1997) and the same is also true for several polysaccharide hydrolases (Šnajdr et al. 2008a). The main reason for this is probably the fact that the large mycelia of saprotrophic basidiomycetes are well adapted to the use of resources with a spatially heterogeneous distribution.

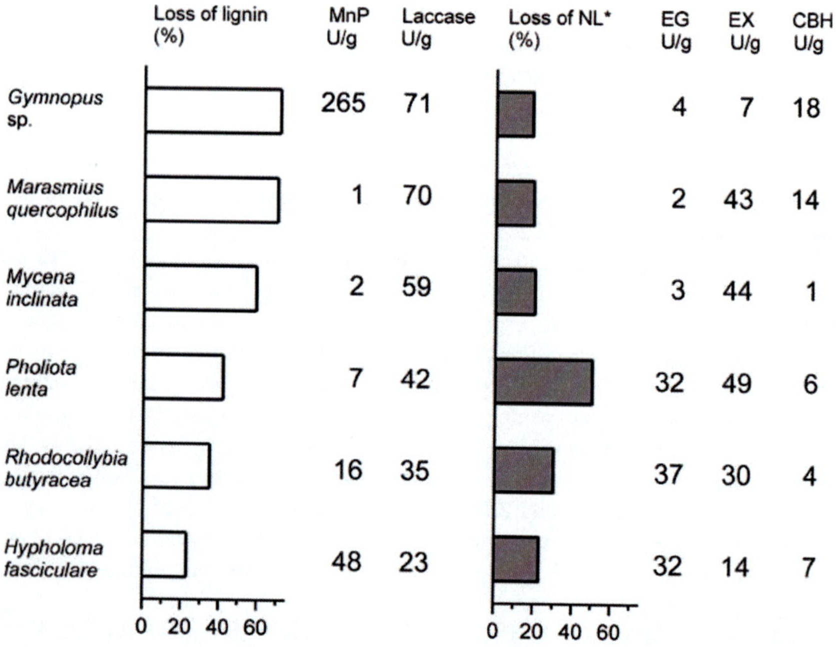

Fig. 9.1 Loss of lignin and nonlignin material (polysaccharides and extractives, NL) and mean lignocellulolytic enzyme activities during 12-week in vitro *Quercus petraea* litter degradation by litter-decomposing basidiomycetes. There is no simple relationship between enzyme activity and mass loss of lignin and polysaccharides. Enzyme activity units in nanomoles per minute per gram dry mass. Abbreviations: *MnP* Mn-peroxidase, *EG* endoglucanase, *EX* endoxylanase, *CBH* cellobiohydrolase. Based on the data of (Steffen et al. 2007a) and (Valášková et al. 2007)

9.5 Lignocellulolytic Enzymes in Different Ecosystems

In a recent meta-analysis of enzymatic activity on a global scale, organic matter and pH turned out to be the most important factors affecting enzyme activity in soils. β-Glucosidase and cellobiohydrolase, as well as phosphatase and chitinase, increased with increasing soil organic matter content, while phenoloxidase and peroxidase did not (Sinsabaugh et al. 2008). Soil organic matter content affects enzyme activity indirectly through its positive effect on soil microbial and especially fungal biomass (Baldrian et al. 2008; Šnajdr et al. 2008b). It is thus clear that land management increasing or decreasing the input of organic matter into soils results in changes in enzymatic activity. Natural ecosystems where dead plant biomass is not removed should thus exhibit higher activity of organic matter decomposing enzymes than ecosystems where plant production is removed or where its transformation rate is increased (e.g., by tillage).

The fact that that land use is probably the major factor affecting enzymatic activity was experimentally confirmed (Tscherko and Kandeler 1999). Amount of organic C and respiration decreased from grassland > poplar plantation > maize field under tillage, as did β-glucosidase activity (Saviozzi et al. 2001). In tropical climates, primary forest soils and pasture soils exhibited higher β-glucosidase activity than plantation soils where plant biomass was removed, and even less activity was found in arable and paddy soils (Salam et al. 1998; Acosta-Martinez et al. 2007b). One probable reason for low enzymatic activity in arable soils is the damage to fungal mycelia by tillage, and resultant relative enrichment of bacteria (van der Wal et al. 2006).

The following text summarizes the most important results on ecosystem-specific factors affecting activity of lignocellulose-degrading enzymes in soils. Selected important papers on this topic are presented in Table 9.1. The enzymology of forest soils is also a subject of Chap. 4.

9.5.1 Polar and Mountainous Soils

Polar and mountainous soils are characterised by short periods of activity during the vegetation season resulting in slow soil profile development and vulnerability.

Table 9.1 Reports on lignocellulose-degrading enzymes and their response to selected environmental factors in contrasting ecosystems

Arable soils	Effect of tillage	Monreal and Bergstrom (2000)
	Soil suppresivity	Rasmussen et al. (2002)
	Fertilization	Řezáčová et al. (2007), Niemi et al. (2008)
Coniferous forests	Postharvest practice	Waldrop et al. (2003)
	Depth gradient, seasonality	Wittmann et al. (2004)
	Tree species	Niemi et al. (2007)
	Fire effect	Waldrop and Harden (2008)
Desert and arid soils	Seasonality	Doyle et al. (2006)
	Management type	Acosta-Martinez et al. (2007a)
Grassland soils	Burning, Fertilization	Ajwa et al. (1999)
	Soil moisture, CO_2	Henry et al. (2005)
	N addition	Zeglin et al. (2007)
Evergreen forests	Seasonality of moisture	Criquet et al. (2000), Criquet et al. (2002), Sardans and Penuelas (2005)
	Succession	Rutigliano et al. (2004)
	Fire effect	Fioretto et al. (2005)
Polar and alpine soils	CO_2	Moorhead and Linkins (1997)
	Seasonality	Lipson et al. (2002)
	reezing / thawing	Yergeau and Kowalchuk (2008)
Wetlands	Seasonality	Bonnett et al. (2006)
	Vegetation effect	Reboreda and Cacador (2008)
Temperate forests	Burning	Eivazi and Bayan (1996), Boerner et al. (2000), Boerner and Brinkman (2003)
	N addition	Carreiro et al. (2000), Saiya-Cork et al. (2002)
	CO_2	Moscatelli et al. (2005)
	Seasonality	Courty et al. (2007), Mosca et al. (2007)
	Depth gradient	Šnajdr et al. (2008b)

Mountain soils exhibit sharp seasonality where snowmelt is an important breakpoint event. Higher specific cellulolytic activity and microbial activity was, however, recorded during winter than in summer in some soils (Lipson et al. 2002). Carbon cycle related enzymes (β-glucosidase and β-xylosidase) are more temperature-regulated than N-cycle enzymes (Koch et al. 2007). The activity in soils at temperatures close to zero might be highly dynamic since both fungal biomass and laccase activity (but not bacterial biomass) respond more to freezing-thawing cycles than to warming, as observed in Antarctic soils (Yergeau and Kowalchuk 2008). Possible results of global climate change were addressed by Moorhead and Linkins (1997) in tussock tundra; increased CO_2 reduced cellulolytic activity, possibly due to root exudation of simple sugars that inhibited cellulose decomposition.

9.5.2 Boreal Forests

Boreal forest soils are also subject to long periods of low temperatures. The fact that aboveground litter is the most important C input into soils in forests leads to the formation of a sharp vertical gradient of soil properties. In the coniferous forests of Northern Europe, activities of cellobiohydrolase and β-glucosidase decreased sharply with soil depth along with the decreased microbial biomass (Wittmann et al. 2004). No dramatic effects of temperature on soil processes, including respiration and β-glucosidase activity, were detected in these soils. A significant part of the annual enzymatic turnover was achieved during the winter when surface temperatures dropped below zero (Kahkonen et al. 2001; Wittmann et al. 2004). Some seasonal effects are, however, apparent even within the warm season (Niemi et al. 2007).

Enzyme activities in boreal forests were also demonstrated to reflect vegetation cover and litter quality. The activities of cellobiohydrolase, β-glucosidase, and β-xylosidase, along with microbial biomass, were higher in *Alnus* forest than in *Pinus* forest planted in the same type of soil (Niemi et al. 2007). The litter of some plants, such as *Kalmia angustifolia,* inhibit β-glucosidase and phosphatase due to high tannin content. Therefore, in native soils, enzymatic activity decreased with *Kalmia* vegetation cover (Joanisse et al. 2007).

Thinning did not have a long-term effect on enzymatic activities, but did change microbial biomass content in spruce forest soil (Maassen et al. 2006). On the other hand, significant effects of post-harvest practices (slashing, burning, and chipping) on the activities of laccase, cellulases, and hemicellulases, was demonstrated. Compared to untreated forests, enzymatic activity usually decreases in the forest floor of treated stands. Laccase was also lower at burned sites in a litterbag experiment (Waldrop et al. 2003). The vulnerability of boreal soils was demonstrated in a study where laccase activity and lignin mineralization were detectably reduced after 5 years following wildfire or permafrost reduction (Waldrop and Harden 2008).

9.5.3 Temperate Forests

Temperate forest soils are not subject to dramatic temperature changes, but summer temperatures may, under certain circumstances, result in temporary droughts. The seasonal changes are thus not primarily related to ambient temperature, but also reflect the litter quality in the litter horizon, which changes dramatically after the autumn litterfall period (Baldrian et al. 2008). Cellobiohydrolase, β-glucosidase, β-xylosidase, and laccase activities in ectomycorrhiza exhibit seasonal changes in temperate forests due to changes in assimilate supply from their tree hosts (Courty et al. 2007; Mosca et al. 2007).

Temperate forest soils are vertically structured and the activity of lignocellulolytic enzymes and fungal biomass decrease sharply with soil depth (Šnajdr et al. 2008b). This is due to the dominance of saprotrophic fungi in the litter horizon and mycorrhizal species in the deeper soil horizons. Saprotrophic basidiomycetes isolated from temperate forest soils produce a rich array of lignocellulose-degrading enzymes and litter turnover is faster in boreal forest soils where ectomycorrhizal fungi are more dominant. Litter incubated with pure cultures of saprotrophic basidiomycetes, however, differs in chemical composition from litter decomposing *in situ* due to contribution of other decomposer taxa to total decay (Steffen et al. 2007a; Valášková et al. 2007).

Thinning was demonstrated to increase microbial activity, as documented by the increase of phosphatase and laccase activities (Giai and Boerner 2007). Contrary to polar soils, increased CO_2 in poplar plantation increased microbial biomass and β-glucosidase significantly (Moscatelli et al. 2005). The reason for this may be increased plant production and litter input.

9.5.4 Evergreen Forests

Temperature variation is of minor importance in evergreen forests and a distinct litterfall period is also missing. Rainfall and drought thus seem to be the key factors in enzyme activity regulation in this ecosystem. The activities of lignocellulose-degrading enzymes increase during moist periods, and MnP activity was detectable only during this time of the year (Criquet et al. 2000, 2002). The effects of drought are considerable, taking into account that drought manipulation excluding runoff and/or rain in *Quercus ilex* forest resulted in an enzymatic activity reduction of tens of percents (e.g., by 10–85% in the case of β-glucosidase) (Sardans and Penuelas 2005).

Due to summer droughts, combined with high temperatures, wildfires are an important issue in this environment. The effects of fires were found to be vegetation-dependent in Mediterranean evergreen forests (Fioretto et al. 2005). The successive changes of vegetation typically occurring at post fire sites also affect enzymatic activity: the highest β-glucosidase activity was found in the middle phase of this succession (Rutigliano et al. 2004).

9.5.5 *Grasslands*

Grasslands and pasture soils are different from forest soils in that they have higher root density in the rhizosphere and have arbuscular mycorrhiza instead of ectomycorrhiza on the roots. Since part of the litter decomposes still standing, a distinct litter horizon is usually not formed. The fungal decomposer community has a lower proportion of saprotrophic basidiomycetes than in forest soils and the ability of cellulose and lignin decomposition in this environment seems to be less frequent (Deacon et al. 2006).

Cellulase activity in grasslands was found to decrease with age, likely due to the changes in SOM quality (Shi et al. 2006). The seasonal effects are more likely due to moisture changes than to temperature, and are less pronounced than in forest soils (Baldrian et al. 2008). In heathlands, temporary drying during summer decreased laccase activity as well as the diversity of soil fungi (Toberman et al. 2008).

Pasture soils are affected by cattle grazing, soil compaction and nitrogen and phosphorus input. Application of cattle slurry leads to higher microbial biomass in soil and a corresponding increase in xylanase activity (Kandeler and Eder 1993). Nitrogen addition also increased cellobiohydrolase and β-glucosidase activities in different grasslands (Zeglin et al. 2007), but the long term N fertilization or burning ultimately decreased microbial biomass in tallgrass prairie areas. β-Glucosidase activity was decreased by burning, and increased in N-fertilized plots (Ajwa et al. 1999).

Elevated water content and CO_2 decreased the activity of polysaccharide hydrolases but increased laccase and peroxidase activity (Henry et al. 2005). In European grasslands, however, CO_2 increased polysaccharide hydrolases in various extents, probably due to rhizodeposition and root litter (Drissner et al. 2007).

9.5.6 *Arable Soils*

Arable soils are characterised by reduced C input from aboveground plant biomass due to harvesting. As a consequence of biomass removal, combined with tillage, litter horizon is virtually missing and fungal biomass is low (van der Wal et al. 2006). Due to the absence of fungi, lignin decomposition is slow and ligninolytic enzyme activities are very low (Řezáčová et al. 2007). The seasonality of enzymatic activity reflects the annual management cycles of individual crops, and the crops themselves significantly influence enzyme production (Bergstrom and Monreal 1998).

Tillage reduces enzymatic activity in general, and β-glucosidase activity in particular (Deng and Tabatabai 1996; Monreal and Bergstrom 2000). Tillage intensity is also an important factor. Minimum tillage in Peru preserved similar enzymatic activity of β-glucosidase as uncultivated soils (Dick et al. 1994). Non-tilled soils tend to stratify, while soil mixing due to tillage may result in the increase of activity of particular cellulolytic enzymes in deeper soil horizons due to burying (Kandeler and Bohm 1996).

To substitute for biomass removal, and to increase yields, arable soils are subject to fertilization. Higher organic matter input (e.g., mulching or fertilization) results in higher organic carbon and nitrogen content and increases cellulolytic activities (Debosz et al. 1999; Bohme et al. 2005; Řezáčová et al. 2007), but clear differences were found among different fertilization treatments (organic/chemical/peat addition) (Niemi et al. 2008). Addition of nitrogen also contributes to higher microbial biomass, endocellulase, cellobiohydrolase, and hemicellulase activities during wheat straw decomposition in soil (Henriksen and Breland 1999).

9.5.7 Wetlands

Temporary drought is the most important factor affecting biopolymer transformation in wetlands. β-Glucosidase was found to be regulated by drought and the amount of dissolved organic carbon (Freeman et al. 1997; Toberman et al. 2008). In addition to β-glucosidase, seasonality also significantly affects laccase activity, and this seems to be due to changes in peat chemistry rather than temperature (Bonnett et al. 2006). Increasing plant root biomass in salt marshes was found to increase β-glucosidase activity (Reboreda and Cacador 2008), probably due to rhizodeposition.

9.5.8 Arid and Desert Soils

Arid soils are characterised by short periods of high microbial activity following the increase of soil moisture content. High activity of laccase and peroxidase in arid grassland soils are stabilised by the binding to other soil components (Stursova and Sinsabaugh 2008). The stabilization may contribute to the persistence of enzymes in the environment and a resulting fast decomposition immediately after soil rewetting. This is particularly important in ecosystems with scarce and irregular rain periods. The most important factors affecting cellulolytic activity in desert soil were seasonality, organic matter content and moisture content (Pavel et al. 2004; Doyle et al. 2006), as well as the type of vegetation (Garcia et al. 2005).

β-Glucosidase in semiarid soils was found to be higher in unmanaged soils and to positively correlate strongly with soil C and N content and negatively with soil pH (Acosta-Martinez et al. 2003). Moreover, unmanaged fields also showed higher fungal biomass (Acosta-Martinez et al. 2007a). Since the turnover of soil organic matter is slow in arid soils, nutrient addition may have a long term effect. The effect on soil C content, microbial biomass, respiration and activity of multiple enzymes, including β-glucosidase was still apparent 8 years after the treatment (Pascual et al. 1999).

9.6 Factors Affecting Lignocellulose Degradation in Soils

There are several factors that affect the activity of lignocellulose-degrading enzymes in soils independently from their producers and ecosystem processes. One of the most important is the interaction of enzymes with other soil components. Enzymes differ in their association with different sizes and types of soil particles. Specific enzyme activity in soil size-fractionated material from grasslands differs with fraction size. Carbohydrate-utilizing enzymes are associated with larger fraction sizes, in contrast to the P and N cycle enzymes (Stemmer et al. 1998; Marx et al. 2005). Adsorption of cellulases to soil components limits their free movement in the soil, but leads to an increase in their stability during freezing/thawing cycles (Lahdesmaki and Piispanen 1992). Also, humic material in soil binds a small, yet significant, part of the total enzyme activity in soils (e.g., 3–21% in case of β-glucosidase) (Ceccanti et al. 2008). Moreover, a significant proportion of lignocellulases is bound to the cell walls of microorganisms, which makes their reaction product more accessible to their producers (Valášková and Baldrian 2006).

Enzymatic activity is also regulated by soil pH, since individual enzymes differ in their catalytic optima (Wittmann et al. 2004; Niemi and Vepsalainen 2005). Higher litter loss was recorded in soils with higher Mn concentrations, possibly due to improved function of MnP (Berg et al. 2007). On the other hand, heavy metals in soils cause a decrease in total microbial biomass frequently accompanied by a shift in the fungal/bacterial biomass ratio and changes in enzyme activities (Baldrian 2010). These are usually inhibitory, except in the case of laccase, whose activity is usually increased in the presence of Cu and Cd (Baldrian 2006).

Enzymatic activity also reflects the gradual changes in soils undergoing successive development or shift. In soils developing from an initial stage with low biomass content, enzymatic activity tends to increase. In soils of a different age, in glacier forelands, enzyme activity including β-glucosidase and β-xylosidase increased for 50 years and remained stable thereafter (Tscherko et al. 2003). The effect of plant species was not apparent in this environment during initial succession, but became important in older, established soils (Tscherko et al. 2005). During the succession following post-mining deposits, fungal biomass and the activity of most extracellular hydrolytic enzymes, including cellobiohydrolase and β-glucosidase, also increased with time but peaked after 21 years of succession and dropped later (Baldrian et al. 2008). The differences in enzymatic activity between grassland and forest soils were also notable during the successive changes in enzymatic activity in transition from meadows to forests where β-glucosidase decreased during succession (Griffiths et al. 2005).

On a global scale, human activity has increased the atmospheric input of NO_3 to many terrestrial ecosystems. Atmospheric NO_3 may potentially affect ecosystem function, especially in temperate forests that are often N-limited. The potential effects of nitrogen were addressed, since data on saprotrophic fungi indicated that excess nitrogen may limit lignin decomposition (Bonnarme et al. 1991). Response to N depends on vegetation and litter quality. The net gain or loss of C is supposedly mediated by the regulation of laccase, peroxidise, and cellobiohydrolase activity and the effect of nitrogen is stronger in the litter horizon than in the soil (Waldrop et al. 2004; Sinsabaugh et al. 2005). The response of cellulases and laccase on nitrogen addition in a spruce forest litter depends on the litter C/N ratio. N addition to pine forest soils resulted in a 30–70% decrease in fungal biomass, and decrease of F/B ratio and laccase activity (Frey et al. 2004). In boreal pinelands, however, N addition in different organic and inorganic forms resulted in a shift in the soil microbial community, but not in detectable alteration of enzyme activities (Lucas et al. 2007).

In hardwood forests, NO_3 addition decreased microbial biomass, β-glucosidase and laccase activities in upper soil layers (DeForest et al. 2004). In another experiment, peroxidase activity was also reduced by NO_3, and the addition resulted in the increase of soil C and decrease of microbial biomass (DeForest et al. 2005). The response of oxidative enzymes to nitrate deposition controls both enzyme activity and dissolved organic carbon fluxes (Waldrop and Zak 2006). Results from different soil types confirmed the inhibition of laccase activity, but showed an increase in cellulolytic enzymes (Carreiro et al. 2000; Saiya-Cork et al. 2002). Thus, laccase seems to be the major regulatory component in NO_3-supplemented soils. Molecular studies demonstrated that nitrate affects the activity, but not the diversity, of this enzyme and its producers (Blackwood et al. 2007; Hofmockel et al. 2007).

9.7 Molecular Biology of Enzymes in Soils

It is obvious that enzyme activity measurement represents the first step in the elucidation of soil processes. Molecular biology methods offer the possibility to link the soil microbial community structure and function, and the modern methods of large throughput sequencing represent a novel opportunity to investigate the identity of enzyme producers in soils. The first attempts in this field were undertaken in studies focusing on laccase gene pools in hardwood forest soils (Luis et al. 2004). Later studies linked the detection of enzyme transcripts to their potential producers, or used sequence data to understand the diversity and function of laccase producers in soils (Luis et al. 2005; Blackwood et al. 2007; Hofmockel et al. 2007). The major hindrance for similar studies seems to be the methodological problems with obtaining suitable primers for gene or transcript detection covering the whole diversity of soil enzymes (Edwards et al. 2008; Morgenstern et al. 2008).

However, we can expect major developments in this field along with the development of molecular methods of gene analysis and the accomplishment of genome sequencing projects.

9.8 Conclusions

The degradation of lignocellulose is of major importance for the understanding of the decomposition of part of the world's carbon in soils – the largest carbon pool on Earth. The importance of this understanding arises now, when the Earth is challenged with the risks of global climate change. The current knowledge of enzymatic activities in soils does not provide enough information for a complete picture of carbon fate in soils. Studies focusing more on the relative importance of individual processes, individual enzymes and taxa of soil organisms are needed. This, however, will only be possible using the tools of molecular biology, which will help to link individual microorganisms with biochemical processes.

Acknowledgment Financial support from the Ministry of Education, Youth and Sports of the Czech Republic (Project LC06066) and from the Ministry of Agriculture of the Czech Republic (Project QH72216) is gratefully acknowledged.

References

Acosta-Martinez V, Klose S, Zobeck TM (2003) Enzyme activities in semiarid soils under conservation reserve program, native rangeland, and cropland. J Plant Nutr Soil Sci 166:699–707

Acosta-Martinez V, Mikha MM, Vigil MF (2007a) Microbial communities and enzyme activities in soils under alternative crop rotations compared to wheat-fallow for the Central Great Plains. Appl Soil Ecol 37:41–52

Acosta-Martinez V, Cruz L, Sotomayor-Ramirez D, Perez-Alegria L (2007b) Enzyme activities as affected by soil properties and land use in a tropical watershed. Appl Soil Ecol 35:35–45

Ajwa HA, Dell CJ, Rice CW (1999) Changes in enzyme activities and microbial biomass of tallgrass prairie soil as related to burning and nitrogen fertilization. Soil Biol Biochem 31:769–777

Baldrian P (2004) Increase of laccase activity during interspecific interactions of white-rot fungi. FEMS Microbiol Ecol 50:245–253

Baldrian P (2006) Fungal laccases – occurrence and properties. FEMS Microbiol Rev 30:215–242

Baldrian P (2008a) Wood-inhabiting ligninolytic basidiomycetes in soils: ecology and constraints for applicability in bioremediation. Fungal Ecol 1:4–12

Baldrian P (2008b) Enzymes of saprotrophic Basidiomycetes. In: Boddy L, Frankland JC, van West P (eds) Ecology of saprotrophic Basidiomycetes. Academic, Amsterdam, pp 19–41

Baldrian P (2010) Effect of heavy metals on saprotrophic soil fungi. In: Sherameti I, Varma A (eds) Soil heavy metals. Springer, New York, pp 263–279

Baldrian P, Valášková V (2008) Degradation of cellulose by basidiomycetous fungi. FEMS Microbiol Rev 32:501–521

Baldrian P, Trögl J, Frouz J, Šnajdr J, Valášková V, Merhautová V, Cajthaml T, Herinková J (2008) Enzyme activities and microbial biomass in topsoil layer during spontaneous succession in spoil heaps after brown coal mining. Soil Biol Biochem 40:2107–2115

Bending GD, Read DJ (1997) Lignin and soluble phenolic degradation by ectomycorrhizal and ericoid mycorrhizal fungi. Mycol Res 101:1348–1354

Berg B, Steffen KT, McClaugherty C (2007) Litter decomposition rate is dependent on litter Mn concentrations. Biogeochemistry 82:29–39

Bergstrom DW, Monreal CT (1998) Increased soil enzyme activities under two row crops. Soil Sci Soc Am J 62:1295–1301

Blackwood CB, Waldrop MP, Zak DR, Sinsabaugh RL (2007) Molecular analysis of fungal communities and laccase genes in decomposing litter reveals differences among forest types but no impact of nitrogen deposition. Environ Microbiol 9:1306–1316

Boerner REJ, Brinkman JA (2003) Fire frequency and soil enzyme activity in southern Ohio oak-hickory forests. Appl Soil Ecol 23:137–146

Boerner REJ, Decker KLM, Sutherland EK (2000) Prescribed burning effects on soil enzyme activity in a southern Ohio hardwood forest: a landscape-scale analysis. Soil Biol Biochem 32:899–908

Bohme L, Langer U, Bohme F (2005) Microbial biomass, enzyme activities and microbial community structure in two European long-term field experiments. Agric Ecosyst Environ 109:141–152

Bonnarme P, Perez J, Jeffries TW (1991) Regulation of ligninase production in white-rot fungi. ACS Symp Ser 460:200–206

Bonnett SAF, Ostle N, Freeman C (2006) Seasonal variations in decomposition processes in a valley-bottom riparian peatland. Sci Total Environ 370:561–573

Burns RG, Dick RP (2002) Enzymes in the environment: activity, ecology and applications. Marcel Dekker, New York

Carreiro MM, Sinsabaugh RL, Repert DA, Parkhurst DF (2000) Microbial enzyme shifts explain litter decay responses to simulated nitrogen deposition. Ecology 81:2359–2365

Ceccanti B, Doni S, Macci C, Cercignani G, Masciandaro G (2008) Characterization of stable humic-enzyme complexes of different soil ecosystems through analytical isoelectric focussing technique (IEF). Soil Biol Biochem 40:2174–2177

Claus H, Filip Z (1997) The evidence of a laccase-like enzyme activity in a *Bacillus sphaericus* strain. Microbiol Res 152:209–216

Colpaert JV, van Laere A (1996) A comparison of the extracellular enzyme activities of two ectomycorrhizal and a leaf-saprotrophic basidiomycete colonizing beech leaf litter. New Phytol 134:133–141

Colpaert JV, van Tichelen KK (1996) Decomposition, nitrogen and phosphorus mineralization from beech leaf litter colonized by ectomycorrhizal or litter-decomposing basidiomycetes. New Phytol 134:123–132

Courty PE, Breda N, Garbaye J (2007) Relation between oak tree phenology and the secretion of organic matter degrading enzymes by *Lactarius quietus* ectomycorrhizas before and during bud break. Soil Biol Biochem 39:1655–1663

Criquet S, Farnet AM, Tagger S, Le Petit J (2000) Annual variations of phenoloxidase activities in an evergreen oak litter: influence of certain biotic and abiotic factors. Soil Biol Biochem 32:1505–1513

Criquet S, Tagger S, Vogt G, Le Petit J (2002) Endoglucanase and β-glycosidase activities in an evergreen oak litter: annual variation and regulating factors. Soil Biol Biochem 34:1111–1120

Dari K, Bechet M, Blondeau R (1995) Isolation of soil *Streptomyces* strains capable of degrading humic acids and analysis of their peroxidase activity. FEMS Microbiol Ecol 16:115–121

de Boer W, Folman LB, Summerbell RC, Boddy L (2005) Living in a fungal world: impact of fungi on soil bacterial niche development. FEMS Microbiol Rev 29:795–811

Deacon LJ, Pryce-Miller EJ, Frankland JC, Bainbridge BW, Moore PD, Robinson CH (2006) Diversity and function of decomposer fungi from a grassland soil. Soil Biol Biochem 38:7–20

Debosz K, Rasmussen PH, Pedersen AR (1999) Temporal variations in microbial biomass C and cellulolytic enzyme activity in arable soils: effects of organic matter input. Appl Soil Ecol 13:209–218

DeForest JL, Zak DR, Pregitzer KS, Burton AJ (2004) Atmospheric nitrate deposition, microbial community composition, and enzyme activity in northern hardwood forests. Soil Sci Soc Am J 68:132–138

DeForest JL, Zak DR, Pregitzer KS, Burton AJ (2005) Atmospheric nitrate deposition and enhanced dissolved organic carbon leaching: test of a potential mechanism. Soil Sci Soc Am J 69:1233–1237

Deng SP, Tabatabai MA (1996) Effect of tillage and residue management on enzyme activities in soils.2. Glycosidases. Biol Fertil Soils 22:208–213

Dick RP, Sandor JA, Eash NS (1994) Soil enzyme activities after 1500 years of terrace agriculture in the Colca Valley, Peru. Agric Ecosyst Environ 50:123–131

Doyle J, Pavel R, Barness G, Steinberger Y (2006) Cellulase dynamics in a desert soil. Soil Biol Biochem 38:371–376

Drissner D, Blum H, Tscherko D, Kandeler E (2007) Nine years of enriched CO_2 changes the function and structural diversity of soil microorganisms in a grassland. Eur J Soil Sci 58:260–269

Edwards IP, Upchurch RA, Zak DR (2008) Isolation of fungal cellobiohydrolase I genes from sporocarps and forest soils by PCR. Appl Environ Microbiol 74:3481–3489

Eivazi F, Bayan MR (1996) Effects of long-term prescribed burning on the activity of select soil enzymes in an oak-hickory forest. Can J For Res 26:1799–1804

Ekschmitt K, Kandeler E, Poll C, Brune A, Buscot F, Friedrich M, Gleixner G, Hartmann A, Kastner M, Marhan S, Miltner A, Scheu S, Wolters V (2008) Soil-carbon preservation through habitat constraints and biological limitations on decomposer activity. J Plant Nutr Soil Sci 171:27–35

Fioretto A, Papa S, Pellegrino A (2005) Effects of fire on soil respiration, ATP content and enzyme activities in Mediterranean maquis. Appl Veg Sci 8:13–20

Freeman C, Liska G, Ostle NJ, Lock MA, Hughes S, Reynolds B, Hudson J (1997) Enzymes and biogeochemical cycling in wetlands during a simulated drought. Biogeochemistry 39:177–187

Frey SD, Knorr M, Parrent JL, Simpson RT (2004) Chronic nitrogen enrichment affects the structure and function of the soil microbial community in temperate hardwood and pine forests. For Ecol Manage 196:159–171

Garcia C, Roldan A, Hernandez T (2005) Ability of different plant species to promote microbiological processes in semiarid soil. Geoderma 124:193–202

Giai C, Boerner REJ (2007) Effects of ecological restoration on microbial activity, microbial functional diversity, and soil organic matter in mixed-oak forests of southern Ohio, USA. Appl Soil Ecol 35:281–290

Gramss G (1997) Activity of oxidative enzymes in fungal mycelia from grassland and forest soils. J Basic Microbiol 37:407–423

Gramss G, Ziegenhagen D, Sorge S (1999) Degradation of soil humic extract by wood- and soil-associated fungi, bacteria, and commercial enzymes. Microb Ecol 37:140–151

Griffiths R, Madritch M, Swanson A (2005) Conifer invasion of forest meadows transforms soil characteristics in the Pacific Northwest. For Ecol Manage 208:347–358

Hatakka A (2001) Biodegradation of Lignin. In: Steinbüchel A, Hofrichter M (eds) Biopolymers 1: lignin, humic substances and coal. Wiley, Weinheim, pp 129–180

Henriksen TM, Breland TA (1999) Nitrogen availability effects on carbon mineralization, fungal and bacterial growth, and enzyme activities during decomposition of wheat straw in soil. Soil Biol Biochem 31:1121–1134

Henry HAL, Juarez JD, Field CB, Vitousek PM (2005) Interactive effects of elevated CO_2, N deposition and climate change on extracellular enzyme activity and soil density fractionation in a California annual grassland. Glob Chang Biol 11:1808–1815

Hofmockel KS, Zak DR, Blackwood CB (2007) Does atmospheric NO_3- deposition alter the abundance and activity of ligninolytic fungi in forest soils? Ecosystems 10:1278–1286
Hofrichter M (2002) Review: lignin conversion by manganese peroxidase (MnP). Enzyme Microb Technol 30:454–466
Joanisse GD, Bradley RL, Preston CM, Munson AD (2007) Soil enzyme inhibition by condensed litter tannins may drive ecosystem structure and processes: the case of *Kalmia angustifolia*. New Phytol 175:535–546
Kahkonen MA, Wittmann C, Kurola J, Ilvesniemi H, Salkinoja-Salonen MS (2001) Microbial activity of boreal forest soil in a cold climate. Boreal Environ Res 6:19–28
Kandeler E, Bohm KE (1996) Temporal dynamics of microbial biomass, xylanase activity, N-mineralisation and potential nitrification in different tillage systems. Appl Soil Ecol 4:181–191
Kandeler E, Eder G (1993) Effect of cattle slurry in grassland on microbial biomass and on activities of various enzymes. Biol Fertil Soils 16:249–254
Kästner M, Hofrichter M (2001) Biodegradation of humic substances. In: Steinbüchel A, Hofrichter M (eds) Biopolymers 1: lignin, humic substances and coal. Wiley, Weinheim, pp 349–378
Kjoller A, Struwe S (2002) Fungal communities, succession, enzymes, and decomposition. In: Burns RG, Dick RP (eds) Enzymes in the environment: activity, ecology and applications. Marcel Dekker, New York, pp 267–284
Koch O, Tscherko D, Kandeler E (2007) Temperature sensitivity of microbial respiration, nitrogen mineralization, and potential soil enzyme activities in organic alpine soils. Global Biogeochem Cycles 21:GB4017. doi:10.1029/2007GB002983
Lahdesmaki P, Piispanen R (1992) Soil enzymology – role of protective colloid systems in the preservation of exoenzyme activities in soil. Soil Biol Biochem 24:1173–1177
Leonowicz A, Cho NS, Luterek J, Wilkolazka A, Wojtas-Wasilewska M, Matuszewska A, Hofrichter M, Wesenberg D, Rogalski J (2001) Fungal laccase: properties and activity on lignin. J Basic Microbiol 41:185–227
Lindahl BD, Ihrmark K, Boberg J, Trumbore SE, Hogberg P, Stenlid J, Finlay RD (2007) Spatial separation of litter decomposition and mycorrhizal nitrogen uptake in a boreal forest. New Phytol 173:611–620
Lipson DA, Schadt CW, Schmidt SK (2002) Changes in soil microbial community structure and function in an alpine dry meadow following spring snow melt. Microb Ecol 43:307–314
Lucas RW, Casper BB, Jackson JK, Balser TC (2007) Soil microbial communities and extracellular enzyme activity in the New Jersey Pinelands. Soil Biol Biochem 39:2508–2519
Luis P, Walther G, Kellner H, Martin F, Buscot F (2004) Diversity of laccase genes from basidiomycetes in a forest soil. Soil Biol Biochem 36:1025–1036
Luis P, Kellner H, Zimdars B, Langer U, Martin F, Buscot F (2005) Patchiness and spatial distribution of laccase genes of ectomycorrhizal, saprotrophic, and unknown basidiomycetes in the upper horizons of a mixed forest cambisol. Microb Ecol 50:570–579
Lynd LR, Weimer PJ, van Zyl WH, Pretorius IS (2002) Microbial cellulose utilization: fundamentals and biotechnology. Microbiol Mol Biol Rev 66:506–577
Maassen S, Fritze H, Wirth S (2006) Response of soil microbial biomass, activities, and community structure at a pine stand in northeastern Germany 5 years after thinning. Can J For Res-Revue Canadienne De Recherche Forestiere 36:1427–1434
Martin F, Selosse MA (2008) The *Laccaria* genome: a symbiont blueprint decoded. New Phytol 180:296–310
Marx MC, Kandeler E, Wood M, Wermbter N, Jarvis SC (2005) Exploring the enzymatic landscape: distribution and kinetics of hydrolytic enzymes in soil particle-size fractions. Soil Biol Biochem 37:35–48
McCarthy AJ (1987) Lignocellulose-degrading actinomycetes. FEMS Microbiol Rev 46:145–163
Monreal CM, Bergstrom DW (2000) Soil enzymatic factors expressing the influence of land use, tillage system and texture on soil biochemical quality. Can J Soil Sci 80:419–428

Moorhead DL, Linkins AE (1997) Elevated CO_2 alters belowground exoenzyme activities in tussock tundra. Plant Soil 189:321–329

Morgenstern I, Klopman S, Hibbett D (2008) Molecular evolution and diversity of lignin degrading heme peroxidases in the *Agaricomycetes*.. J Mol Evol 66:243–257

Mosca E, Montecchio L, Scattolin L, Garbaye J (2007) Enzymatic activities of three ectomycorrhizal types of *Quercus robur* L. in relation to tree decline and thinning. Soil Biol Biochem 39:2897–2904

Moscatelli MC, Lagomarsino A, De Angelis P, Grego S (2005) Seasonality of soil biological properties in a poplar plantation growing under elevated atmospheric CO_{2}.. Appl Soil Ecol 30:162–173

Nagendran S, Hallen-Adams HE, Paper JM, Aslam N, Walton JD (2009) Reduced genomic potential for secreted plant cell-wall-degrading enzymes in the ectomycorrhizal fungus *Amanita bisporigera*, based on the secretome of *Trichoderma reesei*.. Fungal Genet Biol 46:427–435

Niemi RM, Vepsalainen M (2005) Stability of the fluorogenic enzyme substrates and pH optima of enzyme activities in different Finnish soils. J Microbiol Meth 60:195–205

Niemi RM, Vepsalainen M, Erkomaa K, Ilvesniemi H (2007) Microbial activity during summer in humus layers under *Pinus silvestris* and *Alnus incana*.. For Ecol Manage 242:314–323

Niemi RM, Vepsalainen M, Wallenius K, Erkomaa K, Kukkonen S, Palojarvi A, Vestberg M (2008) Conventional versus organic cropping and peat amendment: impacts on soil microbiota and their activities. Eur J Soil Biol 44:419–428

O'Brien HE, Parrent JL, Jackson JA, Moncalvo JM, Vilgalys R (2005) Fungal community analysis by large-scale sequencing of environmental samples. Appl Environ Microbiol 71:5544–5550

Pascual JA, Garcia C, Hernandez T (1999) Lasting microbiological and biochemical effects of the addition of municipal solid waste to an arid soil. Biol Fertil Soils 30:1–6

Pavel R, Doyle J, Steinberger Y (2004) Seasonal patterns of cellulase concentration in desert soil. Soil Biol Biochem 36:549–554

Rasmussen PH, Knudsen IMB, Elmholt S, Jensen DF (2002) Relationship between soil cellulolytic activity and suppression of seedling blight of barley in arable soils. Appl Soil Ecol 19:91–96

Reboreda R, Cacador I (2008) Enzymatic activity in the rhizosphere of *Spartina maritima*: potential contribution for phytoremediation of metals. Mar Environ Res 65:77–84

Řezáčová V, Baldrian P, Hršelová H, Larsen J, Gryndler M (2007) Influence of mineral and organic fertilization on soil fungi, enzyme activities and humic substances in a long-term field experiment. Folia Microbiol 52:415–421

Rutigliano FA, D'Ascoli R, De Santo AV (2004) Soil microbial metabolism and nutrient status in a Mediterranean area as affected by plant cover. Soil Biol Biochem 36:1719–1729

Saiya-Cork KR, Sinsabaugh RL, Zak DR (2002) The effects of long term nitrogen deposition on extracellular enzyme activity in an *Acer saccharum* forest soil. Soil Biol Biochem 34:1309–1315

Salam AK, Katayama A, Kimura M (1998) Activities of some soil enzymes in different land use systems after deforestation in hilly areas of West Lampung, South Sumatra, Indonesia. Soil Sci Plant Nutr 44:93–103

Sardans J, Penuelas J (2005) Drought decreases soil enzyme activity in a Mediterranean *Quercus ilex* L. forest. Soil Biol Biochem 37:455–461

Saviozzi A, Levi-Minzi R, Cardelli R, Riffaldi R (2001) A comparison of soil quality in adjacent cultivated, forest and native grassland soils. Plant Soil 233:251–259

Shi W, Dell E, Bowman D, Iyyemperumal K (2006) Soil enzyme activities and organic matter composition in a turfgrass chronosequence. Plant Soil 288:285–296

Sinsabaugh RL, Gallo ME, Lauber C, Waldrop MP, Zak DR (2005) Extracellular enzyme activities and soil organic matter dynamics for northern hardwood forests receiving simulated nitrogen deposition. Biogeochemistry 75:201–215

Sinsabaugh RL, Lauber CL, Weintraub MN, Ahmed B, Allison SD, Crenshaw C, Contosta AR, Cusack D, Frey S, Gallo ME, Gartner TB, Hobbie SE, Holland K, Keeler BL, Powers JS,

Stursova M, Takacs-Vesbach C, Waldrop MP et al (2008) Stoichiometry of soil enzyme activity at global scale. Ecol Lett 11:1252–1264

Šnajdr J, Valášková V, Merhautová V, Cajthaml T, Baldrian P (2008a) Activity and spatial distribution of lignocellulose-degrading enzymes during forest soil colonization by saprotrophic basidiomycetes. Enzyme Microb Technol 43:186–192

Šnajdr J, Valášková V, Merhautová V, Herinková J, Cajthaml T, Baldrian P (2008b) Spatial variability of enzyme activities and microbial biomass in the upper layers of *Quercus petraea* forest soil. Soil Biol Biochem 40:2068–2075

Soponsathien S (1998) Study on the production of acetyl esterase and side-group cleaving glycosidases of ammonia fungi. J Gen Appl Microbiol 44:389–397

Steffen KT, Hofrichter M, Hatakka A (2000) Mineralisation of C-14-labelled synthetic lignin and ligninolytic enzyme activities of litter-decomposing basidiomycetous fungi. Appl Microbiol Biotechnol 54:819–825

Steffen KT, Hatakka A, Hofrichter M (2002) Degradation of humic acids by the litter-decomposing basidiomycete *Collybia dryophila*. Appl Environ Microbiol 68:3442–3448

Steffen KT, Cajthaml T, Šnajdr J, Baldrian P (2007a) Differential degradation of oak (*Quercus petraea*) leaf litter by litter-decomposing basidiomycetes. Res Microbiol 158:447–455

Steffen KT, Schubert S, Tuomela M, Hatakka A, Hofrichter M (2007b) Enhancement of bioconversion of high-molecular mass polycyclic aromatic hydrocarbons in contaminated non-sterile soil by litter-decomposing fungi. Biodegradation 18:359–369

Stemmer M, Gerzabek MH, Kandeler E (1998) Organic matter and enzyme activity in particle-size fractions of soils obtained after low-energy sonication. Soil Biol Biochem 30:9–17

Stursova M, Sinsabaugh RL (2008) Stabilization of oxidative enzymes in desert soil may limit organic matter accumulation. Soil Biol Biochem 40:550–553

Toberman H, Freeman C, Evans C, Fenner N, Artz RRE (2008) Summer drought decreases soil fungal diversity and associated phenol oxidase activity in upland *Calluna* heathland soil. FEMS Microbiol Ecol 66:426–436

Tscherko D, Kandeler E (1999) Classification and monitoring of soil microbial biomass, N-mineralization and enzyme activities to indicate environmental changes. Bodenkultur 50:215–226

Tscherko D, Rustemeier J, Richter A, Wanek W, Kandeler E (2003) Functional diversity of the soil microflora in primary succession across two glacier forelands in the Central Alps. Eur J Soil Sci 54:685–696

Tscherko D, Hammesfahr U, Zeltner G, Kandeler E, Bocker R (2005) Plant succession and rhizosphere microbial communities in a recently deglaciated alpine terrain. Basic Appl Ecol 6:367–383

Valášková V, Baldrian P (2006) Estimation of bound and free fractions of lignocellulose-degrading enzymes of wood-rotting fungi *Pleurotus ostreatus*, *Trametes versicolor* and *Piptoporus betulinus*.. Res Microbiol 157:119–124

Valášková V, Šnajdr J, Bittner B, Cajthaml T, Merhautová V, Hofrichter M, Baldrian P (2007) Production of lignocellulose-degrading enzymes and degradation of leaf litter by saprotrophic basidiomycetes isolated from a *Quercus petraea* forest. Soil Biol Biochem 39:2651–2660

van der Wal A, van Veen JA, Smant W, Boschker HTS, Bloem J, Kardol P, van der Putten WH, de Boer W (2006) Fungal biomass development in a chronosequence of land abandonment. Soil Biol Biochem 38:51–60

Waldrop MP, Harden JW (2008) Interactive effects of wildfire and permafrost on microbial communities and soil processes in an Alaskan black spruce forest. Glob Chang Biol 14:2591–2602

Waldrop MP, Zak DR (2006) Response of oxidative enzyme activities to nitrogen deposition affects soil concentrations of dissolved organic carbon. Ecosystems 9:921–933

Waldrop MP, McColl JG, Powers RF (2003) Effects of forest postharvest management practices on enzyme activities in decomposing litter. Soil Sci Soc Am J 67:1250–1256

Waldrop MP, Zak DR, Sinsabaugh RL, Gallo M, Lauber C (2004) Nitrogen deposition modifies soil carbon storage through changes in microbial enzymatic activity. Ecol Appl 14:1172–1177

Wittmann C, Kahkonen MA, Ilvesniemi H, Kurola J, Salkinoja-Salonen MS (2004) Areal activities and stratification of hydrolytic enzymes involved in the biochemical cycles of carbon, nitrogen, sulphur and phosphorus in podsolized boreal forest soils. Soil Biol Biochem 36:425–433

Yergeau E, Kowalchuk GA (2008) Responses of Antarctic soil microbial communities and associated functions to temperature and freeze-thaw cycle frequency. Environ Microbiol 10:2223–2235

Zavarzina AG, Zavarzin AA (2006) Laccase and tyrosinase activities in lichens. Microbiology 75:546–556

Zeglin LH, Stursova M, Sinsabaugh RL, Collins SL (2007) Microbial responses to nitrogen addition in three contrasting grassland ecosystems. Oecologia 154:349–359

Chapter 10
Heterophase Synthesis of Humic Acids in Soils by Immobilized Phenol Oxidases

A.G. Zavarzina

10.1 Introduction

Soil organic matter (humus) is one of the largest carbon reservoirs in the biosphere and holds about 1,500 Pg of C_{org} (Batjes 1996). Humus has a vital significance for the development and functioning of terrestrial ecosystems. Two major processes are responsible for C_{org} accumulation in soils: (1) humification, leading to formation of recalcitrant humic substances (HS); (2) organo-mineral interactions leading to chemical (via adsorption) or physical (occlusion within aggregates) stabilization of organic molecules. As a result of organo-mineral surface interactions organic coatings of varying thickness are formed on the mineral grains (Fig. 10.1 - former 10.5). The most stable C_{org} fraction in soils with mean residence time of $n \times 10^2$–10^3 years is represented by adsorption complexes of humic substances with fine mineral particles (Mikutta et al. 2006). Although clay-sized organo-mineral complexes comprise 50–75% of soil organic matter in cold and temperate soils (Christensen 2001), mechanisms of their formation are not fully understood yet. The concept of sorptive preservation implies that organic matter must occur in a dissolved state prior to adsorption (Guggenberger and Kaiser 2003). This is not in contradiction with formation of fulvic acid complexes with minerals. Fulvic acids (FA) are low molecular weight (0.3–2 kDa) water- and acid-soluble humic compounds, capable of downward migration in the soil profile to adsorption sites. A considerable fraction of soil humus is represented by humic acids (HA), which are highly polydisperse (5–100 kDa) and macromolecular by nature (mean average molecular weight is about 50 kDa). Only low molecular weight HA fractions can move as true solutions from the place of synthesis (e.g., litter) to the underlying mineral soil; mobilization of high molecular weight fractions is only possible as colloids. Indeed, the mean average molecular weight of dissolved organic matter in

A.G. Zavarzina
Faculty of Soil Science, Moscow State University, Moscow 119991, Russia
e-mail: zavarzina@mail.ru

G. Shukla and A. Varma (eds.), *Soil Enzymology*, Soil Biology 22,
DOI 10.1007/978-3-642-14225-3_10, © Springer-Verlag Berlin Heidelberg 2011

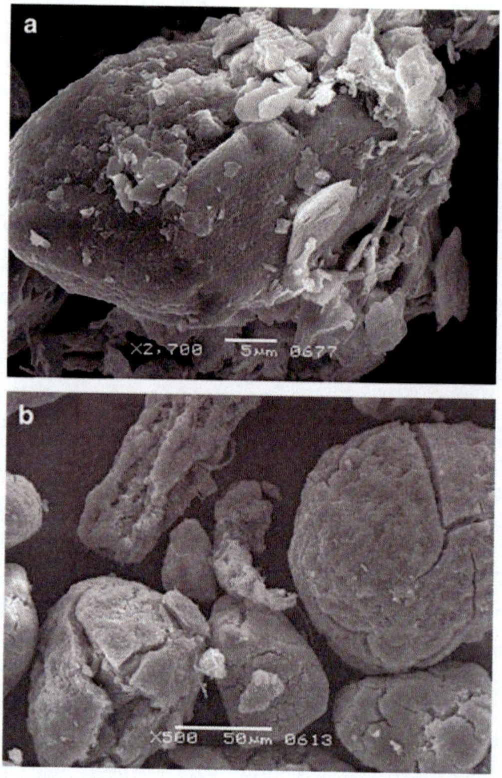

Fig. 10.1 SEM images of organic coatings on the mineral grains: (**a**) A_h horizon of albeluvisol: thin wave-like coatings on silt-sized particles; (**b**) B_{hf} horizon of Al–Fe humic podzol: thick coatings consisting of amorphous Al oxyhydroxide–humic acid complexes on the non-weathered surface of primary minerals. The cracks are formed upon coating drying

soil solutions is 1.7 kDa (Perdue and Ritchie 2004). One can assume that HA polymers are formed in situ in mineral soil horizons. A possible mechanism is heterophase polymerization of low molecular weight (thus soluble and mobile) precursor material in presence of catalytically active solid phases. In this chapter, the available data supporting the concept of surface HS polymerization are summarized, and evidence for the key role of immobilized phenol oxidases and solid matrix in the catalytic synthesis of HAs is provided.

10.2 Synthesis of Humic Substances from Soluble Precursors

Two main humification pathways co-exist in soils: (1) synthesis of HS from polymeric precursors (lignins, melanins) by their partial oxidative degradation or (2) synthesis of HS from low molecular weight precursors by their oxidative coupling (Stevenson 1994). While the first pathway (lignin–protein theory) is more typical for wood, litter, or poorly drained peaty horizons, the second pathway should be important way of HA formation in mineral soil layers (Table 10.1).

Table 10.1 Principal differences in humification processes in litter and humus horizons of forest soils

Property	Litter	Humus horizon
Actual enzyme activity	High	Moderate/Low
Main starting material	Particulate organic matter (foliage, twigs, wood at different stages of decomposition)	Leached-down soluble organic substances, root exudates and root decomposition products, microbial metabolites
Initial molecular weight of precursor material	High molecular weight	Low molecular weight
Dominant solid phase	Organic	Inorganic
Dominant process	Solid-state fermentation	Heterophase synthesis
Reactions, leading to HS formation	Oxidative transformation	Precipitation or surface polymerization
Product	Humic colloids	Humus–mineral adsorption complexes

Synthesis of HS from soluble compounds occurs by: (1) oxidative coupling of polyphenols with nitrogenous compounds and other soluble precursors (polyphenol theory); (2) sugar–amino acid condensation (Maillard reaction). The polyphenol theory is more popular and postulates that soluble phenolic substrates are oxidized into highly reactive phenoxy radicals and quinones, which then undergo non-enzymatic spontaneous coupling reactions. Dark-colored heterogeneous structures of varying composition and molecular weight are formed as a result of the process. Polymerization occurs via C–C and C–O coupling of phenolic reactants and N–N and C–N coupling of aromatic nitrogenous compounds (Sjoblad and Bollag 1981). It is widely accepted that HS formation is a catalytic process, rather than auto-oxidation; however, the role of enzymes and abiotic catalysts in synthesis of HS is still under the discussion (Bollag et al. 1998).

10.2.1 Enzymatic Catalysis

Peroxidases (EC 1.11.1.7), laccases (EC 1.11.1.14), and tyrosinases (EC 1.14.18.1) are the major enzymes that catalyze polymerization of phenolic compounds via a free radical mechanism. Peroxidases are heme-containing oxidases catalyzing one-electron oxidation of a broad spectrum of phenolic substrates by H_2O_2 with formation of phenoxy radicals and H_2O (Fig 10.2a). Laccase is a multicopper oxidase that performs four one-electron oxidations of the wide range of substituted phenols and aromatic amines by O_2 with formation of semiquinones and quinones; O_2 is reduced to H_2O (Fig 10.2b). Tyrosinases contain a copper pair at the active site and catalyze two concomitant reactions: *o*-hydroxylation of monophenols yielding *o*-diphenols (monophenolase activity); $2e^-$ oxidation of *o*-diphenols to *o*-quinones (diphenolase activity); O_2 is reduced to H_2O in the course of the reaction (Fig 10.2c).

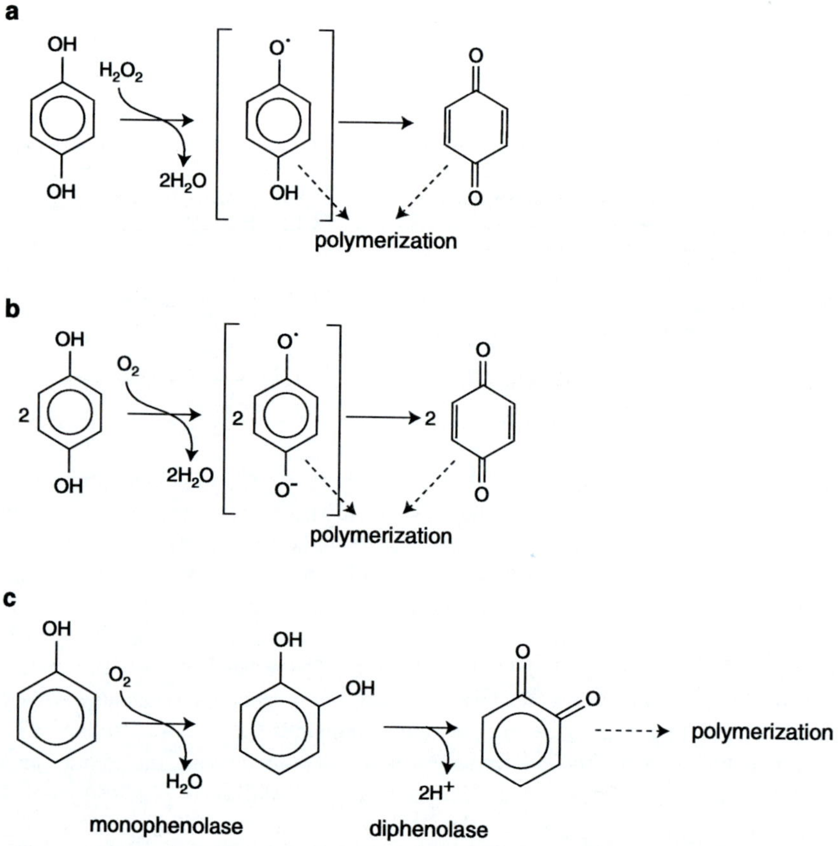

Fig. 10.2 Schematic representation of phenolic substrates oxidation by (**a**) peroxidase, (**b**) laccase, and (**c**) tyrosinase

10.2.1.1 Occurrence of Phenol Oxidases in Soils

Among the enzymes catalyzing humus polymerization, laccases followed by peroxidases are most widespread and common in soils; tyrosinase is less abundant (Criquet et al. 2000; Di Nardo et al. 2004; Snajdr et al. 2008). Fungi are the main source of phenol oxidases in soils (see Chap. 11), although peroxidases and laccases may be also excreted by bacteria and plant routes (Gramss et al. 1998). Phenol oxidase activities in soils exhibit high spatial heterogeneity, more pronounced in litter than in underlying organo-mineral horizons (Snajdr et al. 2008). As a rule, phenol oxidase activities decrease with depth following the decrease in microbial biomass, organic matter content, and its utilizable forms; mineral horizons are characterized by several times lower activities of enzymes than litter (Snajdr et al. 2008). In podzol soils with surface (A_h) and subsurface (B_{hf}) organic-rich horizons two maxima of laccase and peroxidase activities were observed which correlated with distribution of microfungi, organic matter, and Al (Fe) oxyhydroxides (Zavarzina et al. 2007). Phenol oxidase activities in soils may display optimal moisture levels, above which low oxygenation inhibit activity (Fenner et al. 2005) and below which activity declines due to moisture limitations (Toberman et al. 2008).

10.2.1.2 Phenol Oxidase Distribution among Soil Phases

Soil structure is defined as an arrangement of organic, mineral, and organo-mineral particles, forming aggregates of different size and stability with aqueous phase present in macro-, mezo- and micropores between the aggregates or inside them. Soil enzymes can be distributed among soil aqueous and solid phases upon release from their producers. It is widely accepted that free enzymes are unstable in soil environment and are therefore quantitatively insignificant (Nannipieri and Gianfreda 1998). Immobilized enzymes are more resistant to changes in environmental factors, proteolysis, and inhibitory substances, which allow higher enzyme concentrations to persist in soils (Quiquampoix et al. 2002; Tietjen and Wetzel 2003). Binding of enzymes to solid surfaces is determined by the enzyme isoelectric point, surface area, and charge of solid supports. In mineral soil horizons, most phenol oxidase activity is usually found in the silt and clay-sized fractions (Sarkar et al. 1989; Allison and Jastrow 2006). This fraction contains primary minerals, clay minerals, amorphous metal oxyhydroxides, and humus–mineral complexes, which are <0.05 mm in size. Primary minerals are characterized by low surface area and are therefore poor adsorbents for the enzymes and organic matter. Clay minerals such as smectites and illites as well as humus–clay complexes possess large surface areas (100–600 $m^2\ g^{-1}$) but carry an overall negative charge (Tipping and Cooke 1982; Schulze 2002). At pH 4–6, typical for most forest soils, phenol oxidases are also negatively charged (pI 3.0–4.5). This can limit enzyme–mineral interactions due to electrostatic repulsions. Oxyhydroxides of Al and Fe are common and abundant in soils and are present as individual minerals (e.g., gibbsite, goethite) or as coatings on silicates and alumosilicates. Besides possessing large surface area, Al and Fe oxyhydroxides carry a positive charge at pH < 8.0 (Huang et al. 2002a) enabling strong sorption of negatively charged enzymes. Dominant adsorption mechanisms of phenol oxidases onto the oxyhydroxide surfaces include electrostatic attraction (Ahn et al. 2007) and ligand exchange (Naidja et al. 1997). It was found that pure Al hydroxide adsorbed about nine times more laccase from *Trametes villosa* than other non-crystalline minerals such as ferrihydrite ($Fe_5HO_8{\cdot}4H_2O$) or birnessite (δ-MnO_2); laccase activity and kinetic properties remained almost unaffected (Ahn et al. 2007). Coating of clay minerals by Al hydroxides favored adsorption of tyrosinase (Naidja et al. 1997). Kaolinite and illite coated by Al hydroxide adsorbed 5–10 times more *Panus tigrinus* laccase than uncoated minerals (Zavarzina 2006a). Thus, oxyhydroxides of Al and Fe can be considered as major inorganic supports for phenol oxidases in soils due to ubiquity, abundance, and surface characteristics, enabling effective enzyme immobilization.

The role of amorphous Al and Fe compounds as surfaces modifiers is particularly important in sandy soils (e.g., podzols), where primary minerals form the bulk of mineral matrix. The correlation between phenol oxidase activities and depth distribution of Al and Fe oxyhydroxides (Zavarzina et al. 2007) additionally supports this contention.

10.2.1.3 Synthesis of Humic Substances in the Aqueous Phase

Polyphenol theory of humus formation is largely based on experiments with homogeneous (monophase) systems where enzymes and their substrates were in dissolved state and reacted with each other in the solution bulk. In the history of the study of HA synthesis, the following work can be mentioned: synthesis of dark-colored humic-like products in mixtures containing pyrogallol, pepton, H_2O_2, and cell-free culture liquid of *Aspergillus niger* and *Penicillium sp* (Kononova 1966); synthesis of HS in mixtures containing catechin or hydroquinone, amino acids, glucose, and laccase of *Polystictus versicolor* (Trojanowski 1961, as reviewed by Kononova 1966); formation of HS from monophenols, phenolic acids, and N-containing compounds in presence of phenol oxidases (Flaig 1966). Later experiments have shown that condensation in aqueous phase is highly dependent on precursor concentration. At substrate concentrations of 0.5–10 mM, only oligomers (Bollag et al. 1983; Liu et al. 1985; Leontievsky et al. 1999; Zavarzina 2006a), polycondensates with *m*/*z* ranges up to 900 (Naidja et al. 1998), or polymers up to 4.0 kDa (Rittstieg et al. 2002) were formed. At high precursor concentrations (>1 mg ml^{-1}), the molecular weight of the soluble polymer could reach 10 kDa (Zavarzina 2006a); however, further polymerization was terminated by precipitate formation process, which consumed the available monomers. The insoluble product, consisted of high molecular weight fraction (>75 kDa, minor peak) and low molecular weight co-precipitate (10 kDa, major peak).

Although homogeneous catalysis is important for the understanding of principle reaction mechanisms, it has low relevance to the soils where enzymes are mostly bound to solid surfaces and work in heterogeneous system (see Sect. 10.2.1.1). If it is assumed that polymeric HS are formed in the aqueous phase, the possible mechanism can be so-called *precipitation or adsorption polymerization*. The precipitative polymerization mechanism is well known from organic chemistry, for example, for polyaniline formation (Fedorova and Stejskal 2002, Yagudaeva et al. 2007). The factors that favor this reaction are high monomer concentrations and a chemically inert template with high surface area (e.g., silica gel). If one applies precipitation polymerization to the humus synthesis in soils, the following reactions should occur: (1) phenolic compounds are oxidized at the solid surface by immobilized enzymes to phenoxy radicals and quinones which then (2) dissociate from the enzyme active site and undergo spontaneous coupling in equilibrium solution with (3) subsequent deposition and immobilization of the insoluble (polymeric) product on the solid surface (Fig. 10.3). An example of the laboratory study that apparently mimicked this process was that of Naidja et al. (1997): they

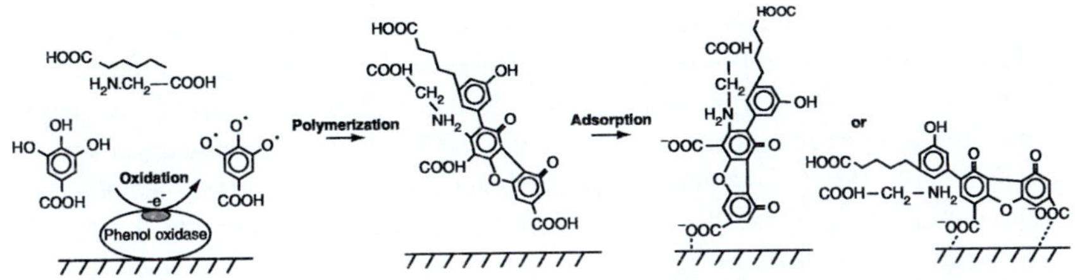

Fig. 10.3 The possible reaction sequence during synthesis of humic substances by precipitation polymerization of monomeric precursors in presence of immobilized laccase

demonstrated that oxidation of dissolved catechol by tyrosinase, immobilized on Al oxyhydroxide-coated montmorillonite resulted in formation of dark-colored products that were adsorbed on the mineral surface and formed organic coating. Infrared spectroscopy revealed similarity of the adsorbed compounds to natural HSs. Precipitation polymerization could also lead to polymeric precipitate formation in some abiotic systems when primary minerals or metal oxides are used as abiotic oxidants (see Sect. 10.2.2). However, in the natural soil environment the formation of polymeric HS on soil minerals by precipitation polymerization is questionable for the following reasons:

1. Monomeric substrate condensation to insoluble products requires high solution concentrations (>1 g L^{-1}). Average concentrations of dissolved organic carbon in natural environment are several orders of magnitude lower: 0.1 mg L^{-1} in groundwater and up to 100 mg L^{-1} in peat bogs (Klavinš 1997; Perdue and Ritchie 2004). No formation of insoluble polymeric product can be expected at such conditions.
2. If it is assumed that the soil solution can be concentrated to appropriate levels (e.g., upon drying), the presence of charged solid surfaces should interfere in polymerization process in the aqueous phase. Radical self-coupling (coupling with each other) dominates in systems that lack appropriate solid surfaces to participate in cross-coupling (Huang and Weber 2004). Charged solid surfaces, in addition to potentially binding phenoxy radicals, can adsorb original phenolic substrates, reducing their concentration in soil solution. Fast adsorption of dissolved organic matter, and especially of phenolic compounds onto soil mineral phases, is a well-known phenomenon (Lehmann et al. 1986; Dalton et al. 1989; Gallet and Pellissier 1997; Kalbitz et al. 2000). Adsorption is largely irreversible (Lehmann and Cheng 1988; Cecchi et al. 2004), resulting in low concentrations of individual phenolic acids in both the aqueous phase and soil extracts. For example, in soddy-podzolic soils, amounts of ethanol-extractable phenolic acids were 15–150 μg per 100 g of soil (Kuvaeva 1980), while average amount of identifiable lignin-derived phenols in soil solutions comprised 0.6% of DOM (Perdue and Ritchie 2004).
3. If it is assumed that temporal increase in concentration of soil solution occurs and the soil mineral phase is inert and does not adsorb enzyme substrates and

monomeric reaction products (e.g., elluvial horizons in podzols, consisting largely of weathered primary minerals), then the polymerization in the solution bulk will be limited by the reaction kinetics. The polymerization process leading to precipitate formation is slow (>24 h) even in homogeneous systems (Kononova 1966; Zavarzina 2006a). In heterogeneous systems (especially in an unstirred medium) the diffusion of substrates to the active site of the enzyme becomes more limiting.

4. And finally, the problem with polymerization in dilute solution lies also in thermodynamics of the polymerization reaction (Lambert 2008). Taking 0.5 M glycine solution as an example, Lambert (2008) has demonstrated that successive polymerization events in solution lie further and further up on the G^o scale, making polymer formation in aqueous phase unfavorable.

It can be thus concluded that the synthesis of high molecular weight HAs is barely possible in the aqueous soil phase under natural soil conditions. Formation of only fulvic acid–like products can be expected. Accepting that polymeric HA (50–100 kDa) do exist as coatings on soil minerals, some other mechanisms than homogeneous catalysis or precipitation polymerization should be responsible for their formation if not only they originate from humic colloids that undergo solubilization (Sect. 11.4.2.1) and subsequent adsorption.

10.2.1.4 Synthesis of Humic Substances on the Solution/Solid Interface

As discussed in previous section, the following factors should be kept in mind when dealing with humus formation from soluble precursors in natural soil systems: (1) substrate concentrations in the bulk soil solution are very low; (2) enzymes are present in an immobilized form; (3) monomeric phenolic compounds leached from the forest floor or excreted by plant roots become rapidly and irreversibly adsorbed onto solid soil matrix. It is generally accepted that low extractability of phenolic compounds from soils is a consequence of their high reactivity at solid surfaces resulting in the oxidative cross-coupling to solid phase or polymerization. Thus, it is reasonable to assume that synthesis of polymeric HS from soluble precursors in mineral soil horizons proceeds on the solid–solution interface and not in the solution bulk. The following experimental data support this concept:

1. At low solution concentrations, the presence of interfaces substantially accelerates the rates of substrate coupling in comparison to solid-free systems (e.g., Huang et al. 2002a, b). This effect is explained by the concentration of monomers on surface due to e.g., electrostatic attraction (Danielewicz-Ferchmin and Ferchmin 2004), which helps to overcome the energetic barrier to polymerization, making polymerization thermodynamically favorable (Lambert 2008).
2. Direct experimental evidence exists that polymerization reactions at the surface of a solid support precede polymerization in the supernatant solution, even at high monomer concentrations (Fedorova and Stejskal 2002, Sapurina et al. 2003). Polymerization in adsorbed state (so-called *surface or boundary*

polymerization) gives rise to organic coatings that consists of polymeric, partially non-extractable compounds.

Although surface polymerization process is well known from polymer science (Sapurina et al. 2002; Boufi and Gandini 2002), experimental data that demonstrate its application to humus chemistry are rather scarce. Surface polymerization instead of precipitation polymerization could well have occurred in the study of Naidja et al. (1997) (see Sect. 2.1.3), but unfortunately molecular weight of the mineral-bound reaction product was not measured. To fill this gap, we have made an attempt to demonstrate the possibility of high molecular weight HAs formation by surface polymerization of monomers in the presence of immobilized fungal laccase, and to elucidate the effect of the nature of the mineral support and the role of biotic catalysts in the polymerization process (Zavarzina 2006a, b). It is necessary to outline briefly the experimental design used, to show that the polymerization process proceeded at solid–solution interface. First, purified fungal laccase was immobilized by adsorption on kaolinite, kaolinite–hydroxyaluminum complex, illite, or montmorillonite. After washing of enzyme–mineral complexes with acetate buffer (pH 4.5), precursor mixture solution containing gallic, caffeic, ferulic, hydroxybenzoic, vannilic acids, tryptophan, and tyrosin was added. On the basis of preceding adsorption experiments, the concentration of precursor solution was selected so as to achieve maximal adsorption of monomers and multilayer formation. After 15 min of the reaction period (no polymerization in the bulk of the reaction mixture occurred according to HPLC analysis of supernatants and monophase controls), the mixtures were centrifuged, the supernatant solutions with unbound monomers were removed, and the pellets containing enzyme and adsorbed precursors were rapidly washed with acetate buffer. Then, fresh buffer was added to the pellets and the mixtures were incubated at room temperature in the dark for 24 h without agitation. Initially white kaolinite-based supports became brown within 15 min of adsorption stage; during further incubation period, the mineral staining became progressively darker in color. After 24 h, the alkali-extractable products were analyzed by spectroscopic methods and gel filtration. They were found to be polymeric (molecular weights from 5 to >75 kDa), resembling soil HA by visible and infrared spectra and molecular weight distributions (Fig. 10.4). Addition of EDTA to the extracts in order to destroy possible metal bridges between HA "subunits" caused only slight reduction in the amount of high molecular weight fraction, suggesting that the extracted reaction products were true macromolecular by their nature. It should also be emphasized that some portion of high molecular weight products could well have been retained by the clay surfaces as the extraction was not complete. The peak of high molecular weight fraction was largest in the extract from Al oxyhydroxide–coated kaolinite. This is possibly because this mineral adsorbed the largest amount of monomers and had the lowest suppressing effect on the laccase activity compared to the other minerals used. HA-like products formed on montmorillonite were most polydisperse, suggesting that surface morphology of mineral supports may be another important factor that determines molecular weight distribution pattern of the polymeric product. No high molecular

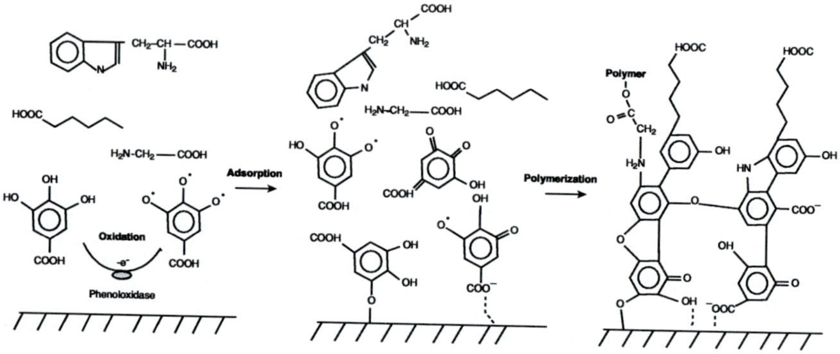

Fig. 10.4 The possible reaction sequence during synthesis of humic substances by surface polymerization of monomeric precursors in presence of immobilized laccase

weight fraction formation was observed in parallel monophase experiments in which the same enzyme activities and substrate concentrations as those reacted on solid surfaces were used. No products with molecular weight larger than 5 kDa were formed on the minerals in absence of immobilized laccase (abiotic controls). Surprisingly, humic-like polymers were formed on hydroxyaluminum kaolinite even at reverse mode of reactants addition i.e., when phenolic compounds were adsorbed first and then laccase was added (unpublished data). It should be mentioned that inorganic nature of solid support was not a necessary pre-condition for the polymeric HS formation. In our experiments on soil HA transformation in submerged culture of laccase-producing fungus *P. tigrinus*, polymerization of low molecular weight HA fractions into higher molecular weight products was observed on the mycelium surface (Zavarzina et al., unpublished data) (Fig. 10.5).

Although the possibility of HS formation by surface polymerization is not in doubt, the reaction mechanisms remain largely unclear. We can speculate that surface humus polymerization is a complicated heterophase process consisting of the following possible steps: (1) electrostatic substrate attraction to the surface; (2) enzyme-catalyzed substrate oxidation to free radicals at solid/solution interface; (3) adsorption of free-radical intermediates together with initial substrate on the mineral surface; (4) spontaneous polymerization/cross-coupling with polymer chain growth perpendicular to the surface (Fig. 10.6). It is important to underline that the concept of surface polymerization implies that polymeric organic films are produced on the surface before polymerization in the solution bulk has started (Sapurina et al. 2003). However, further experiments are needed to define the reaction sequence at solid/solution interface leading to humic polymers formation. For example, it might be that at first substrate adsorption occurs and then its oxidation proceeds directly on the surface by immobilized enzyme molecules. In this case, adsorbed phenolic substrate should interact somehow with enzyme active site (substrate adsorption onto the enzyme molecule is required). The study of Wershaw and Pinckney (1980) apparently supports such pathway as they found that organic matter is often bound to mineral surfaces by amino acids or peptides.

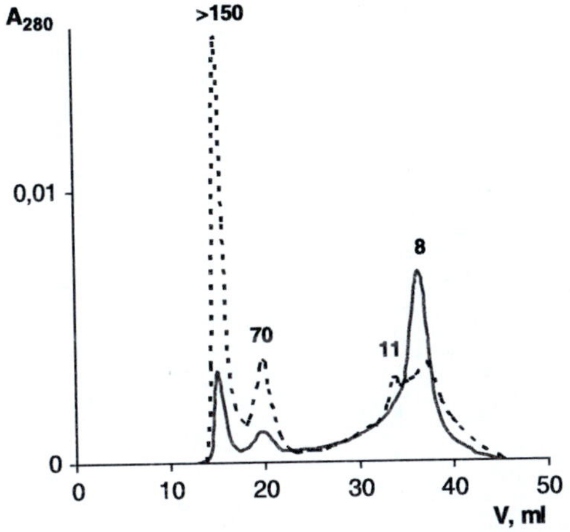

Fig. 10.5 Gel-chromatograms of HA, extracted from mycelium of *Panus tigrinus* on days 9 (solid line) and 12 (dashed line) of submerged cultivation of the fungus in presence of dissolved soil HA. The increase in the amounts of high MW fractions was due to surface oxidation and polymerization of HA by laccase, produced by the fungus. Bold numbers represent MWs in kDa. Column 1×60 cm, Sephadex G-100 gel, elution by 0.025M Tris–HCl buffer (pH 8.2) with 0.05M NaCl, 0.1% SDS, and 0.02% NaN_3 at a flow rate 3 ml h^{-1}

Perpendicular orientation of the polymeric product to the mineral surface is an important condition of surface polymerization process, experimentally confirmed for e.g., polyaniline formation (Sapurina et al. 2002). Vertical orientation of polymers should have a positive effect both on the reaction thermodynamics (e.g., Gerstner et al. 1994) and on enzyme activity, because there is a high probability that at least some enzyme molecules would not be inactivated by the growing polymer (as in the case of planar polymer orientation). The concept of vertical orientation of humic polymers during surface polymerization is in good agreement with the models of organic matter organization on the natural solid surfaces. Recent research has shown that aluminosilicate sediments with the loadings of organic matter <3 mg C m^{-2} have less than 15% of their surface coated (Arnarson and Keil 2001). It was experimentally confirmed that distribution of organic (humic) material on the mineral surface is not uniform and it is organized in discrete spots (patches) with some vertical extension; the coating increases in thickness while retaining practically the same surface coverage (Wang and Xing 2005).

The recent concept of zonal self-organization of adsorbing molecules on mineral surfaces (Wershaw, 1993; Guggenberger and Kaiser 2003; Kleber et al. 2007) can provide an explanation for the effective surface polymerization of monomers when first precursors are adsorbed and then the enzyme is added (see above). It is considered that with increasing surface loading, an increasing portion of the sorbing molecules do not attach to mineral surface, but form organic multilayers as a result of hydrophobic interactions or bridging by polyvalent cations (Guggenberger and Kaiser 2003). The first layer (contact zone) may be more compact due to direct

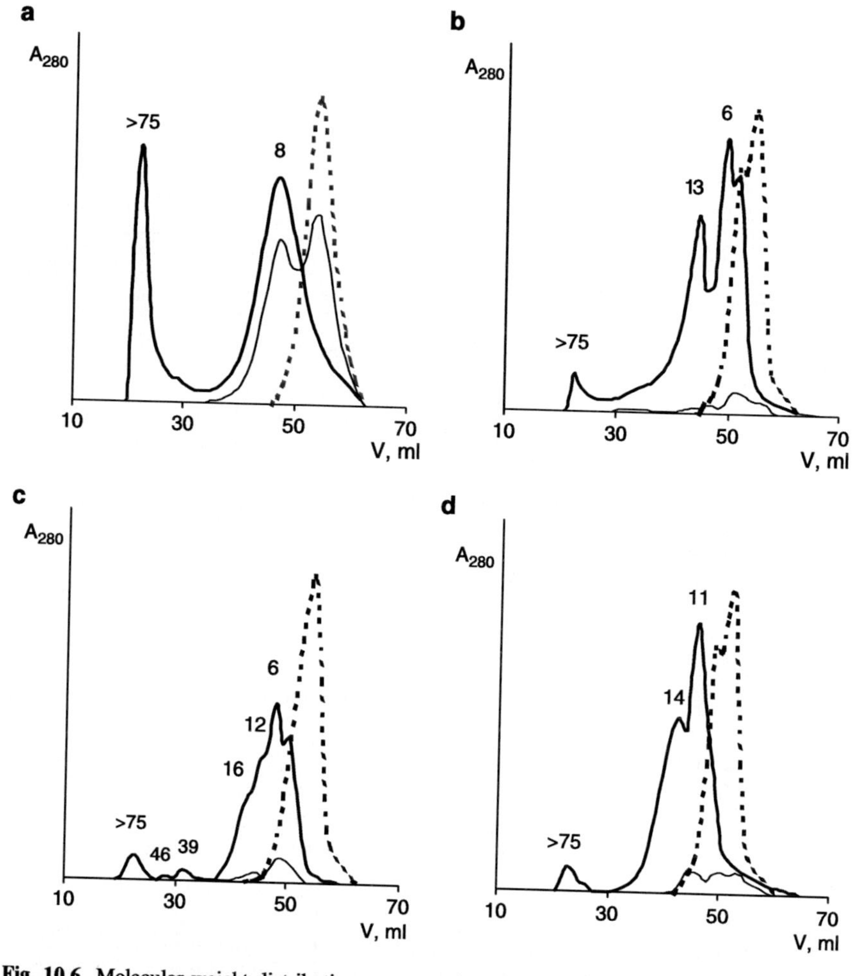

Fig. 10.6 Molecular weight distribution patterns of synthetic humic substances formed on the surface of clay minerals in the presence (*bold line*) or absence (*thin line*) of immobilized laccase: (**a**) hydroxyaluminum-kaolinite; (**b**) kaolinite; (**c**) montmorillonite; (**d**) illite. *Bold numbers* represent molecular weights in kDa. Gel chromatogram of precursor mixture is shown in dashed line. Column 1 × 60 cm, Sephadex G-75 gel, elution by 0.025 M Tris–HCl buffer (pH 8.2) with 0.05 M NaCl, 0.1% SDS, and 0.02% NaN_3 at a flow rate 8 ml h^{-1}

electrostatic attraction to the mineral surface; the components of second layer (hydrophobic zone) are more dynamic and can exchange with soil solution although being retained with considerable force, while molecules in the outer region of hydrophobic zone are loosely retained by cation bridging or hydrogen bonding and form kinetic zone (Kleber et al. 2007). Such organization of phenolic and nitrogenous precursors upon adsorption should enable enzyme and O_2 diffusion within organic "brushes" resulting in the oxidative cross-coupling between the components of contact, hydrophobic, and kinetic zones.

Oxidative polymerization of soluble precursors by immobilized enzymes at solid/solution interface can serve as good explanation for the existence of high molecular weight HA-like coatings on mineral particles. The reaction proceeds despite the reduced catalytic efficiency (lower V_{max} values) and substrate affinity (higher K_m constants) of immobilized enzymes (Nannipieri and Gianfreda 1998). The question arises as to whether presence of biotic catalyst is obligatory or polymer formation on the mineral surface can occur abiotically as well.

10.2.2 *Abiotic Heterogeneous Catalysis*

Numerous studies have shown that various inorganic soil constituents, such as metal oxides (Shindo and Huang 1984; Lehmann et al. 1986; McBride 1987), hydroxides (Liu and Huang 2002), clay minerals (Wang et al. 1978), and even primary minerals (Shindo and Huang 1985) possess oxidative activities and can catalyze transformation of phenolic compounds into humic-like substances. Smectites were even able to catalyze Maillard reaction (Gonzalez and Laird 2004). Manganese (IV) oxides, such as common soil mineral birnessite (δ-MnO_2) are considered as the most powerful oxidants of phenolic compounds. The catalytic power of Fe (III) oxyhydroxides was much lower (Shindo and Huang 1984). As for the clay minerals, montmorillonite and illite were found to be better catalysts than kaolinite because their active sites were located on the planar surfaces and not on crystal edges as in case of kaolinite (Wang and Li 1977). Among the primary minerals studied, the oxidative power of tephroite (Mn-bearing silicate) was the greatest, followed by actinolite, hornblende, fayalite, augite, biotite, and muscovite = orthoclase = microcline = quartz (Shindo and Huang 1985). In general, the presence of transition metals (especially Mn) on the mineral surface or in the crystal lattice is required for efficient abiotic catalysis (Wang et al. 1986; Huang 2000). It was found that the nature of phenolic substrates affected the rate of their transformation into HA-like products. Polyphenols and polyhydroxyphenolic acids with para- and ortho-OH groups were more rapidly converted into HAs by Mn oxides than phenolic compounds with meta-oriented OH groups (Pohlman and McColl 1989; Shindo 1990). Any electron-attracting carboxyl group substituted on the ring reduced the polymerization rate while electron-releasing methyl group increased the rate (Wang et al. 1983).

The widely accepted mechanism of phenol oxidation by soil metal oxides is precipitation polymerization (Stone and Morgan 1984; McBride 1987), which involves the following steps: (1) binding of the organic molecule to the surface via phenolic or carboxylic groups; (2) electron transfer from the adsorbed organic to the oxide (surface oxidation); (3) release of oxidized molecule and reduced metal into solution due to dissociation of the complex; (4) under the aerobic conditions, the reduced metal is quickly re-oxidized, while semiquinones and quinones produced from phenolic substrate oxidation undergo spontaneous polymerization in aqueous phase with subsequent adsorption of the polymeric product on the mineral.

An alternative mechanism has been proposed recently for catechol polymerization, which includes: (1) heterogeneous catechol oxidation on metal oxide surface leading to release of reduced metal in solution; (2) immediate complexation of reduced metal ions by dissolved catechol; (3) homogeneous oxidation of metal–catechol complexes by dissolved oxygen, resulting in the formation of insoluble polymers (Colarieti et al. 2006). If no metal–organic complex dissociation occurs, the insoluble organo-mineral compounds are formed (Wang et al. 1978). In both cases, precursor surface complex formation is prerequisite for the electron transfer and is a rate-limiting step, while the electron transfer within surface complex is rapid (Matocha et al. 2001).

The overall difficulty with analyzing the results of abiotic catalysis in terms of polymeric HS formation is that the molecular weights of reaction products have rarely been measured. Dark-colored compounds were designated as polymeric HS on the basis of spectroscopy data and product insolubility either in aqueous (Liu and Huang 2002) or in the acidic medium (e.g., Wang et al. 1978; Shindo and Huang 1984, 1985; Shindo 1990). Even if we assume that these precipitates were polymeric, their formation under natural soil conditions is hardly possible because very high precursor concentrations (1–10 mg ml^{-1}) and long reaction periods (2–14 days) were used to produce them under "ideal" laboratory conditions. When mass spectrometry and high pressure liquid chromatography were used to analyze molecular weights of soluble and insoluble products of abiotic oxidation, it was found that they were mostly oligomeric by nature (Lehmann and Cheng 1988; Naidja et al. 1998).

Comparative studies on synthesis of humic-like substances using biotic and abiotic catalysts have shown that enzymatic oxidation of phenolic precursors was substantially more rapid than abiotic reaction (Pal et al. 1994; Bollag et al 1995; Naidja and Huang 2002; Ahn et al. 2006). Molecular weights and the degree of aromatic ring condensation were higher in the products of biotic catalysis (Naidja et al. 1998). Interestingly, it was found that heterogeneous catalysis of liquid substrates by solid abiotic catalysts obeyed the Henri–Michaelis–Menten kinetic model (Naidja and Huang 2002). Determination of the kinetic constants for abiotic catalysis allowed direct quantitative comparison of catalytic efficiency between the enzyme (tyrosinase) and the mineral oxide (birnessite). It was found that while the V_{max} of tyrosinase was 2.5–4 times higher than that for birnessite, the turnover frequency (k_{cat}) and the efficiency (k_{cat}/K_m) of the enzyme were three to four orders of magnitude higher than those of the mineral oxide (Naidja and Huang 2002). Higher efficiency of enzymes as oxidative agents than soil minerals is attributed to the ability of continuous oxidation of a substrate (Pal et al. 1994) while inorganic soil constituents can lose their oxidizing ability quite rapidly. The number of active sites on the mineral surface considerably decreases upon mineral aggregation or adsorption of reaction products on mineral surface (Lehmann and Cheng 1988). The abiotic reaction can be thus terminated once organic coating on the mineral surface has been produced.

It can be thus concluded that abiotic catalysis cannot lead to highly polymeric products formation (>10 kDa), at least at solute concentrations close to those in natural soil environment. Absence of polymer formation is not only the result of slow kinetics of the process but also a result of active sites inactivation by reaction products. Nevertheless, abiotic oxidation route can be considered as one of the important mechanisms of organic matter stabilization in soils, especially at primary stages of soil development.

10.3 Conclusions

The experimental data summarized in this chapter allow us to conclude that high molecular weight HAs and their adsorption complexes in mineral soil horizons can originate from surface polymerization of low molecular weight precursor molecules on minerals in the presence of enzymatic catalysts. Polymerization at the surface precedes precipitation polymerization in the solution bulk. The positive effect of interfaces on the polymerization process lies in the energetics of the reaction: substrate concentration at the solution/solid interface is higher than that in dilute solution bulk, making polymerization reaction thermodynamically favorable. The positive effect of enzymatic catalysts on surface polymerization process lies in the reaction kinetics, resulting in accelerated rates of polymer formation in comparison to enzyme-free systems. Surface polymerization is complicated heterophasic multistep process, which is likely to include adsorption/oxidation and polymerization/cross-coupling steps. The concept of surface polymerization implies vertical polymer chain growth on the solid support surface, which is in good agreement with models of humic matter organization at natural solid phases (Kleber et al. 2007). It is very likely that surface polymerization produce polymeric organo-mineral complexes with different strengths of organic component binding to the mineral surface. Molecules attached directly to the mineral surface form humin-like structures upon oxidation. Organic compounds from the next layers can be extracted as HA fraction after oxidative polymerization of adsorbed organic molecules occurred. As a general conclusion, heterophase synthesis of HAs at mineral surfaces by immobilized enzymes may have wide application to soil systems given that enzymes and their substrates are commonly bound to mineral surfaces, and substrate concentrations in the aqueous soil phase are extremely low. Synthesis of humic substances on solid–solution interfaces can be particularly important in soils of cold and temperate humid climate, rich in Al and Fe oxyhydroxides. The mechanisms of heterophase reactions of humus synthesis remain largely unclear and need to be elucidated in future research. The effects of abiotic catalysts on the heterogeneous enzymatic catalysts should be also defined. Experimental evidence exists that enzyme-catalyzed reaction can be inhibited in presence of strong inorganic oxidants (such as birnessite) due to enzyme deactivation by humic-like compounds produced by the mineral (Ahn et al. 2006).

Acknowledgments The work was supported by the Programme No. 15 of the Presidium of the Russian Academy of Sciences "Origin of the Biosphere and Evolution of Geobiological systems" and Russian Foundation for Fundamental Research grant No. 09-04-00570. The author expresses her sincere gratefulness to Prof. Richard P. Beckett for English revision and to Dr. Alexander A. Lisov for the assistance in preparation of Figs. 10.1, 10.3, and 10.4.

References

Ahn MY, Martinez CE, Archibald DD, Zimmerman AR, Bollag JM, Dec J (2006) Transformation of catechol in presence of a laccase and birnessite. Soil Biol Biochem 38:1015–1020

Ahn MY, Zimmerman AR, Martinez CE, Archibald DD, Bollag JM, Dec J (2007) Characteristics of *Trametes villosa* laccase adsorbed on aluminum hydroxide. Enzym Microb Technol 41:141–148

Allison SD, Jastrow JD (2006) Activities of extracellular enzymes in physically isolated fractions of restored grassland soils. Soil Biol Biochem 38:3245–3256

Arnarson TS, Keil RG (2001) Organic-mineral interactions in marine sediments studied using density fractionation and X-ray photoelectron spectroscopy. Org Geochem 32: 1401–1415

Batjes NH (1996) Total carbon and nitrogen in the soils of the world. Eur J Soil Sci 47:151–163

Bollag JM, Minard RD, Liu SY (1983) Cross-linkage between anilines and phenolic humus constituents. Environ Sci Technol 17:72–80

Bollag JM, Meyers C, Pal S, Huang PM (1995) The role of abiotic and biotic catalysts in the transformation of phenolic compounds. In: Huang PM, Berthelin J, Bollag JM, McGill WB, Page AL (eds) Environmental impact of soil component interactions, vol 1. CRS/Lewis, Boca Raton, FL, pp 299–310

Bollag JM, Dec J, Huang PM (1998) Formation mechanisms of complex organic structures in soil habitats. Adv Agron 63:237–265

Boufi S, Gandini A (2002) Formation of polymeric films on cellulosic surfaces by admicellar polymerization. Cellulose 8:303–312

Cecchi AM, Koskinen WC, Cheng HH, Haider K (2004) Sorption-desorption of phenolic acids as affected by soil properties. Biol Fertil Soils 39:235–242

Christensen BT (2001) Physical fractionation of soil and structural and functional complexity in organic matter turnover. Eur Soil Sci 52:345–353

Colarieti ML, Toscano G, Ardi MR, Greco G (2006) Abiotic oxidation of catechol by soil metal oxides. J Hazard Mater 134:161–168

Criquet S, Farnet AM, Tagger S, Le Petit J (2000) Annual variations of phenoloxidase activities in an evergreen oak litter: influence of certain biotic and abiotic factors. Soil Biol Biochem 32:1505–1513

Dalton BR, Blum U, Weed SB (1989) Plant phenolic acids in soils: sorption of ferulic acid by soil and soil components sterilized by different techniques. Soil Biol Biochem 21: 1011–1018

Danielewicz-Ferchmin I, Ferchmin AR (2004) Water at ions, biomolecules and charged surfaces. Phys Chem Liq 42:1–36

Di Nardo C, Cinquegrana A, Papa S, Fuggi A, Fioretto A (2004) Laccase and peroxidase isoenzymes during leaf litter decomposition of *Quercus ilex* in a Mediterranean ecosystem. Soil Biol Biochem 36:1539–1544

Fedorova S, Stejskal J (2002) Surface and precipitation polymerization of aniline. Langmuir 18:5630–5632

Fenner N, Freeman C, Reynolds B (2005) Hydrological effects on the diversity of phenolic degrading bacteria in a peatland: implications for carbon cycling. Soil Biol Biochem 37:1277–1287

Flaig W (1966) The chemistry of humic substances. In: The use of isotopes in soil organic matter studies, Report of FAO/IAEA Technical Meeting. Pergamon, New York, pp 103–127

Gallet C, Pellissier F (1997) Phenolic compounds in natural solutions of the coniferous forest. J Chem Ecol 23:2401–2411

Gerstner JA, Bell JA, Cramer SM (1994) Gibbs free energy of adsorption of biomolecules in ion-exchange systems. Biophys Chem 52:97–106

Gonzalez JM, Laird DA (2004) Role of smectites and Al-substituted goethites in the catalytic condensation of arginine and glucose. Clays Clay Miner 52:443–450

Gramss G, Voigt KD, Kirsche B (1998) Oxidoreductase enzymes liberated by plant roots and their effects on soil humic material. Chemosphere 38:1481–1494

Guggenberger G, Kaiser K (2003) Dissolved organic matter in soil: challenging the paradigm of sorptive preservation. Geoderma 113:293–310

Huang PM (2000) Abiotic catalysis. In: Summer ME (ed) Handbook of soil science. CRC Press LLC, Boca Raton, FL, pp 303–334

Huang Q, Weber WJ (2004) Peroxidase-catalyzed coupling of phenol in the presence of model inorganic and organic solid phases. Environ Sci Technol 38:5238–5245

Huang PM, Wang MK, Kampf N, Schulze DG (2002a) Aluminum hydroxides. In: Dixon JB, Schulze DG (eds) Soil mineralogy with environmental applications. Soil Science Society of America, Madison, WI, USA, pp 261–289

Huang Q, Selig H, Weber WJ (2002b) Peroxidase-catalyzed oxidative coupling of phenols in the presence of geosorbents: rates of non-extractable product formation. Environ Sci Technol 36:596–602

Kalbitz K, Solinger S, Park JH, Michalzik B, Matzner E (2000) Controls on the dynamics of dissolved organic matter in soils: a review. Soil Sci 165:277–304

Klavinš M (1997) Aquatic humic substances: characterization, structure and genesis. University of Latvia, Riga, 234 p

Kleber M, Sollins S, Sutton RA (2007) A conceptual model of organo-mineral interactions in soils: self-assembly of organic molecular fragments into multilayered structures on mineral surfaces. Biogeochem 85:9–24

Kononova MM (1966) Soil organic matter. Pergamon, Oxford

Kuvaeva YV (1980) The content and composition of phenolic acids in some soils of non-chernozemic zone. Sov Soil Sci 1:97–106

Lambert JF (2008) Adsorption and polymerization of amino acids on mineral surfaces: a review. Orig Life Evol Biosph 38:211–242

Lehmann RG, Cheng HH (1988) Reactivity of phenolic acids in soil and formation of oxidation products. Soil Sci Soc Am J 52:1304–1309

Lehmann RG, Cheng HH, Harsh JB (1986) Oxidation of phenolic acids by soil iron and manganese oxides. Soil Sci Soc Am J 51:352–356

Leontievsky AA, Myasoedova NM, Baskunov BP, Pozdnyakova NN, Vares T, Kalkkinen N et al (1999) Reactions of blue and yellow fungal laccases with lignin model compounds. Biochemistry (Moscow) 64:1150–1156

Liu C, Huang PM (2002) Role of hydroxy-aluminosilicate ions (proto-imogolite sol) in the formation of humic substances. Org Geochem 33:295–305

Liu SY, Freyer AJ, Minard RD, Bollag JM (1985) Enzyme-catalyzed complex formation of amino acid esters and phenolic humus constituents. Soil Sci Soc Am J 49:337–342

Matocha CJ, Sparks DL, Amonette JE, Kukkadapu RK (2001) Kinetics and mechanism of birnessite reduction by catechol. Soil Sci Soc Am J 65:58–66

McBride MB (1987) Adsorption and oxidation of phenolic compounds by iron and manganese oxides. Soil Sci Soc Am J 51:1466–1472

Mikutta R, Kleber M, Torn MS, Jahn R (2006) Stabilization of soil organic matter: association with minerals or chemical recalcitrance? Biogeochemistry 77:25–56

Naidja A, Huang PM (2002) Significance of the Henri-Michaelis-Menten theory in abiotic catalysis: catechol oxidation by δ-MnO_2. Surface Sci 506:243–249

Naidja A, Huang PM, Bollag JM (1997) Activity of tyrosinase immobilized on hydroxyaluminum-montmorillonite complexes. J Mol Catal A 115:305–316

Naidja A, Huang PM, Bollag JM (1998) Comparison of reaction products from the transformation of catechol catalyzed by birnessite or tyrosinase. Soil Sci Soc Am J 62:188–195
Nannipieri P, Gianfreda L (1998) Kinetics of enzyme reactions in soil environments. In: Huang PM, Senesi N, Buffle J (eds) Environmental particles – structure and surface reactions of soil particles. Wiley, Chichester, pp 449–479
Pal S, Bollag JM, Huang PM (1994) Role of abiotic and biotic catalysts in the transformation of phenolic compounds through oxidative coupling reactions. Soil Biol Biochem 26:813–820
Perdue EM, Ritchie JD (2004) Dissolved organic matter in freshwaters. In: Holland HD, Turekian KK (eds) Treatise on geochemistry. Elsevier, Amsterdam, pp 273–318
Pohlman AA, McColl JG (1989) Organic oxidation and manganese and organic mobilization in forest soils. Soil Sci Soc Am J 53:686–690
Quiquampoix H, Servagent-Noinville S, Baron M (2002) Enzyme adsorption on soil mineral surfaces and consequences for the catalytic activity. In: Burns RG, Dick RP (eds) Enzymes in the environment. Marcel Dekker, New York, pp 285–306
Rittstieg K, Suurnakki A, Suortti T, Kruus K, Guebitz G, Buchert J (2002) Investigations on the laccase-catalyzed polymerization of lignin model compounds using size-exclusion HPLC. Enzyme Microb Technol 31:403–410
Sapurina I, Osadchev AY, Volchek BZ, Trchova M, Riede A, Stejskal J (2002) In-situ polymerized polyaniline films: 5. Brush-like chain ordering. Synth Met 129:29–37
Sapurina I, Fedorova S, Stejskal J (2003) Surface polymerization and precipitation polymerization of aniline in presence of sodium tungstate. Langmuir 19:7413–7416
Sarkar JM, Leonowicz A, Bollag JM (1989) Immobilization of enzymes on clays and soils. Soil Biol Biochem 21:223–230
Schulze DG (2002) An introduction to soil mineralogy. In: Dixon JB, Schulze DG (eds) Soil mineralogy with environmental applications. Soil Science Society of America, Madison, WI, pp 1–36
Shindo H (1990) Catalytic synthesis of humic acids from phenolic compounds by Mn(IV) oxide (birnessite). Soil Sci Plant Nutr 4:679–682
Shindo H, Huang PM (1984) Catalytic effects of manganese (IV), iron (III), aluminum and silicon oxides on the formation of phenolic polymers. Soil Sci Soc Am J 48:927–934
Shindo H, Huang PM (1985) Catalytic polymerization of hydroquinone by primary minerals. Soil Sci 139:505–511
Sjoblad RD, Bollag JM (1981) Oxidative coupling of aromatic compounds by enzymes from soil microorganisms. In: Paul EA, Ladd JN (eds) Soil biochemistry, vol 5. Marcel Dekker, New York, pp 113–152
Snajdr J, Valaskova V, Merhautova V, Herinkova J, Cajthaml T, Baldrian P (2008) Spatial variability of enzyme activities and microbial biomass in the upper layers of *Quercus petraea* forest soil. Soil Biol Biochem 40:2068–2075
Stevenson FJ (1994) Humus chemistry: genesis, composition, reactions, 2nd edn. Wiley, New York
Stone AT, Morgan JJ (1984) Reduction and dissolution of manganese (III) and manganese (IV) oxides by organics: 1. Reaction with hydroquinone. Environ Sci Technol 18:450–456
Tietjen T, Wetzel RG (2003) Extracellular enzyme-clay mineral complexes: enzyme adsorption, alteration of enzyme activity and protection from photodegradation. Aquat Ecol 37:331–339
Tipping E, Cooke D (1982) The effects of adsorbed humic substances on the surface charge of goethite (α-FeOOH) in freshwaters. Geochim Cosmochim Acta 48:75–80
Toberman H, Evans CD, Freeman C, Fenner N, White M, Emmett BA, Artz RRE (2008) Summer drought effects upon soil and litter extracellular phenol oxidase activity and soluble carbon release in an upland Calluna heathland. Soil Biol Biochem 40:1519–1532
Wang TSC, Li SW (1977) Clay minerals as heterogeneous catalysts in preparation of model humic substances. Z Pfanzenernaehr Bodenkd 140:669–676
Wang K, Xing B (2005) Structural and sorption characteristics of adsorbed humic acid on clay minerals. J Environ Qual 34:342–349

Wang TSC, Li SW, Ferng YL (1978) Catalytic polymerization of phenolic compounds by clay minerals. Soil Sci 126:15–21

Wang TSC, Wang MC, Ferng YL, Huang PM (1983) Catalytic synthesis of humic substances by natural clays, silts and soils. Soil Sci 135:350–360

Wang TSC, Huang PM, Chou CH, Chen JH (1986) The role of soil minerals in the abiotic polymerization of phenolic compounds and formation of humic substances. In: Huang PM, Schnitzer M (eds) Interactions of soil minerals with natural organics and microbes. Soil Science Society of America, Madison, WI, USA, pp 251–281

Wershaw RL (1993) Model for humus in soils and sediments. Environ Sci Technol 27:814–816

Wershaw RL, Pinckney DJ (1980) Isolation and characterization of clay-humic complexes. In: Baker RA (ed) Contaminants and sediments. Ann Arbor Science Publishers, Ann Arbor, Michigan, pp 207–219

Yagudaeva EYu, Muidinov MR, Kapustin DV, Zubov VP (2007) Oxidative polymerization of aniline on the surface of insoluble solid poly(sulfo acids) as a method for the preparation of efficient biosorbents. Russ Chem Bull, Int Edn 56:1166–1173

Zavarzina AG (2006a) A mineral support and biotic catalyst are essential in the formation of highly polymeric soil humic substances. Eurasian Soil Sci 39:48–53

Zavarzina AG (2006b) Key role of mineral support and biotic catalyst in the formation of highly polymeric soil humic acids. In: Proceedings of the 13th meeting of the international humic substances society. Karlsruhe, Germany, July 30–Aug 4 2006, pp 605–608

Zavarzina AG, Semenova TA, Kuznetsova AM, Pogozhev EYu (2007) Synthesis of humic-like substances on the mineral surfaces in presence of oxidases. In: Proceedings International Conference "Humic substances in the biosphere." Moscow, Dec 19–21, 2007, pp 139–145 (in Russian)

Chapter 11
Fungal Oxidoreductases and Humification in Forest Soils

A.G. Zavarzina, A.A. Lisov, A.A. Zavarzin, and A.A. Leontievsky

11.1 Introduction

Humic substances (HS) are ubiquitous and recalcitrant by-products of dead matter hydrolysis and oxidative biotransformation (humification). Their resistance to biodegradation is both a result of structural complexity due to selective preservation of most stable chemical forms during microbial decay (Orlov 1990) and a result of physicochemical protection by interactions with soil minerals (Mikutta et al. 2006). The residence time of HS in soils is 10^2–10^3 years; they comprise up to 90% of soil organic matter (humus), which is the largest carbon reservoir in the biosphere estimated at 1,462–1,548 Pg of C_{org} in the 0–1 m layer excluding litter and charcoal (Batjes 1996). Humification can be thus considered as a key process in Netto Biome production leading to a long-time sink of atmospheric CO_2. About 1/3 (470 Pg) of world soil organic carbon reserves is captured in boreal forests soils and almost half of this amount (224 Pg C) is accumulated in the soils of Russia (Stolbovoi 2006). A better knowledge of humus turnover processes in forests of cold humid climate will allow better predictions of the global carbon dynamics under changing environment. Synthesis, transformation, and mineralization of HS are largely oxidative processes with wood- and soil-inhabiting fungi being a major driving force due to extracellular production of non-specific oxidative enzymes. In this chapter, we provide an over view of the occurrence of oxidoreductases in wood-decomposing,

A.G. Zavarzina (✉)
Faculty of Soil Science, Moscow State University, Moscow 119991, Russia
e-mail: zavarzina@mail.ru

A.A. Lisov and A.A. Leontievsky
Institute of Biochemistry and Physiology of Microorganisms, Russian Academy of Sciences, Pushchino, Moscow Region 142292, Russia

A.A. Zavarzin
Faculty of Biology and Soil Sciences, St. Petersburg State University, St. Petersburg 199034, Russia
e-mail: az@bio.pu.ru

G. Shukla and A. Varma (eds.), *Soil Enzymology*, Soil Biology 22,
DOI 10.1007/978-3-642-14225-3_11,

soil-inhabiting and symbiotic fungi and attempt to elucidate the role of certain fungal groups in humus synthesis and transformation in soil.

11.2 The Origin of HSs

Natural HS comprise operationally defined material extracted from soils by alkali and further separated into humic acids insoluble at pH < 2 (HA, molecular weight, MW, 5–100 kDa) and acid-soluble fulvic acids (FA, MW 1–10 kDa). Non-extractable residues that bind tightly to soil minerals is defined as humin. Distinctive features of HS are a N content of 1–3% (FA) or 2–6% (HA), a C content of 40–50% (FA) and 50–60% (HA), an aromatic C content of 25–35%, a total acidity of 6–14 mmol(−)/g (FA) and 5–8 mmol(−)/g (HA) and characteristic infrared and UV-Vis spectra (Orlov 1990). Humus formation in forest soils begins in the litter: 93–94% of fresh litter mass is utilized by microbiota on the soil surface with the release of CO_2 as a final product (Glazovskaya 1996); only 6–7% of initial C input is leached down as soluble products of decay or/and undergoes transformation into HSs. HSs are formed from highly heterogeneous material comprising dead matter: modified lignin, polyphenols, melanins, chitin, aliphatic compounds (lipids, waxes), carbohydrates, amino acids, proteins, etc. In forest ecosystems lignin serves as the main source of humus precursors due to its quantitative abundance in plant tissues. By contrast, in tundra soils polyphenols, chitin, and melanins from lignin-free lichens and mosses can be important starting material of HS. It is generally recognized that two major humification pathways co-exist in soils (reviewed by Stevenson 1994). (1) Oxidative biodegradation theory (Waksman 1931; Alexandrova 1980) postulates that initial polymeric material (e.g., lignin) is only partially modified by oxidases yielding HA, which can be then oxidized and depolymerized to FA. Changes in lignin include loss of OCH_3 groups with formation of hydroxyphenols and oxidation of aliphatic side chains to form COOH groups. Further oxidation of phenolic groups to semiquinones and quinones enables incorporation of nitrogenous and other compounds into HS structure via free-radical reactions. (2) Polyphenol theory postulates that low MW phenolic aldehydes and acids released during lignin breakdown (or from other sources) are oxidized to reactive semiquinones and quinones and undergo polymerization in presence of nitrogenous compounds and other soluble precursors. First FA are formed and then HA (Kononova 1966; Flaig 1966). The first pathway should predominate in poorly drained soils and peats, while second pathway should be more typical for mineral horizons of well-aerated forest soils, where soluble polyphenols in leachates from litter are main humus precursors. Irrespective of how they are formed the resulting product (HS) are a polydisperse mixture of N-containing molecules composed of substituted aromatic rings, heterocycles, and aliphatic side chains connected by a variety of linkages and bearing functional groups among which carboxylic and phenolic are most abundant (Stevenson 1994). Differentiation of HS from other alkali-soluble compounds (e.g., lignins and melanins) in the soil extracts is quite problematic. Based on NMR signals in HA extracted from peat and soil A_h horizon, Kelleher and Simpson (2006) concluded that the vast majority of humic material is a complex mixture of microbial and plant biopolymers present in soil at the time of extraction. This finding does not rule out the formation of HS closely related to the parent biopolymers or existence of distinct chemical categories of HS. Nevertheless, the presence of non-humic matter in HS preparations is highly possible, especially when HS from organic sources (litter, peat, coal) are under investigation.

11.3 Major Fungal Oxidative Enzymes

The major fungal redox enzymes, involved in oxidative transformation of plant debris, are lignin peroxidase (LiP), Mn-dependent peroxidase (MnP), versatile peroxidase (VP), other peroxidases, laccase, and tyrosinase (Table 11.1). The catalytic action of these enzymes can be divided into two principal stages: (1) an enzymatic substrate oxidation with formation of diffusible, reactive intermediates – phenoxy radicals or quinones from phenolic units, aryl radicals from non-phenolic units; (2) non-enzymatic spontaneous reactions of free radicals, initiating substrate transformation. Two opposite processes may occur: substrate polymerization via radical cross-coupling or substrate degradation via bond cleavage, aromatic ring opening, demethylation, demethoxylation, substituents release, etc. The post-enzymatic step largely extends the substrate range of the enzymes.

11.3.1 Peroxidases

Peroxidases catalyze one-electron substrate oxidation by H_2O_2 with formation of free-radical cation intermediates and H_2O. Fungal peroxidases are quite similar in the structure of their active center and catalytic cycle to plant peroxidases. Enzyme

Table 11.1 Comparative characteristic of fungal phenol oxidases and ligninolytic peroxidases

<table>
<tr><th>Property</th><th>LiP</th><th>VP</th><th>MnP</th><th>Peroxidase</th><th>Laccase</th><th>Tyrosinase</th></tr>
<tr><td>Redox potential</td><td>1.2–1.5 V</td><td>–</td><td>~1.1 V</td><td>~1.0 V</td><td>0.7–0.9 V</td><td>0.26–0.35 V</td></tr>
<tr><td>pH optimum</td><td>2.5–3.5</td><td>–</td><td>4.0–4.5</td><td>~5.5</td><td>4.0–5.0
5.0–6.0</td><td>6.0–7.0</td></tr>
<tr><td>pI</td><td>3.2–4.0</td><td>3.5</td><td>~4.5</td><td>~3.5</td><td>~4.0</td><td>4.5–8.5</td></tr>
<tr><td>MW, kDa</td><td>38–46</td><td>45</td><td>38–50</td><td>40–45</td><td>40–70</td><td>30–50</td></tr>
<tr><td>Active center</td><td colspan="4">Fe-protoporphirin IX</td><td>4 Cu atoms</td><td>2 Cu atoms</td></tr>
<tr><td>Substrate transformation</td><td colspan="5">Depolymerization, mineralization</td><td></td></tr>
<tr><td></td><td colspan="3"></td><td colspan="3">Polymerization</td></tr>
<tr><td>Main producers</td><td colspan="6">White-rot basidiomycetes</td></tr>
<tr><td></td><td colspan="2"></td><td colspan="4">Litter-decomposing basidiomycetes, ectomicorrhizae</td></tr>
<tr><td></td><td colspan="3"></td><td colspan="3">Ascomycetes, lichens</td></tr>
</table>

Literature used: Fakoussa and Hofrichter (1999), Wong (2008), Hammel and Cullen (2008), Mester and Field (1998), Thurston (1994), Baldrian (2006), Makino et al. (1974)

molecules contain a prosthetic heme group-Fe(III) protoporphyrin IX (active center) and two Ca^{2+} ions supporting native structure of the enzyme. The catalytic cycle includes $2e^-$ oxidation of Fe(III) protoporphyrin IX by H_2O_2 to give the radical cation intermediate (compound I) and then two consecutive $1e^-$ reductions of compound I by reducing substrate to give compound II ($1e^-$ reduced enzyme) and the resting enzyme (Welinder 1992).

Lignin-peroxidase (EC 1.11.1.14, diarylpropane:oxygen, hydrogen peroxide oxidoreductase) is unique in its ability to directly oxidize the non-phenolic structures in lignin, which comprise up to 90% of the polymer and have high redox potential (>1.5 V). The major route is *β-O-4* or C_α–C_β cleavage in the propyl side chain to give benzaldehydes. The enzyme also oxidizes phenolic substrates and aromatic amines to phenoxy radicals. LiP has a specific acidic pH optimum (Table 11.1) and is quite rare and unstable enzyme. The "classic" producer is *Phanerochaete crysosporium* (Tien and Kirk 1983), the enzyme was also described in few other white-rot fungi including *Trametes versicolor*, *Phlebia radiata*, *Bjerkandera adusta*, and *Nematoloma frowardii* (Morgenstern 2008). Expression of LiP in other groups of fungi seems to be limited.

Mn-dependent peroxidase (EC 1.11.1.13, Mn(II):hydrogen peroxide oxidoreductase) catalyses oxidation of Mn^{2+} to Mn^{3+}, which is stabilized by bidentate chelators like oxalate, malonate, tartrate, or lactate and acts as non-specific diffusible oxidant of a variety of phenolic compounds and aromatic amines via phenoxy radical intermediates. MnP is produced almost exclusively by basidiomycetes (Hofrichter 2002).

Versatile peroxidase (VP, EC 1.11.1.16) is a hybrid peroxidase that combines catalytic properties of MnP, LiP, and plant peroxidases and has molecular properties similar to MnP (Camarero et al. 1999). The enzyme catalyses the oxidation of Mn^{2+} to Mn^{3+} resembling MnP; it also oxidizes phenolic substrates and aromatic amines in the absence of Mn^{2+} like plant peroxidases; some VPs (e.g., that of *Pleurotus eryngii*) oxidize non-phenolic compounds like LiP due to presence of an invariant tryptophan residue required for long-range e^- transfer from aromatic donors (Perez-Boada et al. 2005). Production is known in few species of white-rot fungi, e.g., *Bjerkandera* spp., *Pleurotus* spp. (Heinfling et al. 1998), and *Panus tigrinus* (Lisov et al. 2003).

Peroxidase (EC 1.11.1.7) has substrate specificity similar to plant peroxidases and laccase. The enzyme oxidizes variety of phenolic compounds via phenoxy radicals. Production is known in different fungal groups including white rots *Pleurotus ostreatus* (Shin et al. 1997) and *Junghuhnia separabilima* (Vares et al. 1992), litter basidiomycetes from the family *Coprinaceae* (Heinzkill et al. 1998), and deuteromycetes (e.g., *Arthromyces ramosus*, Nakayama and Amachi 1999).

11.3.2 Laccase and Tyrosinase

Laccase and tyrosinase are multicopper phenol oxidases that catalyze oxidation of electron-donor substrates by O_2 with formation of free radicals and reduction of O_2

to H_2O (Solomon et al. 1996). Catalytic cycle of laccase contains several $1e^-$ transfers between the 4 Cu atoms. The T1 "blue" Cu site accepts the electrons from the reducing substrate (four $1e^-$ oxidations) and shuttle them to T2/T3 sites where two $2e^-$ reductions of O2 to H_2O occur. So-called yellow laccases have a modified T1 site and thus lack blue color (Leontievsky et al. 1997a). Tyrosinase has a diamagnetic spin-coupled Cu pair in the active centre. Resting enzyme is a met-form that accepts $2e^-$ from diphenol to give quinone and reduced deoxy-form. Deoxy-form reacts with O_2 to give oxy-form, which is a key intermediate, reacting with a monophenol or diphenol (Sanchez-Ferrer et al. 1995).

Laccase (EC 1.11.1.14, benzendiol: oxygen oxidoreductase) preferable substrates are substituted phenols and aromatic amines which are oxidized to semiquinones. These reactive species can be further oxidized to quinones and/or polymerize to form (in)soluble complexes. Alternatively, the semiquinone may react with O_2 to yield superoxide radical that initiate depolymerization processes in a similar way to MnP via alkyl-phenyl and C_α–C_β cleavage of phenolic oligomers (Guillen et al. 2000). Laccases can also degrade non-phenolic lignin model compounds either directly (yellow laccases; Leontievsky et al. 1997a) or in presence of natural or synthetic redox mediators (Eggert et al. 1996). Laccase is almost ubiquitous in white-rot and litter-decomposing basidiomycetes, widespread in ascomycetes and deuteromycetes (Baldrian 2006) and was found in some taxa of micorrhizal fungi (Burke and Cairney 2002) and lichens (Laufer et al. 2006a; Zavarzina and Zavarzin 2006).

Tyrosinase (EC 1.14.18.1, monophenol, o-diphenol: oxygen oxidoreductase) oxidizes some mono- and diphenols but substitution in the aromatic ring decreases enzyme activity. The enzyme catalyses two concomitant reactions: o-hydroxylation of monophenols yielding o-diphenols (monophenolase or cresolase activity) and $2e^-$ oxidation of o-diphenols to o-quinones (diphenolase or catecholase activity). Highly reactive quinones undergo spontaneous coupling to form mixed melanins and heterogeneous polymers (Selinheimo et al. 2007). Tyrosinases were found in basidiomycetes such as *Neurospora crassa*, *Agaricus bisporus*, *Trametes* spp, *Pycnoporus* spp, ascomycetes e.g., *Aspergillus* spp, *Trichoderma* spp (Selinheimo et al. 2007), in micorrhizal fungi (Burke and Cairney 2002), and lichens (Laufer et al. 2006b; Zavarzina and Zavarzin 2006).

11.4 Humification Activities of Fungi in Wood and Soil

HSs are formed in soils by sequential degradative and synthetic processes. The extents to which wood, soil and litter-decomposing fungi participate in each of these processes depend on their enzymes production patterns. Fungal phenol oxidases differ largely by the redox potential and hence by the oxidative power (Table 11.1). High redox potential ligninolytic peroxidases (LiP, MnP, VP) preferentially catalyze degradation of lignin and polymeric phenolic substrates to CO_2 and small soluble fragments (Leonowicz et al. 1999). Non-specific peroxidases and laccase can cause both degradation/polymerization reactions of polyphenols depending on initial substrate MW and environmental conditions. Depolymerization is favored at acidic pH, high oxygen supply, and if initial enzyme substrate is polymeric (Yaropolov et al. 1994; Rabinovich et al. 2004). Low redox potential of tyrosinase suggests that this enzyme does not participate in degradative processes and is involved exclusively in polymerization (Ghosh and Mukherjee 1998). It can be therefore suggested that humification activities of fungi possessing ligninolytic peroxidases (white-rot and litter-decomposing basidiomycetes) should be associated mainly with soluble precursor and FA production via degradation of lignin and humic acids. Fungi producing peroxidases, neutral laccases, and tyrosinases (ascomycetes) should be responsible for synthesis of humic acids.

11.4.1 Wood Decomposers

Wood decomposers do not directly participate in humification processes in soils; however, soluble degradation products (FA-like compounds) can be leached down to the soil with atmospheric depositions, while insoluble products (HA-like compounds) can enter soil at the final stages of wood decay. The initial fungal substrate in wood is lignocellulose (Fig 11.1a). It is generally recognized that enzymes of wood-decomposing basidiomycetes (LiP, MnP, laccase) are too large to diffuse through sound wood and that small non-proteinaceous species initiate the decay (Hammel et al. 2002). These include ·OH radicals, superoxide radicals (O^{2-}), and ferryl ions (Fe^{4+}), derived from Fenton reaction (brown-rot decay); veratryl alcohol cation radical, oxalic acid, or Mn^{3+} (white-rot decay).

11.4.1.1 Soft-rot Ascomycetes: Production of Large Soluble Fragments

Soft-rot ascomycetes are active in the outer layers of wood, altering its mechanical properties and causing wood "softening," resulting in spongy texture of the wood surface (Schwarze 2007). Xylariaceaeous ascomycetes, e.g., *Daldinia, Hypoxylon, Kretzschmaria*, and *Xylaria* are found in standing trees and rotting wood (Schwarze 2007), while microscopic fungi grow on moist wood in contact with soil (see Sect. 5.2). Although ascomycetes preferentially decompose cellulose and hemicellulose, some species have the ability to partially degrade lignin due to laccase activity (type II soft rot). For instance, *Xylaria* spp selectively delignified litter producing bleached areas on fallen leaves (Osono 2007). Unlike the white-rot fungi, which mainly produce low MW decomposition products, wood-colonizing ascomycetes can produce large water-soluble lignocellulose fragments which can serve as HS precursors in soil. Formation of soluble products with MW of 3, 30, and 200 kDa as a result of bond cleavage between lignin and hemicellulose by hydrolytic enzymes was shown for *Xylaria polymorpha* growing on beech wood (Liers et al. 2006). The same fungus produced water- and dioxin-insoluble products from synthetic lignin in presence of laccase. As soft-rot fungi typically start the fungal

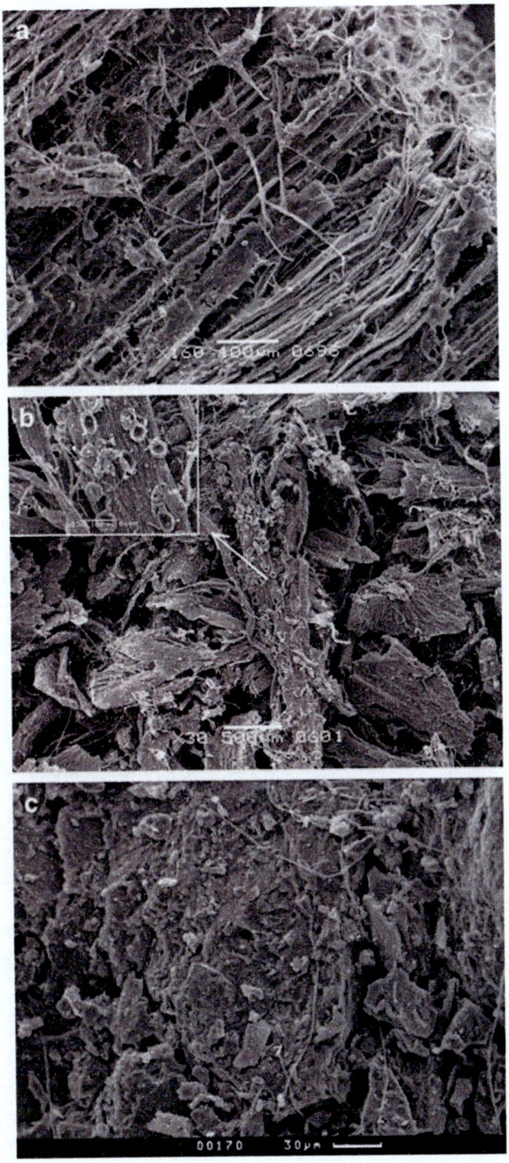

Fig. 11.1 SEM images of decomposing aspen wood (**a**), leaf litter (**b**), and mineral soil particles (**c**) showing difference in organic matter physical state in soil and demonstrating colonization of wood, litter, and soil by fungi (photo by Dr. A.M. Kuznetsova, Biological Faculty, Moscow State University)

succession on moist decaying wood (Rabinovich et al. 2001); they can be considered as "pioneer" humification agents, responsible for release of soluble HA precursors (hydrolytic activity) and their re-polymerization by laccase into HA-like products.

11.4.1.2 Brown-rot Fungi: Formation of Humic Acids from Partially Oxidized Lignin

Brown-rot fungi comprise about 6% of all described wood decay fungi and most of them belong to the *Polyporaceae*. They are predominantly associated with conifers (gymnospermous species) and distributed mostly in northern and southern temperate regions (Schwarze 2007). Representatives comprise *Poria placenta*, *Gloeophyllum trabeum*, and *Coniophora puteana*. Brown-rot fungi effectively depolymerize and metabolize cellulose and hemicelluloses, without altering lignin significantly. Lignin is only partially oxidized, giving the decayed wood the characteristic reddish brown color. The hydroxyl radicals generated by Fenton system ($Fe^{2+} + H_2O_2 + H^+ \rightarrow Fe^{3+} + \cdot OH + H_2O$) are considered to initiate destructive process (Goodell 2003). The ·OH radicals are unable to catalyse cleavage of β-1 and β-*O*-4 bonds in lignin and degradation is limited to demethylation, demethoxylation, aromatic ring hydroxylation, oxidation of initially formed catechol groups, and side-chain oxidation. Participation of lignin-degrading enzymes in oxidative process remains poorly understood. A few reports indicate the presence of extracellular laccase (Lee et al. 2004), LiP (Dey et al. 1991), and MnP (Szklarz et al. 1989) in brown-rots. Intracellular laccase released during hyphae autolysis can cause slight decrease (up to 10%) in lignin content (Rabinovich et al 2004). However the role of laccase in brown-rot decay needs further clarification. The brown-colored modified lignin is enriched in phenolic and carboxylic groups and depleted in methoxyl groups (Kirk 1975), thus approaching HS by the functional groups content and physicochemical properties. High MW HA, formed from polymeric oxidized lignin, are accumulated in decayed wood; absence of destructive ligninolytic peroxidases favors this process (Rypacek and Rypackova 1975). Humified products of decay can easily enter the soil since cubical remnants of brown-rotted wood are abundant on and in the forest floor.

11.4.1.3 White-rot Fungi: Production of Small Soluble Polyphenols (Structural Units) and FAs

White-rot fungi are unique in their ability to completely degrade lignin. They are more frequently found on angiosperm than on gymnosperm wood and can cause simultaneous destruction of lignin, cellulose, and hemicellulose (*Trametes versicolor*, *Phanerochaete chrysosporium*, and *Phlebia radiata*) or selective delignification (*Pleurotus spp.*) giving rise to cellulose-enriched white wood material (Schwarze 2007). The ligninolytic enzyme system consists mainly of MnP, LiP, and laccase. Enzyme production patterns differ between the species. Many white-rot fungi produce either all of the three enzymes (*Nematoloma frowardii*, *Trametes versicolor*) or a combination of any two from them, e.g., LiP and MnP (*Phanerochaete chrysosporium*), MnP (VP), and laccase (*Panus tigrinus*) (Hatakka 1994). In few species (e.g., *Picnoporus cynnabarinus*), solely laccase is produced (Eggert et al. 1996). Production of ligninolytic peroxidases is triggered by starvation in N, C, or S resulting in switching of fungus to secondary metabolism and idiophasic growth. Laccase is mostly a substrate-inducible enzyme, although constitutive expression of laccase can occur (e.g., Koroleva-Skorobogat'ko et al. 1998).

Degradation of lignin and related compounds by white-rot fungi is cometabolic event and occurs only in presence of easily metabolizable carbon source, e.g., glucose. White-rot fungi typically decompose lignin in an acidic medium, producing mainly low MW fulvic acid–like products and CO_2 (Leonowicz et al. 1999). MnP is considered as a key degradative enzyme (Hofrichter 2002), followed by LiP – powerful oxidant of non-phenolic units. The precise role of laccase in ligninolysis remains controversial. Suggestions that laccase acts in synergy with MnP during degradation of macromolecular phenolic substrates (Galliano et al. 1991; Schlosser and Hofer 2002) are contradicted by reports on in vitro depolymerization of lignin (Maltseva et al. 1991) and soil HAs (Zavarzina et al. 2004) by blue laccase in absence of mediators or other enzymes. It seems likely, however, that efficient depolymerizing activity of laccases in vivo require presence of redox mediators. *Picnoporus cinnabarinus* which excretes solely laccase degraded lignin in comparable rate to *P. chrysosporium* due to production of metabolite 3-hydroxyanthranilate (Eggert et al. 1996). Yellow laccases of white-rot fungi, produced exclusively under solid-state fermentation conditions, were found to directly oxidize non-phenolic units in lignin probably due to presence of lignin-generated modifier in their structure that functions as electron-transfer mediator (Leontievsky et al. 1997b, 1999). There is an opinion that laccases in white-rot fungi have polymerizing function and are required for detoxification of low MW lignin breakdown products (Thurston 1994); HA can be formed as a result. Indeed, *Coriolus hirsutus* and/or *Cerrena maxima* grown on oat straw produced humic acid–like substances as a result of laccase activity; HAs were polymeric (23 kDa) and resembled soil HA by elemental composition and spectroscopic data (Yavmetdinov et al. 2003).

The role of white-rot fungi in the synthesis of HA seem to be limited, because under natural conditions FA and not HA accumulate in white-rotted wood and only small amounts of HA are produced on lignified plant material in laboratory experiments. Rypacek and Rypackova (1975) suggested that this can be a result of HA degradation by ligninolytic peroxidases. Indeed, many white-rot fungi were found to decolorize (up to 80%) and depolymerize HA of different origin (reviewed by Grinhut et al. 2007). The decolorization of HA is considered to be caused by splitting of double bonds, resulting in the breakdown of macromolecular mesomeric system and dissipation of the brown color (Fakoussa and Frost 1999). Humic acids with low aromatic C and high carbohydrate C contents (e.g., from litters, low rank coal) are closer by their structure to lignin – original substrate of ligninolytic enzymes – and are therefore more susceptible for oxidative attack than highly oxidized and aromatic soil HA (Almendros and Dorado 1999; Yanagi et al. 2002). For instance, lignite HA were effectively decolorized by laccase in submerged cultures of *P. cinnabarinus* (58% bleaching; Temp et al. 1999) or *Trametes versicolor* (80% bleaching; Fakoussa and Frost 1999), while HA from sod-podzolic soil lost 45% of its initial color under action of *P. tigrinus* laccase *in vivo* (our unpublished data).

11.4.2 Litter and Soil-inhabiting Fungi

Humification in litter and mineral soil horizons differ from that in wood in terms of nature of organic matter and enzymes involved. It was found that laccases, followed by MnP, are dominant redox enzymes in forest soils (Rosenbrock et al. 1995; Snajdr et al. 2008), while activities of LiP and VP have never been reported so far. Fungal substrates in litter are represented by particulate organic matter, consisting of plant debris (leaves, needles, branches), animal remains, root fragments, fungal hyphae, etc. (Fig 11.1b). This organic material is highly heterogeneous but not that compact as lignocellulose in wood that facilitates direct enzymatic attack. Enzymes in litter (MnP, laccase) represent active pool associated mostly with decomposition and mineralization processes (Allison 2006). In underlying soil layers, 90% of organic matter exist as coatings on mineral grains (Fig 11.1c) largely inaccessible to enzymatic attack. Fungi should thus preferentially utilize soluble substrates leaching down from decaying plant litter or excreted by roots. Many of these compounds are potentially toxic polyphenols which are polymerized into HS by laccases, peroxidases, or tyrosinases, immobilized on mineral supports.

11.4.2.1 Microfungi: Lignocellulose and Humus Solubilization, Synthesis of Melanins, and HS

Soil mycelial fungi (micromycetes) are largely ascomycetes, deuteromycetes, and zygomycetes that colonize wood in contact with soil, litter, and also soil up to 1 m depth. They comprise important group of soil microbial community and predominate over other fungal groups on the early stages of litter decomposition (Mirchink 1976; Lindahl et al. 2007). The role of soil microfungi in humification has for long time been attributed to the intracellular production of the high MW polyphenol-like brown pigment melanin (Kang and Felbeck 1965; Kononova 1966; Martin and Haider 1969; Valmaseda et al. 1989). Melanins have certain similarities with humic acids in terms of irregular aromatic structure, stochastic mechanism of synthesis, behavior in solvents, and some physicochemical properties (Zaprometova et al. 1971). This led to assumption that the bulk of unaltered fungal melanins, released upon the cell wall lysis, can form the stable humic acid fraction in soils (Zviagintsev and Mirchink 1986). Recent work has demonstrated that melanins are less resistant to biodegradation than soil HA and before contributing to stable humus fractions undergo oxidative transformations leading to depolymerization, increase in O-content, and optical density (Zavgorodnyaya et al. 2002).

Extracellular humification activity of micromycetes received far less attention than that of basidiomycetes. Soil microfungi are best known for production of cellulase–hemicellulase systems and comprise 60–90% of cellulose degrading microbial population in soils. Representative genera showing high biodiversity are *Aspergillus*, *Chaetomium*, *Ceratocystis*, *Phialophora*, *Trichoderma*, *Fusarium*, *Penicillium*, *Rhizoctonia*, and *Mortierella* (Rabinovich et al. 2001). Although carbohydrates are preferable substrate, representatives of these genera were found to mineralize 5–10%

and solubilize 15–20% of synthetic and wood-derived lignin during *primary growth* (Haider and Trojanowski 1975; Rodriguez et al. 1997; Kluczek-Turpeinen et al. 2003). Moreover, many species including the deuteromycete *Paecylomices inflatus* (Kluczek-Turpeinen et al. 2005), the ascomycetes *Thrichoderma* spp., and *Penicillium* spp. (Laborda et al. 1999), species of *Alternaria, Clonostachys, Phoma,* and *Paecilomyces* (Řezáčová et al. 2006) as well as *Acremonium, Botrytis, Chaetomium,* and *Rhizoctonia* (Gramss et al. 1999) were able to cause partial mineralization (5%), solubilization (6–25%), decolorization (2–30%), and depolymerization of HAs from litters and soil. The deuteromycete *Chalara longipes* isolated from spruce needle litter caused even 75% bleaching of humus extract from O_F litter layer (Koukol et al. 2004). Solubilizing activity is typical for ascomycetes and most likely involves microbial alkaline substances and a synergistic effect of cellulases and hemicellulases (Holker et al. 1999). As a result, some species (e.g., *Trichoderma atroviride*) can grow on HA or coals using them as a sole carbon source (Gramss et al 1999; Silva-Stenico et al. 2007). Oxidative activity is believed to be attributed to production of laccase (Kluczek-Turpeinen et al 2003, 2005), peroxidase (Haider and Trojanowski 1975), peroxidase and tyrosinase (Koukol et al. 2004), phenoloxidase and/or MnP (Laborda et al. 1999; Řezáčová et al. 2006), or combination of laccase, tyrosinase, and peroxidase as in *Botrytis cinerea* (Gramss et al. 1999). The direct involvement of these enzymes in degradation of phenolic compounds remains to be proven, because some species showed bleaching activity in absence of oxidases (Gramss et al. 1999). The extent to which phenoloxidases and peroxidases are produced by microfungi is also poorly understood. Generally, production of ligninolytic peroxidases is rare: LiP-like enzymes were detected so far in *Chrysonilia sitophila* (Duran et al. 1987) and *Penicillium decumbens* (Yang et al. 2005) and MnP-like activity was reported in *Fusarium solani* (Saparrat et al. 2000), *Trichoderma* sp, *Penicillium* sp (Laborda et al. 1999), and some others (Řezáčová et al. 2006).

According to Martin and Haider (1971), microfungi play a significant role in the synthesis of HS in soil. Polymerization activity is mostly associated with laccases, which are widespread in microfungi (Baldrian 2006). Rabinovich et al. (2004) suggested that unlike acidic laccases of white-rot fungi, laccases of the majority of soil micromycetes are predisposed for substrate polymerization, rather than depolymerization due to their more neutral pH optimum (pH 6.0–7.0). Although pH of most forest soils is acidic, some microfungi are able to adjust pH of their microenvironment to neutral values (Stepanova et al. 2003; Kluczek-Turpeinen et al. 2007). Thus, microfungi can be those laccase-producing species responsible for synthesis of HA in litter and mineral soil horizons. Solubilizing activity associated with hydrolytic enzymes can be important prerequisite for lignin and HS modification in litter by more efficient ligninolytic systems of basidiomycetes.

11.4.2.2 Saprotrophic Basidiomycetes: Production of the "White-Rot" Humus and FA-like Compounds

Ligninolytic fungi comprise as much as 10% of the entire fungal decomposer communities in forest litters (Osono 2007): mycelia of these fungi are often concentrated in the interface between the freshly fallen leaves in L layer and near-humus materials in the F layer. Saprotrophic basidiomycetes colonize litter on the later stages of decomposition than ascomycetes and can cause substantial loss of recalcitrant compounds including lignin, HS, tannins, melanins with production of CO_2, soluble fragments, and bleached humus. Representatives of about 20 genera were reported as lignin decomposers: many of them possess MnP activity in addition to laccase, while LiP activity has not been found so far (Steffen et al. 2000; Osono 2007). The most active ligninolytic species belong to the genera *Agrocybe*, *Clitocybe*, *Collybia*, *Marasmius*, *Mycena*, and *Stropharia*. Delignification activity of litter-decomposing basidiomycetes resembles that of white-rot fungi in wood being a cometabolic event (Steffen et al. 2007). Although saprotrophic basidiomycetes decompose lignin at half the rate of white-rot fungi (Gramss et al. 1999; Steffen et al. 2000), they can produce substantial amounts of FA-like compounds (0.9 kDa) from insoluble litter material (Steffen et al. 2002) and can cause decarboxylation (up to 50%) and degradation of HA to CO_2 and lower molecular mass compounds (Rabinovich et al. 2001). A considerable decrease in 30–50 kDa fraction of litter-derived HA and formation of products with mean MWs of 1.0–2.0 kDa were observed in cultures of *Gymnopus* sp., *Hypholoma fasciculare*, *Rhodocollybia butyracea* (Valaskova et al. 2007), and *Collybia dryophila* (Steffen et al. 2002). MnP was considered as a key enzyme in the process. It is important to note that HA depolymerization effect observed in abovementioned studies need careful interpretation because it could indicate the frequent cleavage of lignin polymer present as admixture in alkali extracts from litters rather than HS degradation (Sect. 2).

11.4.3 Symbiotic Fungi

Ectomycorrhiza (ECM) and lichens comprise two groups of symbiotic fungi that are common and abundant in northern forest ecosystems. These fungi obtain all or most of their carbon from the photosynthetic partner, and their saprotrophic activity appears to be limited. Despite this fact, both ECM (e.g., Taylor et al. 2004) and lichens (e.g., Dahlman et al. 2004) are able to assimilate exogenous C by taking up simple organic compounds (glucose, amino acids). Many ECM fungi are known for their abilities to metabolize lignocellulose, hemicellulose and polyphenols (Read and Perez-Moreno 2003). Thus, access to a photosynthate does not preclude facultative saprotrophy, which might be alternative foraging strategy during periods of low photosynthate supply or during massive mycelial production when supplementary resources for growth are needed (Talbot et al. 2008). Among the enzymes involved in organic matter transformation, activities of laccases and tyrosinases have been found in some taxa of ECM and lichens. Irrespective of whether these enzymes are used for saprotrophy-related activities or not, once released or leached into the soil they have a potential to participate in humus synthesis or degradation.

11.4.3.1 Ectomycorrhiza

Ectomycorrhiza is a symbiotic extracellular association of a fungus (usually basidiomycete) with plant fine roots. It is typically formed between the roots of woody plants belonging to *Pinaceaea*, *Betulaceae*, and others. In forest soils ECM are distributed primarily in fragmented litter, humus horizon and mineral soil, being spatially separated from saprotrophs which strongly dominate in the fresh and partially decomposed litter layers rich in labile C (Lindahl et al. 2007). Such prevalence of ECM in specific parts of the soil profile is likely to reflect preferential exploitation of substrates of a particular quality as reflected by their C:N ratios (Read and Perez-Moreno 2003). ECM fungi are considered as nutrient-mobilizing components of the soil fungal community and their hyphae function in the adsorption and translocation of N, P and water to the host plant. Most of N and P in forest soils are present in the colloidal organic material surrounding roots e.g., lignocellulose cell wall structures of dead plant material, HSs, proteins, protein–tannin complexes and HSs. It was found that many ECM fungal taxa can mobilize nutrients from these organic sources via production of lytic enzymes (cellulases, xylanases, proteases, polyphenol oxidases) causing some decrease in lignocellulose and humus in soil (Cairney and Burke 1994; Bending and Read 1996a, b, 1997; Read and Perez-Moreno 2003). When grown on tannic acid as the sole carbon source ECM fungi utilized carbon contained in the substrate and released dark-colored reactive quinone-like compounds (HS precursors) as a by-product (Bending and Read 1996a). Decomposition of lignin and HSs by ECM is limited in comparison with cellulose, hemicellulose and hydrolysable phenols (Durall et al. 1994; Bending and Read 1997; Read and Perez-Moreno 2003). This can be explained by apparent lack in ECM fungi of ligninolytic peroxidases, needed for effective HS and lignin breakdown. There is no convincing evidence for LiP genes in ECM (Cairney et al. 2003) and extracellular production of MnP was confirmed so far only in Tylospora fibrillosa (Chambers et al. 1999). The extents to which ECM fungi produce laccases and tyrosinases and their role in polyphenol transformation is not completely known. Gene fragments with high similarity to laccase from wood-rots were found in ECM species from the genera *Amanita*, *Cortinarius*, *Hebeloma*, *Lactarius*, *Paxillus*, *Piloderma*, *Russula*, *Tylospora*, and *Xerocomus* (Chen et al. 2003). Laccase gene sequences attributed to ECM fungi comprised almost half (45.5%) of investigated laccase genes in forest soil (Luis et al. 2005). However, upregulation of genes is not necessarily mean production of the enzyme. While activities of polyphenol oxidases were detected in many ECM using axenic mycelia, non-sterile mycelia, sporocarp tissue, or ECM root tips, oxidation of the laccase substrate syringaldazine was rare, suggesting that ECM fungi produce tyrosinase rather than laccase (Burke and Cairney 2002). Production of laccase has been confirmed only in few species, e.g., *Cantharellus cibarius* (Ng and Wang 2004) and *Thelephora terestris* (Kanunfre and Zancan 1998). It has been suggested that laccases of ECM might be involved in depolymerization of lignins, release of N from insoluble protein-tannin complexes and HS degradation (Bending and Read 1996, 1997; Gramss et al. 1999). However, an alternative hypothesis suggests high possibility of non-enzymatic oxidation of phenolic substrates in ECM by radicals derived from Fenton reagent (Cairney and Burke 1998). ECM fungi may thus contribute to partial degradation of lignin by the way similar to that of brown rots (Sect. 11.4.1.2.) and can be involved in the HA synthesis rather than degradation. HSs can be also formed as by-products during detoxification of host-defence compounds by laccases and tyrosinases of ECM fungi (Cairney and Burke 2002). Further studies are needed to define the role of ECM fungi in humification.

11.4.3.2 Lichens

Lichens represent bi- or tripartite associations of a fungus (usually ascomycete) with green algae and/or cyanobacteria. They grow on stone, wood, or soil (epilythic, epiphytic, and epigeyic species, respectively) and are characterized by a variety of morphological and chemical adaptations for surviving stressful conditions and for fast restoration of metabolic activity (Beckett et al. 2008; Kranner et al. 2008). Lichens have long been recognized as important agents of soil formation due to their weathering action on rocks (Chen et al. 2000). Being dominant in soil cover of tundra and boreal forest ecosystems, lichens also serve as considerable source of mortmass for humification. Since lichens lack lignin, polymeric HS precursors should be mainly chitin and melanins. Lichens also produce considerable amounts of water-soluble low MW phenolic compounds that can serve as monomeric HS precursors upon leaching from the lichen thalli with rain water. Recent finding of laccases and tyrosinases in lichens of different taxonomic and substrate groups (Laufer et al. 2006a, b; Zavarzina and Zavarzin 2006) opens new perspectives in investigation of pedogenetic role of these symbiotic organisms. Lichens may play important role in humus formation processes as not only the source of the organic compounds for humification, but also producers of enzymes that can synthesize or degrade HS. Representatives of the order *Peltigerales* (genera *Peltigera, Solorina, Nephroma*), which mostly belong to fast-growing epiphytic and epigeyic species in wet microenvironments, were found to be the most active producers of laccase and tyrosinase. One order lower laccase activities and minor peroxidase activities were detected in more xerophytic lichens from the order *Lecanorales* (genera *Cladonia, Cetraria, Stereocaulon*). Laccases in lichens are constitutively expressed and stimulated by desiccation and wounding (Laufer et al. 2006a). Most of laccase activity in studied peltigerous lichens was located intracellularly or in the loosely and hydrophobically bound cell wall fractions, while a greater proportion of tyrosinases occurred intracellularly (Laufer et al. 2006b). Approximately 5–10% of the total extractable laccase activity could be washed out from the intact thalli of *Peltigerous* lichens by distilled water, suggesting possibility of their involvement in extracellular processes of polyphenols transformation (Zavarzina and Zavarzin 2006).

While lichen tyrosinases (e.g., in *Peltigera malacea*, *P. rufescens*, and *Pseudocyphellaria aurata*) had typical MW of about 60 kDa (Laufer et al. 2006b), lichen laccases were found to be unusually large. In the study of Laufer et al. (2009), active laccase isoforms in concentrated water extracts from 13 lichen species belonging to

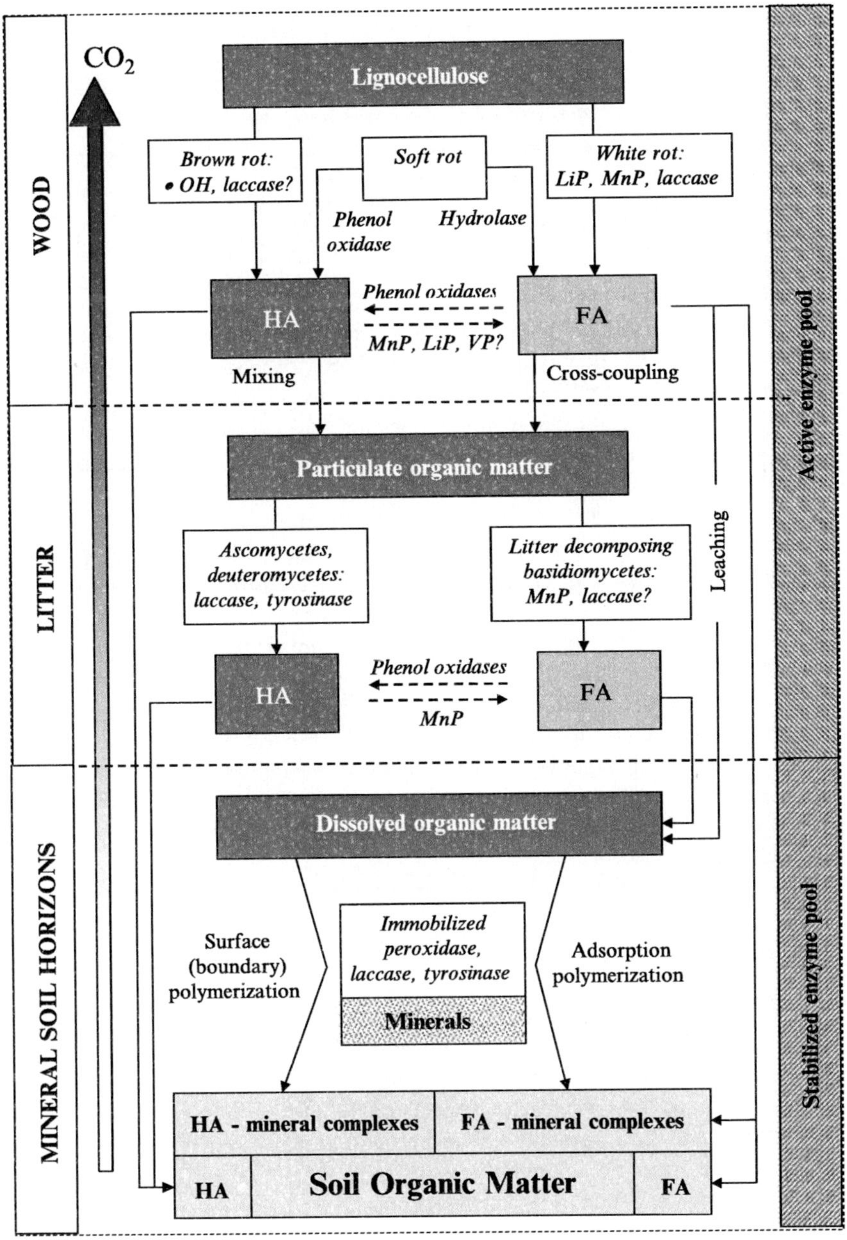

Fig. 11.2 The possible role of fungi and their oxidoreductases in humification process

the suborder *Peltigerineae* had MWs between 135 and 200 kDa, while in 7 lichen species laccases were even larger, 300–350 kDa (e.g., >350 kDa in *P. praetextata*). Lisov et al. (2007) have purified and characterized blue laccase from *Solorina crocea* and yellow laccase from *Peltigera aphthosa* with MWs of 175 kDa and 165 kDa, respectively. The enzymes were typical laccases by their substrate specificity and catalytic properties. In addition to homodimeric laccase forms, monomeric "small" laccases were detected. In *S. crocea* and *P. aphthosa*, "small" laccases had MWs of 45 kDa and 55 kDa, respectively, and consisted of two isoenzymes.

The role of lichen phenol oxidases in humification is unknown. In our preliminary studies we have found that purified laccases from *P. aphthosa* and *Solorina crocea* caused partial depolymerization of soil HA *in vitro*. However, given that washed-out laccase activities are commonly low, involvement of lichen laccases in organic matter degradation in nature is questionable unless lichens produce some metabolites which act as redox mediators. Z. Laufer have found that leachates from lichens were less effective in dye decolorization than intact thalli and that classic laccase mediators speeded up decolorization (personal communication with Prof. RP Beckett). If consider possible effects of laccase alone, the polymerizing activity of leached-out enzyme is more probable, which in turn suggests the possible involvement of lichens in synthesis of HSs. Soil-stabilizing species tightly bound to the mineral substrate, such as *Solorina crocea*, are particularly interesting with this respect. Laccases of such species may be immobilized on mineral grains upon release from the lichen thalli and initiate surface polymerization of water-soluble HS precursors with formation of stable organo-mineral adsorption complexes. Indeed, dark-colored organic coatings (cutanes) are commonly observed on the soil particles or rock fragments under lichen thalli. Further investigations are needed to define the role of lichen enzymes in functioning of the symbiosis and humification.

11.5 Conclusions

The role of fungi in humification processes can be summarized as follows (Fig. 11.2). (1) Humification in wood represents solid-state fermentation of lignocellulose. Soft-rot ascomycetes are pioneers on the surface of wood, while basidiomycetes continue the succession and penetrate in deeper layers by the aid of small non-enzymatic species. Ligninolytic peroxidases (MnP and LiP) of the white-rot fungi cause mineralization of lignin and its breakdown to soluble products (FA-like compounds). Brown-rot fungi and soft-rot fungi cause partial oxidation of lignin with formation of high MW humic acids. Laccases, peroxidases, and OH radicals from Fenton reaction are the oxidants responsible. Soft-rot ascomycetes also produce large soluble lignocellulose fragments by synergistic action of hydrolases. (2) Humification in litter represents transformation of particulate organic matter, which is subjected to direct enzymatic attack. Acidic laccases and MnP of saprotrophic basidiomycetes are considered as main degradative enzymes which decompose complex organic matter to soluble FA-like fragments and CO_2. Microfungi are mainly responsible for synthesis of HA via partial oxidation of lignocellulose, re-polymerization of low MW polyphenols or production of melanins. Laccases and tyrosinases are the main enzymes involved in this process. Humic colloids formed in wood and litter can undergo slow mineralization or oxidation with release of soluble products. MnP and LiP of white-rot fungi are mainly responsible for this process in wood, while MnP of saprotrophic fungi can bleach, depolymerize, and solubilize HA in litters. Soluble products of plant debris and humus decomposition can be re-polymerized *in situ* by phenol oxidases or can be washed down the soil profile. (3) Soluble compounds become the substrate for laccases, peroxidases, and tyrosinases immobilized on inorganic or organo-mineral supports in humus horizon; microfungi, ectomicorhhizae, and lichens are producers of these enzymes. Two possible reactions may occur: oxidative self-coupling in aqueous phase adjacent to soil particles with adsorption/cross-coupling of polymeric products on the support surface ("precipitative polymerization"); adsorption of substrates on solid particles with formation of high MW product directly on the support surface ("surface polymerization"). Both reactions result in irreversible binding of organic matter and formation of organic coatings on mineral grains. Such humus–mineral complexes form the bulk of solid matrix in humus horizons and comprise the most stable fraction of C_{org} in soils.

The extents to which certain fungal groups participate in humification processes in soil need to be further defined. Future studies should concentrate on enzyme producers other than white-rot fungi, which have been most intensively studied over last three decades. White-rot fungi do not colonize soil under natural conditions and thus their contribution to humus synthesis and degradation appear to be limited. Humification activities of microfungi and symbiotic fungi are largely overlooked, and considerable gaps exist in our knowledge of phenol oxidase enzymology in these fungi.

Acknowledgments Financial support from the Russian Foundation for Fundamental Research (grant 09-04-00570) and from the Programme No.15 of the Presidium of Russian Academy of Sciences "Origin of the Biosphere and Evolution of Geobiological systems" is gratefully acknowledged. We are expressing our sincere thanks to Prof. Richard P. Beckett for long lasting cooperation, his valuable comments and revising the language of the manuscript.

References

Alexandrova LN (1980) Soil organic matter and processes of its transformation. Nauka, Leningrad (in Russian)

Allison SD (2006) Soil minerals and humic acids alter enzyme stability: implications for ecosystem processes. Biogeochem 81:361–373

Almendros G, Dorado J (1999) Molecular characteristics related to the biodegradability of humic acid preparations. Eur J Soil Sci 50:227–236

Baldrian P (2006) Fungal laccases: occurrence and properties. FEMS Microbiol Rev 30:215–242

Batjes NH (1996) Total carbon and nitrogen in the soils of the world. Eur J Soil Sci 47:151–163

Beckett RP, Kranner I, Minibaeva F (2008) Stress physiology and the symbiosis. In: Nash TH (ed) Lichen Biology, 2nd edn, Cambridge Univ Press, Cambridge, pp 134–151

Bending GD, Read DJ (1996a) Effects of the soluble polyphenol tannic acid on the activities of ericoid and ectomycorrhizal fungi. Soil Biol Biochem 28:1595–1602

Bending GD, Read DJ (1996b) Nitrogen mobilization from protein-polyphenol complex by ericoid and ectomycorrhizal fungi. Soil Biol Biochem 28:1603–1612

Bending GD, Read DJ (1997) Lignin and soluble-phenolic degradation by ectomycorrhizal and ericoid mycorrhizal fungi. Mycol Res 101:1348–1354

Burke RM, Cairney JWG (2002) Laccases and other polyphenol oxidases in ecto- and ericoid mycorrhizal fungi. Mycorrhiza 12:105–116

Cairney JWG, Burke RM (1994) Fungal enzymes degrading plant cell walls: their possible significance in the ectomycorrhizal symbiosis. Mycol Res 98:1345–1356

Cairney JWG, Burke RM (1998) Do ecto- and ericoid mycorrhizal fungi produce peroxidase activity? Mycorrhiza 8:61–65

Cairney JWG, Taylor AFS, Burke RM (2003) No evidence for lignin peroxidase genes in ectomycorrhizal fungi. New Phytol 160:461–462

Camarero S, Sarcar S, Ruiz-Duenas FJ, Martinez MJ, Martinez AT (1999) Description of a versatile peroxidase involved in the natural degradation of lignin that has both manganese peroxidase and lignin peroxidase substrate interaction site. J Biol Chem 274:10324–10330

Chambers SM, Burke RM, Brooks PR, Cairney JWG (1999) Molecular and biochemical evidence for manganese-dependent peroxidase activity in *Tylospora fibrillosa*. Mycol Res 103:1098–1102

Chen DM, Bastias BA, Taylor AFS, Cairney JWG (2003) Identification of laccase-like genes in ectomycorrhizal basidiomycetes and transcriptional regulation by nitrogen in *Piloderma byssinum*. New Phytol 157:547–554

Chen J, Blume HP, Beyer L (2000) Weathering of rocks induced by lichen colonization – a review. Catena 39:121–146

Dahlman L, Person J, Palmqvist K, Nashholm T (2004) Organic and inorganic nitrogen uptake in lichens. Plants 219:459–467

Dey S, Maiti TK, Bhattacharyya BC (1991) Lignin peroxidase production by a brown rot fungus *Polyporus ostreiformis*. J Ferment Bioeng 72:402–404

Duran N, Ferrer I, Rodriguez J (1987) Ligninases from *Chrysonilia sitophila* (TFB-27441). Appl Microbiol Biotechnol 16:157–167

Eggert C, Temp U, Eriksson KEL (1996) Lignin degradation by a fungus lacking lignin and Mn peroxidase. In: Jeffries T, Vikarii L (eds) Enzymes in the pulp and paper manufacturing. ACS Symp Ser 655: 130–150

Fakoussa RM, Frost PJ (1999) In vivo-decolorization of coal-derived humic acids by laccase-excreting fungus *Trametes versicolor*. Appl Microbiol Biotechnol 52:60–65

Fakoussa RM, Hofrichter M (1999) Biotechnology and microbiology of coal degradation. Appl Microbiol Biotechnol 52:25–40

Flaig W (1966) The chemistry of humic substances. In: The use of isotopes in soil organic matter studies, Report of FAO/IAEA technical meeting. Pergamon, New York, pp 103–127

Galliano H, Gas G, Seris JL, Boudet AM (1991) Lignin degradation by *Rigidoporus lignosus* involves synergistic action of two oxidizing enzymes: Mn-peroxidase and laccase. Enzym Microb Technol 13:478–482

Ghosh D, Mukherjee R (1998) Modeling tyrosinase monooxygenase activity. Spectroscopic and magnetic investigations of products due to reactions between copper(I) complexes of xylyl-based dinucleating ligands and dioxygen: aromatic ring hydroxylation and irreversible oxidation products. Inorg Chem 37:6597–6605

Glazovskaya MA (1996) Role and functions of the pedosphere in geochemical carbon cycles. Pochvovedenije 2:174–186 (in Russian)

Goodell B (2003) Brown-rot fungal degradation of wood: our evolving view. ACS Symp Ser 845:97–118

Gramss G, Ziegenhagen D, Sorge S (1999) Degradation of soil humic extract by wood- and soil-associated fungi, bacteria, and commercial enzymes. Microb Ecol 137:140–151

Grinhut T, Hadar Y, Chen Y (2007) Degradation and transformation of humic substances by saprotrophic fungi: processes and mechanisms. Fungal Biol Rev 21:179–189

Guillén F, Muñoz C, Gómez-Toribio V, Martínez AT, Jesús Martínez M (2000) Oxygen activation during oxidation of methoxyhydroquinones by laccase from *Pleurotus eryngii*. Appl Environ Microbiol 66:170–175

Haider K, Trojanowski J (1975) Decomposition of specifically ^{14}C-labelled phenols and dehydropolymers of coniferyl alcohol as models for lignin degradation by soft and white rot fungi. Arch Microbiol 105:33–41

Hammel KE, Kapich AN, Jensen KA, Ryan AC (2002) Reactive oxygen species as agents of wood decay by fungi. Enzyme Microb Technol 30:445–453

Hammel KE, Cullen D (2008) Role of fungal peroxidases in biological ligninolysis. Curr Opin Plant Biol 11:349–355

Hatakka A (1994) Lignin-modifying enzymes from selected white-rot fungi: production and role in lignin degradation. FEMS Microbiol Rev 13:125–135

Heinfling A, Ruiz-Duenas FJ, Martinez MJ, Bergbauer M, Szewzyk U, Martinez AT (1998) A study on reducing substrates of manganese-oxidising peroxidases from *Pleurotus eryngii* and *Bjerkandera adusta*. FEBS Lett 428:141–416

Heinzkill M, Bech L, Halkier T, Schneider P, Anke T (1998) Characterization of laccases and peroxidases from wood-rotting fungi (family *Coprinaceae*). Appl Environ Microbiol 64:1601–1606

Hobbie EA, Horton TR (2007) Evidence that saprotrophic fungi mobilize carbon and mycorrhizal fungi mobilize nitrogen during litter decomposition. New Phytol 173:447–449

Hofrichter M (2002) Review: lignin conversion by manganese peroxidase (MnP). Enzyme Microb Technol 30:454–466

Holker U, Ludwig S, Scheel T, Hofer M (1999) Mechanisms of coal solubilization by the dueteromycetes *Trichoderma atroviride* and *Fusarium oxysporum*. Appl Microbiol Biotechnol 52:57–59

Kang KS, Felbeck GT (1965) A comparison of the alkaline extract of tissues of *Aspergillus niger* with humic acids from three soils. Soil Sci 99:175–181

Kanunfre CC, Zancan GT (1998) Physiology of exolaccase production by *Thelephora terrestris*. FEMS Microbiol Lett 161:151–156

Kelleher BP, Simpson AJ (2006) Humic substances in soils: are they really chemically distinct? Environ Sci Technol 40:4605–4611

Kirk TK (1975) Effects of a brown-rot fungus *Lenzites trabea* on lignin in spruce wood. Holzforschung 29:99–107

Kluczek-Turpeinen B, Steffen KT, Tuomela M, Hatakka A, Hofrichter M (2005) Modification of humic acids by the compost-dwelling deuteromycete *Paecilomyces inflatus*. Appl Microbiol Biotechnol 66:443–449

Kluczek-Turpeinen B, Tuomela M, Hatakka A, Hofrichter M (2003) Lignin degradation in a compost environment by the deuteromycete *Paecilomyces inflatus*. Appl Microbiol Biotechnol 61:374–379

Kluczek-Turpeinen B, Maijala P, Hofrichter M, Hatakka A (2007) Degradation and enzymatic activities of three *Paecilomyces inflatus* strains grown on diverse lignocellulosic substrates. Intern Biodeterior Biodegrad 59:283–291

Kononova MM (1966) Soil organic matter. Pergamon, Oxford

Koroleva-Skorobogat'ko O, Stepanova E, Gavrilova V et al (1998) Purification and characterization of the constitutive form of laccases from the basidiomycete *Coriolus hirsutus* and effect of inducers on laccase synthesis. J Biotechnol Appl Biochem 28:47–54

Koukol O, Gryndler M, Novak F, Vosatka M (2004) Effect of *Chalara longipes* on decomposition of humic acids from *Picea abies* needle litter. Folia Microbiol 49:574–578

Kranner I, Beckett R, Hochman A, Nash TH (2008) Dessication-tolerance in lichens: a review. The Bryol 111:576–593

Laborda F, Monistrol IF, Luna N, Fernandez M (1999) Processes of liquefaction/solubilization of Spanish coals by microorganisms. Appl Microbiol Biotechnol 52:49–56
Laufer Z, Beckett RP, Minibayeva FV, Luthje S, Bottger M (2006a) Occurrence of laccases in lichenized Ascomycetes in the suborder *Peltigerineae*. Myc Res 110:846–853
Laufer Z, Beckett RP, Minibayeva FV (2006b) Co-occurrence of the multicopper oxidases tyrosinase and laccase in lichens in sub-order *Peltigerineae*. Ann Bot 98:1035–1042
Laufer Z, Beckett RP, Minibaeva FV, Luthje S, Bottger M (2009) Diversity of laccases from lichens in suborder *Peltigerineae*. The Bryol 112:418–426
Lee KH, Wi SG, Singh AP, Kim YS (2004) Micromorphological characteristics of decayed wood and laccase produced by the brown-rot fungus *Coniophora puteana*. J Wood Sci 50:281–284
Leonowicz A, Matuszewska A, Luterek J et al (1999) Biodegradation of lignin by white-rot fungi. Fungal Genet Biol 27:175–185
Leontievsky AA, Vares T, Lankinen P, Shergill JK, Pozdnyakova NN, Myasoedova NM, Kalkkinen N, Golovleva LA, Cammack R, Thurston CF, Hatakka A (1997a) Blue and yellow laccases of ligninolytic fungi. FEMS Microbiol Lett 156:9–14
Leontievsky AA, Myasoedova N, Pozdnyakova N, Golovleva L (1997b) "Yellow" laccase of *Panus tigrinus* oxidizes non-phenolic substrates without electron-transfer mediators. FEBS Lett 413:446–448
Leontievsky AA, Myasoedova NM, Baskunov BP, Pozdnyakova NN, Vares T, Kalkkinen N et al (1999) Reactions of blue and yellow fungal laccases with lignin model compounds. Biochemistry (Moscow) 64:1150–1156
Liers C, Ulrich R, Steffen KT, Hatakka A, Hofrichter M (2006) Mineralization of 14C-labelled synthetic lignin and extracellular enzyme activities of the wood-colonizing ascomycetes *Xylaria hypoxylon* and *Xylaria polymorpha*. Appl Microbiol Biotechnol 69:573–579
Lindahl BD, Ihrmark K, Boberg J, Trumbore SE, Hogberg P, Stenlid J, Finlay RD (2007) Spatial separation of litter decomposition and mycorrhizal nitrogen uptake in a boreal forest. New Phytol 173:611–620
Lisov AV, Leontievsky AA, Golovleva LA (2003) Hybrid Mn-peroxidase from the ligninolytic fungus *Panus tigrinus* 8/18. Isolation, substrate specificity, and catalytic cycle. Biochemistry (Moscow) 68:1027–1035
Lisov AV, Zavarzina AG, Zavarzin AA, Leontievsky AA (2007) Laccases produced by lichens of the order *Peltigerales*. FEMS Microbiol Lett 275:46–52
Luis P, Kellner H, Zimdars B, Langer U, Martin F, Buscot F (2005) Patchiness and spatial distribution of laccase genes of ectomycorrhizal, saprotrophic, and unknown basidiomycetes in the upper horizons of a mixed forest cambisol. Microb Ecol 50:570–579
Makino N, McMahill PHS, Masonthe HS (1974) The oxidation state of copper in resting tyrosinase. J Biol Chem 249:6062–6066
Martin JP, Haider K (1971) Microbial activity in relation to soil humus formation. Soil Sci 111:54–63
Martin JP, Haider K (1969) Phenolic polymers of *Stachybotris atra, Stachybotris chartarum* and *Epicoccum nigrum* in relation to humic acid formation. Soil Sci 107:260–270
Maltseva OV, Niku-Paavola ML, Leontievsky AA, Myasoedova NM, Golovleva LA (1991) Ligninolytic enzymes of the white rot fungus *Panus tigrinus*. Biotechnol Appl Biochem 13:291–302
Mester T, Field JA (1998) Characterization of a novel manganese peroxidase-lignin peroxidase hybrid isozyme produced by Bjercandera species strain BOS55 in the absence of manganese. J Biol Chem 273: 15412–15417
Mikutta R, Kleber M, Torn MS, Jahn R (2006) Stabilization of soil organic matter: association with minerals or chemical recalcitrance? Biogeochemistry 77:25–56
Mirchink TG (1976) Soil mycology. Moscow State University Press, Moscow
Morgenstern I, Klopman S, Hibbett D (2008) Molecular evolution and diversity of lignin degrading heme peroxidases in the Agaricomycetes. J Mol Evol 66:243–257

Nakayama T, Amachi T (1999) Fungal peroxidase: its structure, function, and application. J Mol Catal B – Enzyme 6:185–198
Ng TB, Wang HX (2004) A homodimeric laccase with unique characteristics from the yellow mushroom *Cantharellus cibarius*. Biochem Biophys Res Commun 313:37–41
Orlov DS (1990) Humus acids and the general theory of humification. MSU Press, Moscow (in Russian)
Osono T (2007) Ecology of ligninolytic fungi associated with leaf litter decomposition. Ecol Res 22:955–974
Perez-Boada M, Ruiz-Duenas FJ, Pogni R, Basosi R, Choinowski T, Martinez MJ, Piontek K, Martinez AT (2005) Versatile peroxidase oxidation of high redox potential aromatic compounds: site-directed mutagenesis, spectroscopic and crystallographic investigation of three long-range electron transfer pathways. J Mol Biol 354:385–402
Rabinovich ML, Bolobova AV, Kondrashchenko VI (2001) Theoretical basis of the biotechnology of wood composites. Wood and wood-decaying fungi, vol 1. Nauka, Moscow (in Russian)
Rabinovich ML, Bolobova AV, Vasilchenko LG (2004) Fungal decomposition of natural aromatic structures and xenobiotics: a review. Appl Biochem Microbiol 40:1–17
Read DJ, Perez-Moreno J (2003) Mycorrhizas and nutrient cycling in ecosystems – a journey towards relevance? New Phytol 157:475–492
Řezáčová V, Hršelová H, Gryndlerová H, Mikšík I, Gryndler M (2006) Modifications of degradation-resistant soil organic matter by soil saprobic microfungi. Soil Biol Biochem 38:2292–2299
Rodriguez J, Ferraz A, Nogueira RF, Ferrer I, Esposito E, Duran N (1997) Lignin biodegradation by the ascomycete *Chrisonilia sitophila*. Appl Biochem Biotechnol 62:233–242
Rosenbrock P, Buscot F, Munch JC (1995) Fungal succession and changes in the fungal degradation potential during the initial stage of litter decomposition in a black alder forest (*Alnus glutinosa* (L.) Gaertn.). Eur J Soil Biol 31:1–11
Rypacek V, Rypackova M (1975) Brown rot of wood as a model for studies of lignocellulose humification. Biol Plantarum (Praha) 17:452–457
Sanchez-Ferrer A, Rodriguez-Lopez JN, Garcia-Canovas F, Garcia-Carmona F (1995) Tyrosinase: a comprehensive review of its mechanism. Biochim Biophys Acta 1247:1–11
Saparrat MCN, Martinez MJ, Tournier HA, Cabello MN, Arambarri AM (2000) Production of ligninolytic enzymes by *Fusarium solani* strains isolated from different substrata. World J Microbiol Biotechnol 16:799–803
Schlosser D, Hofer C (2002) Laccase catalyzed oxidation of Mn^{2+} in the presence of natural Mn^{3+} chelators as a novel source of extracellular H_2O_2 production and its impact on manganese peroxidase. Appl Environ Microbiol 68:3514–3521
Schwarze FWMR (2007) Wood decay under the microscope. Fungal Biol Rev 21:133–170
Selinheimo E, Nieidhin D, Steffensen C, Nielsen J, Lomascolo A, Halaouli S, Record E, O'Beirne D, Buchert J, Kruus K (2007) Comparison of the characteristics of fungal and plant tyrosinases. J Biotechnol 130:471–480
Shin KS, Oh IK, Kim CJ (1997) Production and purification of Remazol brilliant blue R decolorizing peroxidase from the culture filtrate of *Pleurotus ostreatus*. Appl Environ Microbiol 63:1744–1748
Silva-Stenico ME, Vengadajellum CJ, Janjua HA, Harrison STL, Burton SG, Cowan DA (2007) Degradation of low rank coal by *Trichoderma atroviride* ES11. J Ind Microbiol Biotechnol 34:625–631
Sklarz G, Antibus RK, Sinsabaugh RL, Linkins AE (1989) Production of phenol oxidases and peroxidases by wood-rotting fungi. Mycologia 81:234–240
Snajdr J, Valaskova V, Merhautova V, Herinkova J, Cajthaml T, Baldrian P (2008) Spatial variability of enzyme activities and microbial biomass in the upper layers of *Quercus petraea* forest soil. Soil Biol Biochem 40:2068–2075
Solomon EI, Sundaram UM, Machonkin TE (1996) Multicopper oxidases and oxygenases. Chem Rev 96:563–2605

Steffen KT, Cajthaml T, Snajdr A, Baldrian P (2007) Differential degradation of oak (*Quercus petraea*) leaf litter by litter-decomposing basidiomycetes. Res Microbiol 158:447–455

Steffen KT, Hatakka A, Hofrichter M (2002) Degradation of humic acids by the litter-decomposing basidiomycete *Collybia dryophila*. Appl Environ Microbiol 68:3442–3448

Steffen KT, Hofrichter M, Hatakka A (2000) Mineralization of 14C-labelled synthetic lignin and ligninolytic enzyme activities of litter-decomposing basidiomycetous fungi. Appl Microbiol Biotechnol 54:819–825

Stepanova EV, Koroleva OV, Vasilchenko LG, Karapetyan KN, Landesman EO, Yavmetdinov IS, Kozlov YP, Rabinovich ML (2003) Fungal decomposition of oat straw during liquid and solid-state fermentation. Appl Biochem Microbiol (Moscow) 39:65–74

Stevenson FJ (1994) Humus chemistry: genesis, composition, reactions, 2nd edn. Wiley, New York

Stolbovoi V (2006) Soil carbon in the forests of Russia. Mitig Adap Strat Gl Change 11:203–222

Talbot JM, Allison SD, Treseder KK (2008) Decomposers in disguise: mycorrhizal fungi as regulators of soil C dynamics in ecosystems under global change. Funct Ecol 22:955–963

Temp U, Meyrahn H, Eggert C (1999) Extracellular phenol oxidase patterns during depolymerization of low-rank coal by three basidiomycetes. Biotechnol Lett 21:281–287

Tien M, Kirk TK (1983) Lignin-degrading enzyme from the hymenomycete *Phanerochaete chrysosporium* Burds. Science 221:661–663

Thurston C (1994) The structure and function of fungal laccases. Microbiol 140:19–26

Valaskova V, Snajdr J, Bittner B, Cajthaml T, Merhautova V, Hofrichter M, Baldrian P (2007) Production of lignocellulose-degrading enzymes and degradation of leaf litter by saprotrophic basidiomycetes isolated from a Quercus petraea forest. Soil Biol Biochem 39:651–660

Valmaseda M, Martinez AT, Almendros D (1989) Contribution by pigmented fungi to P-type humic acid formation in two forest soils. Soil Biol Biochem 21:23–28

Vares T, Lundell TK, Hatakka AI (1992) Novel heme-containing enzyme possibly involved in lignin degradation by the white-rot fungus *Junghuhnia separabilima*. FEMS Microbiol Lett 99:53–58

Waksman SA (1931) Humus. Williams and Wilkins, Baltimore

Welinder KG (1992) Superfamily of plant, fungal and bacterial peroxidases. Curr Opin Struct Biol 2:388–393

Wong DWS (2008) Structure and action mechanism of ligninolytic enzymes. Appl Biochem Biotechnol. doi:10.1007/s12010-008-8279-z

Yanagi Y, Tamaki H, Otsuka H, Fujitake N (2002) Comparison of decolorization by microorganisms of humic acids with different 13C NMR properties. Soil Biol Biochem 34:729–731

Yang JS, Yuan HL, Wang HX, Chen WX (2005) Purification and characterization of lignin peroxidases from *Penicillium decumbens P6*. World J Microbiol Biotechnol 21:435–440

Yaropolov AI, Skorobogat'ko OV, Vartanov SS, Varfolomeyev SD (1994) Laccase. Properties, catalytic mechanism, and applicability. Appl Biochem Biotechnol 49:257–280

Yavmetdinov IS, Stepanova EV, Gavrilova VP, Lokshin BV, Perminova IV, Koroleva OV (2003) Isolation and characterization of humin-like substances produced by wood-degrading white-rot fungi. Appl Biochem Microbiol (Moscow) 39:257–264

Zaprometova KM, Mirchink TG, Orlov DS, Yukhnin AA (1971) Characteristics of black pigments of the dark-colored soil fungi. Soviet Soil Sci 7:22–30

Zavarzina AG, Leontievsky AA, Golovleva LA, Trofimov SY (2004) Biotransformation of soil humic acids by blue laccase of *Panus tigrinus 8/18*: an in vitro study. Soil Biol Biochem 36:359–369

Zavarzina AG, Zavarzin AA (2006) Laccase and tyrosinase activities in lichens. Microbiol (Moscow) 75:546–556

Zavgorodnyaya YA, Demin VV, Kurakov AV (2002) Biochemical degradation of soil humic acids and fungal melanins. Org Geochem 33:347–355

Zviagintsev DG, Mirchink TG (1986) On the nature of soil humic acids. Soviet Soil Sci 5:68–75

Chapter 12
Evolutionary-Economic Principles as Regulators of Soil Enzyme Production and Ecosystem Function

Steven D. Allison, Michael N. Weintraub, Tracy B. Gartner, and Mark P. Waldrop

12.1 Introduction

Extracellular enzymes are ubiquitous in soil environments. Produced by microorganisms and plant roots, these enzymes serve a dual function of degrading complex organic material into simpler forms and acquiring resources for the enzyme producer (Burns 1982; Sinsabaugh 1994). Without extracellular enzymes, microbes and plants would be unable to obtain resources from complex compounds, and cycles of carbon (C) and nutrients would grind to a halt.

Researchers have been measuring and interpreting soil enzyme activities for over 100 years (Eriksson et al. 1974; Skujins 1976). Although extracellular enzymes are clearly important for soil function, the factors regulating enzyme production remain unclear. For example, the mechanisms that determine the composition, timing, spatial location, and quantity of extracellular enzyme production in soil are still poorly known (Wallenstein and Weintraub 2008). Better knowledge of these mechanisms would improve our ability to predict how soil biogeochemical cycles will respond to changes in the environment. Increasing global temperatures, land use change, nutrient deposition, and invasive species are prominent examples of global changes that currently affect soil ecosystems (Trumbore 1997; Swift et al. 1998; Ehrenfeld 2003).

S.D. Allison (✉)
University of California, Irvine, 321 Steinhaus, Irvine, CA 92697, USA
e-mail: allisons@uci.edu

M.N. Weintraub
Department of Environmental Sciences, University of Toledo, Toledo, OH 43606, USA

T.B. Gartner
Department of Biology and the Environmental Science Program, Carthage College, Kenosha, WI 53140, USA

M.P. Waldrop
US Geological Survey, 345 Middlefield Rd, MS 962, Menlo Park, CA 94025, USA

G. Shukla and A. Varma (eds.), *Soil Enzymology*, Soil Biology 22,
DOI 10.1007/978-3-642-14225-3_12, © Springer-Verlag Berlin Heidelberg 2011

Because extracellular enzyme producers are living organisms, they are subject to ecological constraints that may affect growth and enzyme production (Ekschmitt et al. 2005; Allison 2006). Although extracellular enzymes catalyze critical biogeochemical reactions, resource acquisition is the primary function of these enzymes from an organismal perspective. Therefore, enzyme production represents one of several possible foraging strategies, including direct uptake of simple resources, autotrophy, or nitrogen (N) fixation, depending on the resource in question. All these strategies involve costs and benefits that depend on environmental conditions.

The major organismal benefit of enzyme production is the release of organic monomers or mineral nutrients that microbes or plant roots can take up across the cell membrane and assimilate. Extracellular enzymes target nearly every macromolecule on earth, including proteins (proteases), carbohydrates (amylases, cellulases), amino sugar polymers (chitinases), organic phosphates (phosphatases), and lignins (oxidases, peroxidases) (Burns 1978; Allison et al. 2007a). The costs of enzyme production include the metabolic energy required for protein synthesis and excretion, as well as the C and nutrient content of the enzymes themselves. For example, between 50 and 70% of N acquired by microbes may be allocated to amino acids that are the building blocks of enzymes (Friedel and Scheller 2002), and extracellular enzyme production has been reported to consume 1–5% of C and N assimilation by bacteria (Frankena et al. 1988).

The main goal of this chapter is to explore the hypothesis that evolutionary and ecological forces minimize the cost:benefit ratio of extracellular enzyme production and thereby represent an important regulator of enzyme production and activity. We expect that soil physical properties, nutrient availability, and competitive interactions represent strong selective pressures that influence enzyme cost:benefit ratios. For both microbes and plants (the major groups of extracellular enzyme producers), natural selection should favor enzyme production strategies that minimize costs and maximize benefits (Table 12.1). The rationale is that increased costs of enzyme production reduce fitness because those resources cannot be allocated to reproduction. Conversely, the resource benefits of enzyme production can be invested in reproductive effort, thereby increasing fitness. These tradeoffs should apply generally to microbes (and plants) whose fitness often correlates with growth rate, since growth represents the difference between resource inputs and outputs to an organism. We argue that "evolutionary-economic" constraints apply to organisms foraging with extracellular enzymes and provide a mechanistic basis for predicting how they respond to changing environmental conditions.

Table 12.1 Strategies to minimize extracellular enzyme cost: benefit ratio

Inducible enzyme production
End-product inhibition of enzyme production
Binding enzymes to the cell surface
Altering diffusive properties of secreted enzymes
Biofilm formation
Quorum sensing
Antibiotic production

12.2 The Evolutionary Economics of Extracellular Enzyme Production

Components of the evolutionary-economic mechanism of enzyme production have been suggested before across a range of systems, but this work has yet to be unified in a common framework. For example, Sinsabaugh and Moorhead (1994) were among the first to develop an explicit model of microbial allocation to extracellular enzyme production that assumed a tradeoff among enzymes that acquire C, N, and P. This model, called "Microbial Enzyme Allocation during Decomposition" (MEAD), treats microbial communities as economic units that maximize their productivity by allocating resources to extracellular pools of C-, N-, and P-releasing enzymes, depending upon substrate quality and environmental conditions. However, the main goal of this model as well as some more recent models (Schimel and Weintraub 2003; Moorhead and Sinsabaugh 2006) was to predict decomposition rates rather than to examine enzyme production as a foraging strategy.

There are other models that focus more directly on microbial foraging with extracellular enzymes. Based on an analytical model, Vetter et al. (1998) predicted that extracellular enzyme production would be a viable foraging strategy for marine bacteria attached to particulate organic material. The benefits of enzyme production exceeded the costs under the modeled diffusion and substrate conditions, thereby allowing bacterial growth to occur. This foraging concept was recently incorporated into a spatially explicit, individual-based model of enzyme production by bacteria (Allison 2005) that revealed the potential importance of diffusion, nutrient availability, and microbial competition as constraints on extracellular enzyme production. While valuable as a theoretical exercise, this model contains many untested assumptions and has yet to be confronted with experimental data. Therefore, a secondary goal of this chapter is to synthesize empirical and theoretical evidence to test the hypothesis that ecological-economic and evolutionary constraints regulate extracellular enzyme production in soils. In pursuit of this goal, we aim to build a more comprehensive conceptual framework for enzyme production that goes beyond the existing models.

12.3 Controls on Microbial Allocation to Enzyme Production

12.3.1 Microbial Demand

Microbes produce extracellular enzymes that target all essential macronutrients, including C, N, P, and S (enzymes from plant roots target only P and possibly N) (Burns 1978; Allison et al. 2007a). The availability of these nutrients fluctuates in space and time, and nutrient supply does not necessarily match microbial or plant nutrient requirements. The main function of most extracellular enzymes is therefore to bring nutrient supply (from chemically complex resources) more closely in line with nutrient demand. If the supply of available resources is already aligned with microbial and plant requirements, there should be little ecological or evolutionary advantage to enzyme production due to the costs involved. Since nutrient demand effectively determines the relative quantities of resources that organisms need to acquire, understanding the factors that control nutrient demand could help explain patterns in soil enzyme production.

Ultimately, nutrient demand can be traced to constraints on organismal stoichiometry (Sterner and Elser 2002). Macromolecules, such as proteins, nucleic acids, carbohydrates, and cell wall components exhibit well-defined elemental ratios. Since macromolecular composition is a relatively inflexible trait in most organisms, it strongly determines organismal stoichiometry and nutrient demand. For example, because animal tissue is protein rich compared to plant tissue, animals have lower C:N ratios and greater N demand relative to plants (Reiners 1986). At a global scale, stoichiometric constraints on organismal biomass are apparent across a range of taxa. For example, molar ratios of C:N:P for marine phytoplankton are tightly constrained at 106:16:1 (Redfield 1958). Although the ratios differ and the variance is greater, similar patterns hold for plants and microbes in terrestrial systems. Tree foliage ratios average 1,212:28:1 (McGroddy et al. 2004), with a similar ratio of 1,158:24:1 observed in fine root biomass (Jackson et al. 1997). For soil microbial biomass as a whole, the average global ratio is 60:7:1 (Cleveland and Liptzen 2007), which reflects the much lower concentration of structural C and the lack of photosynthetic machinery in microbial cells relative to plant cells.

Within microbes, there are clear stoichiometric differences across taxa that could affect intrinsic demand for different resources. For instance, bacteria have C:N ratios of ~5:1, while fungi show C:N ratios closer to 15:1 (Sterner and Elser 2002). These differences arise because fungi produce cell walls made of C-rich polysaccharide polymers and chitin (Bartnicki-Garcia 1968), while bacterial cell walls are primarily composed of more N-rich peptidoglycans (Schleifer and Kandler 1972). Similarly, C:P and N:P ratios differ across taxa, with P content hypothesized to relate to growth rate because more ribosomes with high P content are required to sustain rapid growth rates (Sterner and Elser 2002; Makino et al. 2003). Although broad groups of microbes (i.e., fungi and bacteria) clearly differ in stoichiometry, more studies of the C:N:P ratios of specific microbial taxa would aid in predicting resource investment in different extracellular enzymes. Based on taxon-specific differences in cell wall chemistry, the stoichiometric variation within bacteria and fungi is likely to be substantial (Bartnicki-Garcia 1968; Schleifer and Kandler 1972).

12.3.2 Enzyme Regulation

Due to the resource costs of enzyme synthesis, microbes (and plant roots) should be under selection to regulate enzyme production. Induction and de-repression are regulatory mechanisms that can potentially mitigate enzyme costs by up-regulating enzyme production only when this strategy will be beneficial to the producer. These mechanisms require additional regulatory machinery at the genetic level, such as promoters that interact with inducer and repressor molecules to signal environmental conditions, such as resource availability. Although not an extracellular enzyme system, the *lac* operon in *E. coli* is a textbook example of how the availability of external resources (lactose and glucose) can interact with a regulatory pathway to control enzyme synthesis (Jacob and Monod 1961).

Regulation of most extracellular enzyme systems has been poorly studied in soil, but there is evidence for regulatory control of extracellular enzyme production in aquatic and laboratory systems. Enzymes that are produced continually with little regulatory control are constitutive, while inducible enzyme activity is produced only under particular environmental conditions. Enzyme production by bacteria can be induced in the laboratory by intermediate degradation products that signal availability of the substrate (Priest 1977). Similarly, extensive work in lake systems has demonstrated that the presence of enzyme substrates can induce the production of alkaline phosphatase and leucine aminopeptidase activity (Chróst 1991). In contrast, high concentrations of low-molecular weight catabolites (e.g., glucose, amino acids) inhibit community enzyme activity either through repression of enzyme gene transcription or through competitive inhibition of the enzyme itself (Hanif et al. 2004).

Although enzyme production is clearly inducible in these systems, some level of constitutive production may be advantageous as a mechanism to detect the presence of substrate. With no baseline level of extracellular enzyme production, it would be difficult to generate intermediates that could act as inducers for additional enzyme synthesis (Chróst 1991; Koroljova-Skorobogatko et al. 1998). Consistent with this idea, Raab et al. (1999) found that protease activities were positively related to soil amino acid concentrations at the low levels typically found in soil. These regulatory mechanisms ensure that enzymes are produced only when substrate is available and the end-products of the enzymatic reaction are scarce.

12.4 Resource Availability in Soil

Although mechanisms of extracellular enzyme regulation were first identified in aquatic ecosystems, there is evidence that the same conceptual models apply in soils. In fact, the model proposed by Sinsabaugh and Moorhead (1994) is effectively a model of end-product inhibition, whereby available forms of N and P suppress the production of N- and P-acquiring enzymes and stimulate microbial allocation to C-degrading enzymes. Empirical evidence provides strong support for the MEAD model. Across a range of sites, the model was able to explain >62% of the variation in decomposition rate of birch wood (Sinsabaugh and Moorhead 1994). In laboratory cultures, proteomic studies have shown that *Bacillus* bacteria produce specific enzymes in response to limitation by C, N, or P (Voigt et al. 2006).

When soil microorganisms are P limited, they produce acid or alkaline phosphatases (depending upon pH and microbial community composition) that release inorganic phosphate from organic matter (Haynes and Swift 1988; Antibus et al. 1992). Moreover, phosphatase activity has been shown to be inversely related to inorganic P availability in both aquatic and soil systems (Chróst 1991; Olander and Vitousek 2000; Treseder and Vitousek 2001; Allison et al. 2007b). This relationship also holds at the global scale, where the ratio of P- to C-acquiring extracellular enzymes increases in tropical ecosystems where P is more likely to limit productivity due to increased P weathering rates (Sinsabaugh et al. 2008). Similarly, the activities of N-acquiring enzymes such as peptidases and chitinases are stimulated by low N availability but inhibited by high concentrations of inorganic N in many systems (Chróst 1991; Olander and Vitousek 2000; Weintraub and Schimel 2005).

As with aquatic systems, there is evidence that soil extracellular enzymes are also inducible in the presence of substrate and that adequate substrate availability may be a requirement for enzyme production. In an infertile Hawaiian soil, addition of available N and P failed to stimulate β-glucosidase activity, but this enzyme activity increased when cellulose substrate was also added in combination with N and P (Fig. 12.1a; Allison and Vitousek 2005). Similarly, glycine aminopeptidase production was unchanged when available C and P were added, but induced when collagen protein was also added (Fig. 12.1b). Thus, extracellular enzyme induction in soil appears to depend on at least two conditions: (1) producers are limited by the resource targeted by the enzyme and (2) a suitable substrate for the enzyme is present in the soil.

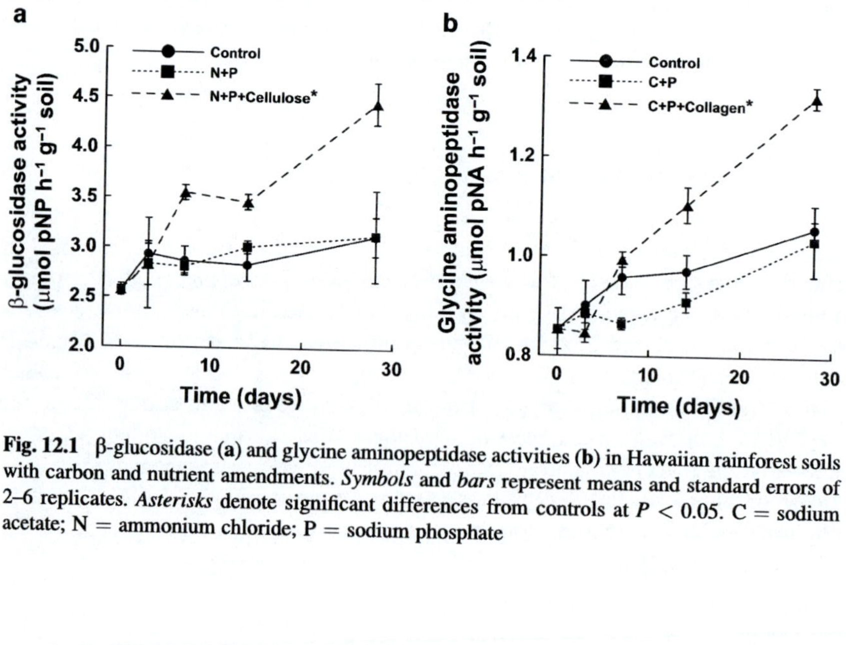

Fig. 12.1 β-glucosidase (**a**) and glycine aminopeptidase activities (**b**) in Hawaiian rainforest soils with carbon and nutrient amendments. *Symbols* and *bars* represent means and standard errors of 2–6 replicates. *Asterisks* denote significant differences from controls at $P < 0.05$. C = sodium acetate; N = ammonium chloride; P = sodium phosphate

12.5 Implications of Enzyme Allocation

12.5.1 Protection of Investment

Because extracellular enzymes act outside the producer cell, other organisms may also derive benefits from the producer investment in enzyme production. If these organisms intercept enough of the resource benefit (or damage the enzyme itself), then the cost:benefit ratio of enzyme production would increase. However, we expect that enzyme producers would evolve strategies to protect their enzyme investment and reduce competition and interference from other organisms. One way of protecting enzymes is to bind them to the cell surface. This helps ensure that neither the direct investment in the enzyme itself nor the products of the enzymatic reaction will be lost. However, the cost of having an enzyme bound to the surface is that distant substrates will not be accessible (Vetter et al. 1998). Alternatively, enzyme-producing microbes may use chemical defenses, such as antibiotics, to eliminate competitors or aggregate with other enzyme producers in a quorum or biofilm (Ekschmitt et al. 2005). This strategy allows microbes to exploit the diffusion losses of their neighbors and increases the chance of taking up reaction products before they diffuse away from the aggregation of cells. Some mycorrhizal fungi employ this strategy by forming dense, hydrophobic mats of hyphae that exude enzymes in water droplets that are later reabsorbed by the fungus, along with the products of decomposition (Sun et al. 1999).

Even if enzymes are not bound to the cell surface, microbes may have strategies to mitigate enzyme loss. The size and structure of an enzyme help determine its diffusivity, and these properties could be altered (as a result of cellular regulation and/or natural selection) to produce different isozymes that optimize enzyme foraging. We used a spatially explicit and individual-based model (Allison 2005) to examine the costs and benefits of changes in enzyme diffusion under different conditions. If substrate availability is high, then the optimal enzyme diffusivity is low because enzymes remain concentrated near the producer and substrate does not become limiting (Fig. 12.2). As substrate availability declines, the model predicts that the optimal enzyme diffusivity increases, allowing the producer to access more distant substrates, once closer substrates become exhausted (Figs. 12.2 and 12.3). However, interception of reaction products by microbial "cheaters" that do not produce enzymes may counter this effect and select for low enzyme diffusivity even under low substrate conditions (Allison 2005).

Secreting extracellular polysaccharides to form a biofilm is another strategy that microbes could employ to restrict the diffusion of extracellular enzymes to an optimal level (Davey and O'Toole 2000). Additionally, microbes may produce autoinducer molecules that allow them to sense the diffusion properties of the environment. Although typically thought to be involved in quorum sensing, Redfield (2002) has proposed that autoinducers are also used by microbes to determine the rate at which secreted molecules move away from the cell. Thus, it is possible that microbes use these autoinducers to sense when diffusion rates are favorable for

Fig. 12.2 Model output of bacterial growth rates as a function of enzyme diffusivity at different substrate (Sub) concentrations (fg μm^{-3})

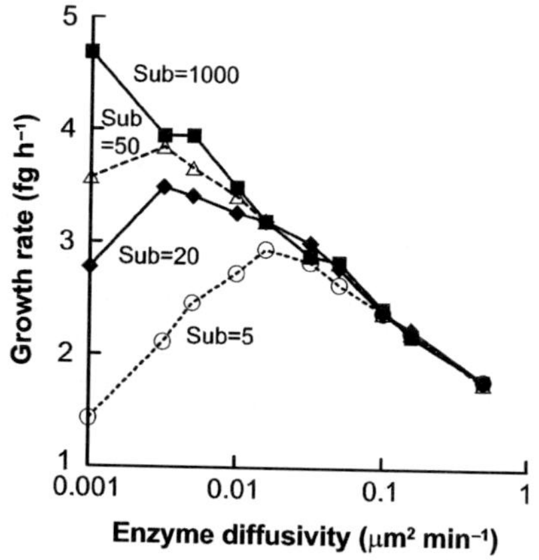

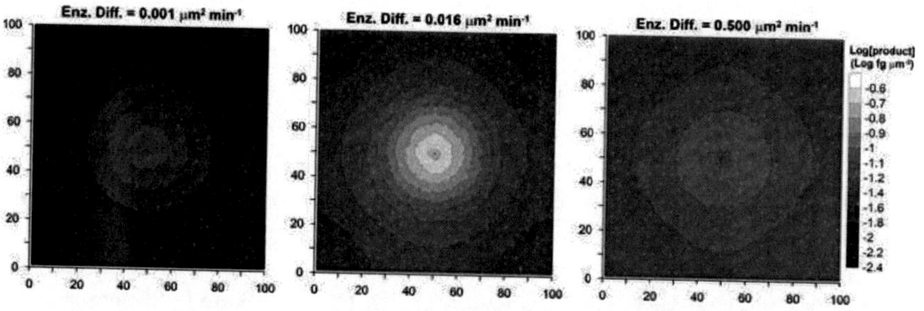

Fig. 12.3 Modeled product concentrations after 50 h on a spatial grid with an enzyme-producing microbe in the center and enzyme diffusivities (EDiff) of 0.001, 0.016, and 0.500 μm^2 min^{-1}. Substrate concentration = 5.0 fg μm^{-3}. Note that product concentrations around the microbe are greatest at intermediate enzyme diffusivity (0.016 μm^2 min^{-1})

enzyme production and extracellular foraging. Autoinducers would be ideal for this regulatory role because they are relatively cheap for cells to produce and are not naturally present in the extracellular environment.

12.5.2 Enzyme Responses to Global Environmental Change

At the ecosystem level, one important consequence of efficient allocation to extracellular enzyme production is a stronger correspondence between resource supply and demand. Microbes (and plant roots) have ecological and evolutionary incentives to use enzymes to extract nutrients from otherwise unavailable organic sources when nutrients are limiting. This allocation pattern suggests that the release of nutrients from complex organic sources will decrease when simple resources are more available. Thus, adding available nutrients should suppress the turnover of complex organic nutrients.

Because extracellular enzymes often control organic matter solubilization – the rate-limiting step in organic matter turnover – shifts in enzyme allocation could have major consequences for rates of C and nutrient cycling under global change. Environmental changes can alter allocation to different enzymes through regulatory pathways as well as shifts in community composition. Altered composition may result from changes in the competitive interactions within microbial communities (Koide et al. 2005). Depending on their competitive abilities, the relative abundances of enzyme producers may increase or decline as the soil environment changes. Although an exhaustive review of enzyme responses to global environmental change is beyond the scope of this chapter, we describe several examples of how enzyme allocation theory can be used to understand ecosystem responses to environmental change.

12.5.2.1 Increase in Atmospheric CO_2

Increasing concentrations of atmospheric CO_2 can alter the microbial production of soil extracellular enzymes through changes in belowground C availability and quality. Elevated CO_2 often stimulates the production of C-rich exudates from plant roots, which increases microbial demand for other elements (Hungate et al. 1997; Hamilton and Frank 2001). In tussock tundra, CO_2 fumigation increased soil phosphatase activity, presumably because plants and microbes were mitigating P deficiency (Moorhead and Linkins 1997). A similar trend has also been observed for N-degrading enzymes at a FACE site in Rhinelander, Wisconsin (Larson et al. 2002). Although the evidence is still somewhat equivocal, there may be changes in plant litter quality under elevated CO_2 that influence substrate availability for C-degrading enzymes (Franck et al. 1997). For example, increasing cellulose concentrations could stimulate cellulase production during litter decomposition if sufficient nutrients are available for enzyme production (Allison and Vitousek 2005).

One important consequence of microbial allocation to nutrient-releasing enzymes under elevated CO_2 is increased mining of nutrients from soil organic sources. Such a response could contribute to enhanced C sequestration if plants gain access to the released nutrients in order to support biomass growth. This mechanism would help alleviate progressive N limitation, which has been hypothesized to constrain plant C sequestration under elevated CO_2 (Johnson 2006). Alternatively, enhanced microbial growth as a result of nutrient-releasing enzyme activity could increase decomposition rates and offset additional C storage. Increased microbial allocation to C-degrading enzymes would have a similar effect; even as greater quantities of litter C enter the soil under elevated CO_2, enzyme-catalyzed decomposition could increase proportionately (Chung et al. 2007; Drissner et al. 2007).

12.5.2.2 Increases in N Deposition

Because extracellular enzymes are N rich and many ecosystems are N limited (at least in terms of plant communities) (Vitousek and Howarth 1991; LeBauer and Treseder 2008), N deposition often has strong impacts on enzyme activity. Based on allocation theory, greater N availability should increase microbial and plant demand for other elements such as C and P. The activities of cellulose-degrading enzymes increase with N addition in deciduous forests (Waldrop et al. 2004a; Sinsabaugh et al. 2005), tallgrass prairies (Ajwa et al. 1999), and California annual grasslands (Henry et al. 2005). N fertilization also increases soil phosphatase activity in grasslands (Ajwa et al. 1999; Phoenix et al. 2004; Henry et al. 2005; Chung et al. 2007), heathlands (Johnson et al. 1998), tropical forests (Olander and Vitousek 2000), and deciduous forests (Saiya-Cork et al. 2002). Furthermore, N fixation by plants has been shown to increase soil phosphatase activity (Zou et al. 1995; Allison et al. 2006). N addition to soil may also inhibit the production of phenol oxidase and peroxidase activities by soil fungi (Fog 1988; Carreiro et al. 2000; Saiya-Cork et al. 2002; Waldrop et al. 2004a). This response is consistent with culture studies, suggesting that oxidative enzymes are typically produced under N limitation and may aid in the acquisition of N from complex polymers such as lignin and humic substances (Fog 1988).

Reallocation of microbial and plant resources under N deposition has strong implications for C and nutrient cycling in ecosystems. Stimulation of cellulase activity can lead to faster decomposition of cellulose-rich litter, whereas inhibition of oxidative enzyme activity may slow the decomposition rate of more recalcitrant litter and soil organic material (Carreiro et al. 2000). Thus, the enzyme-mediated effect of N deposition on soil C cycling depends on the chemical quality of the litter and soil organic matter in a given ecosystem (Carreiro et al. 2000; Neff et al. 2002; Waldrop et al. 2004b). Given that added N may cause secondary P limitation and often increases soil phosphatase activity, enzymes may also contribute to faster rates of P cycling under increased N availability. This pattern may have occurred following invasion of native, nutrient poor Hawaiian ecosystems by the N-fixing tree *Falcataria moluccana* (Allison et al. 2006). The invasion disproportionately stimulated soil phosphatase activities, and P cycling through the ecosystem increased almost as dramatically as N cycling (Hughes and Denslow 2005). In contrast, the degradation of complex organic N probably declines in soil following N addition, as the activities of N-releasing extracellular enzymes decline. N fertilizer suppression of protein- and chitin-degrading enzymes has been observed across a range of ecosystems (Olander and Vitousek 2000; Allison et al. 2008), suggesting that depolymerization of organic N may decline despite increases in N mineralization and the cycling of available N forms.

12.5.2.3 Changes in Temperature and Moisture

There are several mechanisms by which changes in climate could directly or indirectly affect enzyme allocation. One indirect effect could occur through increasing soil temperature, which could result in higher rates of nutrient mineralization (Rustad et al. 2001). Under these conditions, microbes and plant roots may decrease their allocation to nutrient-acquiring enzymes as nutrients become more available. However, this effect could be offset by higher metabolic and growth rates under warmer conditions that could increase rates of constitutive enzyme production.

Changes in soil moisture may impact extracellular enzyme allocation more directly due to alteration of diffusion rates. As soils become drier, the volume of water available to dissolve enzymes and substrates declines and the effective concentrations of these constituents increase. Depending on the initial concentration of substrate and the original diffusion rate of the enzyme, such changes could increase or decrease the return on enzyme investment. Where substrate is not limiting near the enzyme producer, a reduced effective diffusion rate could localize more of the reaction products near the producer for uptake, thereby increasing growth rates. However, if substrate concentrations are low near the producer, then restricting diffusion would reduce enzyme access to more distant substrates (Fig. 12.3). Whether changes in diffusion rates would alter allocation among different enzymes would depend on the relative availabilities of different substrates. It is possible that changes in soil moisture would have similar effects on all enzymes, leaving relative allocation among them unchanged and simply favoring or disfavoring enzyme production relative to other strategies.

12.6 Conclusions

Empirical evidence and existing models support the idea that microbes and plant roots produce soil enzymes according to principles of resource supply and demand. Like the related field of ecological stoichiometry (Elser 2006), these principles are valuable because they link evolutionary theory and ecosystem ecology – the mechanisms that determine resource allocation at the organismal level also scale up to regulate fluxes of elements and energy at the ecosystem level. The stoichiometry of cellular biomass is the major determinant of resource demand, and extracellular enzyme production represents a strategy for acquiring resources to match that demand. There is also good evidence that enzyme producers maximize the benefits of enzyme production while minimizing the costs. Cost reductions can be achieved through regulatory mechanisms, while the benefits can be increased by manipulating enzyme diffusion and suppressing competition for enzyme reaction products. Allocation strategies for soil enzyme production may also help predict ecosystem responses to environmental change. Perturbations to soil resource availability (e.g., N addition, elevated CO_2) cause enzyme producers to shift their allocation patterns and thereby alter rates of C and nutrient cycling. Thus, resource allocation theory based on evolutionary and economic principles can improve our ability to predict ecosystem feedbacks to environmental change.

Acknowledgments We thank B. Caldwell, J. Talbot, K. Treseder, and S. Perakis for valuable comments on the manuscript. This research was supported by a NOAA Climate and Global Change Fellowship to SDA and a workshop grant from the NSF-LTER Network Office.

References

Ajwa HA, Dell CJ, Rice CW (1999) Changes in enzyme activities and microbial biomass of tallgrass prairie soil as related to burning and nitrogen fertilization. Soil Biol Biochem 31:769–777

Allison SD (2005) Cheaters, diffusion, and nutrients constrain decomposition by microbial enzymes in spatially structured environments. Ecol Lett 8:626–635

Allison SD (2006) Brown ground: a soil carbon analogue for the green world hypothesis? Am Nat 167:619–627

Allison SD, Vitousek PM (2005) Responses of extracellular enzymes to simple and complex nutrient inputs. Soil Biol Biochem 37:937–944

Allison SD, Nielsen CB, Hughes RF (2006) Elevated enzyme activities in soils under the invasive nitrogen-fixing tree *Falcataria moluccana*. Soil Biol Biochem 38:1537–1544

Allison SD, Gartner TB, Holland K, Weintraub M, Sinsabaugh RL (2007a) Soil enzymes: linking proteomics and ecological processes. In: Hurst CJ, Crawford RL, Garland JL, Lipson DA, Mills AL, Stetzenbach LD (eds) Manual of environmental microbiology, 3rd edn. ASM, Washington, DC, pp 704–711

Allison VJ, Condron LM, Peltzer DA, Richardson SJ, Turner BL (2007b) Changes in enzyme activities and soil microbial community composition along carbon and nutrient gradients at the Franz Josef chronosequence, New Zealand. Soil Biol Biochem 39:1770–1781

Allison SD, Czimczik CI, Treseder KK (2008) Microbial activity and soil respiration under nitrogen addition in Alaskan boreal forest. Global Change Biol 14:1156–1168

Antibus RK, Sinsabaugh RL, Linkins AE (1992) Phosphatase activities and phosphorus uptake from inositol phosphate by ectomycorrhizal fungi. Can J Bot 70:794–801

Bartnicki-Garcia S (1968) Cell wall chemistry, morphogenesis, and taxonomy of fungi. Annu Rev Microbiol 22:87–108

Burns RG (1978) Soil enzymes. Academic, New York

Burns RG (1982) Enzyme activity in soil: location and a possible role in microbial ecology. Soil Biol Biochem 14:423–427

Carreiro MM, Sinsabaugh RL, Repert DA, Parkhurst DF (2000) Microbial enzyme shifts explain litter decay responses to simulated nitrogen deposition. Ecology 81:2359–2365

Chróst RJ (1991) Environmental control of the synthesis and activity of aquatic microbial ectoenzymes. In: Chróst RJ (ed) Microbial enzymes in aquatic environments. Springer, New York, pp 29–59

Chung H, Zak DR, Reich PB, Ellsworth DS (2007) Plant species richness, elevated CO_2, and atmospheric nitrogen deposition alter soil microbial community composition and function. Global Change Biol 13:908–989

Cleveland CC, Liptzen D (2007) C:N:P stoichiometry in soil: is there a "Redfield ratio" for the microbial biomass? Biogeochemistry 85:235–252

Davey ME, O'Toole GA (2000) Microbial biofilms: from ecology to molecular genetics. Microbiol Mol Biol Rev 64:847–867

Drissner D, Blum H, Tscherko D, Kandeler E (2007) Nine years of enriched CO_2 changes the function and structural diversity of soil microorganisms in a grassland. Eur J Soil Sci 58:260–269

Ehrenfeld JG (2003) Effects of exotic plant invasions on soil nutrient cycling processes. Ecosystems 6:503–523

Ekschmitt K, Liu MQ, Vetter S, Fox O, Wolters V (2005) Strategies used by soil biota to overcome soil organic matter stability – why is dead organic matter left over in the soil? Geoderma 128:167–176

Elser J (2006) Biological stoichiometry: a chemical bridge between ecosystem ecology and evolutionary biology. Am Nat 168:S25–S35

Eriksson KE, Pettersson B, Westmark U (1974) Oxidation: an important enzyme reaction in fungal degradation of cellulose. FEBS Lett 49:282–285

Fog K (1988) The effect of added nitrogen on the rate of decomposition of organic matter. Biol Rev Camb Philos Soc 63:433–462

Franck VM, Hungate BA, Chapin FS III, Field CB (1997) Decomposition of litter produced under elevated CO_2: dependence on plant species composition. Biogeochemistry 36:223–237

Frankena J, Vanverseveld HW, Stouthamer AH (1988) Substrate and energy costs of the production of exocellular enzymes by *Bacillus-licheniformis*. Biotechnol Bioeng 32:803–812

Friedel JK, Scheller E (2002) Composition of hydrolyzable amino acids in soil organic matter and soil microbial biomass. Soil Biol Biochem 34:315–325

Hamilton EW III, Frank DA (2001) Can plants stimulate soil microbes and their own nutrient supply? Evidence from a grazing tolerant grass. Ecology 82:2397–2404

Hanif A, Yasmeen A, Rajoka MI (2004) Induction, production, repression, and de-repression of exoglucanase synthesis in *Aspergillus niger*. Bioresour Technol 94:311–319

Haynes RJ, Swift RS (1988) Effects of lime and phosphate additions on changes in enzyme activities, microbial biomass and levels of extractable nitrogen, sulfur and phosphorus in an acid soil. Biol Fertil Soils 6:153–158

Henry HAL, Juarez JD, Field CB, Vitousek PM (2005) Interactive effects of elevated CO_2, N deposition and climate change on extracellular enzyme activity and soil density fractionation in a California annual grassland. Global Change Biol 11:1808–1815

Hughes RF, Denslow JS (2005) Invasion by a N_2-fixing tree alters function and structure in wet lowland forests of Hawai'i. Ecol Appl 15:1615–1628

Hungate BA, Holland EA, Jackson RB, Chapin FS III, Mooney HA, Field CB (1997) The fate of carbon in grasslands under carbon dioxide enrichment. Nature 388:576–579

Jackson RB, Mooney HA, Schulze E-D (1997) A global budget for fine root biomass, surface area, and nutrient contents. Proc Natl Acad Sci USA 94:7362–7366

Jacob F, Monod J (1961) Genetic regulatory mechanisms in the synthesis of proteins. J Mol Biol 3:318–356

Johnson DW (2006) Progressive N limitation in forests: review and implications for long-term responses to elevated CO_2. Ecology 87:64–75

Johnson D, Leake JR, Lee JA, Campbell CD (1998) Changes in soil microbial biomass and microbial activities in response to 7 years simulated pollutant nitrogen deposition on a heathland and two grasslands. Environ Pollut 103:239–250

Koide RT, Xu B, Sharda J, Lekberg Y, Ostiguy N (2005) Evidence of species interactions within an ectomycorrhizal fungal community. New Phytol 165:305–316

Koroljova-Skorobogatko OV, Stepanova EV, Gavrilova VP, Morozova OV, Lubimova NV, Dzchafarova AN, Jaropolov AI, Makower A (1998) Purification and characterization of the constitutive form of laccase from the basidiomycete *Coriolus hirsutus* and effect of inducers on laccase synthesis. Biotechnol Appl Biochem 28:47–54

Larson JL, Zak DR, Sinsabaugh RL (2002) Extracellular enzyme activity beneath temperate trees growing under elevated carbon dioxide and ozone. Soil Sci Soc Am J 66:1848–1856

LeBauer DS, Treseder KK (2008) Nitrogen limitation of net primary productivity in terrestrial ecosystems is globally distributed. Ecology 89:371–379

Makino W, Cotner JB, Sterner RW, Elser JJ (2003) Are bacteria more like plants or animals? Growth rate and resource dependence of bacterial C:N:P stoichiometry. Funct Ecol 17:121–130

McGroddy ME, Daufresne T, Hedin LO (2004) Scaling of C:N:P stoichiometry in forests worldwide: implications of terrestrial Redfield-type ratios. Ecology 85:2390–2401

Moorhead DL, Linkins AE (1997) Elevated CO_2 alters belowground exoenzyme activities in tussock tundra. Plant Soil 189:321–329

Moorhead DL, Sinsabaugh RL (2006) A theoretical model of litter decay and microbial interaction. Ecol Monogr 76:151–174

Neff JC, Townsend AR, Gleixner G, Lehman SJ, Turnbull J, Bowman WD (2002) Variable effects of nitrogen additions on the stability and turnover of soil carbon. Nature 419:915–917

Olander LP, Vitousek PM (2000) Regulation of soil phosphatase and chitinase activity by N and P availability. Biogeochemistry 49:175–190

Phoenix GK, Booth RE, Leake JR, Read DJ, Grime JP, Lee JA (2004) Simulated pollutant nitrogen deposition increases P demand and enhances root-surface phosphatase activities of three plant functional types in a calcareous grassland. New Phytol 161:279–289

Priest FG (1977) Extracellular enzyme synthesis in the genus *Bacillus*. Bacteriol Rev 41:711–753

Raab TK, Lipson DA, Monson RK (1999) Soil amino acid utilization among species of the Cyperaceae: plant and soil processes. Ecology 80:2408–2419

Redfield AC (1958) The biological control of chemical factors in the environment. Am Sci 46:205–221

Redfield RJ (2002) Is quorum sensing a side effect of diffusion sensing. Trends Microbiol 10:365–372

Reiners WA (1986) Complementary models for ecosystems. Am Nat 127:59–73

Rustad LE, Campbell JL, Marion GM, Norby RJ, Mitchell MJ, Hartley AE, Cornelissen JHC, Gurevitch J (2001) A meta-analysis of the response of soil respiration, net nitrogen mineralization, and aboveground plant growth to experimental warming. Oecologia 126:543–562

Saiya-Cork KR, Sinsabaugh RL, Zak DR (2002) The effects of long term nitrogen deposition on extracellular enzyme activity in an *Acer saccharum* forest soil. Soil Biol Biochem 34:1309–1315

Schimel JP, Weintraub MN (2003) The implications of exoenzyme activity on microbial carbon and nitrogen limitation in soil: a theoretical model. Soil Biol Biochem 35:549–563

Schleifer KH, Kandler O (1972) Peptidoglycan types of bacterial cell walls and their taxonomic implications. Microbiol Mol Biol Rev 36:407–477

Sinsabaugh RL (1994) Enzymic analysis of microbial pattern and process. Biol Fertil Soils 17:69–74

Sinsabaugh RL, Moorhead DL (1994) Resource allocation to extracellular enzyme production: a model for nitrogen and phosphorus control of litter decomposition. Soil Biol Biochem 26:1305–1311

Sinsabaugh RL, Gallo ME, Lauber C, Waldrop MP, Zak DR (2005) Extracellular enzyme activities and soil organic matter dynamics for northern hardwood forests receiving simulated nitrogen deposition. Biogeochemistry 75:201–215

Sinsabaugh RL, Lauber CL, Weintraub MN, Ahmed B, Allison SD, Crenshaw C, Contosta AR, Cusack D, Frey S, Gallo ME, Gartner TB, Hobbie SE, Holland K, Keeler BL, Powers JS, Stursova M, Takacs-Vesbach C, Waldrop MP, Wallenstein MD, Zak DR, Zeglin LH (2008) Stoichiometry of soil enzyme activity at global scale. Ecol Lett 11:1252–1264

Skujins JJ (1976) History of abiontic soil enzyme research. In: Burns RG (ed) Soil enzymes. Academic, London, pp 1–49

Sterner RW, Elser JJ (2002) Ecological stoichiometry: the biology of elements from molecules to the biosphere. Princeton University Press, Princeton, NJ

Sun Y-P, Unestam T, Lucas SD, Johanson KJ, Kenne L, Finlay R (1999) Exudation-reabsorption in a mycorrhizal fungus, the dynamic interface for interaction with soil and soil microorganisms. Mycorrhiza 9:137–144

Swift MJ, Andrén O, Brussaard L, Briones M, Couteaux M-M, Ekschmitt K, Kjoller A, Loiseau P, Smith P (1998) Global change, soil biodiversity, and nitrogen cycling in terrestrial ecosystems: three case studies. Global Change Biol 4:729–743

Treseder KK, Vitousek PM (2001) Effects of soil nutrient availability on investment in acquisition of N and P in Hawaiian rain forests. Ecology 82:946–954

Trumbore SE (1997) Potential responses of soil organic carbon to global environmental change. Proc Natl Acad Sci USA 94:8284–8291

Vetter YA, Denning JW, Jumars PA, Krieger-Brockett BB (1998) A predictive model of bacterial foraging by means of freely released extracellular enzymes. Microb Ecol 36:75–92

Vitousek PM, Howarth RW (1991) Nitrogen limitation on land and in the sea: How can it occur? Biogeochemistry 13:87–115

Voigt B, Schweder T, Sibbald MJJB, Albrecht D, Ehrenreich A, Bernhardt J, Feesche J, Maurer K-H, Gottschalk G, van Dijl JM, Hecker M (2006) The extracellular proteome of *Bacillis licheniformis* grown in different media and under different nutrient starvation conditions. Proteomics 6:268–281

Waldrop MP, Zak DR, Sinsabaugh RL (2004a) Microbial community response to nitrogen deposition in northern forest ecosystems. Soil Biol Biochem 36:1443–1451

Waldrop MP, Zak DR, Sinsabaugh RL, Gallo M, Lauber C (2004b) Nitrogen deposition modifies soil carbon storage through changes in microbial enzyme activity. Ecol Appl 14:1172–1177

Wallenstein MD, Weintraub MN (2008) Emerging tools for measuring and modeling in situ activity of soil extracellular enzymes. Soil Biol Biochem 40:2098–2106

Weintraub MN, Schimel JP (2005) Seasonal protein dynamics in Alaskan arctic tundra soils. Soil Biol Biochem 37:1469–1475

Zou X, Binkley D, Caldwell BA (1995) Effects of dinitrogen-fixing trees on phosphorus biogeochemical cycling in contrasting forests. Soil Sci Soc Am J 59:1452–1458

Chapter 13
Controls on the Temperature Sensitivity of Soil Enzymes: A Key Driver of *In Situ* Enzyme Activity Rates

Matthew Wallenstein, Steven D. Allison, Jessica Ernakovich, J. Megan Steinweg, and Robert Sinsabaugh

13.1 Introduction

Soil microorganisms are surrounded by organic matter that is rich in carbon and nutrients that are required for growth and cell maintenance. However, microbes cannot directly transport these macromolecules into the cytoplasm. Rather, they rely on the activity of a myriad of enzymes that they produce and release into their environment. These enzymes depolymerize organic compounds and generate soluble oligomers and monomers that are then recognized by cell wall receptors and transported across the outer membrane and into the cell. Thus, the activities of extracellular enzymes are critical to soil functioning and for maintenance of the vast biodiversity of organisms in soils.

The activity of glucosidases, phosphatases, phenol oxidases, and other enzymes that degrade the principal components of detrital organic matter have been extensively studied from many perspectives. In early studies, the physical and kinetic characteristics of soil enzymes were a major topic (Bremner and Zatua 1975; McClaugherty and Linkins 1990; Frankenberger and Tabatabai 1991b). More recent studies have concentrated on the ecological significance of soil enzyme activity as a mediator of nutrient cycling. Yet, the fundamental role of temperature

M. Wallenstein (✉)
Natural Resource Ecology Laboratory, Colorado State University, Fort Collins, CO 80523, USA
University of California, Irvine, 321 Steinhaus, Irvine, CA 92697, USA
e-mail: mawallen@nrel.colostate.edu

S.D. Allison
University of California, Irvine, 321 Steinhaus, Irvine, CA 92697, USA

J. Ernakovich and J.M. Steinweg
Natural Resource Ecology Laboratory, Colorado State University, Fort Collins, CO 80523, USA
Graduate Degree Program in Ecology, Colorado State University, Fort Collins, CO 80523, USA

R. Sinsabaugh
Department of Biology, University of New Mexico, Aluquerque, NM 87131, USA

G. Shukla and A. Varma (eds.), *Soil Enzymology*, Soil Biology 22,
DOI 10.1007/978-3-642-14225-3_13, © Springer-Verlag Berlin Heidelberg 2011

in regulating enzyme activities under field conditions has been examined in relatively few studies. In theory, the temperature sensitivity of enzyme activities can be described from first principles of thermodynamics. In this chapter, we consider the utility and limitations of thermodynamic extracellular enzyme activity models for understanding the dynamics of ecosystem processes, and we review our current understanding of the thermal ecology of extracellular enzymes in soils.

13.2 What Controls Enzyme Temperature Sensitivity?

13.2.1 Enzyme Conformation

Enzymes are proteins that catalyze reactions by lowering the activation energy of biochemical reactions. There are two aspects to the temperature sensitivity of enzymes. The first aspect is their thermal stability or their ability to maintain their structure across a range of temperatures. The structure of proteins is the main determinant of their thermal stability, and is defined by the terms primary, secondary, and tertiary elements. The primary structure of a protein or enzyme is its linear amino acid sequence in the secondary structure; protein folding conformations are created by interactions between amino acid side chains that are dispersed along the polypeptide; the tertiary or native structure is the fully folded state of the globular protein, which may include multiple secondary elements (Straub 1964). The folding of proteins is driven by differences in Gibbs free energy of various conformations between the unfolded and the folded protein. The term "macrostability" defines the energy released by the enzyme taking on its 3D tertiary structure.

The second aspect of enzyme temperature sensitivity is the temperature sensitivity of catalytic *activity*, which is determined primarily by the accessibility of the active site of the enzyme. "Microstability" refers to the energy associated with reversible, local changes in structure. The microstability is responsible for the flexibility or rigidity of the *active site* of an enzyme, which is the location of substrate or ligand binding that leads to catalysis (Privalov and Tsalkova 1979). The overall stability of an enzyme is the free energy difference between the macrostability and the microstability (Zavodszky et al. 1998). The active site of enzymes is more flexible than the whole enzyme due to weaker molecular interactions (Tsou 1993), and the active site of an enzyme loses its function (with the addition of chemical denaturants or heat) more quickly than the whole protein can be denatured (Tsou 1993), (Fig. 13.1).

The conformation of enzymes has been described by many models, the most influential of which are Koshland's "induced fit" model and Straub's "fluctuation fit" model. The "induced fit" model postulates that enzyme activation is induced by a change in the conformation of the active site due to interactions of the active site with the substrate. The "fluctuation fit" model of enzyme function says that the

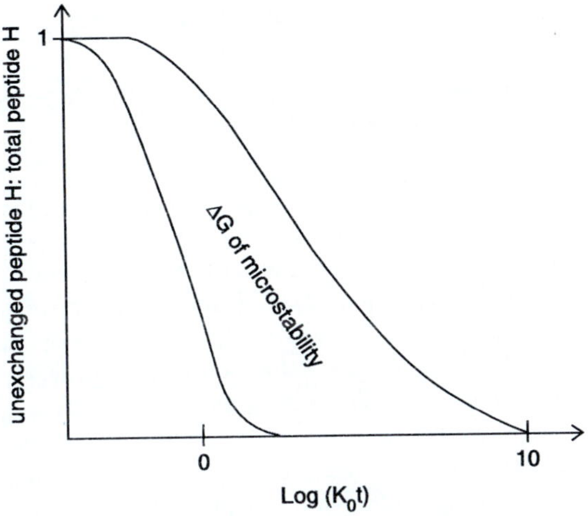

Fig. 13.1 The relationship between the relative proportion of unexchanged peptide hydrogen to the total number of hydrogen in the protein and the Log (K_0t). K_0 is the rate constant for the exchange of hydrogen from the primary structure of a protein. This is the tendency of a protein to undergo micro-unfolding, and the ΔG of microstability is the energy required for the exchange of one peptide hydrogen (adapted from Privalov and Tsalkova 1979)

native conformation of an enzyme can exist in many states in solution. These general principles of enzyme conformation apply across all thermal regimes, but specific adaptations are also made in order to maintain enzyme function across different temperature conditions.

Cold-adapted enzymes have more flexible active sites (Hochachka and Somero 1984) created by a weakening of the intramolecular forces in the active site of the enzyme (Gerday et al. 1997). In contrast, the active sites of heat-adapted enzymes are more rigid (Zavodszky et al. 1998). The changes in flexibility in cold-adapted and thermophilic enzymes are due to one or more changes in the amino acid structure of the active site (Zavodszky et al. 1998). Cold-adapted microorganisms may produce cold-adapted enzymes that catalyze reactions at lower temperatures than their mesophilic counterparts by adjusting their chemical structure (Gerday et al. 1997). The activation energy of these enzymes is lower than that of mesophilic enzymes. Cold-adapted enzymes may have limited range of thermal stability around their temperature optima due to the conformational changes that increase the activity of the enzyme at low temperatures, resulting in a loss of function at the active site before the 3D structure of the protein denatures (Gerday et al. 1997). Thermophilic enzymes have more rigid active sites below their temperature optimum. Zavodszky et al. (1998) found that 3-isopropylmalate dehydrogenase (IPMDH) isolated from the thermophilic bacteria *Thermus thermophilus* did not denature until 17°C higher than the mesophilic homolog, indicating that macrostructure plays an important role in thermal adaptation as well. At their optimum temperature, the flexibility of the active site of a thermophilic enzyme is equal to that of its mesophilic counterpart at its temperature optimum (Secundo et al. 2005; Zavodszky et al. 1998). Thus, it appears that the flexibility and activity of enzymes are closely related, and that organisms have evolved to create enzymes with thermal optima at their habitation temperature.

13.2.2 Modeling Enzyme Kinetics

Thermodynamic models of biochemical and ecological processes begin with the Arrhenius equation:

$$V(T) = V(T_0)e^{-E_a/k(1/T_0-1/T)},$$

where *V(T)* is the reaction rate at temperature *T*, $V(T_0)$ is the reaction rate at a reference temperature T_0, E_a is the activation energy in eV, and *k* is the Boltzmann constant (8.62×10^{-5} eV K^{-1}). E_a is the energy differential between reactants and the transitional species that subsequently decay into products. This activation energy determines the rate of change in reaction rate with temperature. A reaction rate that doubles with a 10°C rise in temperature [$Q_{10} = 2$] has an E_a of 0.5 eV [48 kJ mol^{-1}]. Enzymes facilitate reactions by stabilizing transition states, thereby lowering E_a. Although the Arrhenius model was originally developed to describe simple reactions, it is used to describe the apparent temperature dependency of metabolic processes at every scale of biological organization. In metabolic scaling theory, the model is often considered in relation to area: volume ratios that limit rates of environmental exchange, e.g., oxygen or carbon dioxide in the case of respiration or photosynthesis. This normalization, usually represented as a fractal scalar with a value of 0.75 rather than a surface area to volume ratio, allows the temperature dependence of metabolic processes, e.g., growth and respiration, to be compared on common basis across organizational and body size scales (Gillooly et al. 2001; Allen et al. 2005). Meta-analyses of the temperature dependence of organismal metabolism yield a mean E_a of 0.62 eV (Gillooly et al. 2001).

When considering biochemical processes such as soil EEA that occur over large ranges of temperature, E_a is not the only parameter that influences responses. Enzymes have an effective range and an optimal temperature (T_{opt}) of operation that is determined by their size and composition. At low temperature, enzymes or the matrix with which they are associated freeze. At temperatures only a few degrees above their optimum, the tertiary structure of enzymes begins to denature, unless it is stabilized by interactions with particle surfaces or humic complexes. In ecological systems, enzyme classes, such as laccase or β-glucosidase, are populations of isoenzymes of diverse origin (Di Nardo et al. 2004) that differ in size, polypeptide sequence, and post-translational modifications. Consequently, estimates of E_a and optimal temperature become statistical distributions. These measures are further extenuated by environmental interactions with organic and inorganic particles that can substantially alter both E_a and T_{opt} values by compromising the conformational flexibility of the enzyme. Thus, at the ecosystem scale, organic matter abundance and composition, and soil texture and mineralogy also influence EEA response to temperature fluctuations. In this context, E_a is used as an empirical measurement of system response to temperature change and T_{opt}, which is usually much greater than in situ temperature, is a measure of the relative stability, i.e., turnover rate, of the soil enzyme pools.

These enzyme–environment interactions combined with succession or variation in the composition of the enzyme producer community introduces a scale dependence to apparent EEA temperature response. The T_{opt} and E_a estimates determined for a single sample will not be the same as those calculated by comparing samples collected over spatial or temporal gradients. As the scale of comparison expands, T_{opt} and apparent E_a are conflated and apparent temperature response becomes increasingly a function of resource inputs and community composition to an extent that rates may even be inversely related to temperature, i.e., have an apparent E_a that is negative.

The mean E_a for extracellular enzymes that are not conformationally constrained is about 0.3 eV. In contrast, the E_a for microbial [and plant] metabolism, and presumably extracellular enzyme production, is around 0.62 eV (Gillooly et al. 2001). Reported values of apparent E_a for the metabolism of bacterial communities range from 0.41 to 1.14 eV (Kristensen et al. 1992; Sagemann et al. 1998; Thamdrup et al. 1998; Price and Sowers 2004; Lopez-Urrutia and Moran 2007). The high value of 1.14 eV comes from an analysis by Price and Sowers (2004) that includes data from several types of ecosystems and spans a temperature range from −20°C to 30°C. Enzymes that are conformationally compromised by sorption to elements of the soil matrix may have E_a values similar to that of microbial growth. As a result, the E_a for a particular extracellular enzymatic activity, e.g., phosphatase, is broadly distributed within a soil or litter matrix and some fraction of activity may have an apparent $E_a > 1$ eV.

Superficially, thermal control of EEA seems like a simple thing to model. But temperature responses are conflated with myriad other ecological variables. As a result, estimates of apparent E_a are specific to the enzymes, systems, temperature range, and spatiotemporal scale under consideration.

13.2.3 Michaelis–Menten Model

The kinetics of simple enzymes (i.e., enzymes with one active site that interact with a single substrate) is described by the Michaelis–Menten model as a hyperbolic function:

$$V = \text{Vmax}(S/(S + \text{Km})),$$

where V is reaction rate, S is substrate concentration, Vmax is the rate of substrate conversion when all enzymes are operating at maximum capacity, and Km is a half-saturation constant (i.e., the substrate concentration at which the rate of substrate conversion is equal to Vmax/2). Km is also a measure of the binding affinity of substrate and enzyme.

When the Michaelis–Menten model is applied to ecological systems, its assumptions do not apply and Vmax and Km no longer reflect the biochemical attributes defined in its original context (Williams 1973). In such cases, these parameters are more accurately described as apparent Vmax (^{App}Vmax) and apparent Km (^{App}Km) with ^{App}Vmax, a relative measure of enzyme abundance, and ^{App}Km, a relative measure of substrate concentration. Spatiotemporal variation in ^{App}Vmax most likely reflects differences in the concentration of rate-limiting enzyme, rather than the replacement of one enzyme by another of different structure, particularly when the rate-limiting "enzyme" in question is actually a population of enzymes of similar function produced in many versions, by multiple organisms, under different controls, and heterogeneously dispersed within the environment. Similarly, spatiotemporal variation in ^{App}Km most likely reflects differences in the concentration of the substrate pool because natural substrates act as competitive inhibitors for reactions measured by adding labeled or artificial substrates to environmental samples (Chrost 1990):

$$^{App}\mathrm{Km} = \mathrm{Km}(1 + I/\mathrm{Ki}),$$

where I is the concentration of inhibiter, in this case the background concentration of environmental substrate, and Ki is the half-saturation constant for the enzyme-inhibiter reaction.

In the Michaelis–Menten model, Vmax and Km are independent parameters. However, in ecological systems, ^{App}Vmax and ^{App}Km may be correlated because EEA is tightly controlled by a hierarchy of positive and negative feedback processes linked to substrate availability that operate at the molecular, cellular, and population levels (Chrost 1990; Chróst and Siuda 2002; Lugtenberg et al. 2002; Vial Ludovic et al. 2007). At the molecular level, the activity of individual enzymes is affected by competitive and non-competitive inhibition reactions as well as substrate concentration; at the cellular level, enzyme expression is controlled by induction and repression pathways linked to environmental cues; at the population level, enzyme expression may be coordinated by quorum signals. This correlation between environmental substrate concentration, estimated as ^{App}Km, and enzymatic capacity (^{App}Vmax) can be obscured, particularly at fine spatiotemporal scales, because the extracellular enzyme pool is, to varying extent, spatiotemporally decoupled from the organisms that produced them. Substantial fractions of the pool, particularly for soils, may be stabilized by sorption to humic or mineral colloids, or associated with cell fragments and extracellular polysaccharides creating lags in EEA response to changes in bacterial metabolism (Wilczek et al. 2005).

Activation energies are parameters that mechanistically link enzyme kinetics and temperature responses through the Arrhenius function. In the Michaelis–Menten function, the temperature sensitivity of Vmax is directly related to the activation energy for the enzyme reaction (Davidson and Janssens 2006). In addition, the Km parameter of the Michaelis–Menten function also increases with temperature, which reduces the substrate binding affinity of the enzymes. When substrate concentrations are near Km, this effect can offset the temperature effects on Vmax, resulting in little temperature dependence of the enzyme reaction (Davidson et al. 2006). A mechanistic model that includes temperature sensitivities of Vmax and Km would be superior to non-mechanistic empirical relationships, such as $Q10$.

Biological responses to temperature are often characterized in terms of the parameter Q_{10}, which is the factor by which a biological process changes in response to a 10°C temperature increase (Lloyd and Taylor 1994). Although many biological processes show a Q_{10} of approximately 2, this factor varies with temperature and is not based on a particular biological mechanism, in contrast to the Michaelis–Menten model. As a purely empirical parameter, Q_{10} values cannot be reliably extrapolated beyond measured response ranges or applied to novel systems.

13.2.4 Enzyme Binding to Soil Particles

Within the mineral matrix of the soil, organic matter–mineral binding and physical occlusion of organic matter within soil aggregates both act to limit the mixing of enzymes with otherwise decomposable OM (Tisdall and Oades 1982; Sollins et al. 1996; Jastrow and Miller 1997; Six et al. 2002). The turnover times of free, or bio-available, soil organic matter compounds can be orders of magnitude less than those for the same compounds found in association with soil minerals (Sørensen 1972). Such physical isolation of reactants violates a precept of kinetic theory (Davidson and Janssens 2006; Ågren and Wetterstedt 2007). But organic matter adsorption to mineral surfaces is a chemical process too. Organic matter binds with mineral particles *via* several types of non-covalent bonds (e.g., van der Waals forces, hydrogen bonding). Rates of formation (adsorption) and breakdown (desorption) of those bonds both tend to increase with increasing temperature. But because adsorption reactions are exergonic and have lower E_a's, the equilibrium between adsorption and desorption shifts toward desorption with increasing temperature – more compounds are in solution at warmer temperatures (ten Hulscher and Cornelissen 1996). Thus, increased desorption at higher temperatures could contribute to the temperature sensitivity of in situ enzyme activity. Since enzyme activities are typically measured in lab assays where substrate is non-limiting, temperature sensitivity of in situ enzymes may be under-predicted.

Davidson et al. (Davidson and Janssens 2006; Davidson et al. 2006) have argued that a conceptual framework based on activation energies and substrate availabilities would be a more useful alternative to Q_{10} models. Sinsabaugh and Shah (2010) developed a modeling approach that combines thermal scaling with resource availability. Using estimates of apparent Km, Vmax, and E_a for six extracellular enzymes that mediate nutrient acquisition from carbohydrate, protein, lipid, and organic phosphate pools, they were able to predict variation in bacterial production rates over an annual cycle in two rivers that experience seasonal changes in both temperature and the supply of multiple resources.

13.3 Indirect Effects of Temperature on Enzyme Activities

In general, the overall metabolic rate of enzyme-producing organisms increases with temperature with a mean E_a of 0.62 eV over the range 5–40°C. Thus, the rate of extracellular enzyme production is more responsive to temperature than the kinetics of the enzymes themselves. It is not currently possible to directly measure enzyme production rates in soils (Wallenstein and Weintraub 2008), and data from pure cultures are scarce and are likely to far exceed field rates where resources are limited.

Microorganisms utilize carbon for processes such as growth, maintenance, and enzyme production. Carbon allocation varies based on substrate availability, temperature, moisture, and other environmental factors. The type of carbon assimilated by microbes results in different amounts of energy available for growth and maintenance. Carbon utilization efficiency (CUE) is a measure of how efficiently microorganisms metabolize versus mineralize carbon. In aquatic systems, CUE has been shown to be relatively insensitive to changes in temperature (del Giorgio and Cole 1998; Seto and Misawa 1982); however, in soils it has been demonstrated that carbon utilization efficiency can be temperature dependent (Devevre and Horwat 2000; Steinweg et al. 2008). CUE is currently a fixed parameter in ecosystem models such as CENTURY (Parton et al. 1987); however, it has been demonstrated that CUE is lower at warmer temperatures regardless of the quality of soil organic matter (Steinweg et al. 2008). Low CUE results in more CO_2 produced per unit of substrate incorporated into biomass. Thus, temperature can affect the relative allocation of resources toward enzyme production.

Changes in temperature not only affect enzyme production rates by microorganisms but also affect enzyme degradation rates in the environment. Enzyme turnover is the result of proteolytic enzyme from activity as well as abiotic reactions. Both these processes should increase with temperature, but may show different temperature sensitivities due to differences in activation energy. Enzyme-catalyzed reactions generally show lower activation energies than uncatalyzed reactions, so the temperature sensitivity of the abiotic reactions may be higher (Tabatabai 1982). However, the rates of these reactions are also lower, so the net impact on enzyme activity may be small.

13.4 Temperature Sensitivity of Extracellular Enzymes under Field Conditions

Most contemporary studies of extracellular enzymes focus on spatial or temporal patterns in potential activities, which are typically measured at a single reference temperature in lab assays. This approach neglects the importance of temperature in controlling in situ activities (Wallenstein and Weintraub 2008). In most ecosystems, soil temperatures vary on diel to seasonal time scales, and change in response to long-term climate trends. If we assume that enzyme activity roughly doubles for every 10°C increase in temperature, then the effect of temperature clearly may have a greater impact on in situ activity rates than seasonal fluctuations in enzyme potential at most sites. For example, Wallenstein et al. (2009) developed a quantitative model of in situ B-glucosidase activities based on seasonal lab-based measurements of B-glucosidase potential activities at two temperatures, and using daily soil temperature data from an Arctic tundra site. They found that temperature explained 72% of the variation in predicted in situ activities. Temperature had a larger influence on modeled in situ enzyme activity than seasonal changes in enzyme pools. Clearly, temperature controls on in situ enzyme activities needs to be further explored in other biomes.

The assumption that all enzymes are equally sensitive to temperature, or even that the same class of enzyme exhibits a consistent temperature sensitivity within a single site, has not been borne out in the literature. In fact, several studies have demonstrated that the temperature sensitivity of extracellular enzymes changes seasonally (Fenner et al. 2005; Koch et al. 2007; Trasar-Cepeda et al. 2007; Wallenstein et al. 2009). The most likely explanation is that the measured enzyme pool consists of different isoenzymes (enzymes with the same function, but different structure) through time, which may be produced by different organisms or by a single species capable of producing multiple isoenzymes (Loveland et al. 1994; Sanchez-Perez et al. 2008). Consistent with this hypothesis, Di Nardo et al. (2004) found temporal changes in laccase and peroxidase isoenzymes during leaf litter decomposition. There is also some evidence for biogeographical patterns in enzyme temperature sensitivity. For example, many studies have observed that enzymes from microbes inhabiting cold environments have unusually low temperature optima (Huston et al. 2000; Coker et al. 2003; Feller 2003). Nonetheless, these observations suggest that microbes producing enzymes that maintain optimal activity under native soil conditions are favored. Thus, soil microbial community composition is likely controlled to some extent through feedbacks with enzyme efficacy.

It is widely assumed that enzyme activity roughly doubles with a 10°C increase in temperature ($Q_{10} = 2$); however, the accumulated evidence of numerous studies suggests a wide range in temperature sensitivities for different enzymes, and measured Q_{10}'s are often <2 (McClaugherty and Linkins 1990; Frankenberger and Tabatabai 1991a, b; Lai and Tabatabai 1992; Wirth and Wolf 1992; Criquet et al. 1999; Parham and Deng 2000; Elsgaard and Vinther 2004; Trasar-Cepeda et al. 2007). For example, Trasar-Cepeda et al. (2007) measured the Q_{10} of nine different enzymes in three different soils and found that the Q_{10} at 20°C exceeded 2.0 only for B-glucosidase in one of the soils. Most of the enzymes in that study had a Q_{10} closer to 1.5, corresponding to an E_a of 0.3 eV. However, the apparent temperature sensitivity of enzymes in lab assays with unlimited amounts of substrate and without constraints to diffusion may differ markedly from in situ temperature sensitivities. Even despite methodological concerns, there are insufficient data to assess the degree to which enzyme temperature sensitivity varies across spatial gradients or in response to other environmental factors.

Although there are insufficient data to establish generalized patterns in the temperature sensitivity of specific classes of enzymes, it is clear that their temperature sensitivities differ within a single environment (Koch et al. 2007; Wallenstein et al. 2009). For example, Koch et al. (2007) found that at low temperatures, the relative temperature sensitivity of C-degrading enzymes was greater than aminopeptidases (which degrade N-rich proteins), suggesting that relative N availability could be decreased directly by temperature. Similarly, in the study by Wallenstein et al. (2009), N-degrading enzymes tended to have a lower Q_{10} (overall mean of 1.59) than C-degrading enzymes (overall mean of 2.07). Thus, without any changes in enzyme pools, the relative in situ activity of these enzymes would change along with temperature, resulting in higher rates of C-mineralization relative to N-mineralization. Because different enzymes have different temperature sensitivities, changes in soil temperature may also alter the relative rates of decomposition of different components of soil organic matter. Therefore, seasonal changes in temperature can alter the balance of SOM (Soil Organic Matter) components contributing to soil respiration *without any changes in soil enzyme pools* (or measured enzyme potentials). Natural or human-driven changes in climate could also alter the relative rate of decomposition of SOM components, and ultimately, the quantity and composition of SOM.

One of the most important reasons for understanding the temperature sensitivity of soil enzymatic reactions is to improve predictions of soil respiration responses to temperature. Although there is considerable variability, about half of the CO_2 respired from soil can be attributed to heterotrophic respiration of soil organic matter (Ryan and Law 2005; Czimczik et al. 2006; Scott-Denton et al. 2006). A large fraction of this amount depends on the activity of extracellular enzymes, since much of the dead plant material entering soils are in polymeric form. However, the temperature sensitivity of heterotrophic soil respiration may be decoupled from enzyme temperature sensitivity because enzymatic products must undergo diffusion, uptake, and intracellular metabolism before CO_2 is produced (Ågren and Wetterstedt 2007). Only if the activity of extracellular enzymes is currently limiting the entire flux of carbon to a microbe would heterotrophic respiration rates reflect the temperature sensitivity of extracellular enzymes. This situation is unlikely to occur because microbes also respire labile compounds derived from non-polymeric sources, such as root exudates (Bader and Cheng 2007). Even polymer-derived metabolites may have been released by extracellular enzymes at considerable spatial and temporal distance from the respiring microbe. This decoupling along with the considerable influence of autotrophic respiration on total soil CO_2 efflux suggests that time-integrated models will be needed to link the temperature sensitivity of extracellular enzymes with the temperature response of soil respiration.

13.5 Conclusions

Given the critical roles of enzymes in ecosystem functioning, our limited understanding of enzyme activities under field conditions is an important limitation to our ability to model ecosystem processes under current and future climates. Temperature is among the most important drivers of enzyme activities, yet its role is often neglected, in part due to the challenges involved in measuring or modeling temperature effects. However, there is no easy solution to this challenge. We do not presently have the ability to measure in situ enzyme activities directly and thus cannot simply develop empirical relationships between temperature and enzyme activity. Theoretical models of enzyme behavior under substrate limiting conditions provide a good starting point for modeling approaches. Yet, these simple models cannot account for other temperature-sensitive biological, chemical, and physical processes that affect enzyme activity. Biological responses include changes in enzyme production rates, shifts in isoenzyme production, which affect the statistical distribution of enzyme temperature sensitivities within a population, and changes in microbial community composition which could affect the resource efficiency of enzyme production. Physiochemical processes that affect enzyme activities include absorption–desorption reactions, substrate diffusion rates, and enzyme degradation rates. Clearly, the range and complexity of factors that affect in situ enzyme temperature sensitivity make for a most challenging problem.

References

Ågren G, Wetterstedt JÅM (2007) What determines the temperature response of soil organic matter decomposition? Soil Biol Biochem 39:1794–1798

Allen AP, Gillooly JF, Brown JH (2005) Linking the global carbon cycle to individual metabolism. Funct Ecol 19:202–213

Bader NE, Cheng W (2007) Rhizosphere priming effect of *Populus fremontii* obscures the temperature sensitivity of soil organic carbon respiration. Soil Biol Biochem 39:600–606

Bremner JM, Zatua MI (1975) Enzyme activity in soils at subzero temperatures. Soil Biol Biochem 7:383–387

Chrost RG (1990) Microbial ectoenzymes in aquatic environments. In: Overbeck J, Chrost RJ (eds) Aquatic microbial ecology; biochemical and molecular approaches. Springer, New York, pp 47–78

Chróst RJ, Siuda W (2002) Ecology of microbial enzymes in lake ecosystems. In: Dick RP, Burns RG (eds) Enzymes in the environment: activity, ecology and applications. Marcel Dekker, New York, pp 35–72

Coker JA, Sheridan PP, Loveland-Curtze J, Gutshall KR, Auman AJ, Brenchley JE (2003) Biochemical characterization of a beta-galactosidase with a low temperature optimum obtained from an Antarctic Arthrobacter isolate. J Bacteriol 185:5473–5482

Criquet S, Tagger S, Vogt G, Iacazio G, Le Petit J (1999) Laccase activity of forest litter. Soil Biol Biochem 31:1239–1244

Czimczik CI, Trumbore S, Carbone MS, Winston GC (2006) Changing sources of soil respiration with time since fire in a boreal forest. Global Change Biol 12:1–15

Davidson EA, Janssens IA (2006) Temperature sensitivity of soil carbon decomposition and feedbacks to climate change. Nature 440:165–173

Davidson EA, Janssens IA, Luo Y (2006) On the variability of respiration in terrestrial ecosystems: moving beyond Q_{10}. Global Change Biol 12:154–164

del Giorgio PA, Cole JJ (1998) Bacterial growth efficiency in natural aquatic systems. Annu Rev Ecol Syst 29:503–541

Devevre OC, Horwat WR (2000) Decomposition of rice straw and microbial carbon use efficiency under different soil temperatures and moistures. Soil Biol Biochem 32:1773–1785

Di Nardo C, Cinquegrana A, Papa S, Fuggi A, Fioretto A (2004) Laccase and peroxidase isoenzymes during leaf litter decomposition of *Quercus ilex* in a Mediterranean ecosystem. Soil Biol Biochem 36:1539–1544

Elsgaard L, Vinther FR (2004) Modeling of the fine-scale temperature response of arylsulfatase activity in soil. J Plant Nutr Soil Sci-Zeitschrift Fur Pflanzenernahrung Und Bodenkunde 167:196–201

Feller G (2003) Molecular adaptations to cold in psychrophilic enzymes. Cell Mol Life Sci 60:648–662

Fenner N, Freeman C, Reynolds B (2005) Observations of a seasonally shifting thermal optimum in peatland carbon-cycling processes; implications for the global carbon cycle and soil enzyme methodologies. Soil Biol Biochem 37:1814–1821

Frankenberger WT, Tabatabai MA (1991a) L-Asparaginase activity of soils. Biol Fertil Soils 11:6–12

Frankenberger WT, Tabatabai MA (1991b) L-Glutaminase activity of soils. Soil Biol Biochem 23:869–874

Gerday C, Aittaleb M, Arpigny JL, Baise E, Chessa JP, Garsoux G, Petrescu I, Feller G (1997) Psychrophilic enzymes: a thermodynamic challenge. Biochimica Et Biophysica Acta-Protein Structure and Molecular Enzymology 1342:119–131

Gillooly JF, Brown JH, West GB, Savage VM, Charnov EL (2001) Effects of size and temperature on metabolic rate. Science 293:2248–2251

Hochachka PW, Somero GN (1984) Biochemical adaptation. Princeton University Press, Princeton, NJ

Huston AL, Krieger-Brockett BB, Deming JW (2000) Remarkably low temperature optima for extracellular enzyme activity from Arctic bacteria and sea ice. Environ Microbiol 2:383–388

Jastrow JD, Miller RM (1997) Soil aggregate stabilization and carbon sequestration: feedbacks through organo-mineral associations. In: Lal R, Kimble JM, Follett RF, Stewart BA (eds) Soil processes and the carbon cycle. CRC, Boca Raton, FL, pp 207–224

Koch O, Tscherko D, Kandeler E (2007) Temperature sensitivity of microbial respiration, nitrogen mineralization, and potential soil enzyme activities in organic alpine soils. Global Biogeochem Cy 21:GB4017. doi:10.1029/2007GB002983

Kristensen E, Andersen FO, Blackburn TH (1992) Effects of benthic macrofauna and temperature on degradation of macroalgal detritus – the fate of organic-carbon. Limnol Oceanogr 37:1404–1419

Lai CM, Tabatabai MA (1992) Kinetic-parameters of immobilized urease. Soil Biol Biochem 24:225–228

Lloyd J, Taylor JA (1994) On the temperature dependence of soil respiration. Funct Ecol 8:315–323

Lopez-Urrutia A, Moran XAG (2007) Resource limitation of bacterial production distorts the temperature dependence of oceanic carbon cycling. Ecology 88:817–822

Loveland J, Gutshall K, Kasmir J, Prema P, Brenchley JE (1994) Characterization of psychrotrophic microorganisms producing *Beta*-galactosidase activities. Appl Environ Microbiol 60:12–18

Lugtenberg BJJ, Chin-A-Woeng TFC, Bloemberg GV (2002) Microbe-plant interactions: principles and mechanisms. Antonie Leeuwenhoek Int J Gen Mol Microbiol 81:373–383

McClaugherty CA, Linkins AE (1990) Temperature responses of enzymes in two forest soils. Soil Biol Biochem 22:29–33

Parham JA, Deng SP (2000) Detection, quantification and characterization of beta-glucosaminidase activity in soil. Soil Biol Biochem 32:1183–1190

Parton WJ, Schimel DS, Cole CV, Ojima DS (1987) Analysis of factors controlling soil organic-matter levels in great-plains grassland. Soil Sci Soc Am J 51:1173–1179

Price PB, Sowers T (2004) Temperature dependence of metabolic rates for microbial growth, maintenance, and survival. Proc Natl Acad Sci USA 101:4631–4636

Privalov PL, Tsalkova TN (1979) Micro-stabilities and macro-stabilities of globular-proteins. Nature 280:693–696

Ryan MG, Law BE (2005) Interpreting, measuring, and modeling soil respiration. Biogeochemistry 73:3–27

Sagemann J, Jorgensen BB, Greeff O (1998) Temperature dependence and rates of sulfate reduction in cold sediments of Svalbard, Arctic Ocean. Geomicrobiol J 15:85–100

Sanchez-Perez G, Mira A, Nyiro G, Pasic L, Rodriguez-Valera F (2008) Adapting to environmental changes using specialized paralogs. Trends Genet 24:154–158

Scott-Denton LE, Rosenstiel TN, Monson RK (2006) Differential controls by climate and substrate over the heterotrophic and rhizospheric components of soil respiration. Global Change Biol 12:205–216

Secundo F, Russo C, Giordano A, Carrea G, Rossi M, Raia CA (2005) Temperature-induced conformational change at the catalytic site of *Sulfolobus solfataricus* alcohol dehydrogenase highlighted by Asn249Tyr substitution. A hydrogen/deuterium exchange, kinetic, and fluorescence quenching study. Biochemistry 44:11040–11048. doi:10.1021/bi050469c

Seto M, Misawa K (1982) Growth rate, biomass production, and carbon balance of *Pseudomonas aeruginosa* in a glucose-limited medium at temperature and osmotic pressure extremes. Jpn J Ecol 32:365–371

Sinsabaugh RL, Shah JJ (2010) Integrating resource utilization and temperature in metabolic scaling of riverine bacterial production. Ecology 91:1455–1465

Six J, Conant RT, Paul EA, Paustian K (2002) Stabilization mechanisms of soil organic matter: implications for C-saturation of soils. Plant Soil 241:155–176

Sollins P, Homann P, Caldwell BK (1996) Stabilization and destabilization of soil organic matter: mechanisms and controls. Geoderma 74:65–105

Sørensen LH (1972) Stabilization of newly formed amino acid metabolites in soil by clay minerals. Soil Sci 114:5–11

Steinweg JM, Plante AF, Conant RT, Paul EA, Tanaka DL (2008) Patterns of substrate utilization during long-term incubations at different temperatures. Soil Biol Biochem 40:2722–2728

Straub FB (1964) Formation of the secondary and tertiary structure of enzymes. Adv Enzymol Relat Subj Biochem 26:89–114

Tabatabai MA (1982) Soil enzymes. In: Page AL, Miller RH, Keeney DR (eds) Methods of soil analysis. Part 2. Chemical and microbiological properties, 2nd edn. American Society of Agronomy-Soil Science Society of America, Madison, WI

ten Hulscher TEM, Cornelissen G (1996) Effect of temperature on sorption equilibrium and sorption kinetics of organic micropollutants – A review. Chemosphere 32:609–626

Thamdrup B, Hansen JW, Jorgensen BB (1998) Temperature dependence of aerobic respiration in a coastal sediment. FEMS Microbiol Ecol 25:189–200

Tisdall JM, Oades JM (1982) Organic-matter and water-stable aggregates in soils. J Soil Sci 33:141–163

Trasar-Cepeda C, Gil-Sotres F, Leiros MC (2007) Thermodynamic parameters of enzymes in grassland soils from Galicia, NW Spain. Soil Biol Biochem 39:311–319

Tsou CL (1993) Conformational flexbility of enzyme active- sites. Science 262:380–381

Vial Ludovic MC, Dekimpe Groleau V, Deziel E (2007) Burkholderia diversity and versatility: an inventory of the extracellular products. J Microbiol Biotechnol 17:1407–1429

Wallenstein MD, Weintraub MN (2008) Emerging tools for measuring and modeling the in situ activity of soil extracellular enzymes. Soil Biol Biochem 40:2098–2106

Wallenstein MD, McMahon SK, Schimel JP (2009) Seasonal variation in enzyme activities and temperature sensitivities in Arctic tundra soils. Global Change Biol 15:1631–1639

Wilczek S, Fischer H, Pusch MT (2005) Regulation and seasonal dynamics of extracellular enzyme activities in the sediments of a large lowland river. Microb Ecol 50:253–267. doi:10.1007/s00248-004-0119-2

Williams PJ (1973) The validity of the application of simple kinetic analysis to heterogeneous microbial populations. Limnol Oceanogr 18:159–165

Wirth SJ, Wolf GA (1992) Microplate colourimetric assay for endoacting cellulase, xylanase, chitinase, 1, 3-Beta-glucanase and amylase extracted from forest soil horizons. Soil Biol Biochem 24:511–519

Zavodszky P, Kardos J, Svingor A, Petsko GA (1998) Adjustment of conformational flexibility is a key event in the thermal adaptation of proteins. Proc Natl Acad Sci USA 95:7406–7411

Chapter 14
Actinomycetes: Sources for Soil Enzymes

V. Suneetha and Zaved Ahmed Khan

14.1 Introduction

The classification of the Prokaryotes is a complex issue. Their major subdivisions are the bacteria (Schizophyta), the blue green algae (Cyanophyta), and the Actinomycetales. The latter are sometimes called the "higher bacteria," organisms possessing properties intermediate between the fungi and the bacteria. The Actinomycetes are gram-positive organisms that tend to grow slowly as branching elements. They are prokaryotes, sporulated, powdery growth organisms, and show similarity to fungi in the formation of branched aerial mycelium, which profusely sporulate. But these Actinomycetes differ from fungi in the composition of cell wall; they do not have chitin and cellulose that are commonly found in the cell wall of fungi. The number of Actinomycetes increases in the presence of decomposing organic matter. Depending on the abundance in the soil, the common genera of Actinomycetes are *Streptomyces* (70%), *Nocardia*, and *Micromonospora* although *Actinomyces*, *Actinoplanes*, and *Streptosporangium* have been encountered. The term Actinomycetes is used to indicate organisms belonging to the Actinomycetales, a major subdivision of the Prokaryotae, the kingdom that comprises all organisms with a prokaryotic cell. They were long regarded as fungi, as is reflected in their name: aktino (gr) = ray, mykes (gr) = mushroom (=fungus).

Actinomycetes are divided into several families:

- The family Actinomycetaceae comprises two genera: *Actinomyces* and *Nocardia*. A few species are pathogenic. *A. israelii* causes actinomycosis in man and *A. bovis* in cattle.

V. Suneetha (✉) and Z.A. Khan
School of Biotechnology, Chemical and Biomedical Engineering, VIT University, Vellore 632014, Tamil Nadu, India
e-mail: vsuneetha@vit.ac.in, vuppusuneetha@yahoo.co.in, zavedkhan@vit.ac.in, khan.zaved@gmail.com

G. Shukla and A. Varma (eds.), *Soil Enzymology*, Soil Biology 22,
DOI 10.1007/978-3-642-14225-3_14, © Springer-Verlag Berlin Heidelberg 2011

- The family *Mycobacteriaceae* has a single genus, Mycobacterium, which contains several pathogenic species causing diseases such as leprosy and tuberculosis.
- The family *Streptomyceteae* comprises several organisms found in the soil. They are rarely pathogenic. In contrast, several species of the genus *Streptomyces* produce antibiotics. Erythromycin is produced by *Streptomyces erythreus.*

The characteristic earthy smell of compost is caused by Actinomycetes. Actinomycetes are a form of fungi-like bacteria that form long, thread-like branched filaments that look like gray spider web stretching through the compost. Different species of Actinomycetes predominate during each phase of the composting process (the mesophillic, thermophillic, and maturation phases) but are most easily seen during the early stages of the composting process in the outer 10–15 cm of the pile. Actinomycetes are the primary decomposers of tough plant materials like bark, newspaper, and woody stems. They are especially effective at attacking tough, raw plant tissues (cellulose, chitin, and lignin), softening them up for their less enterprising relatives. The morphology of an Actinomycete growing on agar can provide useful clues to its identity but viewing isolated colonies can give little worthwhile information. Examine the organism streaked in a cross-hatched pattern on the surface of the agar, first with a stereomicroscope and then with a transmitted light microscope with a 40× long-working distance objective to avoid water condensation on the front lens. The appearance of hyphae within the agar and the nature of the spores on the aerial hyphae are important but can only be observed when growth is thin and the medium promotes differentiation.

Sometimes the name "Actinomycetes" is used restrictively for members of the genus Actinomyces only, instead of all members of the Actinomycetales. The industrial enzyme sector in India is growing fast for meeting the need of pharmaceutical, food processing, leather, detergents, pulp, paper, and textile industries. In view of vigorously growing demand, the enzyme industry and sector is poised for high growth rates, necessitating increase in the production or import. The product range and services are growing rapidly as the use of enzymes is gaining widespread acceptance. In addition to the Indian market, export opportunities are also there for the manufacturers. Studies on specific enzyme producing Actinomycetes other than fungi are rather fragmentary and limited in our country. In view of the importance of soil enzymes producing Actinomycetes and scarce information available in our country, it was interesting to carry out studies in this direction (Suneetha and Lakshmi 2004).

The use of enzyme-mediated processes can be traced back to ancient civilizations. Today nearly 4,000 enzymes are known, of which many are commercially produced. The majority of the industrial enzymes are of microbial origin. Until the 1960s, the total sales of enzymes were only a few million dollars annually, but the market has since grown spectacularly. Because of improved understanding of production biochemistry, the fermentation processes, and recovery methods, an increasing number of enzymes can be produced affordably. Also, advances in the methods of using enzymes have greatly expanded the demand. Furthermore, because of the many different transformations that enzymes can catalyze, the number of enzymes used commercially continues to multiply. Enzyme-mediated reactions are attractive alternatives to tedious and expensive chemical methods, and they find great use in a large number of fields such as food, dairy, pharmaceutical, detergent, textile, and cosmetic industries. In the above scenario, enzymes such as proteases and amylases have dominated the world market owing to their hydrolytic reactions for proteins and carbohydrates. With the realization of the biocatalytic potential of microbial lipases in both aqueous and nonaqueous media in the last one and a half decades, industrial fronts have shifted toward utilizing these enzymes for a variety of reactions of immense importance.

For commercialization of any valuable product, emphasis is laid on the use of relatively low-cost raw materials that are converted into commercially important and useful products or services. This process requires coordinated coupling of several units and operations. The efficiency depends on adequate upstream operations and the process of recovery of the product, which includes a series of careful and meticulous steps collectively referred to as downstream processing. Important organisms and their natural products come into light mainly when new screening systems are utilized or when samples from different sources are examined. Further the isolated native organisms are improved upon by conventional techniques or by modern techniques involving rDNA technology. Due to the availability of powerful expression systems in various microbes, the large-scale commercial production of new and useful enzymes is becoming increasingly attractive. The focus of enzyme technology is thus the microorganism, the Actinomycetes itself, that has the capacity to produce the soil enzymes. Though there are many advantages and applications of microbial enzymes, these are still not being exploited as biocatalysts to their optimum capacity for various reasons.

14.2 Keratinases

The microbes like *Actinomycetes* hold prominent position in microbial world because of their diversity and proven ability to produce soil enzymes. They are also potential sources of proteases with typical substrate specificities such as Keratinases that attack normally unreactive protein called keratin. Microbial Keratinases reported till date are produced from organisms like *Streptomycetes*, *Bacillus* species, and fungal members. The studies on isolation, characterization, and strain improvement of keratinase producing organisms have both economic and ecological value but have not been fully exploited yet. The limiting aspect in the wide scale usage of Keratinases is mainly an efficient and cost-effective method for fermentative production of keratinase. Though in recent years patents are being obtained internationally for strains and process of fermentation of Keratinases, there is very little effort in this direction (Table 14.1).

To develop indigenous product technology in India, as great economical and ecological value is associated with the production of Keratinases, any effort to develop the process will be of great potential.

Table 14.1 Some of the reported keratinase producing Actinomycetes (Suneetha 2006)

Strains	References
Actinomycetes: *Doratomyces microsporus*	Vignardet et al. (1999), Gradisar et al. (2000)
Streptomyces spp. *S. fradiae, S raminofaciens* *S. pactum, Streptomyces* spp. A_{11}, *1349, 1382, BA7, etc.*	Noval and Nickerson (1959), Nickerson et al. (1963), Tatsushi et al. (1967), Elmayergi and Smith (1971), Young and Smith (1975), Kunert and Stransky (1988), Kunert (1989), Mukhopadhyay and Chandra (1990, 1993), Galas and Kaluzewska (1992), Bockle et al. (1995), Letourneau et al. (1998), Szabo et al. (2000), Ichida et al. (2001), Ivanko et al. (2002), Korkmaz et al. (2003), Suneetha (2003)
Thermoactinomyces spp. *Thermoactinomyces candidus*	Ignatova et al. (1999)

Tirumala hills is a famous pilgrim center in Andhra Pradesh. It is located in Eastern Ghats on Seshachalam hill range, with North latitude of 13–14′ and East longitude of 70–21′. It is 2,820 feet above the sea level and 100 square miles in extent. Tirumala is visited by thousands of pilgrims every day from all over India as well as the world, throughout year to worship Lord Venkateswara (Hindu mythology). The place has relatively high deposits of keratin as most of the pilgrims coming to Tirumala consider head tonsuring as the most sacred offering to God. Further the prevailing climatic conditions also facilitate the growth of thermo-tolerant organisms. Thus, this place was presumed to provide good enrichment for potential keratinophilic and keratinolytic organisms, and soil samples collected from various locations in Tirumala hills and Tirupati were analyzed to isolate potential keratinolytic organisms. It is expected that the present emphasis on keratinase producing organisms may lead to a step forward in the process of development of indigenous kerazyme technology (Suneetha and Lakshmi 2004).

14.3 Pectinases

The structural unit of pectin has a complex structure. Preparations consist of substructural entities that depend on their source and extraction methodology. Commercial extraction causes extensive degradation of the neutral sugar-containing side chains. The majority of the structure consists of homopolymeric partially methylated poly-α-(1 → 4)-D-galacturonic acid residues (Fig.14.1), but there are substantial "hairy" non-gelling areas (Fig.14.2) of alternating α-(1 → 2)-L-rhamnosyl-α-(1 → 4)-D-galacturonosyl sections containing branch points with mostly neutral side chains (1–20 residues) of mainly L-arabinose and D-galactose (rhamnogalacturonan I). Pectins may also contain rhamnogalacturonan II side chains containing other residues such as D-xylose, L-fucose, D-electronic acid, D-apipose, 3-deoxy-D-manno-2-octulosonic acid (Kdo), and 3-deoxy-D-lyxo-2-heptulosonic acid (Dha) attached to

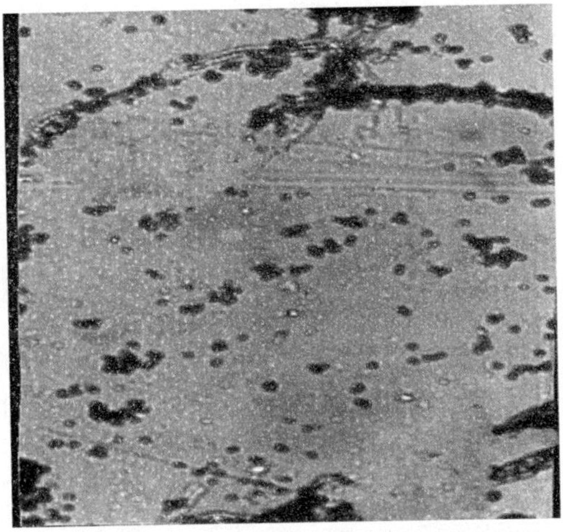

Fig 14.1 Micromorphology of Actinomycetes (Suneetha 2006)

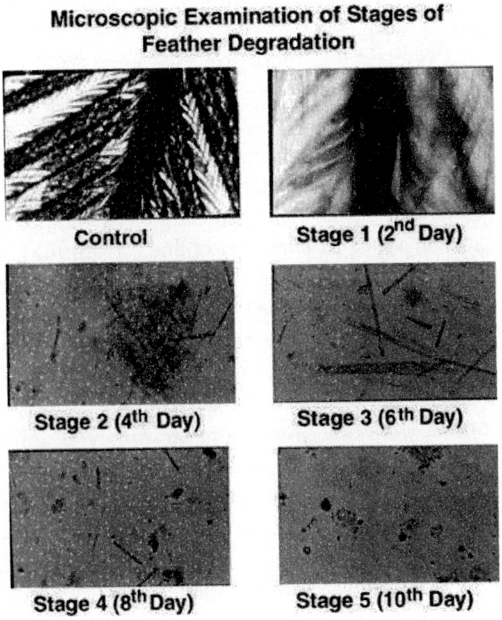

Fig 14.2 Keratin (feather) degradation (Suneetha 2006)

poly-α-(1 → 4)-D-galacturonic acid regions. Enzymes are bio-catalysts. They speed up the rate of chemical reactions taking place in living cells. The reactants of enzyme-catalyzed reactions are termed as substrates and each enzyme is specific in character, acting on a particular substrate(s) to produce particular product or products. Enzyme usage in the fruit industry is a recent innovation. Application of biotechnology to industrial operations is no longer an academic or potentially useful alternative

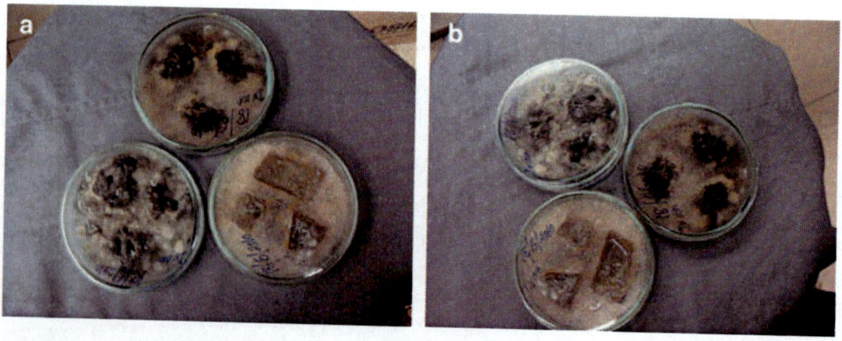

Fig 14.3 Pectin degradation (Suneetha and Zaved 2008)

proposition for the future. Enzymes have become big business, with a wide range of industries using commercial enzymes, in addition to the feed industry. The world annual sales of industrial enzymes were recently valued at $1 billion. Three-quarters of the market is for enzymes involved in the hydrolysis of natural polymers. Food-processing enzymes including pectinases account for about 45% of enzyme usage (Fig 14.3).

Pectinase was found to be the most efficient commercialized enzyme in degrading the fruit waste. Pectinolytic enzymes or pectinases is a group of enzymes that hydrolyze the pectic substances, present mostly in plants. Enzymes are usually offered as "cocktails" of several activities rather than a single enzymatic activity. However, in many cases, the different enzyme activities can still act on the same composition, as the composition can have a complex chemical structure having various types of chemical bounds, requiring different enzyme activities for break down. An example of this is the enzyme cocktails offered as "pectinase." Such pectinase composition often contains one or more of the following activities: polygalacturonase, pectin lysase, pectin methyl esterase. Pectinase preparations are often used in fruit juice processing. It is preferred in the present invention that the enzyme preparation used contains at least one of these three activities mentioned, preferably two, more preferably all three.

Pectinases are a group of enzymes that catalyze the degradation of pectic polymers present in plants. Pectinolytic enzymes are widely distributed in higher plants and microorganisms. Based on their catalytic action, pectin-degrading enzymes have been classified into two major groups: the first group is represented by pectin esterase (PE) and the second by polygalacturonase (PG) and pectin lyase (PL) (Table 14.2).

14.4 Xylanases

Xylans are hydrolyzed mainly by ß-1,4-endoxylanases (1,4- ß-D-xylan xylanohydrolases EC 3.2.1.8.) and xylosidases (1,4-ß-D-xylan xylohydrolases EC 3.2.1.37). The interests in xylan-degrading enzyme and its application in the pulp and paper

Table 14.2 Some of the reported petinase producing microorganism (Suneetha and Zaved 2008)

Source	Organism	References
Actinomycetes	*Streptomyces* spp.	Suneetha et al. (2003)
	Streptomyces spp.	Suneetha et al. (2003)
	Streptomyces lydicus	Jacob and Prema (2006)
	Streptomyces sp. *RCK-SC*	Chander Kuhad et al. (2004)
	Streptomyces sp. *QG-11-3*	Qasim Khalil et al. (2000)
	Thermomonospora flisca	Stutzenberger (1987)
	Streptomyces viridochromogenes	Agate et al. (1962)

industries have advanced significantly over the past few years (Agate et al. 1962). Mainly, cellulose-free xylanase is of great importance in the paper industry. The cost of carbon source plays another major role in the economics of xylanase production. Hence, an approach to reduce the cost of xylanase production is by the use of lignocellulosic materials as substrates such as wheat bran, rice straw, corn cob, agricultural wastes, sugar cane baggase rather than opting for the expensive pure xylans. For the development of suitable xylanase as a pre-bleaching agent, the stability of enzyme at higher optimum pH and temperature is desirable (Bockle et al. 1995). Alkaline xylanase also finds a number of applications. For example, because of high solubility of xylan at alkaline pH, alkaline xylanases may have good potential in the conversion of lignocellulosic wastes to fermentable sugars.

Many xylanases producing alkaliphilic microbial strains have been reported from different laboratories. However, the xylanase from many of the alkaliphilic strains have their optimum pH around neutrality. Naturally occurring habitats for the isolation of alkaliphilic microbial strain are scattered in different parts of the world (Bockle and Muller 1997). Most of the reported literature on xylanase has concentrated on the characterization of the enzyme and its application as a pre-bleaching agent (Nickerson 1947). Relatively less attention was paid to the optimization of xylanase production. In the present study, an attempt was made to describe the optimization studies related to xylanase production from Actinomycetes, which is isolated from soil near paper industry (Fig. 14.4, Table 14.3).

14.5 Lipases

Although bacteria have the highest total population in soil followed by Actinomycetes and fungi, the frequencies of lipase producing bacteria, Actinomycetes, and fungi were 8.5, 55.9, and 23.3%, respectively. Although little research has been published on lipolytic activity in Streptomyces considering their widespread use within antibiotic production, lipase activity has been detected. Lipase production in a wide range of *Streptomycetes* has been investigated. However, lipolytic activity was found in only a few strains tested (15). Lipases (triacylglycerol acylhydrolases, E.C. 3.1.1.3) are ubiquitous enzymes of considerable physiological significance and industrial

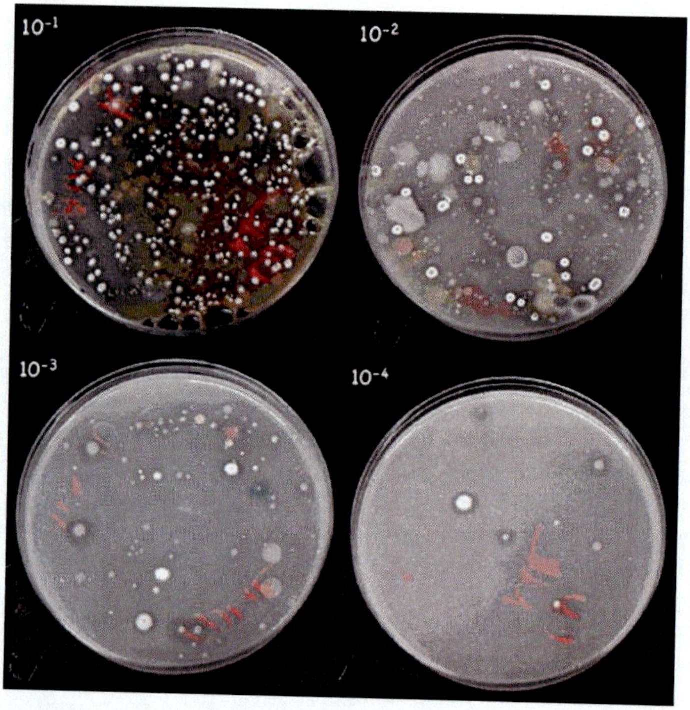

Fig 14.4 Xylan degradation by Actinomycetes (Suneetha 2008)

Table 14.3 Some reported Actinomycetes producing xylanase (Suneetha 2008)

Actinomycetes	References
Actinomyces from soil	Suneetha (2008)
Streptomyces olivaceoviridis E-86	Ding (2004)
Streptomyces sp.	Rawashdeh (2005), *Streptomyces* sp. (strain Ib 24D)

potential. Lipases catalyze the hydrolysis of triacylglycerols to glycerol and free fatty acids. In contrast to esterases, lipases are activated only when adsorbed to an oil–water interface and do not hydrolyze dissolved substrates in the bulk fluid. A true lipase will split emulsified esters of glycerin and long-chain fatty acids such as triolein and tripalmitin. Lipases are serine hydrolases and display little activity in aqueous solutions containing soluble substrates. In eukaryotes, lipases are involved in various stages of lipid metabolism including fat digestion, absorption, reconstitution, and lipoprotein metabolism. In plants, lipases are found in energy-reserve tissues. How lipases and lipids interact at the interface is still not entirely clear and is a subject of intense investigation. The natural substrates of lipases are triacylglycerols, having very low solubility in water. Under natural conditions, they catalyze the hydrolysis of ester bonds at the interface between an insoluble substrate phase and the aqueous

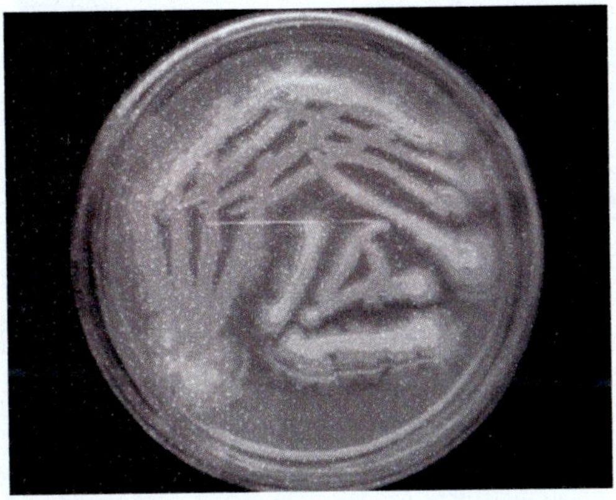

Fig 14.5 Lipid degradation by Actinomycetes (Suneetha and Seena 2008)

Table 14.4 Reported lipase producing Actinomycetes (Suneetha and Seena 2008)

Source	Genus	Species	References
Actinomycetes	Streptomyces	*Streptomyces fradiae NCIB* 8233	Sztajer et al. (1988)
		Streptomyces sp. *PCB27*	Sztajer et al. (1988)
		Streptomyces sp. *CCM 33*	Sztajer et al. (1988)
		S. Coelicolor	Hou (1994)

phase in which the enzyme is dissolved. Under certain experimental conditions, such as in the absence of water, they are capable of reversing the reaction. The reverse reaction leads to esterification and formation of glycerides from fatty acids and glycerol. The occurrence of the lipase reaction at an interface between the substrate and the aqueous phase causes difficulties in the assay and kinetic analysis of the reaction. The usual industrial lipases are special classes of esterase enzymes that act on fats and oils, and hydrolyze them initially into the substituted glycerides and fatty acids, and finally on total hydrolysis into glycerol and fatty acids (Table 14.4).

The focus of soil is thus the habitat for microorganisms, especially Actinomycetes, the wonderful organisms that have the capacity to produce the enzymes. Though there are many advantages and applications of microbial enzymes, these have still not being exploited as biocatalysts to their optimum capacity (Fig. 14.5).

14.6 Conclusion

The use of Actinomycetes for industrial purposes has a long history, which is long before the realization of activities of other microorganisms. We have discussed the production of enzymes like keratinases, pectinases, xylanases, and lipases for commercial applications and screening methods. With the improvement of screening methods, we will be able to produce other commercial products from Actinomycetes in near future. It also suggests new priorities for research based on an integrative approach that combines biochemistry and biophysics, and microbial exploitation for the production of soil enzymes is highly attractive for applications in fruit, detergent, textile, tanning, meat paper industries, waste water treatment, biomedical, food and other bioremediation technology.

References

Agate AD, Bilimoria MH, Bhat JV (1962) Pectin transeliminase activity in *Streptomyces viridochromogenes*. Curr Sci 31:462–463

Bockle B, Muller R (1997) Reduction of disulfide bonds by *Streptomyces pactum* during growth on chicken feathers. Appl Environ Microbiol 63:790–792

Bockle B, Galunsky B, Muller R (1995) Characterization of a keratinolytic serine proteinase from *Streptomyces pactum* DSM 40530. Appl Environ Microbiol 61:3705–3710

Chander Kuhad R, Kapoor M, Rustagi R (2004) Enhanced production of an alkaline pectinase from *Streptomyces* sp RCK-SC by whole-cell immobilization and solid-state cultivation. World J Microbiol Biotechnol 20:257–263

Ding CH, Jiang ZQ, Li XT, Li LT, Kusakabe I (2004) High activity xylanase production by *Streptomyces olivaceoviridis*. E-86 20(1):7–10

Elmayergi HH, Smith RE (1971) Influence of growth of *Streptomyces fradiae* on pepsin-HCL digestibility and methionine content of feather meal. Canadian Journal of Microbiology 17:1067–1072

Galas E, Kaluzewska M (1992) Proteinases of *Streptomyces fradiae*: catalytic and some physicochemical properties of keratinolytic proteinase. Acta Microbiologia Pol 41:169–177

Gradisar H, Kern S, Fredrich J (2000) Keratinase of Doratomyces microsporus. Applied Microbial Biotechnology 53:196–200

Hou CT (1994) pH dependence and thermostability of lipases from cultures from ARS culture collection. J Ind Microbiol 13:242–248

Ignatova Z, Gousterova A, Spassov G, Nedkov P (1999) Isolation and partial characterisation of extracellular keratinase from a wool degrading thermophilic actinomycete strain *Thermoactinomyces candidus*. Canadian Journal of Microbiology 45:217–222

Ivanko OV, Varbanet's LD, Valahurova OV, Nahorna SS, Redchyts TI, Zhdanova NM (2002) Screening of collagenase and keratinase producers. Mikrobiol Z 64:31–36

Jacob N, Prema P (2006) Influence of mode of fermentation on production of polygalacturonase by a novel strain of *Streptomyces lydicus*. Food Technol Biotechnol 44:263–267

Khalil Beg Q, Bhushan B, Kapoor M, Hoondal GS (2000) Effect of amino acids on production of xylanase and pectinase from *Streptomyces* sp. QG-11-3. World J Microbiol Biotechnol 16:211–213

Korkmaz H, Unaldi MN, Asian B, Coral G, Arikan B, Dincer S, Colak O (2003) Keratinolytic activity of *Streptomyces strain* BA-7, a new isolate from Turkey. Ann Microbiol 53:85–93

Kunert J, Stransky Z (1988) Thiosulfate production from cystine by the keratinolytic prokaryote *Streptomyces fradiae*. Arch Microbiol 15:600–601

Letourneau F, Soussote V, Bressollier P, Branland P, Verneuil B (1998) Keratinolytic activity of *Streptomyces sp.S.KL-02:* A new isolated strain. Lett Appl Microbiol 26:77–80

Mukhopadhyay RP, Chandra AL (1990) Keratinase of a *Streptomycete*. Indian J Exp Biol 28:575–577

Mukhopadhyay RP, Chandra AL (1993) Protease of a keratinolytic *Streptomycete* to unhair goat skin. Indian J Exp Biol 31:557–558

Nickerson WJ (1947) Biology of pathogenic fungi. Mycologia. 14. Waltham, MA, pp 10–28

Nickerson WJ, Noval JJ, Robison RS (1963) Properties of the enzyme conjugate elaborated by *Streptomyces fradiae*. Biochem Biophys Acta 77:73–86

Noval JJ, Nikerson WJ (1959) Decomposition of native keratin by Streptomyces fradiae. Journal of Bacteriology 77:251–263

Rawashdeh L (2005) Effect of cultural conditions on xylanase production by Streptomyces sp. (strain Ib 24D) and its potential to utilize tomato pomace. African Journal of Biotechnology 4(3):251–255

Stutzenberger FJ (1987) Inducible thermoalkalophilic polygalacturonate lyase from *Thermornonospora fusca*. J Bacteriol 169:2774–2780

Suneetha V (2005) Optimisation of parameters for feather degrading Actinomycetes. Journal of microbial world pp. 268–276

Suneetha V (2006) Isolation and charaterization of keratinase producing organisms from Trimula and Tirupati. Unpublished PhD thesis. University of SPMVV (Women University), Tirupati

Suneetha (2008) Screening, characterization and optimisation of cellulase free xylanase. Journal of Industrial Biotechnology 1:32–37

Suneetha V, Khan ZA (2008) Screening, characterization and optimization of pectinase enzyme from actinomycetes. Soil Biology and Biochemistry 8(1):57–67

Suneetha V, Lakshmi VV (2004) Fermentative production of kerazyme with isolates from Tirumala hills. In: Deshmukh AM (ed) Biotechnology in agriculture industry and environment. Microbiology Society, Karad, pp 198–203

Suneetha V, Seena Anthony (2008) Screening and characterization of lipase producing actinomycetes. Unpublished M Tech thesis. VIT University, Vellore

Szabo I, Benedek A, Szabo IM, Barabas GY (2000) Feather degradation with a thermotolerant *Streptomyces graminofaciens* strain. World J Microbiol Biotechnol 16:253–255

Sztajer H, Maliszewska I, Wieczorek J (1988) Production of exogenous lipase by bacteria, fungi and actinomycetes. Enzyme Microb Technol 10:492–497

Tatsushi M, Hiroshige T, Tsuzuki T (1967) Multiple proteolytic enzymes of *Streptomyces fraidae:* production, isolation and preliminary characterization. Biochem Biophys Acta 139:382–397

Vignardet C, Guillaume YC, Friedrich J, Millet J (1999) A first order experimental design to assess soluble proteins released by a new keratinase from *Doratomyces microsporus* on human substrates. Int J Pharm (Amsterdam) 191:95–102

Young RA, Smith RE (1975) Degradation of feather keratin by culture filtrates of *Streptomyces fradiae*. Canadian Journal of Microbiology 21:583–586

Chapter 15
Organo-Mineral–Enzyme Interaction and Soil Enzyme Activity

Andrew R. Zimmerman and Mi-Youn Ahn

15.1 Introduction

Both microbially produced enzymes and mineral surfaces can be considered catalysts of chemical transformation in soils and other geological environments such as sediments and subsurface aquifers. While both are important direct agents of organic matter (OM) remineralization and transformation, mineral surfaces can also act as "heterogeneous co-catalysts," influencing the kinetic properties of biological enzymes by providing surfaces upon which reactions can take place. Enzymes may be categorized, according to their location as "intracellular," those present in living and proliferating cells, or "abiontic," i.e. all others (Skujins 1976). Within the latter group, extracellular enzymes may be leaked or lysed from dead cells or actively secreted by living bacteria and fungi cells (Burns 1982). They are of particular importance to the biogeochemistry of soils and sediments, in that they hydrolyze large polymeric organic compounds into small monomers, which then can be passed through cell walls and fuel microbial respiration and growth. Thus, their activity may be the rate-limiting step in governing the degradation of OM and the remineralization of carbon and other nutrients.

While classical enzyme experiments have typically been carried out in buffered solutions and under easily controllable conditions, the soil environment represents a wholly different set of conditions and extracellular enzyme activity cannot be expected to conform to the "ideal." In addition to the variables often cited as influencing enzyme activity such as temperature, pH, and substrate availability, the activity and stability of a particular enzyme may be enormously influenced by its interaction

A.R. Zimmerman (✉)
Department of Geological Sciences University of Florida, 241 Williamson Hall, PO Box 112120, Gainesville, FL 32611, USA
e-mail: azimmer@ufl.edu

M.-Y. Ahn
Soil and Water Science Department University of Florida, 110290, Gainesville, FL 32611, USA

G. Shukla and A. Varma (eds.), *Soil Enzymology*, Soil Biology 22,
DOI 10.1007/978-3-642-14225-3_15, © Springer-Verlag Berlin Heidelberg 2011

with surfaces. Whether in soil, groundwater, or sediment systems, the interactions of mineral or indigenous OM with both enzymes and enzyme-targeted organic substrates may largely determine the effectiveness of enzymatic degradation, and thus, microbial accessibility to organic carbon and nutrient sources. These geological environments are extremely heterogeneous in their biotic (microorganisms, enzymes, micro- and macrofauna) and abiotic components (labile organic compounds, humic substances, mineral types, organo-mineral aggregates). Thus, the actual activity and stability that an extracellular enzyme would be expected to exhibit in a particular system might be considered a sum of the enzyme and enzyme-modulating component pairs in the system. However, this does not account for the interactive and synergistic effects between multiple components that we are now beginning to realize exist. We have begun to examine, in isolation or combination, many of the factors that influence extracellular enzyme activity and, thus, may become better able to estimate the behavior of these enzymes in the natural environment.

A number of papers have reviewed specific aspects of enzyme–organo-mineral interaction such as that of protein (enzyme) adsorption to minerals (Quiquampoix et al. 2002; Quiquampoix and Burns 2007) and enzyme–clay interaction (Boyd and Mortland 1990; Naidja et al. 2000). This chapter outlines the broader range of organo-mineral interactions in soils and their varied effects on enzyme activity, with special focus on providing historical prospective and recent developments in our understanding of this research area. For the purposes of this discussion, OM will be used here to refer to generic environmental OM that is neither enzyme, nor enzyme-targeted substrate.

15.2 History of Enzyme–Organo-Mineral Interaction Studies

Our understanding of the mechanisms and importance of enzyme–organo-mineral interaction has proceeded from a number of fields including biochemical and environmental engineering, material science, microbiology, and soil and sediment organic geochemistry. Techniques and concepts developed in each of these fields have provided insight into the effects of these interactions on soil enzyme activity and microbial ecology. Among the earliest reference to the measurement of surface-associated enzyme activity is the work of Griffin and Nelson (1916) and Nelson and Griffin (1916) who found no change in yeast-derived invertase activity whether sorbed to charcoal, aluminum hydroxide, or even colloidal proteins such as egg albumin, as long as pH was held constant using an appropriate buffer. Since then, these so-called "immobilized enzymes" have seen wide commercial application in processes that require a biocatalyst such as in the food and pharmaceutical industries. The benefits of immobilized enzymes are that, on a solid support, an enzyme can be easily removed from the reaction solution following its use and then reused and can, in some cases, be stabilized (i.e., made to have increased longevity relative to the unbound enzyme).

By one count, more than one thousand reports on the immobilization of about two hundred different enzymes had been published by the mid-1970s (Mosbach 1976). The methods of immobilization can be categorized as involving adsorption, entrapment, covalent attachment, or co-polymerization or cross-linking (Srere and Ueda 1976), and myriad solid supports have been used ranging from natural minerals and ceramics to organic polymer beads and gels. The goal of the immobilization method chosen for any particular application is to prevent the loss of enzyme activity while strongly binding the enzyme. In general, any attachment that changes the nature (structure or chemistry) of reactive groups in the binding site of the enzyme will lead to activity loss. While physical adsorption is the simplest and most economical of these methods (and the most analogous to "bound" extracellular soil enzymes), it is the weakest and least controllable binding method, and offers the greatest possibility of enzyme release and loss of activity. In addition, the understanding derived from work in this field that the chemical nature of the support material determines the amount, stability, and activity of bound enzymes (Goldstein and Katchalski-Katzir 1976), is directly applicable to our understanding of extracellular soils enzymes.

Early work on enzyme immobilization often focused on enzyme–clay, or more generally, protein–clay interaction (e.g., Ensminger and Gieseking 1939; Zittle 1953). Clay-adsorbed proteins were found to degrade more slowly than their "free" counterparts, both by microbes (Estermann and McLaren 1959; Pinck and Allison 1951; Pinck et al. 1954; Skujins et al. 1959) and by proteinases (Ensminger and Gieseking 1941; Ensminger and Gieseking 1942; McLaren 1954; McLaren and Estermann 1956; Sorensen 1969). In other words, the stability of enzymes, i.e., resistance to denaturization *via* proteinases or other physical factors including high and low temperatures, dehydration, and radiation, was found to be increased in the adsorbed state.

From the earliest detection and study of soil enzymes at the beginning of the twentieth century (reviewed in Skujins 1978), workers noted, in addition to microbial factors, the importance of both organic and inorganic soil fractions in controlling catalytic and, specifically, enzyme activity. However, perhaps due to methodological obstacles and a lack of cross-disciplinary expertise, these relationships did not become a widespread focus of research until the 1960–1970s. Though early work on soil enzymology failed to produce its highly sought after "fertility index," it found, instead, strong correlations between enzyme activities and physical and chemical soil characteristics. For example, a number of workers noted a predominance of enzymatic activity in the clay and silt soil size fractions relative to sand, and surmised preferential adsorption to be at play (McLaren and Packer 1970). Durand (1963) reported that clay sorption of either uricase or the substrate urate resulted in lower activity than that of the free enzyme.

At the same time, a line of research developed investigating the association between enzymes and humic substances and other organic components of soils. For example, Ladd and Butler (1969) discovered that protease activity was, in many cases, dramatically inhibited by the presence of soil humic acids (HA). At solution concentrations of 1 and 10 μg/ml, respectively, 80 and 10% of pronase activity, 32 and 14% of trypsin activity, and 74 and 48% of carboxypeptidase A activity remained. An effect on these enzymes similar in magnitude was also observed with other polymeric anions such as a benzoquinone and a catechol polycondensate. However, phaseolain, tyrosinase, and chemotrypsin showed little effect from the presence of HA, and the activity of papain, ficin, and thermolysin actually increased with increasing HA concentrations (by 30, 120, and 115% at 10 μg/ml HA, respectively, depending upon the substrate used). In addition, working with these same proteases, these workers found that HA inhibited enzyme activity to a greater extent than fulvic acids and ascribed this difference to the relatively greater molecular weight and rigidity of HA, which may have led to more structural deformation of certain enzymes (Butler and Ladd 1971). Additional hypotheses that could explain the differential effects of HA on various enzymes, such as metal ion requirements (Ladd and Butler 1969) and the presence of carboxyl groups in the inhibitory organic component (Butler and Ladd 1969), proved not to be universally true. For example, although Kunze (1970, 1971) showed that although a number of acidic organic compounds including syringic and gallic acids inhibited catalase activity, tannins also had a similar effect. Mayaudon and Sarkar (1974, 1975) found that diphenol oxidase enzymes were associated with HA and protein complexes extracted from soils. Further, these enzymes, when complexed with OM, retained significant activity.

In the field of aquatic science, examinations of extracellular enzymatic initially drew from concepts and techniques developed in soil science. During the 1980s, the first measurements of enzymatic activity in mineral-based marine snow (Amy et al. 1987), and marine (Meyer-Reil 1986), freshwater (King and Klug 1980), and intertidal sediments (King 1986; Rego et al. 1985) were made. Though with the primary focus of understanding the degradation of aquatic OM, these studies have led to important contributions by marine scientists, to the understanding of enzyme–organo-mineral interactions, particularly with regard to OM–mineral sorption and surface protection (discussed below).

As research in this field progresses through the development of analytical tools and the cross-fertilization from a variety of disciplines, we are beginning to understand the modes of mineral and OM interaction with enzymes and the chemical mechanisms by which this interaction may affect enzyme activity. However, we are also beginning to see that there are many ways in which soil OM, mineral, and enzyme may interact (Fig. 15.1), and these interactions are both complex and synergistic, and not simply additive. In the following, the processes and mechanisms of enzyme–mineral, enzyme–OM, and ternary interactions between all three are outlined.

15.3 Enzyme–Mineral Interactions

Enzymes, as with most proteins, exhibit a strong affinity for mineral surfaces of many types. Because they are made up of amino acids of a range of properties including hydrophilic, hydrophobic, and negatively, positively, and neutrally charged, and because they may change their shape in response to environmental changes, the mechanisms of protein–mineral adsorption are varied and complex.

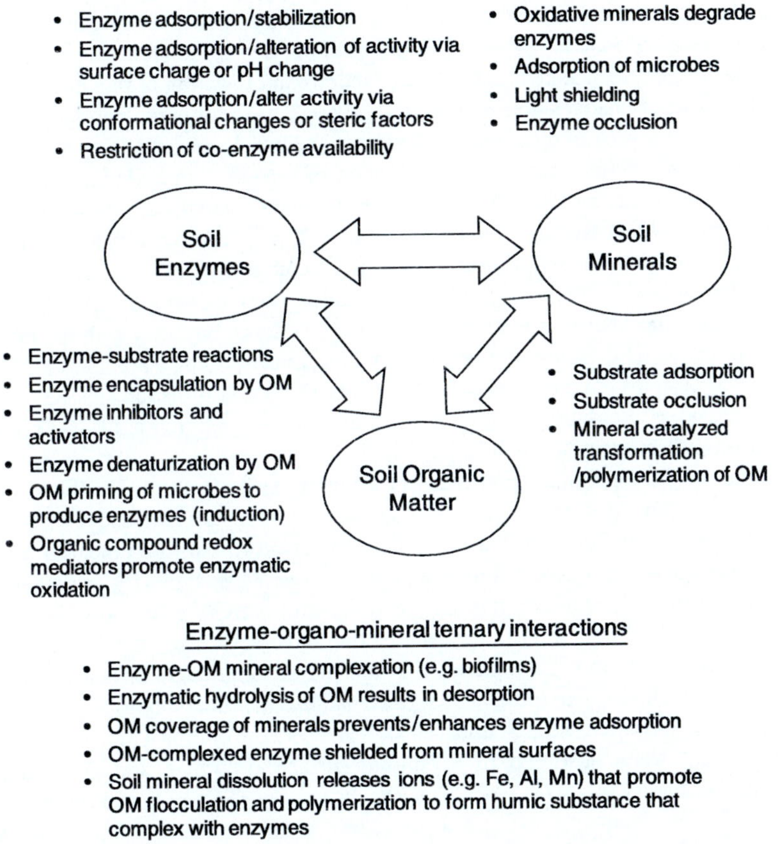

Fig. 15.1 Summary of soil enzyme-mineral-organic matter interactions

Early work examining the adsorption of proteins to clays put forward ionic exchange and electrostatic attraction as the primary mode of bonding (Boyd and Mortland 1990; Naidja et al. 2000). Evidence for this was cited as the pH change that occurs when protein and clay are mixed. However, though the surface of many minerals is negatively charged, and proteins, though amphiphilic, usually carry some net charge, this may not be the dominating factor in controlling protein adsorption. There is no significant relationship between the net charge of adsorbed protein and the charge density of a mineral sorbent's surface, though one does exist for amino acid monomers and dimers (Zimmerman et al. 2004b). Clay interlayer adsorption of protein was thought to occur for expandable clays such as smectites, and especially montmorillonite, but this mechanism has been shown to occur only partially and only in certain cases such as with albumin (De Cristofaro and Violante 2001), glucose oxidase (Garwood et al. 1983), and aspartase (Naidja and Huang 1996).

While some studies have found electrostatic interactions to dominate protein–mineral interaction (e.g., Ding and Henrichs 2002; Servagent-Noinville et al. 2000), entropic effects such as hydrophobic interactions and lowering of free energy due to conformational changes are, increasingly, thought to be a predominant adsorption mechanism for protein onto mineral surfaces. This mechanism explains a number of common observations such as the adsorption of protein onto hydrophobic surfaces such as that of silica oxides even under non-favorable electrostatic conditions (Arai and Norde 1990; Fusi et al. 1989; Giacomelli and Norde 2001), as well as the maximal adsorption of protein occurring at or near the isoelectric point of a mineral surface (e.g., Barral et al. 2008; McLaren 1954; Staunton and Quiquampoix 1994). Though the relative importance of enthalpic and entropic effects may vary with the protein–sorbent pair, and these are not completely independent (Quiquampoix et al. 2002), adsorbed proteins have been clearly shown to exhibit conformation changes such as internal and external unfolding upon mineral adsorption (Baron et al. 1999; Naidja et al. 2002; Servagent-Noinville et al. 2000).

Changes in the kinetic properties of enzymes (activity) are usually observed upon adsorption to mineral surfaces. Table 15.1 is a non-exhaustive list of the changes in enzyme activity that have been observed in enzyme potential activity upon adsorption to various common soil sorbents. Most typically, a decrease in enzyme activity is recorded though the amount of activity inhibition varies from non-existent to complete. The hypotheses that have been proposed to explain this activity decrease are (1) a difference in the pH or ionic strength that an enzyme experiences close to a mineral's surface *versus* the bulk solution (Claus and Filip 1988; Skujins 1976; Skujins et al. 1959), (2) a conformational change in the enzyme when adsorbed to a mineral surface (Leprince and Quiquampoix 1996; Quiquampoix 1987a), or (3) steric hindrance (Baron et al. 1999; Naidja et al. 2000; Quiquampoix and Burns 2007). However, the first hypothesis cannot explain the shift toward a higher pH of maximal activity that is commonly observed when an enzyme is adsorbed (Baron et al. 1999; Pflug 1982; Quiquampoix 1987b; Skujins et al. 1974). Evidence for the second hypothesis is found in more recent studies that have shown pH-dependent structural changes in enzyme conformation upon mineral adsorption that may cause enzyme deactivation (Servagent-Noinville et al. 2000), and the nonreversibility of enzyme activity loss upon desorption (Leprince and Quiquampoix 1996; Quiquampoix 1987a). In further support of this theory, cases in which there is no loss of enzyme activity also show no sign of enzyme secondary structure alteration (e.g., laccase on Al hydroxide; Ahn et al. 2007).

Substrate accessibility to an enzyme's active site, or vice versa (steric restriction), may also play a role in reduced sorbed enzyme activity. The orientation of enzyme attachment has also been shown to be pH dependent (Baron et al. 1999) and this would explain why some minerals exhibit enhanced activity in the bound state and why mineral edge-bound versus face-bound enzymes may display varying activity. Variations in bound enzyme orientation, and differing mineral surface morphology may also explain why enzymes adsorbed to minerals similar in chemistry may widely differ in adsorbed enzyme activity (Table 15.1). For example, the

Table 15.1 Changes in enzyme activity following mineral interaction

Study	Enzyme	Sorbent	Activity change (%)
Morgan and Corke (1976)	Glucose oxidase	Ca-montmorillonite	−57 to −96
Ross and McNeilly (1972)	Glucose oxidase	Kaolinite	−17
Ross and McNeilly (1972)	Glucose oxidase	Illite	−21
Ross and McNeilly (1972)	Glucose oxidase	Ca-montmorillonite	−77
Ross and McNeilly (1972)	Glucose oxidase	Allophane	−52
Hughes and Simpson (1978)	Arylsulphatase	Ca-montmorillonite	−52
Hughes and Simpson (1978)	Arylsulphatase	Kaolinite	−18
Haska (1981)	Mixed endopeptidases	Montmorillonite	−66
Claus and Filip (1988)	Laccase	Bentonite	−11 to −100
Claus and Filip (1988)	Tyrosinase	Bentonite	−84 to 100
Claus and Filip (1988)	Laccase	Kaolinite	−30 to −83
Claus and Filip (1988)	Tyrosinase	Kaolinite	0 to −75
Quiquampoix (1987a)	Glucosidase	Na-montmorillonite	−35 to −100
Quiquampoix (1987a)	Glucosidase	Kaolinite	−13 to −100
Quiquampoix (1987a)	Glucosidase	Goethite	0 to 40
Skujins et al. (1974)	Chitinase	Kaolinite	−95
Naidja et al. (1997)	Tyrosinase	$Al(OH)_x$-coated-montmorillonite	−24 to −62
Naidja and Huang (1996)	Aspartate	Ca-montmorillonite	−20
Rao et al. (2000)	Acid phosphatase	Montmorillonite	−80
Rao et al. (2000)	Acid phosphatase	Al hydroxide	−55
Calamai et al. (2000)	Catalase	Ca-montmorillonite	−81 to −99
Gianfreda et al. (1991)	Invertase	Na-montmorillonite	−88 to −96
Gianfreda et al. (1991)	Invertase	$Al(OH)_x$-montmorillonite	−89 to −95
Gianfreda et al. (1991)	Invertase	$Al(OH)_x$	−94 to −99
Gianfreda et al. (1992)	Urease	Na-montmorillonite	−41
Gianfreda et al. (1992)	Urease	$Al(OH)_x$-montmorillonite	−49 to −67
Gianfreda and Bollag (1994)	Laccase	Montmorillonite	0
Gianfreda and Bollag (1994)	Laccase	Kaolinte	−14
Gianfreda and Bollag (1994)	Peroxidase	Montmorillonite complex	0
Gianfreda and Bollag (1994)	Kaolinite	Montmorillonite complex	−21
Gianfreda and Bollag (1994)	Acid phosphatase	Montmorillonite	−68
Gianfreda and Bollag (1994)	Acid phosphatase	Kaolinite	−64
Huang et al. (2005)	Acid phosphatase	Goethite	−32
Huang et al. (2005)	Acid phosphatase	Kaolinite	−43
(Ahn et al. 2007)	Tyrosinase	Al hydroxide	−11[a]
(Lozzi et al. 2001)	Peroxidase	Ca-montmorillonite	0 to −69
(Lozzi et al. 2001)	Peroxidase	Na-montmorillonite	−88 to −99
Tietjen and Wetzel (2003)	Alkaline phosphatase	Elledge clay	−75
Tietjen and Wetzel (2003)	Alkaline phosphatase	Montmorillonite	−67

(*continued*)

Table 15.1 (continued)

Study	Enzyme	Sorbent	Activity change (%)
Tietjen and Wetzel (2003)	Glucosidase	Elledge clay	+55
Tietjen and Wetzel (2003)	Glucosidase	Montmorillonite	+50
Tietjen and Wetzel (2003)	Protease	Elledge clay	−82 to 82
Tietjen and Wetzel (2003)	Protease	Montmorillonite	−100
Tietjen and Wetzel (2003)	Xylosidase	Elledge clay	+50
Tietjen and Wetzel (2003)	Xylosidase	Montmorillonite	+200
Quiquampoix (1987b)	Glucosidase	Montmorillonite	0 to −82
Quiquampoix (1987b)	Glucosidase	Natural clay fraction	0 to −99
Quiquampoix (1987a)	Glucosidase	Kaolinite	0 to −88
Rao et al. (2000)	Phosphatase	Na-montmorillonite	−80
Rao et al. (2000)	Phosphatase	Al hydroxide	−55
Rao et al. (2000)	Phosphatase	$Al(OH)_x$-montmorillonite	−42

[a]Calculated from V_{max}

enhanced catalytic activity of nanoparticle-bound enzymes relative to their "free" counterpart (Serefoglou et al. 2008) may be due to favorable active site orientation on these surfaces or to alleviation of reactant or product diffusion limitation that may occur on particles of larger sizes. Of course, it is recognized that the nature of the specific sorbent–enzyme pair (e.g., "hard" or "soft" enzyme, hydrophobicity of mineral surface) may also determine which of these possible mechanisms play a dominating role in altering enzyme activity.

As pointed out by Quiquampoix and Burns (2007), the study of enzyme interaction with mineral surfaces is difficult because changes in conformation both result from and drive enzyme adsorption to mineral surfaces and because the structure of enzymes in the adsorbed state cannot be fully examined by the methods currently available. Spectroscopic methods have only permitted elucidation of protein secondary structure, though advances in this regard are being made (Smith et al. 2009). Further, neither the chemical environment at the surface of a mineral nor the orientation of a sorbed enzyme can be completely known. An additional impediment to the advancement of our understanding is that enzyme activity experiments are not often conducted with associated enzyme adsorption and substrate adsorption experiments. Without these, it cannot be known what portion of the activity measured is produced by adsorbed enzymes versus "free" enzymes. Variation in the enzyme activity with substrate concentration, as described, for example, by the Michaelis–Menten equation is not commonly determined for mineral-adsorbed enzymes (discussed below). Even less often is a range of enzyme-mineral surface area loadings tested. Lastly, soils are commonly composed of a heterogeneous assemblage of minerals. But pure clays have been, by far, the main focus of attention with regard to enzyme– mineral interaction experiments. In many clay-poor soils, amorphous or poorly crystalline minerals such as Al and Fe oxyhydroxides may play a dominant role in regulating enzyme stability and activity and require greater study.

15.4 Enzyme–Organic Matter Interactions

It is likely that, particularly in soils with higher OM or sand contents, enzyme–OM interaction plays a more important role in regulating soil enzyme activity and stability than enzyme–mineral interaction. Research in this area has utilized two general methods. The first is to examine the association of enzymes with different natural soil OM fractions through various extraction schemes and, perhaps, to measure the residual potential activity present in each fraction. The other approach is to compare the kinetic properties of free enzyme forms with those of OM–enzyme complexes prepared in the laboratory. Each approach has yielded important insights.

Most soil extracellular enzymes are thought to be intimately associated with soil OM as they cannot be efficiently extracted without also extracting a great deal of humic substances (Ladd 1972). Further, many purification techniques fail to separate enzymes from their associated OM, and enzymatic activity and organic-C contents of the extracts are generally correlated (Ceccanti et al. 1978). It can be assumed that enzymes will form complexes with the OM components of soil much as do proteins, which are associated and react chemically with phenols, quinones, tannins, lignin components, and HA (Ladd and Butler 1975 and references therein). Both hydrogen and other electrostatic bonds and covalent bonds are known to occur, and the natures of both are strongly dependent upon pH.

Studies of extracted HA–enzyme complexes and attempts at purification have not revealed any single mode of association, but rather, a combination of mechanisms have been suggested including ion exchange, physical entrapment within three-dimensional structures, hydrogen and covalent bonding. For example, the involvement of inorganic cations such as Ca^{+2} in encouraging protein complexation and complexed-enzyme inactivation would suggest the influence of electrostatic forces (Ladd and Butler 1969; Mayaudon and Sarkar 1975; Pflug 1981). However, the effectiveness of sodium pyrophosphate in extracting active soil OM-bound extracellular enzymes (Ceccanti et al. 1978; Nannipieri et al. 1988) suggests that a large fraction of enzymes are covalently bound or, at least, strongly associated with humic substances rather than ionically bound. Another prospective is provided by the observation of a direct relationship between molecular weight of the fractionated soil humic extracts and its resistance to thermal denaturation and proteolysis (Butler and Ladd 1971; Nannipieri et al. 1988). This would suggest a HA porous structure that has to permit the diffusion of substrates toward and products away from the enzyme, but restricts the movement of larger molecules such as proteases. Some amount of diffusion limitation would be expected to occur, however, no matter the size of the substrate, and this may play a role in the usually observed reduced activity of OM-bound enzymes (Table 15.2). This type of encapsulation may be maintained mainly by hydrophobic interactions, which is supported by recent calorimetric measurements of protein–HA complexation reactions (Tan et al. 2008) and ^{15}N and ^{13}C NMR spectrographic monitoring of OM degradation (Zang et al. 2000, 2001).

Table 15.2 Changes in enzyme activity following organic matter interaction

Author	Enzyme	Sorbent/Complexant	Activity change (%)
Grego et al. (1990)	Pronase	Resorcinol	−80
Grego et al. (1990)	Pronase	Pyrogallol	−100
Grego et al. (1990)	Pronase	Catechol	−100
Rowell et al. (1973)	Trypsin	Benzoquinone	−56 to −98
Rowell et al. (1973)	Pronase	Benzoquinone	−32 to −96
Serban and Nissenbaum (1986)	Peroxidase	Humic acid	+18 to +125
Serban and Nissenbaum (1986)	Catalase	Humic acid	+100
Ruggiero and Radogna (1988)	Tyrosinase	Humic acid	−5 to +19
Pflug (1981)	Malate dehydrogenase	Humic acid	−16 to −57
Butler and Ladd (1969)	Pronase	Humic acid	−40 to −73
Butler and Ladd (1969)	Trypsin	Humic acid	−44 to −64
Butler and Ladd (1969)	Papain	Humic acid	+88 to +160
Butler and Ladd (1969)	Carboxypeptidase	Humic acid	−35 to −76
Gianfreda et al. (1995b)	Urease	Tannic acid	−29 to −72
Rao et al. (2000)	Acid phosphatase	Tannic acid	−49 to −93
Criquet et al. (2000)	Laccase	Humic acid	−95 to +60
Allison (2006)	Glucosidase	Humic acid	−70 to −100
Allison (2006)	Acid phosphatase	Humic acid	−20 to −65
Allison (2006)	Urease	Humic acid	−10 to −35
Allison (2006)	Glucosaminidase	Humic acid	−75 to −90
Allison (2006)	Polyphenol oxidase	Humic acid	−35 to −70
Boavida and Wetzel (1998)	Phosphatase	Humic acid	+10 to +100
Canas et al. (2007)	Laccase	Acetosyringone	−10 to −50
Canas et al. (2007)	Laccase	Syringaldehyde	0 to −50
Canas et al. (2007)	Laccase	Vanillin	−50 to −60
Canas et al. (2007)	Laccase	Acetovanillone	−70 or +50
Canas et al. (2007)	Laccase	2,4,6-Trimethylphenol	0 to −30
Canas et al. (2007)	Laccase	*p*-coumaric acid	−85 or +250
Canas et al. (2007)	Laccase	Ferulic acid	−30 or +80
Canas et al. (2007)	Laccase	Sinapic acid	−50 or +100
Freeman et al. (2004)	Sulphatase	Phenolic compound	−47[a]
Freeman et al. (2004)	Phosphatase	Phenolic compound	−18[a]
Freeman et al. (2004)	Glucosidase	Phenolic compound	−26[a]
Freeman et al. (2004)	Xylosidase	Phenolic compound	−16 [a]
Freeman et al. (2004)	Chitinase	Phenolic compound	−22[a]
Ladd and Butler (1969)	Pronase	Humic acid	−25 to −35
Freeman et al. (2004)	Trypsin	Humic acid	−18 to −52
Freeman et al. (2004)	Carboxypeptidase A	Humic acid	−33 to −49
Freeman et al. (2004)	Papain	Humic acid	+49 to +226

[a]Not pH buffered

HA–enzyme complexes have been prepared in vitro through simple adsorption techniques, such as with trypsin and pronase adsorbed to benzoquinone (Rowell et al. 1973) and pronase adsorbed to polyphenolic HA precursors (Grego et al. 1990). However, these complexes retained little of their pre-interaction activity. In contrast, better activity retention was achieved when phenoloxidase enzymes, such as peroxidases and laccases, were combined with HA precursors and lignin subunits (Canas et al. 2007; Sarkar and Burns 1983, 1984), which carry out oxidative coupling reactions much like those thought to occur during the formation of natural soil humic substances. Recently, ^{15}N-^{1}H HSQC NMR spectrometry has provided evidence for protein–quinone group covalent bonding of HAs using a tetrapeptide

(Hsu and Hatcher 2005). But while these synthetic model studies, along with those utilizing preformed humic materials, generally show greater stability and decreased activity of OM-associated enzymes (Table 15.2), a wide range of magnitude of changes have been demonstrated with no clear pattern emerging. It seems that, much as with enzyme–mineral interactions, bound enzyme activity is highly specific to the enzyme–sorbent pair and deactivation may occur due to the orientation of an enzyme's active site relative to each specific substrate's accessibility or to distortion of an enzyme's structure due to intermolecular forces in its bound chemical environment.

One way that the effects and mechanisms of enzyme inactivation due to mineral versus OM interaction can be directly compared is through the use of Michaelis–Menten parameters, which model enzyme kinetic properties. Though most of the enzyme interaction experiments (Tables 15.1 and 15.2) have recorded enzyme activity at a given substrate and enzyme concentration, a range of substrate concentrations must be tested to derive K_m, the Michaelis–Menten constant, and V_{max}, the maximum enzyme reaction velocity. The former is a measure of the affinity of an enzyme for a substrate (a higher K_m value indicates lower affinity), and the latter, describes the rate of substrate conversion activity that occurs when enzyme active sites are saturated with substrate (Fig. 15.2). These are useful because they are independent of the enzyme concentration used and also provide insight into the operative inhibition mechanism.

While they can only strictly be applied to simple enzyme–substrate systems with homogeneous reactions, the "apparent" kinetic values obtained for bound enzymes, i.e., heterogeneous catalysts with diffusion limitations, may have mechanistic implications. For example, reversible competitive inhibition, which occurs when a substance competes for the same active site as the substrate, will be characterized by an increase in K_m without a change in V_{max} (Fig. 15.2). This type of inhibition may be analogous to the temporary binding of OM to an enzyme's active site.

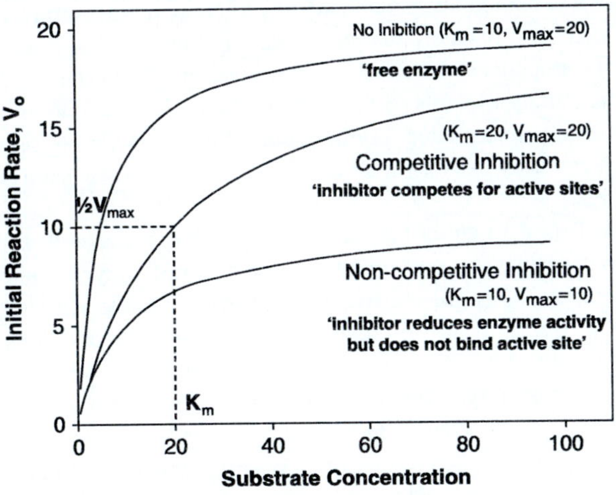

Fig. 15.2 Model Michaelis-Menton plots showing predicted kinetic changes with enzyme inhibition

Noncompetitive inhibition, indicated by a decrease in V_{max} without a change in K_m, occurs when an inhibitor binds to either the enzyme–substrate complex or the enzyme itself, resulting in lowered enzymatic efficiency. This might be likened to an enzyme conformational change, which occurs upon mineral or OM interaction. With irreversible inhibition, one would expect to see time-dependent changes in both K_m and V_{max}. Among those soil mineral and OM interaction experiments that have calculated Michaelis–Menten parameters (Table 15.3), mixed inhibition was most common, indicating that both competitive and noncompetitive inhibition may be operative. That is, substrate binding to both active and nonactive sites of the enzyme, conformation alteration, and diffusion, charge, and steric effects may decrease enzyme affinity for substrate. Decreased K_m values, indicating increased enzyme–substrate affinity, have been observed, mainly for montomorillonite. This may be attributable to the formation of an adsorption bond orientation, which causes enhanced active site accessibility to substrate. And increased enzyme activity has been observed in the presence of urea that has been attributed to loosening of the enzyme's structure, which confers increased conformational flexibility at the active site (Fan et al. 1996; Zhang et al. 1997). Other kinetic constants such as turnover frequency and enzyme efficiency may provide additional information with which to compare the effects of complexents. However, the extent to which these parameters are applicable to the extremely heterogeneous nature of soil systems is debatable.

15.5 Mineral-Organic Matter–Enzyme Synergistic Interactions

While the enzyme complexation experiments of the type reviewed, above, provide some insight into the fundamental processes affecting enzyme activity in soils, an understanding of the actual effect of minerals and OM on soil enzyme activity cannot be achieved without consideration of the possible synergistic interactions between enzymes and these complexants. Soil OM and soil minerals are intimately associated, so it is likely to be only rarely that an enzyme is associated with one or the other, exclusively. That being so, the relatively low number of experiments that have been carried out in ternary systems or examining alternative enzyme–substrate-sorbent relationships is rather surprising. Furthermore, of the wide variety of ways in which minerals, OM and enzymes may interact (Fig. 15.1), only a few have been studied to any great extent.

One method used to study the ternary interaction between enzyme, OM, and mineral is to measure the activity of OM-complexed enzymes fixed to mineral surfaces. In one case, β-D-glucosidase, which had been made somewhat resistant to protease attack over 24 h due to copolymerization with phenolic compounds, was completely resistant when these copolymers were fixed to bentonite (Sarkar and Burns 1984). Interestingly, while the copolymer immobilization was accompanied by increases in K_m and decreases in V_{max}, changes in these parameters were not as large as those that would be expected had enzyme–phenol and enzyme–bentonite changes been additive. These data imply that association with OM can, to a degree, shield an enzyme from the conformational or steric limitations that might otherwise

Table 15.3 Changes in Michaelis–Menten kinetic parameters following enzyme complexation with mineral or organic matter

Study	Enzyme	Sorbent/Complexant	K_m[a] (% change from free to complexed)	V_{max}[a]
Minerals				
Makboul and Ottow (1979a)	Acid phosphatase	Montorillonite	+5779	−46
Makboul and Ottow (1979a)	Acid phosphatase	Illite	+500	−56
Makboul and Ottow (1979a)	Acid phosphatase	Kaolinite	+443	−69
Makboul and Ottow (1979b)	Urease	Montorillonite	+191	−50
Makboul and Ottow (1979b)	Urease	Illite	+88	−50
Makboul and Ottow (1979b)	Urease	Kaolinite	+161	−48
Dick and Tabatabai (1987)	Acid phosphatase	Montorillonite	0	−38 to −67
Dick and Tabatabai (1987)	Acid phosphatase	Illite	0	−21 to −46
Dick and Tabatabai (1987)	Acid phosphatase	Kaolinite	+68 to +307	0
(Dick and Tabatabai 1987)	Pyrophosphatase	Montorillonite	0	−39 to −48
(Dick and Tabatabai 1987)	Pyrophosphatase	Illite	0	−18 to −41
(Dick and Tabatabai 1987)	Pyrophosphatase	Kaolinite	+71 to +138	0
Serefoglou et al. (2008)	β-glucosidase	Montmorillonite	−60	−3
Gianfreda et al. (1992)	Urease	Montorillonite	−73	−30
Sarkar and Burns (1984)	β-glucosidase	Bentonite	+45	−98
Sarkar and Burns (1984)	β-glucosidase	$Al(OH)_x$	−9.4	−81
Sarkar and Burns (1984)	β-glucosidase	Montorillonite-$Al(OH)_x$	−48	−36
Marzadori et al. (1998b)	Urease	Hydroxyapatite	−6.7	−34
Ahn et al. (2007)	Laccase	$Al(OH)_x$	−16	−11
Rosas et al. (2008)	Acid phosphatase	Allophonic clay	−14	+42
Shindo et al. (2002)	Acid phosphatase	Fe oxide	+67	−77
Shindo et al. (2002)	Acid phosphatase	Al oxide	−20	−87
Shindo et al. (2002)	Acid phosphatase	Mn oxide	+120	−87
Organic matter				
Rosas et al. (2008)	Acid phosphatase	Tannic acid	+135	−35
Gianfreda et al. (1995b)	Urease	Tannic acid (0.02 mM)	−11	−25
Gianfreda et al. (1995b)	Urease	Tannic acid (0.1 mM)	−63	−71
Vuorinen and Saharinen (1996)	Phosphomonoesterase	Soil OM extract (0.05 mg/L)	+124	+2.6
Vuorinen and Saharinen (1996)	Phosphomonoesterase	Soil OM extract (0.5 mg/L)	+22	−0.7
Marzadori et al. (1998a)	Acid phosphatase	Ca-polygalacturonate	+102	−59
Rao et al. (2000)	Acid phosphatase	Tannic acid	+100	−75
Rao et al. (2000)	Acid phosphatase	Tannic acid + Fe	+233	−75
Rao et al. (2000)	Acid phosphatase	Tannic acid + Mn	+167	−63
Sarkar and Burns (1984)	β-glucosidase	Resorcinol	+13	−45
Sarkar and Burns (1984)	β-glucosidase	Tyrosine	+52	−75

[a]Michaelis–Menton constant and maximum conversions rate, K_m and V_m, respectively, calculated using Linweaver–Burk equation

accompany mineral boundedness. In another study (Ahn et al. 2006), laccase activity in the ternary system of laccase with catechol and humic-like polymers, produced through birnessite catalysis of catechol, was less than the sum of enzyme activity in binary systems of laccase and birnessite or laccase and catechol. Enzyme inactivation both by birnessite-generated humic-like polymers and by Mn^{+2} ions released from the mineral were considered possible causes of the inhibition. An effect of metal ions, which are commonly released from minerals especially when redox changes occur, on enzyme activity has been observed in other cases. For example, Fe^{+2} has been shown to stimulate phenol oxidase activity (Van Bodegom et al. 2005) and soluble Fe^{+3}, Mn^{+2}, and Al^{+3} species stimulated the activity of tannate–urease complexes (Gianfreda et al. 1995a; Gianfreda et al. 1995b). Interestingly, these species had little effect on the activity of phosphatase–tannic acid complexes unless montomorillite was also present (Rao et al. 2000).

Perhaps due to the possibility of industrial and contaminant remediation applications, while many studies have examined the activity of "free" and mineral-adsorbed enzymes, only a few studies have examined enzyme activity when provided mineral- or OM-adsorbed substrates. This may be carried out by performing substrate adsorption prior to enzyme addition; however, the possibility of substrate desorption and enzyme adsorption must be considered. In one study (McLaren and Estermann 1957), the activity of chymotrypsin, when supplied kaolinite-sorbed lysozyme as a substrate, exhibited no loss of activity relative to "free" lysozyme. However, an increase in the pH of maximum activity was observed, much as occurs with most mineral-adsorbed enzymes. In another study (Skujins et al. 1974), an 84% decrease in chitinase activity occurred when provided with kaolinite-sorbed chitin as a substrate, but this was less than the 94% decrease observed when the chitinase was first adsorbed to kaolinite. Zimmerman et al. (2004a) found that oxidation of an amorphous alumina-adsorbed phenol by laccase was somewhat less (by a mean of 7%) than that of the free compound, while that of amorphous silica-adsorbed phenol was enhanced (by a mean of 20%). In this same study, almost complete loss of laccase activity occurred when the substrate was occluded within alumina and silica nanopores, showing that mineral surface morphology can play a major role in substrate accessibility to enzymes. Thus, bound substrates may or may not be completely protected from enzymatic digestion, depending upon the situation, but sorbed-enzyme degradation of sorbed substrate is likely to be extremely limited.

Native OM can also control the adsorption of enzymes to mineral surfaces. For example, HAs added before urease resulted in less urease adsorption to hydroxyapatite and less urease activity than when urease was added first (Borghetti et al. 2003). Interestingly, the order of addition, in this case, had no effect on long-term enzyme stability. Conversely, minerals, particularly short-range-ordered minerals such as Fe, Al, and Mn oxyhydroxides promote the immobilization of OM and enzyme-associated OM via polymerization and copolymerization reactions (Ahn et al. 2006; Gianfreda et al. 1995a; Gianfreda et al. 1995b; Rao et al. 1996).

Soil OM may play an additional role in enzyme activity regulation by providing microbes with nutrients needed to produce enzymes (induction), providing co-enzymes, activator or inhibitor molecules, or substrate competitors. Measurements of enzyme activity following experimental additions of various substrates have been carried out, but often produce complex results that vary with enzyme and substrate type, soil mineralogy, and response period. For example, acid phosphatase activity was only stimulated by citrate in a clayey soil, versus oxalate, glutamate, and citrate in a sandy soil (Renella et al. 2007). In the same study, urease activity was only stimulated by glucose and citrate in the clayey soil, and by no substrates in the sandy soil. In another study (Nannipieri et al. 1983), additions of glucose or rye grass to a clay-loam increased phosphatase and urease activities in proportion to increases in bacterial biomass, whereas casein hydrolysis was delayed. Often, enzymatic remineralization out of proportion to the amount of substrate added is observed, i.e. "OM priming" (Kuzyakov et al. 2000). Although some of the results of these types of addition experiments conform to prediction, i.e., increased enzyme activity when with labile substrates are abundant and decreased enzyme activity when only complex substrates are present (Allison and Vitousek 2005), many times they do not, suggesting that factors other than substrate supply influence soil enzyme activity. Among them, surely, is enzyme–organo-mineral interaction. In addition, the common observation that most enzyme activities returned to their pre-OM addition levels (e.g., Dilly and Nannipieri 2001; Kuzyakov et al. 2000; Nannipieri et al. 1983) is support for the hypothesized homeostatic level of microbial/enzymatic activity for each particular soil, first proposed by Burns (1982), controlled perhaps by the enzyme–organo-mineral interaction present in each soil type. Future work should combine substrate adsorption experiments with controlled incubations of mixed microbial populations in binary and ternary OM-mineral–enzyme systems to identify the mechanisms responsible for these interactions.

15.6 Consequences of Organo-Mineral–Enzyme Interactions

Aside from the obvious importance of soil enzymes in supporting all microbial life and, thus, making nutrients available to the plants upon which we all depend, organo-mineral–enzyme interactions, in particular, are of fundamental importance in explaining diverse phenomena such as the bioavailability and migration of organic contaminants (Luthy et al. 1997), the sequestration of organic carbon in soils and coastal and marine sediments (Mayer 1994; Mayer 1999; Torn et al. 1997), and the global carbon balance regulating atmospheric CO_2 concentrations. Although most biomolecular OM is inherently labile, some portion of it is preserved in soils and sediments over long timescales and remains apparently unavailable to microbial decomposers (Hedges and Keil 1995; Luthy et al. 1997). Direct correlations between soil and sediment surface area and organic carbon (Baldock and Skjemstad 2000; Kennedy et al. 2002) suggest that organo-mineral complexation stabilizes labile forms of OM against enzymatic attack, and this has been shown experimentally (Jastrow and Miller 1997; Kaiser and Guggenberger 2000; Zimmerman et al. 2004a). The mechanisms responsible for this OM protection are the same ones outlined, above, including enzyme activity limitation via adsorption or complexation within OM or on mineral surfaces. Enzymes or substrates may be occluded within mineral micro- and nanofabric structures such as aggregates and clay domains, sorbed to mineral surfaces (particularly clays, poorly crystalline minerals or micro- or nanopores), or complexed within organic polymer structures.

15.7 Conclusion

Advances in enzyme technology will allow the examination of enzyme–organo-mineral interactions in greater detail than previously allowed. For example, advances in spectroscopic techniques will allow better detection of changes in enzyme tertiary structure that occur with OM and mineral interaction (Smith et al. 2009). Experimental and analytical techniques developed by biochemists and chemical engineers have recently been creatively employed to probe the enzyme–organo-mineral relationship. For example, Ziervogel et al. (2007) tethered a fluorescently labeled polysaccharide (covalently bonded pullulan) to agarose beads to examine the accessibility of mobility-restricted substrates to a "free" and clay-adsorbed enzyme (pullanase). In this study, apportionment of the products of enzymatic hydrolysis of the substrate into different molecular weight classes was monitored by gel permeation chromatography. In another study, decreases in the molecular size of "free" and montmorillonite-sorbed spin-labeled polysaccharides were monitored in real time using electron paramagnetic resonance spectroscopy (Steen et al. 2006), allowing high-resolution determinations of enzyme sorption and degradation kinetics. Finally, in a novel application of histochemical staining, Curry et al. (2007) were able to visualize OM protected from enzymatic digestion in nanoscale clay structures using acid-thiosemicarbazide-silver proteinate-tagged polysaccharides and transmission electron microscopy. With analytical and visualization techniques such as these, we will be able to examine enzyme–organo-mineral relationships in ever greater detail. While past research has mainly examined binary interaction of enzymes with certain minerals (mostly clays) or specific OM fractions, future work should focus on ternary enzyme–organo-mineral synergistic interactions, including those of mineral catalysis, OM priming, and mineral and OM protective structures.

References

Ahn MY, Martínez CE, Archibald DD, Zimmerman AR, Bollag J-M, Dec J (2006) Transformation of catechol in the presence of a laccase and birnessite. Soil Biol Biochem 38:1015–1020

Ahn MY, Zimmerman AR, Martinez CE, Archibald DD, Bollag JM, Dec J (2007) Characteristics of *Trametes villosa* laccase adsorbed on aluminum hydroxide. Enzyme Microb Technol 41:141–148

Allison SD, Vitousek PM (2005) Responses of extracellular enzymes to simple and complex nutrient inputs. Soil Biol Biochem 37:937–944

Allison SD (2006) Soil minerals and humic acids alter enzyme stability: implications for ecosystem processes. Biogeochem 81:361–373

Amy PS, Caldwell BA, Soeldner AH, Morita RY, Albright LJ (1987) Microbial activity and ultrastructure of mineral-based marine snow from Howe Sound, British-Columbia. Can J Fish Aquat Sci 44:1135–1142

Arai T, Norde W (1990) The behavior of some model proteins at solid-liquid interfaces 1. Adsorption from single protein solutions. Colloids Surf 51:1–15

Baldock JA, Skjemstad JO (2000) Role of soil matrix and minerals in protecting natural organic materials against biological attack. Org Geochem 31:697–710

Baron MH, Revault M, Servagent-Noinville S, Abadie J, Quiquampoix H (1999) Chymotrypsin adsorption on montmorillonite: enzymatic activity and kinetic FTIR structural analysis. J Colloid Interf Sci 214:319–332

Barral S, Villa-Garcia MA, Rendueles M, Diaz M (2008) Interactions between whey proteins and kaolinite surfaces. Acta Mater 56:2784–2790

Boavida MJ, Wetzel RG (1998) Inhibition of phosphatase activity by dissolved humic substances and hydrolytic reactivation by natural ultraviolet light. Freshw Biol 40:285–293

Borghetti C, Gioacchini P, Marzadori C, Gessa C (2003) Activity and stability of urease-hydroxyapatite and urease-hydroxyapatite-humic acid complexes. Biol Fertil Soils 38:96–101

Boyd SA, Mortland M (1990) Enzyme interactions with clays and clay-organic matter complexes. In: Stotzky G, Bollag J-M (eds) Soil biochemistry, vol 6. Marcel Dekker, New York, pp 1–28

Burns RG (1982) Enzyme activity in soil: location and a possible role in microbial ecology. Soil Biol Biochem 14:423–427

Butler JHA, Ladd JN (1969) The effect of methylation of humic acids on their influence of proteolytic enzyme activity. Aust J Soil Res 7:263–268

Butler JHA, Ladd JN (1971) Importance of the molecular weight of humic and fulvic acids in determining their effects on protease activity. Soil Biol Biochem 3:249–257

Calamai L, Lozzi I, Stotzky G, Fusi P, Ristori GG (2000) Interaction of catalase with montmorillonite homoionic to cations with different hydrophobicity: effect on enzymatic activity and microbial utilization. Soil Biol Biochem 32:815–823

Canas AI, Alcalde M, Plou F, Martinez MJ, Martinez AT, Camarero S (2007) Transformation of polycyclic aromatic hydrocarbons by laccase is strongly enhanced by phenolic compounds present in soil. Environ Sci Technol 41:2964–2971

Ceccanti B, Nannipieri P, Cervelli S, Sequi P (1978) Fractionation of humus-urease complexes. Soil Biol Biochem 10:39–45

Claus H, Filip Z (1988) Behavior of phenoloxidases in the presence of clays and other soil-related adsorbents. Appl Microbiol Biotechnol 28:506–511

Criquet S, Farnet AM, Tagger S, Le Petit J (2000) Annual variations of phenoloxidase activities in an evergreen oak litter: influence of certain biotic and abiotic factors. Soil Biol Biochem 32:1505–1513

Curry KJ, Bennett RH, Mayer LM, Curry A, Abril M, Biesiot PM, Hulbert MH (2007) Direct visualization of clay microfabric signatures driving organic matter preservation in fine-grained sediment. Geochim Cosmochim Acta 71:1709–1720

De Cristofaro A, Violante A (2001) Effect of hydroxy-aluminium species on the sorption and interlayering of albumin onto montmorillonite. Appl Clay Sci 19:59–67

Dick WA, Tabatabai MA (1987) Kinetics and activities of phosphatase-clay complexes. Soil Sci 143:5–15

Dilly O, Nannipieri P (2001) Response of ATP content, respiration rate and enzyme activities in an arable and a forest soil to nutrient additions. Biol Fertil Soils 34:64–72

Ding X, Henrichs SM (2002) Adsorption and desorption of proteins and polyamino acids by clay minerals and marine sediments. Mar Chem 77:225–237

Durand G (1963) Microbiologie des sols – sur la degradation des bases puriques et pyrimidiques dans le sol – fixation de ces composes par les argiles – etude en fonction du pH et de la concentration. C R Hebd Seances Acad Sci 256:4126

Ensminger LE, Gieseking JE (1939) The adsorption of proteins by montmorillonitic clays. Soil Sci 48:467–473
Ensminger LE, Gieseking JE (1941) The absorption of proteins by montmorillonitic clays and its effect on base-exchange capacity. Soil Sci 51:125–132
Ensminger LE, Gieseking JE (1942) Resistance of clay-adsorbed proteins to proteolytic hydrolysis. Soil Sci 53:205–209
Estermann EF, McLaren AD (1959) Stimulation of bacterial proteolysis by adsorbents. J Soil Sci 10:64–78
Fan YX, Ju M, Zhou JM, Tsou CL (1996) Activation of chicken liver dihydrofolate reductase by urea and guanidine hydrochloride is accompanied by conformational change at the active site. Biochem J 315:97–102
Freeman C, Ostle NJ, Fenner N, Kang H (2004) A regulatory role for phenol oxidase during decomposition in peatlands. Soil Biol Biochem 36:1663–1667
Fusi P, Ristori GG, Calamai L, Stotzky G (1989) Adsorption and binding of protein on "clean" (homoionic) and "dirty" (coated with Fe oxyhydroxides) montmorillonite, illite and kaolinite. Soil Biol Biochem 21:911–920
Garwood GA, Mortland MM, Pinnavaia TJ (1983) Immobilization of glucose-oxidase on montmorillonite clay – hydrophobic and ionic modes of binding. J Mol Catal 22:153–163
Giacomelli CE, Norde W (2001) The adsorption-desorption cycle: reversibility of the BSA-silica system. J Colloid Interface Sci 233:234–240
Gianfreda L, Rao MA, Violante A (1991) Invertase (beta-fructosidase) – effects of montmorillonite, Al-hydroxide and $Al(OH)_X$-montmorillonite complex on activity and kinetic-properties. Soil Biol Biochem 23:581–587
Gianfreda L, Rao MA, Violante A (1992) Adsorption, activity and kinetic-properties of urease on montmorillonite, aluminum hydroxide and $Al(OH)_X$-montmorillonite complexes. Soil Biol Biochem 24:51–58
Gianfreda L, Bollag JM (1994) Effect of soils on the behavior of immobilized enzymes. Soil Sci Soc Am J 58:1672–1681
Gianfreda L, Decristofaro A, Rao MA, Violante A (1995a) Kinetic-behavior of synthetic organo-complexes and organo-mineral-urease complexes. Soil Sci Soc Am J 59:811–815
Gianfreda L, Rao MA, Violante A (1995b) Formation and activity of urease-tannate complexes affected by aluminum, iron, and manganese. Soil Sci Soc Am J 59:805–810
Goldstein L, Katchalski-Katzir E (1976) Immobilized enzymes – a survey. In: Wingard LB Jr, Goldstein L, Katchalski-Katzir E (eds) Immobilized enzyme principles, vol 1. Academic, London, pp 1–22
Grego S, Dannibale A, Luna M, Badalucco L, Nannipieri P (1990) Multiple forms of synthetic pronase phenolic copolymers. Soil Biol Biochem 22:721–724
Griffin EG, Nelson JM (1916) The influence of certain substances on the activity of invertase. J Am Chem Soc 38:722–730
Haska G (1981) Activity of bacteriolytic enzymes adsorbed to clays. Microb Ecol 7:331–341
Hedges JI, Keil RG (1995) Sedimentary organic matter preservation: an assessment and speculative synthesis. Mar Chem 49:81–115
Hsu PH, Hatcher PG (2005) New evidence for covalent coupling of peptides to humic acids based on 2D NMR spectroscopy: a means for preservation. Geochim Cosmochim Acta 69:4521–4533
Huang QY, Liang W, Cai P (2005) Adsorption, desorption and activities of acid phosphatase on various colloidal particles from an Ultisol. Colloid Surf B 45:209–214
Hughes JD, Simpson GH (1978) Arylsulfatase-clay interactions. 2. Effect of kaolinite and montmorillonite on arylsulfatase activity. Aust J Soil Res 16:35–40
Jastrow JD, Miller RM (1997) Soil aggregate stabilization and carbon sequestration: feedbacks through organo-mineral associations. CRC, Boca Raton, FL
Kaiser K, Guggenberger G (2000) The role of DOM sorption to mineral surfaces in the preservation of organic matter in soils. Org Geochem 31:711–725

Kennedy MJ, Pevear DR, Hill RJ (2002) Mineral surface control of organic carbon in black shale. Science 259:657–660
King GM, Klug MJ (1980) Sulfhydrolase activity in sediments of Wintergreen Lake, Kalamazoo-County, Michigan. Appl Environ Microbiol 39:950–956
King GM (1986) Characterization of beta-glucosidase activity in intertidal marine-sediments. Appl Environ Microbiol 51:373–380
Kunze C (1970) The effect of streptomycin and aromatic carbonic acid on the catalase activity in soil samples. Zentralbl Bakter Par 124:658–61
Kunze C (1971) Modulation of catalase activity in soil samples with tannic, gallic and para hydroxybenzoic acids. Oecolog Plantar 6:197–202
Kuzyakov Y, Friedel JK, Stahr K (2000) Review of mechanisms and quantification of priming effects. Soil Biol Biochem 32:1485–1498
Ladd JN, Butler JHA (1969) Inhibition and stimulation of proteolytic enzyme activities by soil humic acids. Aust J Soil Res 7:253–261
Ladd JN (1972) Properties of proteolytic enzymes extracted from soil. Soil Biol Biochem 4:227–237
Ladd JN, Butler JHA (1975) Humus-enzyme systems and synthetic, organic polymer-enzyme analogs. In: Paul EA, McLaren AD (eds) Soil biochemistry, vol 4. Marcel dekker, New York, pp 143–194
Leprince F, Quiquampoix H (1996) Extracellular enzyme activity in soil: effect of pH and ionic strength on the interaction with montmorillonite of two acid phosphatases secreted by the ectomycorrhizal fungus *Hebeloma cylindrosporum*. Eur J Soil Sci 47:511–522
Lozzi I, Calamai L, Fusi P, Bosetto M, Stotzky G (2001) Interaction of horseradish peroxidase with montmorillonite homoionic to Na^+ and Ca^{2+}: effects on enzymatic activity and microbial degradation. Soil Biol Biochem 33:1021–1028
Luthy RG, Aiken GR, Brusseau ML, Cunningham SD, Gschwend PM, Pignatello JJ, Reinhard M, Traina SJ, Weber WJ Jr, Westall JC (1997) Sequestration of hydrophobic organic contaminants by geosorbents. Environ Sci Technol 31:3341–3347
Makboul HE, Ottow JCG (1979a) Michaelis constant (K_m) of acid phosphatase as affected by montmorillonite, illite, and kaolinite clay-minerals. Microb Ecol 5:207–213
Makboul HE, Ottow JCG (1979b) Alkaline-phosphatase activity and Michaelis constant in the presence of different clay-minerals. Soil Sci 128:129–135
Marzadori C, Gessa C, Ciurli S (1998a) Kinetic properties and stability of potato acid phosphatase immobilized on Ca-polygalacturonate. Biol Fertil Soils 27:97–103
Marzadori C, Miletti S, Gessa C, Ciurli S (1998b) Immobilization of jack bean urease on hydroxyapatite: urease immobilization in alkaline soils. Soil Biol Biochem 30:1485–1490
Mayaudon J, Sarkar JM (1974) Chromatography and purification of diphenol oxidases of soil. Soil Biol Biochem 6:275–285
Mayaudon J, Sarkar JM (1975) Laccases of polyporus versicolor in soil and litter. Soil Biol Biochem 7:31–34
Mayer LM (1994) Relationships between mineral surfaces and organic carbon concentrations in soils and sediments. Chem Geol 114:347–363
Mayer LM (1999) Extent of coverage of mineral surfaces by organic matter in marine sediments. Geochim Cosmochim Acta 63:207–215
McLaren AD (1954) The adsorption and reactions of enzymes and proteins on kaolinite. J Phys Chem 58:129–137
McLaren AD, Estermann EF (1956) The adsorption and reactions of enzymes and proteins on kaolinite. 3. The isolation of enzyme-substrate complexes. Arch Biochem Biophys 61:158–173
McLaren AD, Estermann EF (1957) Influence of pH on the activity of chymotrypsin at a solid-liquid interface. Arch Biochem Biophys 68:157–160
McLaren AD, Packer L (1970) Some aspects of enzyme reactions in heterogeneous systems. Adv Enzymol Relat Areas Mol Biol 33:245–308

Meyer-Reil LA (1986) Measurement of hydrolytic activity and incorporation of dissolved organic substrates by microorganisms in marine-sediments. Mar Ecol Prog Ser 31:143–149
Morgan HW, Corke CT (1976) Adsorption, desorption, and activity of glucose oxidase on selected clay species. Can J Microbiol 22:684–693
Mosbach K (1976) Introduction. In: Mosbach K (ed) Immobilized enzymes, vol 44. Academic, New York, pp 3–7
Naidja A, Huang PM (1996) Deamination of aspartic acid by aspartase-Ca-montmorillonite complex. J Mol Catal A: Chem 106:255–265
Naidja A, Huang PM, Bollag JM (1997) Activity of tyrosinase immobilized on hydroxyaluminum-montmorillonite complexes. J Mol Catal A: Chem 115:305–316
Naidja A, Huang PM, Bollag JM (2000) Enzyme-clay interactions and their impact on transformations of natural and anthropogenic organic compounds in soil. J Environ Qual 29:677–691
Naidja A, Liu C, Huang PM (2002) Formation of protein-birnessite complex: XRD, FTIR, and AFM analysis. J Colloid Interface Sci 251:46–56
Nannipieri P, Muccini L, Ciardi C (1983) Microbial biomass and enzyme-activities - production and persistence. Soil Biol Biochem 15:679–685
Nannipieri P, Ceccanti B, Bianchi D (1988) Characterization of humus phosphatase complexes extracted from soil. Soil Biol Biochem 20:683–691
Nelson JM, Griffin EG (1916) Adsorption of invertase. J Am Chem Soc 38:1109–1115
Pflug W (1981) Inhibition of malate dehydrogenase by humic acids. Soil Biol Biochem 13:293–299
Pflug W (1982) Soil enzymes and clay-minerals. 2. Effect of clay-minerals on the activity of polysaccharide cleaving soil enzymes. Pflanz Bodelkunde Z 145:493–502
Pinck LA, Allison FE (1951) Resistance of a protein-montmorillonite complex to decomposition by soil microorganisms. Science 114:130–131
Pinck LA, Dyal RS, Allison FE (1954) Protein-montmorillonite complexes, their preparation and the effects of soil microorganisms on their decomposition. Soil Sci 78:109–118
Quiquampoix H (1987a) A stepwise approach to the understanding of extracellular enzyme-activity in soil.1. Effect of electrostatic interactions on the conformation of a beta-D-glucosidase adsorbed on different mineral surfaces. Biochimie 69:753–763
Quiquampoix H (1987b) A stepwise approach to the understanding of extracellular enzyme-activity in soil. 2. Competitive effects on the adsorption of a beta-D-glucosidase in mixed mineral or organo mineral systems. Biochimie 69:765–771
Quiquampoix H, Servagent-Noinville S, Baron M-H (2002) Enzyme adsorption on soil mineral surfaces and consequences for the catalytic activity. In: Burns RG, Dick RP (eds) Enzymes in the environment. Marcel Dekker, New York, pp 285–306
Quiquampoix H, Burns RG (2007) Interactions between proteins and soil mineral surfaces: environmental and health consequences. Elements 3:401–406
Rao MA, Gianfreda L, Palmiero F, Violante A (1996) Interactions of acid phosphatase with clays, organic molecules and organo-mineral complexes. Soil Sci 161:751–760
Rao MA, Violante A, Gianfreda L (2000) Interaction of acid phosphatase with clays, organic molecules and organo-mineral complexes: kinetics and stability. Soil Biol Biochem 32:1007–1014
Rego JV, Billen G, Fontigny A, Somville M (1985) Free and attached proteolytic activity in water environments. Mar Ecol Prog Ser 21:245–249
Renella G, Landi L, Valori F, Nannipieri P (2007) Microbial and hydrolase activity after release of low molecular weight organic compounds by a model root surface in a clayey and a sandy soil. Appl Soil Ecol 36:124–129
Rosas A, Mora MD, Jara AA, Lopez R, Rao MA, Gianfreda L (2008) Catalytic behaviour of acid phosphatase immobilized on natural supports in the presence of manganese or molybdenum. Geoderma 145:77–83
Ross DJ, McNeilly BA (1972) Some influences of different soils and clay minerals on the activity of glucose. Soil Biol Biochem 4:9–18

Rowell MJ, Ladd JN, Paul EA (1973) Enzymically active complexes of proteases and humic acid analogues. Soil Biol Biochem 5:699–703

Ruggiero P, Radogna VM (1988) Humic acids tyrosinase interactions as a model of soil humic enzyme complexes. Soil Biol Biochem 20:353–359

Sarkar JM, Burns RG (1983) Immobilization of beta-D-glucosidase and beta-D-glucosidase-polyphenolic complexes. Biotechnol Lett 5:619–624

Sarkar JM, Burns RG (1984) Synthesis and properties of beta-D-glucosidase phenolic copolymers as analogs of soil humic-Enzyme complexes. Soil Biol Biochem 16:619–625

Serban A, Nissenbaum A (1986) Humic-acid association with peroxidase and catalase. Soil Biol Biochem 18:41–44

Serefoglou E, Litina K, Gournis D, Kalogeris E, Tzialla AA, Pavlidis IV, Stamatis H, Maccallini E, Lubomska M, Rudolf P (2008) Smectite clays as solid supports for immobilization of beta-glucosidase: synthesis, characterization, and biochemical properties. Chem Mater 20:4106–4115

Servagent-Noinville S, Revault M, Quiquampoix H, Baron MH (2000) Conformational changes of bovine serum albumin induced by adsorption on different clay surfaces: FTIR analysis. J Colloid Interface Sci 221:273–283

Shindo H, Watanabe D, Onaga T, Urakawa M, Nakahara O, Huang QY (2002) Adsorption, activity, and kinetics of acid phosphatase as influenced by selected oxides and clay minerals. Soil Sci Plant Nutr 48:763–767

Skujins J, Pukite A, McLaren AD (1974) Adsorption and activity of chitinase on kaolinite. Soil Biol Biochem 6:179–182

Skujins J (1976) Extracellular enzymes in soils. CRC Crit Rev Microbiol 4:383–421

Skujins J (1978) History of abiontic soil enzyme research. In: Burns RG (ed) Soil enzymes. Academic, London, pp 1–49

Skujins JJ, Estermann EF, McLaren AD (1959) Proteolytic activity of *Bacillus subtilis* in a clay protein paste system analogous to soil. Can J Microbiol 5:631–634

Smith JR, Cicerone MT, Meuse CW (2009) Tertiary structure changes in albumin upon surface adsorption observed via fourier transform infrared spectroscopy. Langmuir 25:4571–4578

Sorensen LH (1969) Fixation of enzyme protein in soil by the clay mineral montmorillonite. Experientia 25:20–21

Srere PA, Ueda K (1976) Functional groups on enzymes suitable for binding matrices. In: Mosbach K (ed) Immobilized enzymes, vol 44. Academic, New York, pp 3–7

Staunton S, Quiquampoix H (1994) Adsorption and conformation of bovine serum-albumin on montmorillonite – modification of the balance between hydrophobic and electrostatic interactions by protein methylation and pH variation. J Colloid Interface Sci 166:89–94

Steen AD, Arnosti C, Ness L, Blough NV (2006) Electron paramagnetic resonance spectroscopy as a novel approach to measure macromolecule-surface interactions and activities of extracellular enzymes. Mar Chem 101:266–276

Tan WF, Koopal LK, Weng LP, van Riemsdijk WH, Norde W (2008) Humic acid protein complexation. Geochim Cosmochim Acta 72:2090–2099

Tietjen T, Wetzel RG (2003) Extracellular enzyme-clay mineral complexes: enzyme adsorption, alteration of enzyme activity, and protection from photodegradation. Aquat Ecol 37:331–339

Torn MS, Trumbore SE, Chadwick OA, Vitousek PM, Hendricks DM (1997) Mineral control of soil organic carbon storage and turnover. Nature 389:170–173

Van Bodegom PM, Broekman R, Van Dijk J, Bakker C, Aerts R (2005) Ferrous iron stimulates phenol oxidase activity and organic matter decomposition in waterlogged wetlands. Biogeochem 76:69–83

Vuorinen AH, Saharinen MH (1996) Effects of soil organic matter extracted from soil on acid phosphomonoesterase. Soil Biol Biochem 28:1477–1481

Zang X, van Heemst JDH, Dria KJ, Hatcher PG (2000) Encapsulation of protein in humic acid from a histosol as an explanation for the occurrence of organic nitrogen in soil and sediment. Org Geochem 31:679–695

Zang X, Nguyen RT, Harvey HR, Knicker H, Hatcher PG (2001) Preservation of proteinaceous material during the degradation of the green alga *Botryococcus braunii*: a solid-state 2D N-15 C-13 NMR spectroscopy study. Geochim Cosmochim Acta 65:3299–3305

Zhang HJ, Sheng XR, Pan XM, Zhou JM (1997) Activation of adenylate kinase by denaturants is due to the increasing conformational flexibility at its active sites. Biochem Biophys Res Commun 238:382–386

Ziervogel K, Karlsson E, Arnosti C (2007) Surface associations of enzymes and of organic matter: consequences for hydrolytic activity and organic matter remineralization in marine systems. Mar Chem 104:241–252

Zimmerman AR, Chorover J, Goyne KW, Brantley SL (2004a) Protection of mesopore-adsorbed organic matter from enzymatic degradation. Environ Sci Technol 38:4542–4548

Zimmerman AR, Goyne KW, Chorover J, Komarneni S, Brantley SL (2004b) Mineral mesopore effects on nitrogenous organic matter adsorption. Org Geochem 35:355–375

Zittle CA (1953) Adsorption studies of enzymes and other proteins. Adv Enzymol Rel S Bi 14:319–374

Chapter 16
The Influence of Pesticides on Soil Enzymes

Liliana Gianfreda and Maria A. Rao

16.1 Introduction

Pesticides are one of the major products developed by man in the last century. They have had a beneficial impact not only on agricultural productivity along with the reduction of costs but also on the quality of life and improvement of longevity. For instance, they have allowed various pests like insects, weeds, worms, and rodents to be controlled or even eliminated. Moreover, they have combated the diffusion of insects linked to the transmission of diseases such as malaria, yellow fever, and typhus with consequent manifest advantages of human health.

Occasionally, application of pesticides can be unsafe, as not the entire applied chemical reaches the target site. The reduction of weed control, the damage of other plants as well as the possible pollution of soil and water may occur. Unfortunately, pesticides belong to the group of xenobiotics, i.e., man-made organic chemicals that mimic organic chemicals important for the life, but with properties and features extraneous to the living organisms and therefore not recognized by them.

Many processes may happen to pesticides and affect their behavior in the environment. They may accumulate and persist into the environment with serious and severe threats to wildlife, animals, and humans. Among the environmental compartments, soil is undoubtedly the ecosystem, more exposed to pesticides and their possible negative effects. The biological activity of soil, comprising microbial and plant activity, can be seriously compromised with negative, very often irreversible effects, on the life not only of soil biota but also on the whole living population, including humans.

In the soil, enzyme activities play a predominant role in nutrient cycling and soil fertility. Indeed, microbial and plant enzymes, intra-, extra- or ecto-cellular, are

L. Gianfreda (✉) and M.A. Rao
Dipartimento di Scienze del Suolo della Pianta, dell'Ambiente e delle Produzioni Animali, Università di Napoli Federico II, Via Università 100, 80055 Portici, Napoli, Italy
e-mail: liliana.gianfreda@unina.it

G. Shukla and A. Varma (eds.), *Soil Enzymology*, Soil Biology 22,
DOI 10.1007/978-3-642-14225-3_16, © Springer-Verlag Berlin Heidelberg 2011

the main effectors of the biochemical and biological activity occurring at plant, rhizosphere, and bulk soil sites. To understand and elucidate the potential effects that pesticides may have on the activity and catalytic behavior of soil enzymes, a better comprehension of possible processes undertaken by a pesticide when entering the soil is needed.

This chapter is aimed at analyzing the effects of pesticides on the activity of enzymes involved in the biogeochemical cycles of the main nutrients. Some brief preliminary, general notes on what are pesticides and what can be their fate in soil will be also provided. The effects of pesticides at more general levels on enzyme-producing microbial and plant cells will be mentioned.

16.2 Pesticides and Soil

According to the Environmental Protection Agency (http://www.epa.gov/pesticides), pesticides are substances intended to repel, kill, or control any species designated a "pest." They are man-made organic chemicals, synthesized and used to control or even eliminate weeds in fields and grass, and unwanted or harmful pests, such as insects and mites that feed on crops, rodents, fungi, bacteria, or other organisms. Pesticides can be classified in many different ways: according to the target organisms, they are planned to control (Table 16.1) the chemical structure of the compound used (Gevao et al. 2000) or the degree or type of health hazard involved. Pesticides can be also classified, according to their potential toxicity to humans and other animals and organisms, as restricted use (can only be purchased and applied by certified persons who have had training in pesticide application) and general use (may be purchased and applied by any person).

Due to their indiscriminate and uncontrolled use in agricultural practices, dramatically increased in the last century, pesticides are also considered xenobiotic pollutants of great concern. In particular, the presence of pesticides in soil can adversely impact human and animal health, and beneficial plants and soil organisms (Madhun and Freed 1990). The capacity of the soil to filter, buffer, degrade,

Table 16.1 Categories of pesticides according to EPA (http://www.epa.gov/pesticides)

Category	Control type
Algicides	Used to control algae in swimming pools and water tanks
Bactericides	Used to control or destroy bacteria, typically in the home, schools, or on hospital equipment
Fungicides	Used to control or destroy fungi on food or grain crops
Herbicides	Designed to control or kill plants, weeds, or grasses. Almost 70% of all pesticides used by farmers and ranchers are herbicides. These chemicals have wide-ranging effects on non-target species
Insecticides	Compounds specifically used to kill or prevent the growth of insects
Miticides	Used to kill mites that feed on plants and animals
Nematicides	Used to kill nematodes (microscopic, worm-like organisms that feed on plant roots)
Rodenticides	Used to kill rats and other rodents or to prevent them from damaging food, crops, or forage

immobilize, and detoxify pesticides is a function or quality of the soil and the extent of processes undergone by the pesticide in the environment.

When a pesticide is released deliberately or accidentally (i.e., accidental spillage, industrial and domestic wastes) into the environment, it may interest all the environmental compartments, i.e., water, air, soil, sediments, and biomass (Cheng 1990). Many processes affect what happens to pesticides in the environment (Fig. 16.1). They are basically *transfer* processes (e.g., volatilization, spray drift, runoff in dissolved or sorbed state, leaching, absorption and crop removal, and mainly sorption and desorption to soil colloid surfaces), where the pesticide is moved away from the target site without any alteration of its chemical structure, and *degradation* or *breakdown* processes, where the chemical structure of the pesticide is profoundly altered or even completely mineralized to simple, nontoxic compounds (Bollag and Liu 1990). Therefore, pesticides in soil may be degraded rapidly or exist in widely varying concentrations at different soil microsites.

Among the processes depicted in Fig. 16.1, some can be beneficial and others can be harmful. For instance, if some herbicides are leached into the root zone, a better weed control can occur. By contrast, not all the applied pesticide may reach the target site, because it can be moved away by runoff. It can be partially wasted, with reduction of the target control and potential damage of other plants or pollution of soil and water. Similarly, some of the pesticide may drift downwind and outside of the intended application site.

Transfer processes are strictly dependent on several characteristics linked to the chemical nature and features of both the pesticide and the system (soil, plant or

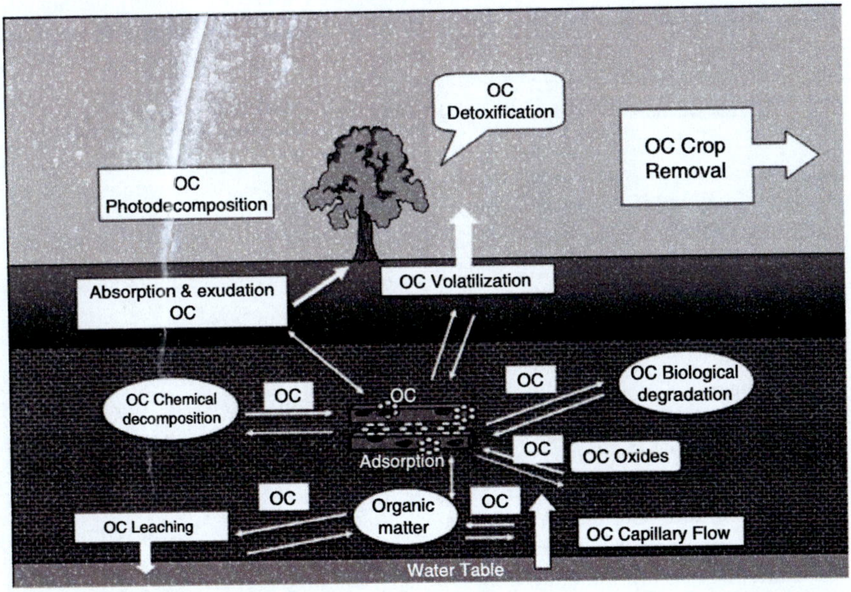

Fig. 16.1 Fate of an organic chemical (OC) in the environment

water) involved in the process. The pesticide may be associated to soil inorganic and organic colloids through adsorption or entrapping mechanisms. It can be absorbed by plant roots or volatilized, depending on the soil zone where it is located and its chemical and physical properties. When it accumulates on "soil surface," it can undergo photodecomposition, and the process will be strongly influenced by intrinsic and extrinsic properties of soil (e.g., abundance of organic matter, color, climatic, and geographical position) as well as the inherent chemical structure of the compound. Chemical decomposition of the chemical can also occur if soil contains some components (i.e., oxides) and properties (i.e., Redox state, pH) capable of sustaining such reaction. Part of the pesticide can be distributed in ground water by leaching processes.

Adsorption is one of the key processes affecting the ultimate fate of pesticides in soils (Bailey and White 1970; Adams 1973). It is the primary process of how the soil retains a pesticide and is defined as the accumulation of a pesticide on the soil particle surfaces. In particular, the interactions at the interface between organic and inorganic soil colloids and pesticides through adsorption–desorption mechanisms may influence movements of pesticides, and thus their availability for plant or microbial uptake, or their transformation by abiotic or biotic agents. The extent of adsorption is related to various soil properties, including organic matter content, type and content of clay, pH, exchange capacity, surface charge characteristics, permeability, and physical and chemical parameters of the pesticide such as water solubility and polarity (Bailey and White 1970; Adams 1973). For most pesticides, organic matter content is the most important soil property controlling the degree of adsorption.

As summarized by Gianfreda and Rao (2008 and references herein), degradation of organic chemicals may occur by direct or indirect mechanisms and lead to the partial or total degradation of the chemical. A complete mineralization of the pesticide may occur if the pesticide produces selective pressure for microorganisms able to degrade it or stimulate the production of extracellular enzymes capable of detoxifying the pesticide molecule. In this case, the pesticide could be degraded so rapidly that it no longer performs its planned function of pest control. Major enzymes involved in pesticide transformation and degradation are oxido-reductases, catalyzing a large array of oxidative and reductive reactions, hydrolases, catalyzing the break down of different chemical bonds, and synthetases, catalyzing synthetic reactions.

However, to be a pesticide efficiently degraded in soil, several conditions must simultaneously occur. Soil must contain organisms capable not only to metabolizing and degrading the pesticide but also to surviving into the soil; the chemical must have properties to be suitable for degradation, i.e., low persistence, relatively high solubility, relatively high bioavailability; the degrading organism or the extracellular enzyme and the chemical must be located sufficiently close to each other to allow their interaction; the environmental conditions in terms of pH, temperature, and soil properties must allow the proliferation of the degrading organisms and/or the catalytic action of the degrading enzymes. When some of these conditions are not met, the pesticide will not be degraded. It may accumulate and persist in the soil for long periods of time with possible negative influence on soil biological activity and therefore also on soil enzyme performance.

Without going in detail two main properties will strongly affect the fate of a pesticide in soil and in turn its effects: bioavailability and persistence. They are mutually related to each other and depend not only on the intrinsic and inherent properties of the pesticide molecule but also on the properties of soil. As clearly underlined by Ruggiero et al. (2002), two concepts of bioavailability can be defined as "mass bioavailability," i.e., "an actual decrease in the contaminant mass over time as a consequence of the mass transfer to biologically accessible regions," and a "toxicological bioavailability," i.e., when "the bioavailable fraction is regarded as the carrier of toxicity" and "the organism exposure to the contaminant leads to toxic response" (Ruggiero et al. 2002). Bioavailability is a consequence of the persistence or the "lasting power" of the pesticide, usually measured in "half-life," e.g., the amount of time the pesticide takes for half of the pesticide to be converted into something else, or its concentration is half of its initial level. The half-life of a pesticide depends on soil type, its formulation, and environmental conditions (e.g., temperature, moisture). For instance, some pesticides like chlordane, paraquat, or bromacil may have a half-life greater than 100 days being thus classified as persistent. For some others, this value is lower than 30 days and the pesticides are non-persistent. Moreover, some of their metabolites produced by microbial transformation may have a different behavior from their parent compound and behave as persistent.

Persistence and bioavailability are, however, also dependent on possible ageing processes undergone by the chemical molecule and affecting the amount of the chemical really available for interacting with both enzymes and microorganisms. Ageing processes usually lead to the adsorption and/or sequestration of organic compounds into the soil matrix. The chemical can be sequestered from diffusion into sites within sorbing matrices or entry in nanopores, where they are not easily accessible to microbial cells (Reid et al. 2000; Nam and Alexander 2001). Chemical bonds with soil components or physical entrapment of the compound in the soil organic matter or the soil mineral lattice may occur. In this case, the pesticide can be indicated as "bound residue" (Gevao et al. 2000; Barraclough et al. 2005). This term applies to those compounds not easily extractable from soil without altering their chemical structure. As a consequence, the organic compound become less, or even not bioavailable. With time, unpredictable release of the chemical may, however, occur and lead to undesired effects on soil biological activity.

If a partial degradation of the pesticide takes place with formation of intermediate, possibly toxic metabolites, these latter may be involved in similar phenomena and all the same affect the soil biological activity.

16.3 Soil Enzymes

The biological activity of soil is guaranteed by the coordinated and integrated activity of a multitude of enzymes acting within and outside the living cells inhabiting the soil. They are involved in energy transfer, nutrient cycling, environmental quality, and crop productivity (Dick 1994, 1997), and therefore, their activity is related to soil fertility and sustainability (Nannipieri 1994).

The determination of the quality-related properties of soil (which are sensitive to changes caused by management practices and environmental stress) may help to monitor the changes in its sustainability and environmental quality. This is especially true for the agricultural management involving, for instance, application of pesticides and to assist into the establishment of policies for an appropriate use of these compounds. Soil enzymes activities have been suggested as suitable indicators of soil quality, because they can be easily assayed and nowadays several drawbacks in the determination of soil enzyme activities are going to be overcome (Gianfreda and Ruggiero 2006). Soil enzymes are strictly related to the nutrient cycles and transformations being a measure of the soil microbial activity, and they may rapidly respond to the changes caused by anthropogenic factors such as the presence of xenobiotic substances (Gianfreda and Bollag 1996; Gianfreda and Bollag 2002; Nannipieri et al. 2002; Gianfreda et al. 2005). Moreover, soil enzyme activities may be considered early and sensitive indicators to measure the degree of soil degradation in both natural and agro-ecosystems, being thus well suited to measure the impact of pollution on the quality of soil (Dick and Tabatabai 1993; Dick 1997; van Beelen and Doelman 1997; Trasar-Cepeda et al. 2000).

Enzymes belonging to different classes have been found in soil. However, not all those usually present in a living organism have been assayed in soil samples. Different intracellular and extracellular enzymatic components contribute to the overall enzymatic activity of soils. They are from microbial, animal, or plant origin, may exist in different states (Fig. 16.2), may present different features and

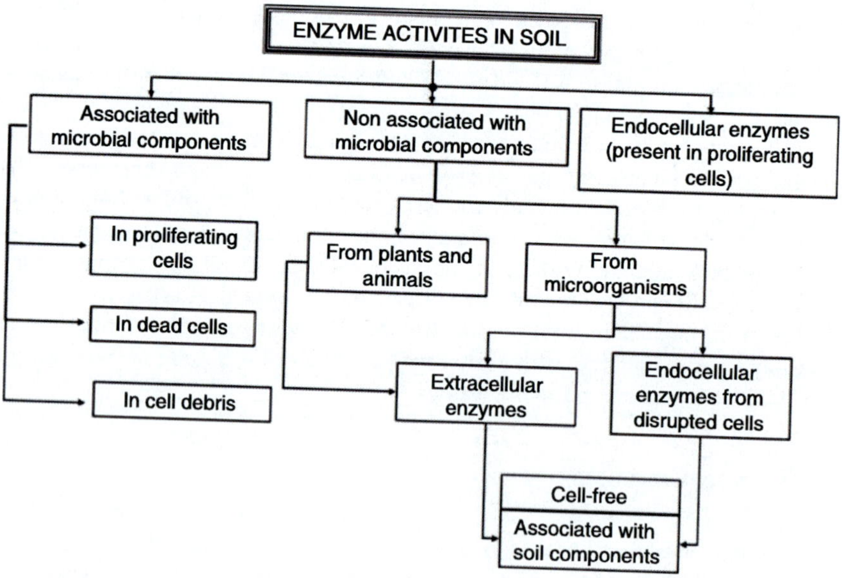

Fig. 16.2 Categories of enzyme activities in soil (adapted from Gianfreda and Bollag 1996)

properties, and act under a range of microenvironmental conditions that affect their catalytic behavior (Burns 1982; Gianfreda and Bollag 1996; Nannipieri and Gianfreda 1998; Gianfreda et al. 2002; Nannipieri et al. 2002; Gianfreda and Ruggiero 2006). Due to their complexity, it is still not easy to define their exact originating organisms, their spatial location, or their temporal relationship with them.

Among the distinct categories depicted in Fig. 16.2 for classifying enzymes in soil, enzymes associated with clay minerals and/or humic colloids as a result of adsorption/entrapment are defined "soil-bound enzymes" or "naturally immobilized enzymes" (Gianfreda and Bollag 1996). In soil, many enzymatic molecules are associated with the solid phase of the soil and behave similarly to the enzymes of the living cell operating within the framework of a highly organized structure. Therefore, soil enzymes show a typical behavior strictly dependent on the particular environment of such soil, where they act as catalysts. The finding that soil can show a catalytic activity attributable to enzyme action even in the absence of active living organisms has led soil enzymologists to assume that a consistent fraction of enzymes present in soil is constituted by soil-bound enzymes or naturally immobilized enzymes (Gianfreda and Bollag 1996).

When compared with enzymes from different sources, soil enzymes commonly show particular and peculiar features. In particular, they are very often more resistant to deactivating agents such as temperature, presence of proteases, irradiation, or prolonged storage. In most cases, no significant or reduced loss of activity occurs after exposure to most of those agents (Gianfreda and Bollag 1996; Gianfreda et al. 2002; Gianfreda and Ruggiero 2006). Moreover, soil enzymes typically have a kinetic behavior different from that exhibited by the same enzyme in a purified form; lower and higher values of the kinetic parameters V_{max} and K_m are usually measured, respectively, indicating that they have a lower catalytic efficiency and a reduced substrate affinity. Similarly, different activity–pH or activity–temperature profiles are found for soil enzymes (Nannipieri and Gianfreda 1998 and references therein). These peculiar behaviors can, at least in part, be explained by the presence of enzymes immobilized on soil supports and, consequently, working in a heterogeneous system. However, the exact mechanism by which enzymes are immobilized and consequently stabilized in soils still remains not completely clear.

As approached by several authors and revised recently by Gianfreda et al. (2010 and references therein), soil-bound enzymes may play an important ecological role being capable of immediately responding and transforming available substrates, and render them easily up taken by cells without these latter waiting for induction or synthesis of competent enzymatic molecules. Moreover, they may mediate the cleave-off of large substrates with production of intermediate metabolites easily usable by cells even as possible inducers. The activity of extracellular enzymes, as free and mainly as soil-bound enzymes, can also have a pivotal importance not only in the carbon cycle, i.e., in the decomposition and/or storage of carbonaceous substrates, and, in turn, on their potential effects on global climate, but also in the cycle of P- and N-substrates being P and N essential nutrients for the growth and enzyme synthesis.

Soil enzyme activities may be involved in the transformation and degradation of pollutants, including pesticides, thereby contributing to the restoration and recovery of polluted soils (Nannipieri and Bollag 1991; Sutherland et al. 2002; Gianfreda and Rao 2004). However, the reduced catalytic activity, often displayed by the soil-bound enzyme category, can hinder an effective restoration of polluted soils. Conversely, the increased stability of this soil enzyme fraction and the possible involvement of soil components in the pollutant transformation may improve and stabilize the active pollutant degradative capability of soil.

16.4 Pesticides and Soil Enzymes

Pesticides and in general xenobiotic substances may present direct, either reversible or irreversible, and indirect effects on soil enzyme activities. Although pesticide molecules are not deliberately synthesized to inhibit enzymes, a direct reversible inhibition of soil enzyme activities may occur as the result of reversible interactions of the pesticide with intra- and/or extracellular enzymes. Possible competitive and/or noncompetitive substrate inhibition or alteration of the protein conformation may result. If the pesticide molecule is degraded in intermediate metabolites by biotic or abiotic transformation, a similar effect could be shown by its degradation products. When a covalent binding occurs with catalytic groups involved in the enzyme function, a direct irreversible effect is the supposed consequence.

Indirect effects are, instead, the consequence of the influence that pesticides may have on soil microbial growth and activity. Pesticides may induce detectable changes in size, structure, and functionality of the microbial community, thereby altering life functions, dynamics, and biodiversity of soil organisms.

As previously reviewed by Cervelli et al. (1978) and later summarized by Gianfreda and Rao (2008), several studies have been performed on the impact of pesticides on soil microbial growth and activity (Johnsen et al. 2001). Indirectly, these studies have provided information on the effect that a pesticide may have on soil enzymes that are the ultimate expression and actors of microbial activity. In most of these studies, the response of dehydrogenase activity in soils applied with pesticides was investigated (Xu et al. 2000), being this enzyme a measurement of the active biomass in soil and related to the metabolic state of soil microbial population. A consequent influence on other enzymatic soil activities was also evidenced when the effect of the herbicide (brominal) and the insecticide (selecron) on the microbial soil population was studied (Omar and Abdel-Sater 2001). For instance, a corresponding decrease of cellulase activity was measured along with the reduction of the number of cellylolytic fungi. Similarly to what observed with bacterial and actinomycetes population, the activity of phosphatase was promoted by both the pesticides at field application rates and inhibited at higher levels (Omar and Abdel-Sater 2001). A different behavior was observed with alkaline phosphatase that was promoted even at high application rates.

The main conclusions drawn from these studies on the influence of pesticides on soil microbial activity were summarized by Gianfreda and Rao (2008). The life functions of soil organisms may be altered by pesticides at different levels. In particular, pesticides may modify and regulate the genesis and synthesis of proteins and enzymes by repression or induction. The production of new intra- and extracellular enzymes as well as changes in the ratio between intra- and extracellular enzyme amounts will result. Pesticides may also affect some physiological functions like membrane's functionality or influence the dynamics of the soil populations and consequently the soil biodiversity. Indeed, they may induce the death of responsive organisms with the consequent formation of organic products usable by the surviving organisms. Pesticides may be directly degraded or metabolized by competent organisms, i.e., they may behave as energy and C source for microbial growth, or be co-metabolized, produce intermediate metabolites, and promote the growth of microbes depending on secondary nutrient sources (Gianfreda and Rao 2008). In all cases, an influence on the activity of enzymes involved in the processes will occur.

Since the activity displayed by an enzyme in soil is the result of complex processes of synthesis, persistence, stabilization, regulation, and catalytic behavior of the involved enzymatic protein, soil enzyme response to an applied disturbance such as the presence of a pesticide will be a complex event, also function by the changes of physical, chemical, and biological soil composition. Therefore, a truthful comprehension of such events requires methods capable not only of accurately measuring the activity of the chosen enzyme but also of detecting its changes following the application of the disturbing factor. As underlined by Gianfreda and Ruggiero (2006), it is not easy to decide which and how many enzymes are important to assess the impact of a given alteration on soil quality. In other words, it can be very difficult to decide which enzymes should be measured to that purpose. A suitable set of more than one enzyme should be preferentially assayed to have a more detailed picture of the occurred disturbance.

The most exhaustive, timely review of effects of pesticides on soil enzyme activities was provided by Schaffer (1993). The author evaluated almost all the findings so far published and tried to summarize the responses of a large number of enzymes (17), belonging to different classes and to different pesticides in soils with different properties. Mainly oxido-reductases and hydrolases and more than 100 pesticides from different chemical categories were considered. The main conclusion drawn by Schaffer (1993) was that the response of a soil enzyme to a given pesticide is unpredictable, because a pesticide that behaves as inhibitor of an enzyme in a soil may be activator or has no effect on the same enzyme in a different soil. Moreover, the concentration of the pesticide found toxic for an enzyme can be not influent or even beneficial for another enzyme. The contrasting and contradictory results often found by Schaffer (1993) are well exemplified by data shown in Table 16.2, where the responses of some enzymes to the presence of different pesticides are summarized. Furthermore, very often results from laboratory studies are not comparable to those obtained from field experiments, and general conclusions cannot be easily drawn.

Table 16.2 Variable results obtained with pesticides on soil enzyme activities

Enzyme	Linuron	Atrazine	Simazine	Oxamyl	Diazinon	Fenamiphos	2,4-D	Tefluthrin
Dehydrogenase	+/o	–/o/+	–/+,o	o/o,+	o/o,+		+/o,+	+/–,+
Phosphatase	–/o	+,o/–	+,o/–	o,–	o/–	o	o/–	+/–
Urease	–	o	o	o/+	+	–,o/o		o/+
Sulfatase			–/o	–,o/o		–	+	
Invertase		–,+	–,+	o	+,o	o	o	–,o/o

+ (stimulated); – (reduced); o (unchanged)
Symbols separated by a comma (,) indicate changes during the observation period
Symbols separated by a slash (/) refer to differences in experimental conditions
Adapted from Schaffer (1993)

Further studies on the effects of old and new pesticides on soil enzymes often confirmed what earlier observed by Schaffer (1993). Investigations have also been performed to evaluate the impact on soil enzyme activity by pesticide residues, the repeated application of the same pesticide, the possible interactive effects of more than one pesticide applied in combination, or of the simultaneous presence of organic amendments or inorganic pollutants. Most of these studies were performed under field conditions, though much information was also achieved by laboratory experiments.

The application of botanical insecticides, like pyrethrins and Neemix-4E, brought about only transitional effects on urease activity, thus prolonging the duration of N availability to plants (Antonius 2003). By contrast, experiments performed with dimehypo (He et al. 2003), and chlorimuron-ethyl and furadan (Yang et al. 2006) on the activity and kinetic parameters of urease in different soils (meadow brown and black soils from different ecological zones) indicated that significant correlations occurred between the pesticide concentrations and the urease activity. Dimehypo significantly inhibited the soil urease activity, and urease kinetic parameters were decreased with increasing dimehypo concentrations. By contrast, chlorimuron-ethyl and furadan activated the urease activity of soils within a large range of concentrations (Yang et al. 2006). Moreover, a quadratic polynomial equation was found that better described the relationships between the concentrations of chlorimuron-ethyl and furadan and the activity of soil urease ($P < 0.05$). These results suggest to a certain extent that the urease activity can be used as an index to characterize the soil polluted by dimehypo or chlorimuron-ethyl and furadan.

The sensitivity of urease to pesticide impact was also confirmed by studies performed with increasing concentrations of chlorpyrifos (commercial formulation) (Rani et al. 2008). When the pesticide was applied in loamy sand soil under laboratory conditions, the activities of urease and dehydrogenase were more responsive to the pesticide present, while cellulase, amylase, and protease showed recovery tendency around 21st day of incubation in chlorpyrifos-treated soil (Rani et al. 2008). No significant influence on urease activity in four soils from Australia and Ecuador was shown by fenamiphos, a widely used organophosphorus pesticide, up to 100 mg kg^{-1} soil (Cáceres et al. 2009).

While a decrease of phosphatase activity was observed with chlorpyrifos for the entire period of study (Rani et al. 2008), previous studies demonstrated that the activity of the enzyme of soil samples receiving 5.0 kg ha^{-1} of the pesticide increased up to 20 days of incubation and decreased progressively with increasing the incubation period (Madhuri and Rangaswamy 2002). The same response was observed with the same amounts of methyl parathion and with 2.5 kg ha^{-1} of the insecticides dichlorovos, phorate, and methomyl (Madhuri and Rangaswamy 2002). The repeated application for two years of chloropyrifos on the activity of phosphomonoesterase was investigated in groundnut (*Arachis hypogaea* L.) field as seed treatment at 25 ml kg^{-1} and soil treatment at 4 L ha^{-1} (Pandey and Singh 2006). Similar studies were performed also with quinalphos. The residues of these insecticides were monitored during the entire crop season and findings showed that dissipation of quinalphos in comparison to chlorpyrifos was slow in both seed-treated and soil-treated field. The activity of the enzyme was significantly inhibited by up to 25.2% as compared to the control, and the effects were transitory. Differently, the phosphomonoesterase activity was inhibited by quinalphos till the end of the period in the soil-treated field, whereas it recovered within 30–60 days of treatment in the seed-treated field (Pandey and Singh 2006).

A strong negative influence on phosphatase activity and soil respiration in upland soil was also detected with a new pesticide, acetamipirid, applied at normal and high field concentration (0.5, 5, and 50 mg kg^{-1} dried soil) (Yao et al. 2006). The other tested enzymes dehydrogenase, urease, catalase, and proteinase differently responded to the acetamipirid application. Dehydrogenase was activated after acetamiprid application for 2 weeks, while variance of urease and catalase had no distinct relationship with the applied concentration, and proteinase was inhibited only from the fourth week after acetamiprid application. Overall, the pesticide at normal field dose did not pose a toxicological threat to soil enzymes, but only a certain potential threat to soil respiration.

By contrast, dehydrogenase activity in Mollic Gleysol, Eutric Fluvisol, and Eutric Histosol soil samples was inhibited by 5–21% up to 17–44% with increase of fonofos doses from 1 μg g^{-1} to ten times higher doses but only in the first stage after application (Stepniewska et al. 2007). Substantial decreases of the dehydrogenase activity were also observed by Xie et al. (2004) when increasing amounts of triazophos, bensulforon-methyl, and chlorobenthiazone were supplied to a paddy soil under controlled moisture and temperature conditions. The relative toxicity of the three chemicals was in the order triazophos > chlorobenthiazone > bensulforon-methyl irrespective of the rates of application (Xie et al. 2004). Differently, dehydrogenase was not significantly affected by fenamiphos in the Australia and Ecuador soils even up to 100 mg kg^{-1} soil (Cáceres et al. 2009). However, potential nitrification was found to be highly sensitive to fenamiphos with a significant inhibition recorded even at 10 mg kg^{-1} soil, thus suggesting that fenamiphos is likely to be detrimental to nitrification at field application rates.

The simultaneous influence of the pesticide doses and the application time was investigated with validamycin, a non-systemic fungicide poorly studied for its effects on soil enzymatic activities (Qian et al. 2007). Promotion and/or inhibition were detected with catalase, urease, or acid phosphatase with high valymadicin doses, but the effects were transitory.

Recently, an integrated methodological approach has been used by Niemi et al. (2009). The authors have investigated the impacts of the herbicides metribuzin and linuron and the fungicide fluazinam on ten different soil enzyme activities (arylsulphatase, phosphomonoesterase, phosphodiesterase, leucine-aminopeptidase, alanine-aminopeptidase, chitinase, ß-D-xylosidase, cellobiosidase, ß- and α-D-glucosidase) and soil ATP content in microcosm, mesocosm, and field experiments in potato cultivation. In the mesocosm tests, the separate addition of each pesticide and the simultaneous use of all the pesticides were investigated. Their hypothesis was that micro- and mesocosm experiments can differentiate direct impacts of pesticides on microbiota and indirect impacts due to plant growth and that the sensitivity of different microbial processes varies and depends on the pesticide and exposure time. Increases of several enzyme activities were observed in microcosms supplied with metribuzin and linuron. Soil toxicity testing with luminescent bacteria indicated bioavailability of fluazinam and severe toxic effects throughout the experiments. Conversely, in the mesocosm, some enzyme activities decreased only with combined use of pesticides, being less impacted by their separate use. Decreases of enzyme activities were also monitored in the field experiment. In conclusion, herbicides increased, though temporarily, some enzyme activities in unplanted soil, i.e., without rhizosphere. The higher activities of some enzymes in mesocosms and the field measured in the control soil as compared with herbicide-treated soil was justified by the presence of the denser plant cover. There was no serious influence of the herbicides on soil biodegradation or fertility. Therefore, the changes in enzyme activities caused by the three pesticides were probably the result of changes in the microbial composition (Niemi et al. 2009).

Modern agriculture practices utilize, very often, application of different groups of pesticides, at the same time or in succession for effective control of a variety of pests. The accumulation of pesticide residues in soil with injurious effects on the environment may result.

When diazinon, imidacloprid, and lindane were applied for three consecutive years (1997–1999) in groundnut (*Arachis hypogaea* L.) field, differentiate responses of dehydrogenase and alkaline phosphomonoesterase enzyme activities were observed (Singh and Singh 2005). Both activities significantly decreased, increased, and did not change after lindane, imidacloprid, and diazinon seed treatment, respectively (Singh and Singh 2005). A detectable inhibition in dehydrogenase activity was observed in 2-year seed and soil treatments with chlorpyrifos (Pandey and Singh 2006). A 17% reduction was measured after 60 days of seed treatment in comparison to control, whereas the inhibitory effect increased up to 63% after 15 days of quinalphos seed treatment. In the second year of treatment and also after soil treatment, similar trends were observed (Pandey and Singh 2006).

Detectable reductions of several enzymatic activities (acid phosphatase, alkaline phosphatase, urease, catalase, and invertase) as well as bacterial, fungal, and actinomycete populations were measured only after the first and second applications of chlorothalonil (Yun et al. 2006). In particular, the most marked inhibition occurred after the second pesticide application. However, after initial variations, all the measured microbial and enzymatic parameters adapted gradually to the presence of the pesticides, and the negative effects became transient and weaker following the third and fourth treatment (Yun et al. 2006). Interestingly, 21 days after the fourth treatment with chlorothalonil, three bacterial strains, capable of utilizing chlorothalonil as a sole carbon and energy source for growth, were isolated. Therefore, soil microorganisms adapted to the pesticide and capable of degrading it developed during the experiment (Yun et al. 2006).

Complex phenomena inhibiting the degradation of xenobiotics may arise when more polluting compounds were simultaneously present. The repeated application of chlorpyrifos, fenamiphos, and chlorothalonil and their combination suppressed in many cases their own rates of degradation (Singh et al. 2002a, b). The dynamics of residues of major metabolites of the three pesticides were also influenced by the pesticide combinations (Singh et al. 2002a). Enzyme activities and total microbial biomass were all adversely affected by chlorothalonil, but very small or insignificant effects were observed with chlorpyrifos and fenamiphos (Singh et al. 2002b).

When tested for 30 days under laboratory conditions, combinations of monocrotophos or quinalphos with cypermethrin yielded synergistic, antagonistic, and additive interaction effects on cellulase and amylase in two agricultural soils: black vertisol soil and red alfinsol soil. In contrast, the activities of the two enzymes were increased by individual application of the three insecticides at 5, 10, and 25 mg g^{-1} soil. The interaction responses were persistent even for 30 days and relationships with the populations of cellulolytic and amylolytic organisms in the two soils were found (Gundi et al. 2007).

A long-term experiment, known as the Chemical Reference Plots, started in 1974 on a silty clay loam soil at Rothamsted. Up to five pesticides (aldicarb, benomyl, chlorfenvinphos, glyphosate, and chlorotoluron or triadimefon) were applied to plots, each receiving the same treatment annually for up to 20 years. Indicator of soil fertility was considered the yield of spring barley, grown each year (Bromilow et al. 1996). Crop productivity was not negatively affected by these pesticide applications, and no differences were found in microbial processes in soils sampled in April 1992. Moreover, in August 1994, 17 months after the last experimental treatment, no pesticide residues were detected in soil samples (Bromilow et al. 1996).

By contrast, residues of organophosphorus and organochlorine pesticides (chlorpyrifos, ethion, α endosulfan, β endosulfan, and endosulfan sulfate) were found in six tea garden soils and two adjacent forest soils (control) in West Bengal, India, and had a strong impact on some soil microbial and biochemical components (MBC, BSR, SIR, FDAH, and β-glucosidase activity) (Bishnu et al. 2008).

Attention was also devoted to the possible joint effects of pesticides and heavy metals or organic amendments on soil enzymatic activities. For instance, urease activity of four soils (meadow burozem and phaeozem), exposed to various concentrations of chlorimuron-ethyl and furadan and mercury (Hg) individually and simultaneously, was activated by either chlorimuron-ethyl (14–18%) or furadan (up to 13–21%) but markedly inhibited by Hg (Yang et al. 2007). The combined effect of Hg and chlorimuron-ethyl was synergistic and depended on the investigated soil.

Conversely, the interactive effect between Hg and furadan was synergistic or antagonist depending on the Hg+furadan concentrations.

The potential ecological risk of the combined effect of copper and pyrethroids on the soil ecosystem was evaluated by monitoring soil catalase in soil supplied with five concentrations of each pollutant alone or in combination (Liu et al. 2008). According to literature, the inhibition of catalase by single Cu addition was explained by the possible interaction of the metal with the substrate alone, the enzymatic active site, or the enzyme–substrate complex. Cypermethrin and Cu in combination affected catalase activity stronger and weaker than single cypermethrin or Cu, respectively. This effect was attributed to the possible interaction between the two pollutants or to their bioavailability (Liu et al. 2008).

Perucci et al. (1999, 2000) examined, under laboratory conditions, the combined effects of rimsulfuron, a sulfonylurea herbicide, or imazethapyr, an imidazolinone herbicide, with a vermi-compost, from sewage sludge on several enzymatic activities (dehydrogenase, global hydrolytic capacity, catalase, nitrogenase, acid, and alkaline phosphatase) as well as on soil respiration, microbial biomass-C and -N, and ATP contents of two soils (a silty clay loam and a Vertic Aquic Ustorthent), at varying conditions of temperature and humidity. Slight and transitory increases of some properties were monitored at the highest applied pesticide rates (tenfold field rate), whereas others such as microbial biomass-C content, ATP contents, dehydrogenase, acid, and alkaline phosphatase activities decreased. These detrimental effects seemed to be enhanced by organic amendments. Moreover, the author proposed a new synthetic index, the specific hydrolytic activity (qFDA), for assessing microbial activity in reply to xenobiotic treatments.

The beneficial effects shown by vermi-composting on the activities of soil dehydrogenase, phosphatase, and urease (increases up to 128, 30, and 31%, respectively, as respect to the control soil) were annulled in the presence of propiconazole, profenos, or pretilachlor, and decreases of enzyme activities were measured. Different inhibition at different application rates was measured for each pesticide (Kalam et al. 2004). By contrast, beneficial effect of urban compost in ameliorating the toxic effects of metalaxyl and pendimethalin on phosphatase activity in the rhizosphere of wheat was demonstrated by Setty and Magu (1996).

The contrasting and contradictory results very often obtained in studies so far commented might be the results of the rather complicated interplay of phenomena that may occur when enzymes, soil colloids, and pesticides are simultaneously present in soil, thereby leading to a very complex sequence of pathways and products being formed (Fig. 16.3). As reported above, enzymes in soil exist as free or immobilized forms, and pesticides may interact with soil colloids giving rise to stable soil–colloid–pesticide complexes (Fig. 16.3).

In order to take into account these processes and overcome the confusing effects due to the various categories of enzymes in soil, Gianfreda and co-workers (Gianfreda et al. 1993, 1994, 1995, 2002; Sannino and Gianfreda 2001) studied pesticide effects with synthetic, enzymatic model systems, which closely resemble those present in soil, and whole soils. Three herbicides (atrazine, paraquat, and glyphosate), one insecticide (carbaryl), and three enzymes (invertase, urease and phosphatase) were

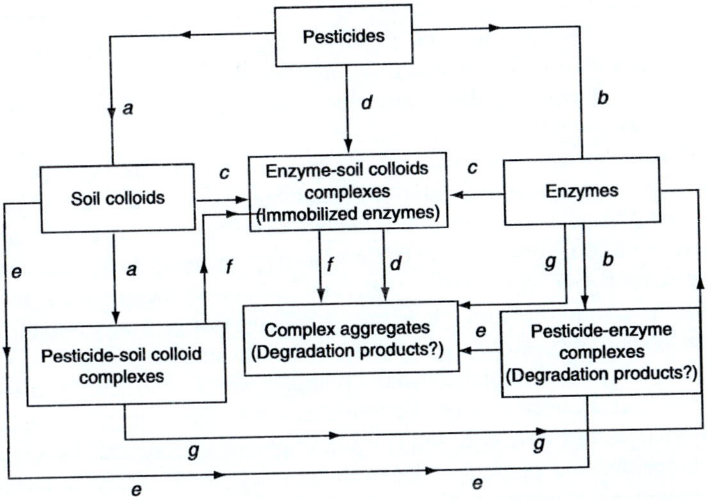

Fig. 16.3 Interactions occurring between pesticides, enzymes, and soil colloids: *a*: interaction between pesticides and soil colloids with formation of pesticide–soil colloid complexes; *b*: interaction between pesticides and enzymes with formation of pesticide–enzyme complexes (and degradation products); *c*: interaction between enzymes and soil colloids with formation of enzyme–soil colloid complexes; *d*: interaction between pesticides and immobilized enzymes with formation of complex aggregates (and degradation products); *e*: interaction between soil colloids and pesticide–enzyme complexes with formation of complex aggregates (and degradation products); *f*: interaction between pesticides–soil colloid complexes and immobilized enzymes with formation of complex aggregates (and degradation products); *g*: interaction between pesticides–soil colloid complexes and enzymes with formation of complex aggregates (and degradation products)

used. The responses of three enzymatic states were investigated: (1) free enzymes, which should simulate the fractions of enzymes free in soil solution, if any; (2) synthetic clay–, organo–, and organo-clay–enzyme complexes, which should simulate enzyme–soil colloid associations; and (3) whole soils, which represent natural systems. The synthetic enzymatic complexes used were enzyme–montmorillonite, enzyme–tannic acid, and enzyme–$Al(OH)_x$–tannic acid–montmorillonite complexes, previously characterized for their catalytic properties (Gianfreda et al. 2002 and references therein).

Activation, inhibition, or no influence were observed, thus suggesting that the responses were enzyme and pesticide specific, depending on both the "state" of the enzyme and the type of pesticide. Therefore, no generalizations could be made. For instance, the activity of free invertase was markedly activated by glyphosate and paraquat, whereas the performance of urease was not influenced by both pesticides (Gianfreda et al. 1993). An activation effect by the two pesticides and by only paraquat was also detected on invertase and urease immobilized on montmorillonite, respectively. By contrast, both pesticides did not affect or caused their decrease the activity of invertase and urease immobilized on tannic acid (Gianfreda et al. 1994, 1995), while glyphosate strongly inhibited the activity of all phosphatase complexes (Sannino and Gianfreda 2001). A general inhibition of enzyme activity was measured with methanol, used as solvent for atrazine and carbaryl solubilization, partially removed by the presence of the two pesticides. This partial recovery of activity varied with both the involved enzyme and its state (Gianfreda et al. 2002 and references therein).

Contrasting results (increases, decreases, and no effects) and no univocal responses with the four pesticides were obtained for the activity of invertase, urease, and phosphatase in 22 soils sampled in different sites of Italy and characterized by different physical–chemical properties (Sannino and Gianfreda 2001). Increases of soil invertase activity from 4 to 204% were measured by glyphosate and paraquat addition, whereas phosphatase activity was generally inhibited (up to 98%) by glyphosate. As for the synthetic systems, atrazine and carbaryl effects were appreciably affected by methanol. General inhibition and activation effects of atrazine on soil invertase and urease activities, respectively, were recognized. No reliable relationships were obtained when multiple regression analysis was applied to soil properties and pesticide effects. By contrast, some satisfying conclusions were reached by comparing the results obtained with the synthetic enzymatic systems and whole soils. For examples, the response of the enzyme activity of some soils to the presence of the four pesticides was very similar to that obtained with one of the model systems, thus suggesting the prevalence of that enzymatic fraction in the soils (Sannino and Gianfreda 2001).

The heterogeneity of results obtained with the synthetic enzymatic systems and even more with whole soils confirm the complexity of the systems and processes occurring between enzymes and pesticides in soil (Fig. 16.3). When free enzymes interact with pesticides, the system is homogeneous, and direct interactions at the molecular level may occur. Conversely, when enzymes, soil colloids, and pesticides are considered, the system is heterogeneous and several interactions may be generated: direct interactions of the pesticide with the enzyme molecules, which could have varied their catalytic feature if immobilized on soil colloids, but also indirect effects, deriving from the interactions between pesticide and inorganic and organic supports. Competition phenomena between immobilized enzyme and pesticide molecules could occur and result in a possible release of free enzymatic molecules from matrices. For instance, the marked increased activity of urease–montmorillonite complex measured with paraquat (Gianfreda et al. 1994) was possibly due to paraquat adsorption on external and internal montmorillonite surfaces with consequent displacement of immobilized urease molecules from specific sites and partial recover of their activity, lost during the immobilization process (Gianfreda et al. 1994).

16.5 Conclusions

Studies of pesticide influence on soil enzyme activities have provided often contradictory and contrasting not easily explainable results. Indeed, the complexity of soil enzyme categories as well as the complex processes undergone by pesticides in soil have contributed to enhance this difficulty. Direct, unequivocal evidence of the interactions between pesticides and all the possible forms of enzymes existing in a soil are still lacking and is a challenge for future research. Furthermore, other decisive factors (e.g., the concentration of the chemical, its persistence and bioavailability, its toxicity, the mode of inhibition) should be considered to evaluate whether and how much the soil biological ecosystem will be significantly impaired by the presence of the pesticide, in either a reversible or an irreversible mode. Obviously, some of them are directly affected by the peculiar chemical properties of the pesticide and the processes occurring in the soil system.

In general, pesticides do not have much effect on the soil enzymatic activities, except at concentrations greatly exceeding normal recommended field rates. Indeed, if recommended field application rates are used, inhibition of some microbial species may be temporary; others may rapidly develop and replace the sensitive species. Consequently, enzyme activities will return to levels similar to those in untreated soils but in a few weeks or months. A baseline or background level of enzyme activity is likely contained in soils and very hardly it will be permanently changed (Zantua and Bremner 1977).

In 2002, Speir and Ross questioned "Is it therefore worthwhile to continue to test for pesticide effects on soil enzyme activities?" The studies undertaken from 2002 onwards seem to positively answer this question and support what also claimed by the two authors, i.e. "On balance, it probably is important that newly registered pesticides continue to be subjected to tests for effects on soil enzymes, even though they have already passed registration criteria that are arguably more stringent than these tests. It is conceivable that a new chemical or a metabolite may, by design or accident, be a particularly potent inhibitor of a soil enzyme." Moreover, "Further investigations may be profitable only if they concentrate on degradative enzymes from the perspective of understanding mechanisms of pesticide metabolism in soil and/or for the cleanup of contaminated soils" (Speir and Ross 2002).

Knowledge in this direction is still not complete and deserves additional research efforts.

References

Adams RS Jr (1973) Factors influencing soil adsorption and bioactivity of pesticides. Residue Rev 47:1–54

Antonius GF (2003) Impact of soil management and two botanical insecticides on urease and invertase activity. J Environ Sci Health Part B: Pestic, Food Contam Agric Wastes 38:479–488

Bailey GW, White JL (1970) Factors influencing the adsorption, desorption, and movement of pesticides in soil. Residue Rev 32:29–92

Barraclough D, Kearney T, Croxford A (2005) Bound residues: environmental solution or future problem? Environ Pollut 133:85–90

Bishnu A, Saha T, Mazumdar D, Chakrabarti K, Chakraborty A (2008) Assessment of the impact of pesticide residues on microbiological and biochemical parameters of tea garden soils in India. J Environ Sci Health Part B: Pestic, Food Contam Agric Wastes 43:723–731

Bollag J-M, Liu S (1990) Biological transformation processes of pesticides. In: Cheng HH (ed) Pesticides in the soil environment Processes, impacts, and modeling, vol 2, SSSA Book Series. SSSA, Madison, WI, pp 169–211
Bromilow RH, Evans AA, Nicholls PH, Todd AD, Briggs GG (1996) The effect on soil fertility of repeated applications of pesticides over 20 years. Pestic Sci 48:63–72
Burns RG (1982) Enzyme activity in soil: location and a possible role in microbial ecology. Soil Biol Biochem 14:423–427
Cáceres TP, He W, Megharaj M, Naidu R (2009) Effect of insecticide fenamiphos on soil microbial activities in Australian and Ecuadorean soils. J Environ Sci Health B 44:13–17
Cervelli S, Nannipieri P, Sequi P (1978) Interactions between agrochemicals and soil enzymes. In: Burns RG (ed) Soil enzymes. Academic, London, pp 251–293
Cheng HH (1990) Pesticides in the soil environment: processes, impacts, and modeling, vol 2. SSSA Inc, Madison, WI
Dick RP (1994) Soil enzymes activities as indicators of soil quality. In: Doran JW, Coleman DC, Bezdicek DF, Stewart BA (eds) Defining soil quality for a sustainable environment. SSSA special publication, vol 35, SSSA. Madison, Wl, pp 107–124
Dick RP (1997) Enzyme activities as integrative indicators of soil health. In: Pankhurst CE, Doube BM, Gupta VVSR (eds) Biological indicators of soil health. CAB International, Oxon, UK, pp 121–156
Dick WA, Tabatabai MA (1993) Significance and potential uses of soil enzymes. In: Metting FB (ed) Soil microbial ecology: application in agricultural and environmental management. Marcel Dekker, New York, pp 95–125
Gevao B, Semple KT, Jones KC (2000) Bound pesticide residues in soils: a review. Environ Pollut 108:3–12
Gianfreda L, Bollag J-M (1996) Influence of natural and anthropogenic factors on enzyme activity in soil. In: Stosky G, Bollag J-M (eds) Soil biochemistry, vol 9. Marcel Dekker, New York, pp 123–194
Gianfreda L, Bollag J-M (2002) Isolated enzyme for the transformation and detoxification of organic pollutants. In: Burns RG, Dick RP (eds) Enzyme in the environment. Activity, ecology and applications. Marcel Dekker, New York, pp 495–538
Gianfreda L, Rao MA (2004) Potential of extra cellular enzymes in remediation of polluted soils: a review. Enzym Microb Technol 35:339–354
Gianfreda L, Rao MA (2008) Interaction between xenobiotics and microbial and enzymatic soil activity. Critical Rev Environ Sci Technol 38:269–310
Gianfreda L, Ruggiero P (2006) Enzyme activities in soil. In: Nannipieri P, Smalla K (eds) Soil biology, vol 8, Nucleic acids and proteins in soil. Sprinter, Berlin, pp 257–311
Gianfreda L, Sannino F, Filazzola MT, Violante A (1993) Influence of pesticides on the activity and kinetics of invertase, urease and acid phosphatase enzymes. Pestic Sci 39:237–244
Gianfreda L, Sannino F, Ortega N, Nannipieri P (1994) Activity of free and immobilized urease in soil: effects of pesticides. Soil Biol Biochem 26:777–784
Gianfreda L, Sannino F, Violante A (1995) Influence of pesticides on the behaviour of free, immobilized and soil invertase. Soil Biol Biochem 27:1201–1208
Gianfreda L, Rao MA, Saccomandi F, Sannino F, Violante A (2002) Enzymes in soil: properties, behavior and potential applications. In: Violante A, Huang PM, Bollag J-M, Gianfreda L (eds) Soil mineral-organic matter-microorganism interactions and ecosystem health, development in soil science 28B. Elsevier, London, pp 301–328
Gianfreda L, Rao MA, Piotrowska A, Palumbo G, Colombo C (2005) Enzyme activities of soils affected by different anthropogenic alterations. Sci Tot Environ 341:265–279
Gianfreda L, Rao MA, Mora M (2010) Enzymatic activity as influenced by soil and humic colloids: the impact on the ecosystem. In: Huang PM (ed) Section E "Soil physical, chemical, and biological interfacial interactions". Handbook of Soil Sciences. 2nd edn (in press)
Gundi VAKB, Viswanath B, Chandra MS, Kumar VN, Reddy BR (2007) Activities of cellulase and amylase in soils as influenced by insecticide interactions. Ecotoxicol Environ Safety 68:278–285

He WX, Jiang X, Yu GF, Bian YR (2003) Effect of dimehypo on soil urease activity. Acta Pedol Sinica 40:750–755 (in Chinese)

Johnsen K, Jacobsen CS, Torsvik V, Sørensen J (2001) Pesticide effects on bacterial diversity in agricultural soils - a review. Biol Fertil Soils 36:443–453

Kalam A, Tah J, Mukherjee AK (2004) Pesticide effects on microbial population and soil enzyme activities during vermicomposting of agricultural waste. J Environ Biol 25:201–208

Liu J, Xie J, Chu Y, Sun C, Chen C, Wang Q (2008) Combined effect of cypermethrin and copper on catalase activity in soil. J Soils Sediments 8:327–332

Madhun YA, Freed VH (1990) Impact of pesticides on the environment. In: Pesticides in the soil environment: processes, impacts, and modeling. SSSA Inc. Vol 2, Madison, WI, pp 429–466

Madhuri RJ, Rangaswamy V (2002) Influence of selected insecticides on phopshatase activity in groundnut (*Arachis hypogeae* L.) soils. J Environ Biol 23:393–397

Nam K, Alexander M (2001) Role of nanoporosity and hydrophobicity in sequestration and bioavailability: test with model soils. Environ Sci Technol 32:71–74

Nannipieri P (1994) The potential use of soil enzymes as indicators of productivity, sustainability and pollution. In: Pankhurst CE, Doube BM, Gupta VVSR, Grace PR (eds) Soil biota: management in sustainable farming systems. CSIRO, East Melbourne, pp 238–244

Nannipieri P, Bollag J-M (1991) Use of enzymes to detoxify pesticide-contaminated soils and waters. J Environ Qual 20:510–517

Nannipieri P, Gianfreda L (1998) Kinetics of enzyme reactions in soil environments. In: Huang PM, Senesi N, Buffle J (eds) Environmental particles – structure and surface reactions of soil particles. Wiley, New York, pp 449–479

Nannipieri P, Kandeler E, Ruggiero P (2002) Enzyme activities and microbiological and biochemical processes in soil. In: Burns RG, Dick RP (eds) Enzymes in the environment. Activity, ecology and applications. Marcel Dekker, New York, pp 1–33

Niemi RM, Heiskanen I, Ahtiainen JH, Rahkonen A, Mäntykoski K, Welling L, Laitinen P, Ruuttunen P (2009) Microbial toxicity and impacts on soil enzyme activities of pesticides used in potato cultivation. Appl Soil Ecol 41:293–304. doi:10.1016/j.apsoil.2008.12.002

Omar SA, Abdel-Sater MA (2001) Microbial populations and enzyme activities in soil treated with pesticides. Water Air Soil Pollut 127:49–63

Pandey S, Singh DK (2006) Soil dehydrogenase, phosphomonoesterase and arginine deaminase activities in an insecticide treated groundnut (*Arachis hypogea* L.) field. Chemosphere 63:869–880

Perucci P, Vischetti C, Battistoni F (1999) Rimsulfuron in a silty clay loam soil: effects upon microbiological and biochemical properties under varying microcosm conditions. Soil Biol Biochem 31:195–220

Perucci P, Dumontet S, Bufo SA, Mazzatura A, Casucci C (2000) Effects of organic amendment and herbicide treatment on soil microbial biomass. Biol Fertil Soils 32:17–23

Qian H, Hu B, Wang Z, Xu X (2007) Effects of validamycin on some enzymatic activities in soil. Environ Monit Assess 125:1–8

Rani MS, Lakshmi KV, Devi PS, Madhuri RJ, Devi SH, Jyothi K (2008) Impact of chlorpyrifos on soil enzyme activities in agricultural soil. Asian J Microbiol Biotechnol Environ Sci 10:295–300

Reid BJ, Jones KC, Semple KT (2000) Bioavailability of persistent organic pollutants in soils and sediments - a perspective on mechanisms, consequences and assessment. Environ Pollut 108:103–113

Ruggiero P, Pizzigallo MDR, Crecchio C (2002) Effects of soil abiotic processes on the bioavailability of anthropogenic organic residues. In: Violante A, Huang PM, Bollag J-M, Gianfreda L (eds) Soil mineral-organic matter-microorganism interactions and ecosystem health development in soil science, vol 28B. Elsevier, Amsterdam, pp 95–133

Sannino F, Gianfreda L (2001) Pesticide influence on soil enzymatic activities. Chemosphere 22:1–9

Schaffer A (1993) Pesticide effects on enzyme activities in the soil ecosystem. In: Bollag J-M, Stotzky G (eds) Soil biochemistry, vol 8. Marcel Dekker, New York, pp 273–340
Setty PK, Magu SP (1996) Influence of metalaxyl and pendimethalin on soil phosphatase activity in the rhizosphere of wheat. J Teaching Res Chem 3:53–59
Singh J, Singh DK (2005) Dehydrogenase and phosphomonoesterase activities in groundnut (*Arachis hypogaea* L.) field after diazinon, imidacloprid and lindane treatments. Chemosphere 60:32–42
Singh BK, Walker A, Wright DJ (2002a) Degradation of chlorpyrofos, fenamiphos, and chlorothalonil alone and in combination and their effects on soil microbial activity. Environ Toxicol Chem 21:2600–2605
Singh BK, Walker A, Wright DJ (2002b) Persistence of chlorpyrofos, fenamiphos, chlorothalonil and pendimethalin in soil and their effects on soil microbial characteristics. Bull Environ Contam Toxicol 69:181–188
Speir TW, Ross DJ (2002) Hydrolytic enzyme activities to assess soil degradation and recovery. In: Burns RG, Dick R (eds) Enzyme in the environment. Activity, ecology and applications. Marcel Dekker, New York, pp 407–431
Stepniewska Z, Wolińska A, Lipińska R (2007) Effect of fonofos on soil dehydrogenase activity. Intern Agrophysics 21:101–105
Sutherland T, Russel R, Selleck M (2002) Using enzymes to clean pesticide residues. Pestic Outlook 13:149–151
Trasar-Cepeda C, Leirós MC, Seoane S, Gil-Sotres F (2000) Limitation of soil enzymes as indicators of soil pollution. Soil Biol Biochem 32:1867–1875
Van Beelen PV, Doelman P (1997) Significance and application of microbial toxicity tests in assessing ecotoxicological risks of contaminants in soil and sediments. Chemosphere 43:455–499
Xie X-M, Liao M, Huang C-Y, Liu W-P, Abid S (2004) Effects of pesticides on soil biochemical characteristics of a paddy soil. J Environ Sci 16:252–255
Xu B, Zhang Y, Chen M, Zhu N, Ming H (2000) Impact of repeated insecticide application on soil microbial activity. EU Pestic Environ Meeting, Brussel
Yang C, Sun T, He W, Chen S (2006) Effects of pesticides on soil urease activity. Chinese J Appl Ecol 17:1354–1356
Yang C, Sun T, He W, Zhou Q, Chen S (2007) Single and joint effects of pesticides and mercury on soil urease. J Environ Sci 19:210–216
Yao X, Min H, Lü Z, Yuan H (2006) Influence of acetamiprid on soil enzymatic activities and respiration. Eur J Soil Biol 42:120–126
Yun LY, Shan M, Fang H, Wang X, Xiao QC (2006) Responses of soil microorganisms and enzymes to repeated applications of chlorothalonil. J Agric Food Chem 54:10070–10075
Zantua MI, Bremner JM (1977) Stability of urease in soil. Soil Biol Biochem 9:135–140

Chapter 17
Behavior of Enzymatic Activity in Chilean Volcanic Soil and Their Interactions with Clay Fraction

Analí Rosas, Ada López, and Roxana López

17.1 Introduction

Soils derived from volcanic ashes are divided into two great taxonomic orders: Andisols and Ultisols. Andisols derive from recent volcanic ashes, which in Chile are denominated *Trumaos*, corresponding to mapudungun, the indigenous name for this kind of soil, which means dust accumulation. Ultisols, derived from Latin *ultimus*, which means "last", indicates by this very fact that it is a very advanced state of weathering.

We will consider the properties of soils derived from volcanic ash and their effects on soil enzyme behavior related to management practices that have been poorly studied. In addition, we will consider the frequent management use of these soils, which impacts their biologic properties.

A. Rosas (✉)
Facultad de Agronomía, Departmento de Ciencia del Suelo y Recursos Naturales, Universidad de Concepción, 537, Av. Vicente, Méndez 595 Chillan, Chile
e-mail: arosas@udec.cl

A. López,
Instituto de Biología Vegetal y Biotecnología, Universidad de Talca, Casilla 747, Talca, Chile

R. López
Facultad de Ciencias Agrarias, Departamento de Producción Agrícola, Universidad de Talca, Casilla 747, Talca, Chile

G. Shukla and A. Varma (eds.), *Soil Enzymology*, Soil Biology 22,
DOI 10.1007/978-3-642-14225-3_17, © Springer-Verlag Berlin Heidelberg 2011

17.2 Chilean Volcanic Soil

17.2.1 Characteristics of Chilean Volcanic Soils

Soils in Chile, except a recent volcano-free zone between Atacama and Aconcagua (27° 30′–32° 20′), originate from intense volcanic activity, which causes large surfaces of the territory to be covered with piroclastic sediments, constituted mainly by volcanic ash (Besoaín 1985). In fact, nearly 50% of the soils in the south-central zone present a volcanic surface a few metres deep (Besoaín 1985). In the case of Andisols, they correspond to recent soils, formed during the Holocene. The age of these soils is between 15,000 and 20,000 years , as determined by the main mineral constituent, allophane.

On the other hand, in Chilean Ultisols, the predominant clay is halloysite. This clay is formed from allophane and directly from volcanic ashes and pomes of varied composition as well (Sieffermann and Millot 1969; Aomine and Wada 1962). In Chile, it is frequent to find laminar halloysite in the superficial horizons, while tubular halloysite predominates in the sub-superficial horizons (Besoaín 1985). The length of halloysite nanotubes varies from 0.02 to 30 μm and the external diameter from ca. 30 to 190 nm, with an internal diameter range of ca. 10–100 nm (Bates et al. 1950; Churchman et al. 1995). The other secondary mineral in Ultisols is kaolinite, which corresponds to the final phyllosilicate in the sequence of volcanic ash weathering, preceded by the allophane–halloysite series (Fieldes 1955). Kaolinite presents the hexagonal morphology, with a particle size of 2 μm, specific surface area of 6–22 $m^2\ g^{-1}$, and density of 2.6 $g\ cm^{-3}$ (Fig.17.1c).

17.2.2 Properties of Volcanic Ashes-derived Soils

Andisols are characterized by a variable charge that gives them chemical and physical properties completely different from those soils with permanent charge minerals. Characteristics such as organic matter (OM) accumulation, high biological activity, high aluminum content, phosphate fixation, and aggregate formation, are closely related to the nature and properties of non-crystalline minerals (Galindo and Escudey 1985). It would be interesting to understand the properties that govern OM accumulation and also allow a high biological activity, which helps the mineralization processes.

In Andisol, the allophane particles are aggregates. These aggregates are amorphous alumino-silicates and have physical features quite close to those in synthetic silica gels (Woignier et al. 2006): fractal geometry described as clusters, formed by limited diffusion aggregation (Yasuhisa and Karube 1999), and a low bulk density (close to 0.5 $g\ cm^{-3}$). In addition, the allophane particles forming a solid network can be described as an assembly of fractal cluster, built by the aggregation of these particles. In this natural gel, the fractal structure is in the mesopore range

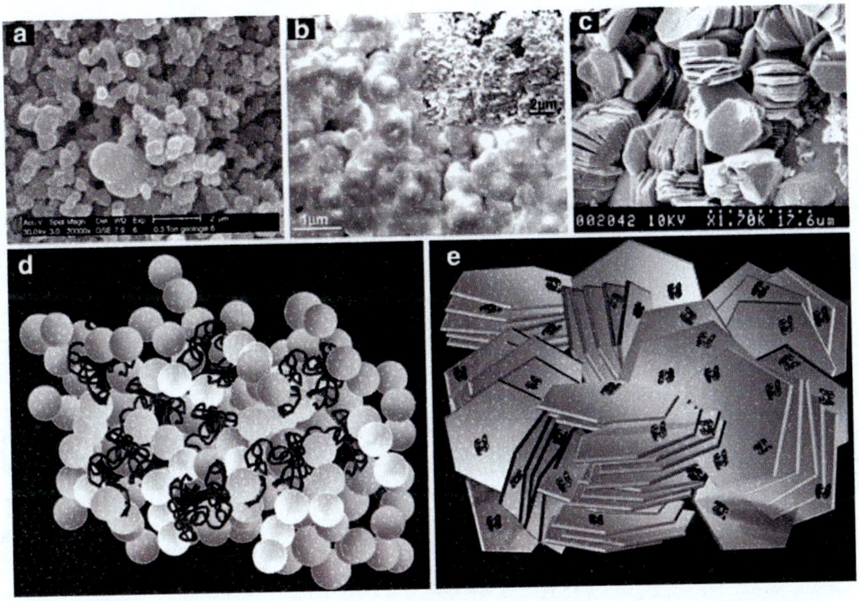

Fig. 17.1 Natural or synthetic particles used for the enzyme immobilization. (**a–c**) Transmission electron microscopy of allophane, silica nanoparticles (adopted from Luckarift 2004) and kaolinite. (**d–e**) Schematic representation of enzyme immobilization on natural clay; (**a**) Entrapment of enzyme in allophane and (**b**) chemical immobilization of acid phosphatase on kaolinite

(2–50 nm). The allophane particles have a high specific surface area (Denaix et al. 1999), which varies from 700 to 1,100 $m^2\ g^{-1}$ and an approximate size from 3.5 to 5.5 nm (Parfitt 1990). Due to this, allophane is considered a natural nanoparticle (Fig. 17.1a).

A study demonstrated as well that the allophane aggregates have a fractal structure very similar to that of the synthetic gels too. Chevallier et al. (2008) evaluated the importance of the aggregate fractal structure, described both for allophane and for synthetic gels related to C sequestration (Woignier et al. 2007; Schaeffer and Keefer 1986; Vacher et al. 1988; Emmerling and Fricke 1992; Dietler et al. 1986). Initially, six soils were incubated for 28 days at 28°C to evaluate the soil organic C transformation through the microorganism respiration. The results showed that the bioavailable C (g of C transformed in CO_2 per gram of C in the soil) is negatively correlated to the content of soil allophane.

Allophanic soils are well known by their organic carbon (OC) retention capability, accumulating 3 or 4 times more carbon than soils with other origin mineral materials (Wada 1985; Boudot et al. 1986). This fact would explain the impact of these soils in the carbon sequester and the greenhouse gas emissions mitigation. OM accumulation in soils has been attributed to Fe and Al-humus stable complexes formation, which has been reaffirmed with numerous studies. Some works indicate that the Andisols organic C content extension cannot be explained solely through Fe, Al-humus complexes formation (Boudot et al. 1986). Carbon sequester has also been related to allophanic soils' water content, but the reason of this relationship is not yet understood (Feller et al. 2001). One of the most recent studies realized with 11 Chilean Andisols showed a high correlation between allophane and the C associated to the silt and clay fraction of soils ($R^2 = 0.82$). This was consistent with the fact that the C, in this fraction, explained most of the OC variation (Matus et al. 2008). Other authors have identified through 13C-NMR spectroscopy that the Al monomeric and polymeric forms interact with the carboxylic groups of the OM (Parfitt et al. 1999). However, this organic C retention mechanism has also been described for different types of soil such as Ultisols and Alfisols (Matus et al. 2008). Moreover, a study performed by Zunino et al. (1982) demonstrated that allophane addition in an Alfisol soil would diminish the microbial products transformation, so that the mineral fraction would be mainly responsible for the OM accumulation. In Andisol, the presence of allophane may hinder the transformation of OM in CO_2 due to their structure and the pores shape. There are some insights that the fractal structure could be involved within the C sequestration. The tortuous allophane aggregates porosity would be involved in the low oxygen diffusion and in the microorganisms' limited access to the OM (Chevallier et al. 2008). Finally, this effect results in a microbial respiration reduction, as defined in the mesopore protection hypothesis (Mayer et al. 2004; Wang et al. 2003).

17.2.3 Mechanisms of Enzyme Immobilization in Soil Derived of Volcanic Ash

17.2.3.1 Enzyme Immobilization on Andisol Clay Fraction

The enzymes are considered amongst the most important OM compounds, and their activity strongly influences the content and quality of this fraction. In the soil, the enzymes are mainly immobilized in the organic and mineral fraction. The mineral fraction is able to stabilize proteins and protect them from proteolytic degradation (Rao et al. 1996). However, it has been informed that enzymatic immobilization in natural mineral supports modifies the protein conformational structure, which finally alters its catalytic properties and reduces its activity (Rao et al. 2000). In general, it is known that allophane and other reactive minerals can complex and stabilize organic material (Schwertmann and Taylor 1989; Wada 1989). The soil mineral may have particularly strong effects on the stability and activity of proteins, including enzymes. The enzymes are found principally complexed with the organo-mineral soil compounds. A common strategy to understand the relationships between enzyme and clay of soil is the use of enzyme–clay synthetic complex as a model system (Rao et al. 1996; Huang and Shindo 2000; Rao et al. 2000; Gianfreda et al. 2002; Kelleher et al. 2004). One of the first studies regarding the enzyme immobilization in allophane was performed by Shindo et al. (2002). The results showed a low retention of alkaline phosphatase enzyme in allophane.

However, comparing the activity of the enzyme immobilized in different types of clays (allophane, kaolinite, and montmorillonite), the highest enzymatic activity was observed when the enzyme was immobilized in allophane. At the moment, there are significant evidences, which indicate that the immobilization on allophanic materials increase the enzymes activity and catalytic efficiency. One of the first studies that evidenced this effect was carried out by Allison (2006), in which a synthetic allophane was added to an Andisol with a suppressed biological activity. The results showed that allophane addition had a positive effect on most of the evaluated enzymatic activities. Another study, in which acid phosphatase was immobilized on an Andisol clay fraction (Eutric Pachic Fulvudands) from southern Chile, an increment of 33% of the enzymatic activity and catalytic efficiency was observed (López 2006; Rosas 2006). In addition, the presence of heavy metals, such as molybdenum (Mo) and manganese (Mn), did not affect the enzymatic activity in comparison with the free enzyme. In order to explain this result, phosphatase immobilized in allophane were analyzed by FTIR (Fig. 17.2). The spectrum of acid clay phosphatase did not show new signals, which indicates that new bindings do not exist. Therefore, the mechanism of immobilization of the enzyme does not appear to be the adsorption. In this immobilization process, acetate buffer was used. This carboxylate was bonded to the clay showing a band of around 1,600 cm^{-1}, which corresponds to enzyme carboxylate and acetate C–O symmetric stretching. The presence of these new bands indicates that part of the acetate was adsorbed on clay. Clay and enzyme isoelectric points are 8.35 and 5.2, respectively, and thus both are positively charged at pH 5.0. The acetate buffer (0.5 M) is negatively charged at pH 5.0 and might be forming an extern sphere complex with the mineral surface (Sparks 1995; Violante et al. 2002). This way, acetate present in the clay extern surface might have contributed to the process of enzyme immobilization through the electrostatic forces. The electrostatic interaction has already been described as a mechanism of enzyme immobilization (Stauton and Quiquampoix 1994; Gianfreda and Scarfi 1991; Huang et al. 2005). However, electrostatic forces are unstable, generate protein loss during the immobilization process, and generally

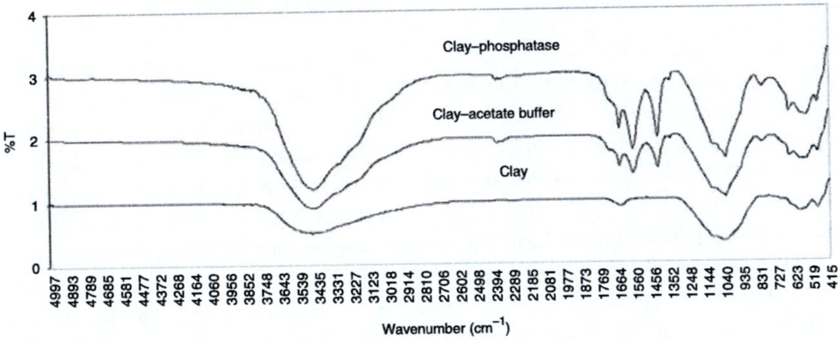

Fig. 17.2 FTIR spectra of clay, clay–acetate buffer and clay–acid phosphatase. Clay extracted from Andisol

diminish the enzyme catalytic activity (Rao et al. 2000; Rao and Gianfreda 2000; Huang and Shindo 2000; Kelleher et al. 2004; Huang et al. 2005). In addition, in a study performed by Eggers and Valentine (2001), it has been shown that the activity of immobilized enzymes on silica gel is unaltered by changes in electrostatic forces. Specifically, they explained that the secondary structure of immobilized protein was unaltered by changes in pH and ionic strength of KCl.

Then, what explains the increase in the activity and the catalytic efficiency of the enzyme immobilized on allophane and silica gel?

The analogy between allophane and silica gel particles has been widely studied by Woignier et al. (2005, 2006, 2007) and Chevallier et al. (2008). In particular, Woignier et al. (2005) performed an experiment for determined physical properties of allophane and silica gel demonstrating that both have pores with similar physical behaviors (Fig. 17.1a, b). Some studies demonstrated an activation and high stability of the enzymes immobilized in silicated nanomaterials obtained from sol-gel processes (Gill 2001; Reetz et al. 2003; Shchipunov et al. 2004). This nanomaterial has both particles and pores with a size similar to allophanic clay aggregates. The immobilization mechanism described for sol-gel has been described as encapsulation (Eggers and Valentine 2001). In this regard, Wei et al. (2001) demonstrated that the enzyme catalytic efficiency was increased due to the pore size and the superficial area of support that facilitates both the substrate and product diffusion (Wei et al. 2001). Then, the increase of catalytic efficiency of the enzymes immobilized in allophane might be attributed to an encapsulation of the enzyme on clay pores (Fig. 17.1d). Thus, due to the similitude between the allophane aggregates and the sol-gel structure, the enzyme immobilization process in both supports is produced by encapsulation in the pores (Rosas 2006; López 2006). The acid phosphatase has a three-dimensional structure with dimensions of approximately $4 \times 6 \times 7.5$ nm. Since the molecular size ranges of biopolymers such as proteins are approximately 1–20 nm, it is to be expected that allophane could act as biomolecular trapper materials.

17.2.3.2 Enzyme Immobilization on Ultisol Clay Fraction

On the other hand, clays of Ultisols have been used as enzymatic support for different hydrolytic enzymes. The use of these clays to immobilize enzymes has shown diverse effect on their activity and catalytic properties. In a study in which the specific activity of an acid phosphatase immobilized in the Ultisol clay fraction and commercial kaolinite were compared, the results showed that the immobilized enzyme in these clays had a lower specific activity than the free enzyme (Huang et al. 2005). Different results have been obtained when immobilizing other enzymes in modified kaolinite. In a study about immobilization of lacasse on kaolinite with an activator and glutaraldehyde, the enzyme exhibited a very high activity. In addition, in this support, the enzyme showed a high stability as a response to inhibitors and long time of storage (Dodor et al. 2004). Similar results were obtained when immobilizing a β-glucoronidase in a Ca homoionic kaolinite

(Fiorito et al. 2008). This result seems to indicate that the increase of activity is due to a conformational change of the enzyme that allowed a better access of substrate to the active site. In studies with other laminar clays or iron and aluminum oxides, frequently found in Ultisols, a diminution in the catalytic activity of the immobilized enzymes was reported (Rao et al. 2000; Rao and Gianfreda 2000; Huang and Shindo 2000; Kelleher et al. 2004).

In order to explain the mechanism of enzyme immobilization in Ultisol clay, Huang et al. (2005) studied the adsorption process of an acid phosphatase in kaolinite. The results indicated that in this clay, the retention of the enzyme is mainly due to electrostatic forces such as Van der Waals force and hydrogen bonding. Meanwhile, between 13 and 18% of the enzyme was retained by ligand interchange. Schematic representations of acid phosphatase immobilization on kaolinite are shown in Fig. 17.1e.

17.2.3.3 Some Considerations about Model Using Synthetic Complexes Clay–Enzyme

To study the behavior of soil enzyme, it is possible to use synthetic complexes between enzyme and mineral or organic soil constituents. For example, it has been demonstrated that acid phosphatase complexed with montmorillonite, kaolinite, and tannic acid have been used as models of soil (Gianfreda and Bollag 1994; Rao et al. 1996; Rao and Gianfreda 2000; Gianfreda et al. 2002).

According to some authors' statements, the diversity of results from studies with different clay-bound enzyme indicates that no generalizations can exist regarding the cause of changes in the activity. This may be clarified when comparing three studies in which an acid phosphatase was immobilized in allophane (Rosas et al. 2008; López and Rosas 2008; Shindo et al. 2002). In the study carried out by Shindo et al. (2002), the allophanic clay was treated with dithionite-citrate-bicarbonate and cold 5% Na_2CO_3 before phosphatase immobilization. This treatment dissolves the most reactive components and layers of the greater part of the organized allophane (Farmer et al. 1977). The activity of the enzyme immobilized in this support was lower than that in the free enzyme. In a first study, carried out by Rosas et al. (2008), the Andisol clay fraction was saturated with 0.1 M KCl, according to the methodology proposed by Jara et al. (2006), previous to the enzyme immobilization. The immobilized enzyme showed an activation of 33%. Later, in a second study performed by López and Rosas (2008), the immobilized phosphatase exhibited a significant activation, with an increase higher than 100% when the clay was not saturated with cations.

Although in these three studies, there are some differences related to the enzyme origin and the buffer pH, the immobilization was performed with the same Andisol clay and acid phosphatase. The significant differences in the enzyme activity can be explained by the clay preparation process. All these treatments, described above, could have affected the structure or physico-chemical properties of the clay, and then, its capacity both to retain the enzymes as well as to keep its native structure.

The results seem to be more confusing when trying to elucidate the effect of the Ultisol clays on the enzymatic activities. Regarding kaolinite, the most studied of the synthetic enzymatic complexes, the clay preparation is also the main difference in the methodologies. The results show that when the clay is saturated with cations, especially Na, the enzyme is weakly adsorbed or shows a high rate of inhibition (Lozzi et al. 2001; Huang et al. 2005; Fiorito et al. 2008). On the other hand, when clay was prepared without cations but with glutaraldehyde and activator compounds, a high stability and activity of the immobilized enzymes was obtained (Dodor et al. 2004). Therefore, in order to compare the true effect of the enzymatic immobilization on clays and to simulate the natural processes of the soil in model systems, it could be necessary to maintain the natural structure and physico-chemical properties of the clay.

17.3 Soil of Enzymatic Activities in Relation with Some Management Practices

17.3.1 Soil Enzymatic Activities in Andisols and Ultisols of Southern Chile

Some references about enzyme activity and microbial biomass in volcanic soil of southern Chile are shown in Table 17.1 Compared to other soil, the effect of agricultural management on enzyme activities has not been systematically investigated in Chilean volcanic soils.

Amongst the Chilean volcanic kinds of soil, Ultisols are the most affected area for physical and biological erosion. Due to their susceptibility, the effect of conservation practices has been evaluated (Alvear et al. 2006) through enzyme activities. The range of enzyme activities in response to different tillage practices are detailed in Table 17.1. Thus, no tillage system increased dehydrogenase, acid phosphatase, arysulphatase, and urease activities compared with conventional tillage. Similar results were obtained for phosphatase activity under no tillage practices in interaction with lupine-wheat rotation (Redel et al. 2007).

Meanwhile, regarding Chilean Andisol, there is a report in which the most important enzyme activities, related with the soil quality, were evaluated (Alvear et al. 2006). In this study, the enzyme activity by effect of application of herbicides was evaluated. The acid phosphatase, dehydrogenase, and urease activities showed a slight and temporal decrease by application of simazine, trifluralin, and MCPA and metsulfuron-methyl MCPA herbicides. In another study on Andisol, acid phosphatase activity where only different Nothofagus rainforest ecosystems were compared was determined (Redel et al. 2008). Interestingly, the activity of pristine and deciduous forest ecosystem showed a great value compared with cultivated soils. According to Redel et al. (2008), the higher phosphatase indicated an enhanced P cycling under this type of forest, as response to a greater labile and moderate labile organic P content in these soils.

Table 17.1 Microbial biomass and enzymatic activities in Chilean volcanic soil

Soil	Season	Treatments	C-Biomass (mg C g^{-1})	N-Biomass (µg N g^{-1})	D-ase (µg RF g^{-1})	S-ase	β-glucosidase	P-ase	Urease µmol NH_3g^{-1}	Reference
						µmol PNF $g^{-1}h^{-1}$				
Ultisols	Winter	No tillage	412	78	143	0.10	0.39	3.60	18.49	Alvear et al. (2006)
		Conventional tillage	296	54	67	0.11	2.08	2.69	15.10	
	Summer	No Tillage	159	88	388	0.14	1.16	6.14	11.67	
		Conventional tillage	258	56	350	0.12	0.89	3.66	4.98	
	Autumn	No tillage, rotation oat-wheat	–	–	–	–	–	5.47	–	Redel et al. (2007)
		No tillage, rotation lupine-wheat	–	–	–	–	–	7.59	–	
		Conventional tillage, rotation oat-wheat	–	–	–	–	–	4.96	–	
		Conventional tillage, rotation lupine-wheat	–	–	–	–	–	6.73	–	
Andisol	Spring 15 days	Control	320	82	290	–	–	4.4	42	Alvear et al. (2006)
		Simazine	340	84	220	–	–	3.5	118	
		Trifluralin	310	75	280	–	–	3.8	54	
		MCPA and metsulfuron-methyl	290	73	290	–	–	3.8	50	
	Summer 150 days	Control	540	65	410	–	–	5.7	16	
		Simazine	360	72	405	–	–	5.8	6	
		Trifluralin	300	95	400	–	–	4.7	6	
		MCPA and metsulfuron-methyl	340	100	455	–	–	4.8	8	
	Winter	Pristine forest ecosystem	–	–	–	–	–	43.2	–	Alvear et al. (2007)
	Winter	Deciduous forest ecosystem	–	–	–	–	–	51.8	–	

17.3.1.1 Evaluation of the Effect of Mn and Mo on Acid Phosphatase in a Model System of Andisol

Other management practice that can affect the enzyme activity is the application of fertilizer. In Chilean volcanic soil, the effect of Mo has been studied, that is applied as fertilizer since Andisols have a deficiency of this micronutrient. Due to its acidic pH, these soils are characterized by high Al and Mn content. However, although Mn and Mo are micronutrients for plant, they are heavy metals that in high concentrations may affect the soil enzyme activities. Previous reports regarding the effect of metals on free enzyme activity indicate that acid phosphatase can be inhibited (Bozzo et al. 2002; Yenugün and Güvenilir 2003) or activated (Tso and Chen 1997; Bozzo et al. 2002) by Mn. In contrast, Mo has been described as a competitive acid phosphatase activity inhibitor; however, the inhibitory effects observed at similar concentrations of the metal are very wide (Tso and Chen 1997; Yenugün and Güvenilir 2003; Schützendübel and Polle 2002). In studies performed with free acid phosphatase, it has been demonstrated that, at levels usually present in Andisols, the activity showed a decrease of approximately 29% by effect of Mn at 1.17 mM. Meanwhile, the enzyme activity was competitively inhibited by Mo (López et al. 2007a).

Now, with regard to the effect of these metals on soil enzymatic activity of volcanic soil, it has been demonstrated that additions on 200 μg Mo kg^{-1} dry soil have a detrimental effect on the L-asparaginase and L-glutaminnase activity (Table 17.1) of four Andisols (López et al. 2007b). A more specific research to determine the effect of Mn and Mo on enzyme activity was performed in a model system of Andisol (Rosas et al. 2008). To simulate enzymatic reactions occurring in Andisols, synthetic complexes were formed by interaction between acid phosphatase and either OM or natural allophanic clay (Table 17.2). In this study, the effect of 0.58 mM and 1.17 mM of Mn, levels similar to those present in acid soil, and of Mo, at doses of 0.002 and 0.01 mM, at levels used frequently in agriculture, was evaluated. These metals were applied during and after the immobilization of acid phosphatase on clay and OM. Specifically, as a representative of organic soil components, tannic acid was used, a precursor of fulvic acid that can form synthetic complexes with enzymatic proteins (Gianfreda and Bollag 1994). The results showed that by addition of Mn, the residual activity of the enzyme in interaction with tannic acid decreased around 33 and 41% as compared with the same complex without Mn. According to chemical characterization of enzyme–OM model system, only around 5% of the metal is present in the solution of phosphatase immobilized on tannic acid. Therefore, Mn exerted its negative effect on enzyme activity during the immobilization process probably by interactions occurring between Mn^{2+} with groups and/or other metallic ions usually present in the active site of phosphatases (Olczack et al. 2003; Zambonelli and Roberts 2003).

On the other hand, the presence of Mo affects significantly the activities of the enzyme during the immobilization process, with a decrease of about 53% at the highest Mo concentration. Significant effects were also detected on the kinetic parameters of enzyme immobilized on tannic acid in the presence of Mn and Mo.

Table 17.2 Kinetic parameters of acid phosphatase immobilized on tannic acid or Andisol clay in the presence and absence of different levels of Mn and Mo (adopted from Rosas et al. 2008)

	V_{max} (μmol min^{-1} ml^{-})	K_m (mM)	V_{max}/K_m	R^2
Free P	0.312	0.085	3.7	0.999
Organic support				
[T–P]	0.201	0.198	1.0	0.999
[T–P–Mn] 0.58 mM	0.136	0.140	1.0	0.999
[T–P–Mn] 1.16 mM	0.139	0.080	1.7	0.999
[T–P–Mo] 0.002 mM	0.170	0.316	0.5	0.996
[T–P–Mo] 0.010 mM	0.224	0.343	0.7	0.999
Mineral support				
[C–P–Mn] 0.58 mM	0.403	0.064	6.3	0.998
[C–P–Mn] 1.16 mM	0.431	0.065	6.6	0.999
[C–P–Mo] 0.002 mM	0.381	0.140	2.7	0.999
[C–P–Mo] 0.010 mM	0.280	0.223	1.3	0.998

The maximum velocity and substrate affinity decreased by effect of Mn or Mo compared with the free enzyme (Table 17.2). The changes of both V_{max} and K_m values resulted in lower catalytic efficiencies, but this effect was the greatest in the presence of Mo (Table 17.2).

Moreover, when the immobilization process was performed on allophanic clay, no significant effects on the activity of the enzyme were observed (Table 17.2) by addition of Mn, even though 59 and 48% of the initial Mn added were found in the complexes (Fig. 17.3). These results seem to indicate that the immobilization of the enzyme on the clay prevented the enzyme from the negative effects of Mn.

However, the Mo addition to the clay–enzyme mixture strongly influenced the activity and the kinetic parameters of the immobilized enzyme. Compared with the same complex without Mo, a decrease of 38% in enzyme activity was observed by application of 0.01 mM of Mo. Moreover, a significant decrease in the substrate affinity can be explained by the 100% of Mo anions retained in the enzyme–clay complexes (Rosas et al. 2008). However, according to results obtained by López et al. (2007), when concentration of 0.01 mM of Mo was applied to free acid phosphatase, a decrease of 71% on enzyme activity was observed. These overall results show that both metals had a less inhibitory effect on the activity and kinetics of acid phosphatase immobilized on allophane compared with the free enzyme. Thus, the immobilization process protects the enzyme against the inhibitory effect of metals, probably by protecting its catalytic site.

17.4 Conclusion

The studies performed in soil model system allow elucidating the mechanism of enzyme immobilization in soil. Thus, the differences in the behavior of enzyme activity in Chilean volcanic soil seem to be related to the kind of clay. Andisol has

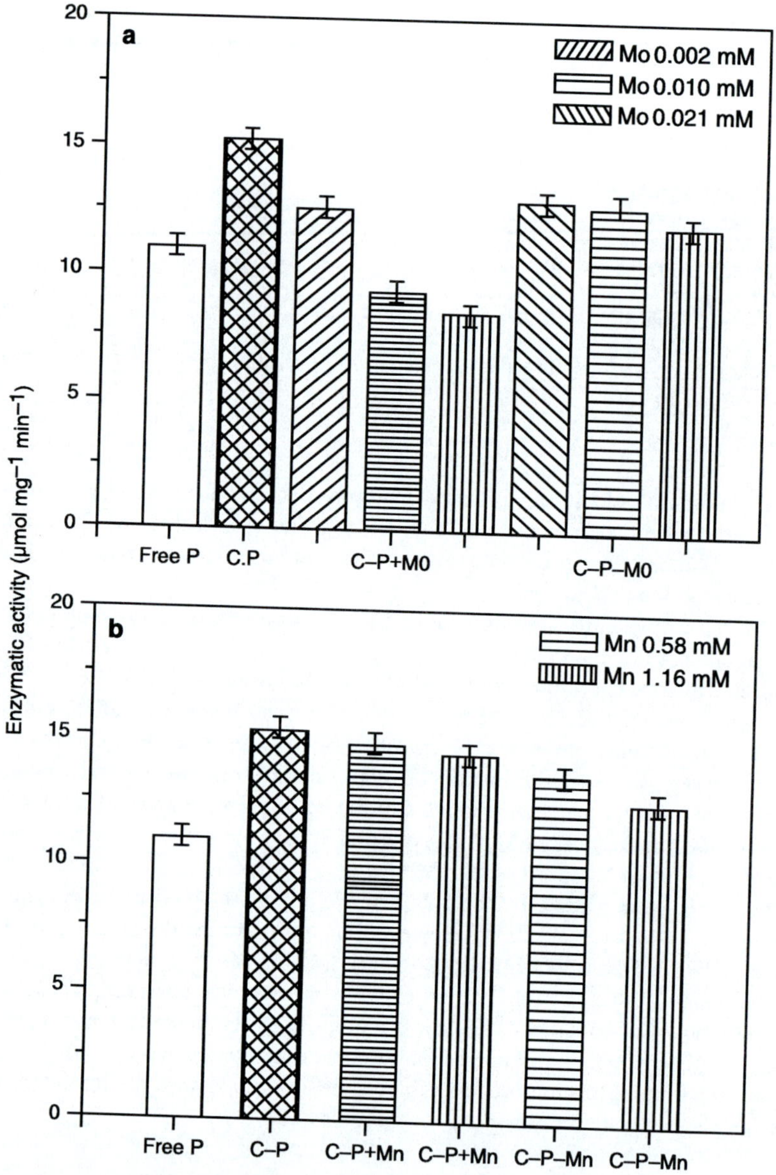

Fig. 17.3 Specific activities of tannic acid–enzyme and clay–enzyme complexes in the presence and absence of different levels of Mn and Mo (adopted from Rosas et al. 2008)

allophanic clay, a nanoparticle that forms solid network, where the enzyme can be retained. While Ultisol shows laminar clay, such as kaolinite that, due to the size and natural staking of its layers, affects negatively the enzymes retention and activity.

In fact, we can unveil that the natural clay can be enhanced by the catalytic efficiency of the enzyme due to the physico-chemical properties of clay. Moreover, we proposed that the encapsulation in the pores of clay aggregates could occur by the mechanism used for enzyme immobilization on alloplane, which explains the increased catalytic efficiency. By contrast, the laminar clay has an opposite effect on enzyme, probably due to the mechanism used for enzyme retention is by electrostatic forces.

On the other hand, the behavior of the enzyme in soil is strongly influenced by the management practices. There are greater enzyme activities in a Chilean pristine forest ecosystem, although these activities are decreased by effect of the intensive agricultural system.

References

Allison S (2006) Soil minerals and humic acids alter enzyme stability: implications for ecosystem processes. Biogeochemistry 81:361–373

Alvear M, Pino M, Castillo C, Trasar-Cepeda C, Gil-Sotres F (2006) Effect of non-tillage on some biological activities in an Alfisol from Southern Chile. J Soil Sci Plant Nutr 6:38–53

Alvear M, Urra C, Huaiquilao R, Astorga M, Reyes F (2007) Actividades biológicas y estabilidad de agregados en un suelo del bosque templado chileno bajo dos etapas sucesionales y cambios estacionales. R C Suelo Nutr Veg 7(3):38–50, J Soil Sc Plant Nutr 7(3):38–50

Aomine S, Wada K (1962) Differential weathering of volcanic ash and pumice resulting in formation of hydrated halloysite. Am Mineral 47:1024–1048

Bates T, Hildebrand F, Swineford A (1950) Morphology and structure of endelite and halloysite. Am Mineral 35:463–484

Besoaín E (1985) Los suelos. In: Tosso J (ed) Suelos Volcánicos de Chile. Instituto de Investigaciones Agropecuarias (INIA), Santiago, pp 23–106

Boudot JP, Hadj BAB, Chrone T (1986) Carbon mineralization in Andosols and aluminium-rich highland soils. Soil Biol Biochem 18:457–461

Bozzo G, Raghothama K, Plaxton W (2002) Purification and characterization of two secreted purple acid phosphatase isozymes from phosphate-starved tomato (*Lycopersicon esculentum*) cell cultures. Eur J Biochem 269:6278–6286

Chevallier T, Woignier T, Toucet J, Blanchart E, Dieudonné P (2008) Fractal estructure in natural gels: effect on carbon sequestration in volcanic soil. Sol-Gel Sci Technnol 48:231–238

Churchman G, Davy T, Aylmore L, Gilkes R, Self P (1995) Characteristics of fine pores in some halloysites. Clay Miner 30:89–98

Denaix L, Lamy I, Botero JY (1999) Structure and affinity towards Cd2+, Cu2+, Pb2+ of synthetic colloidal amorphous aluminosilicates and their precursors. Colloid Surf A 158:315–325

Dietler G, Aubert C, Cannell DS, Wiltzius LP (1986) Gelation of colloidal silica. Phys Rev Lett 57:3117

Dodor D, Hwang H, Ekunwe S (2004) Oxidation of anthracene and benzo[a]pyrene by immobilized laccase from *Trametes versicolor*. Enzym Microb Technol 35:210–217

Eggers DK, Valentine JS (2001) Molecular confinement influences protein structure and enhances thermal protein stability. Protein Sci 10:250–261

Emmerling A, Fricke J (1992) Small angle scattering and the structure of aerogels. J Non-Cryst Solids 145:113–120

Farmer VC, Smith BFL, Tait JM (1977) Alteration of allophane and imogolite by alkaline digestion. Clay Miner 12:195–198

Feller C, Albrecht A, Blanchart E, Cabidoche YM, Chevallier T, Hartmann C, Eschenbrenner V, Larre-Larrouy MC, Ndandou JF (2001) Soil organic carbon sequestration in tropical areas. General considerations and analysis of some edafic determinants for Lesser Antilles soils. Nutr Cycl Agroecosys 61:19–31

Fieldes M (1955) Allophane and related mineral colloids. N Z J Sci Tech 37:336–350

Fiorito T, Icoz I, Stotzky G (2008) Adsorption and binding of the transgenic plant proteins, human serum albumin, β-glucuronidase, and Cry3Bb1, on montmorillonite and kaolinite: microbial utilization and enzymatic activity of free and clay-bound proteins. Appl Clay Sci 39:142–150

Galindo G, Escudey M (1985) Interacciones superficie-solución en suelos volcánicos y sus componentes. In: Tosso J (ed) Suelos Volcánicos de Chile. Instituto de Investigación Agropecuaria (INIA). Ministerio de Agricultura, Santiago, Chile, pp 303–333

Gianfreda L, Bollag JM (1994) Effect of soils on the behavior of immobilized enzymes. Soil Sci Soc Am J 58:1672–1681

Gianfreda L, Scarfi MR (1991) Enzyme stabilization: state of the art. Mol Cell Biochem 199:97–128

Gianfreda L, Rao MA, Saccomandi F, Sannino F, Violante A (2002) Enzymes in soil: properties, behavior and potential applications. In: Violante A, Huang PM, Bollag JM, Gianfreda L (eds) Soil mineral-organic matter-microorganism interactions and ecosystem health. Development in soil science 28B. Elsevier, London, pp 301–328

Gill I (2001) Bio-doped nanocomposite polymers: sol-gel bioencapsulates. Chem Mater 13:3404–3421

Huang Q, Shindo H (2000) Effects of copper on the kinetics of free and immobilized acid phosphatase. Soil Biol Biochem 32:1885–1892

Huang Q, Liang W, Cai P (2005) Adsorption, desorption and activities of acid phosphatase on various colloidal particles from an Ultisol. Colloid Surf B 45:209–214

Jara A, Violante A, Pigna M, Mora M (2006) Mutual interactions of Sulfate, oxalate, citrate, and phosphate on synthetic and natural allophanes. Soil Sci Soc Am J 70:337–346

Kelleher BP, Simpson AJ, Willeford OK, Simpson MJ, Stout R, Rafferty A, Kingery WL (2004) Acid phosphatase interactions with organo-mineral complexes: influence on catalytic activity. Biogeochemistry 71:285–297

López R (2006) Evaluación del efecto de molibdeno sobre algunos parámetros bioquímicos del suelo y la planta en andisoles del sur de Chile Tesis para optar al grado de Doctor en Ciencias de Recursos Naturales. Universidad de La Frontera, Temuco, Chile

López R, Alvear M, Gianfreda L, Mora M (2007) Molybdenum availability in Andisols and its effect on biological parameters of soils and red clover (*Trifolium pratense* L.). Soil Sci 172(11):913–924

López R, Rosas A (2008) Effect of acid phosphatase immobilized in allophanic clay on P availability in soil. J Soil Sci Plant Nutr 8:201–202

López R, Rosas A, Rao M, Mora ML, Alvear M, Gianfreda L (2007) Manganese and molyldenum affect acid phophatases from potato. Acata Agr Scand BS P 57:65–73

Lozzi I, Calamai L, Fusi P, Bosetto M, Stotzky G (2001) Interaction of horseradish peroxidase with montmorillonite homoionic to Na+and Ca2+: effects on enzymatic activity and microbial degradation. Soil Biol Biochem 33:1021–1028

Luckarift HR, Spain JC, Naik RR, Stone M (2004) Enzyme immobilization in a biomimetic silica support. Nat Biotechnol 22:211–213

Matus F, Garrido E, Sepúlveda N, Cárcamo I, Panichini M, Zagal E (2008) Relationship between extractable Al and organic C in volcanic soil of Chile. Geoderma 148:180–188

Mayer LM, Schick LL, Hardy K, Wagai R, McCarthy J (2004) Organic matter content of small mesopores in sediments and soils. Geochim Cosmochim Acta 68:3863–3872

Olczack M, Morawiecka B, Watorek W (2003) Plant purple acid phosphatases-genes, structures and biological function. Acta Biochim Pol 50:1245–1257

Parfitt RL (1990) Allophane in New Zealand – a review. Aust J Soil Res 28:343–360

Parfitt RL, Yuan G, Theng BKG (1999) A ^{13}C-NMR study of the interactions of soil organic matter with aluminium and allophane in podzols. Eur J Soil Sci 50:695–700

Rao MA, Gianfreda L (2000) Properties of acid phosphatase-tannic acid complexes formed in the presence of Fe and Mn. Soil Biol Biochem 32:1921–1926

Rao MA, Gianfreda L, Palmiero F, Violante A (1996) Interaction of acid phosphatase with clays, organic molecules and organo-mineral complexes. Soil Sci 11:751–760

Rao MA, Violante A, Gianfreda L (2000) Interaction of acid phosphatase with clays, organic molecules and organo-mineral complexes: kinetics and stability. Soil Biol Biochem 32:1007–1014

Redel Y, Rubio R, Rouanet J, Borie F (2007) Phosphorus bioavailability affected by tillage and crop rotation on a Chilean volcanic derived Ultisol. Geoderma 139:388–396

Redel Y, Rubio R, Godoy R, Borie F (2008) Phosphorus fractions and phosphatase activity in an Andisol under different forest ecosystems. Geoderma 145:216–221

Reetz MT, Tielman P, Wisenhöfer W, Könen W, Zonta A (2003) Second generation sol-gel encapsulated lipases: robust heterogeneous biocatalysts. Adv Synth Catal 345:717–728

Rosas A (2006) Evaluación del efecto del manganeso sobre parámetros bioquímicos de la planta y del suelo en sistemas modelo. Tesis para optar al grado de Doctor en Ciencias de Recursos Naturales. Universidad de La Frontera, Temuco, Chile

Rosas A, Mora M, Jara A, López R, Rao M, Gianfreda L (2008) Catalytic of acid phosphatase immobilized on natural supports in the presence of mangase or molybdenum. Geoderma 145:77–83

Schaeffer W, Keefer KD (1986) Structure of Random Porous Materials: Silica Aerogel. Phys Rev Lett 56:2199–2202

Schützendübel A, Polle A (2002) Plant responses to abiotic stresses: heavy metal-induced oxidative stress and protection by mycorrhization. J Exp Bot 53:1351–1365

Schwertmann U, Taylor RM (1989) Iron oxides. In: Dixon JB, Weld SB (eds) Minerals in soil environments. Soil Science Society of America, Madison, WI, pp 379–438

Shchipunov YA, Karpenko TY, Bakunina IY, Burtseva YV, Zvyagintseva TN (2004) A New precursor for the immobilization of enzymes incide sol-gel derived hibrid silica nanocomposites containing polysaccharides. J Biochem Bioph Meth 58:25–38

Shindo H, Watanabe D, Onaga T, Urakawa M, Nakahara O, Huang Q (2002) Adsorption, activity and stability of acid phosphatase as influenced by selected inorganic soil components. Soil Sci Plant Nutr 48:763–767

Sieffermann G, Millot G (1969) Equatorial and tropical weathering of recent basalts from cameroun allophanes, halloysita, metahalloysita, kaolinite and gibbsite. Intern Clay Conf 1:417–430

Sparks DL (1995) Environmental soil chemistry. Academic, San Diego

Stauton S, Quiquampoix H (1994) Adsorption and conformation of bovine serum albumin on montmorillonite: Modification of the balance between hydrophobic and electrostatic interactions by protein methylation and pH variation. J Colloid Interf Sci 166:89–94

Tso S, Chen Y (1997) Isolation and characterization of a group III isozyme of acid phosphatase from rice plants. Bot Bull Acad Sinica 38:245–250

Vacher R, Woignier T, Pelous J (1988) Structure and self-similarity of silica aerogels. Phy Rev B 37:6500–6503

Violante A, Krishnamurti GSR, Huang PM (2002) Impact of organic substances on the formation of metal oxides in soil environments. In: Huang PM (ed) Interactions between soil particles and microorganism and their impact on the terrestrial environment. Wiley, New York, pp 133–188

Wada KJ (1985) The distinctive properties of Andosols. Springer, Heidelberg

Wada K (1989) Allophane and immogolite. In: Dixon JB, Weld SB (eds) Minerals in soil environments. Soil Science Society of America, Madison, WI, pp 1051–1087

Wang WJ, Dalal RC, Moody PW, Smith CJ (2003) Relationships of soil respiration to microbial biomass, substrate availability and clay content. Soil Biol Biochem 35:273–284

Wei Y, Xu J, Feng Q, Lin M, Dong H, Zhang W, Wang C (2001) A novel method for enzyme immobilization: direct encapsulation of acid phosphatase in nanoporus silica host materials. J Nanosci Nanotechnol 1:83–93

Woignier T, Braudeau E, Doumenc H, Rangon L (2005) Supercritical drying applied to natural "gels": allophanic soils. J Sol-Gel Sci Technol 36:61–68

Woignier T, Primera J, Hashmy A (2006) Application of the DLCA model to "natural" gel: the allophanic soils. J Sol-Gel Sci Technol 40:201–207

Woignier T, Pochet G, Doumenc H, Dieudonné P, Duffours L (2007) Allophane: a natural gel in volcanic soils with interesting environmental properties. J Sol-Gel Sci Technnol 41:25–30

Yasuhisa Y, Karube J (1999) Application of a scaling law to the analysis of allophane aggregates. Colloid Surface A 151:43–47

Yenugün B, Güvenilir Y (2003) Partial purification and kinetic characterization of acid phosphatase from garlic seedling. Appl Biochem Biotech 107:677–687

Zambonelli C, Roberts MF (2003) An iron-dependent bacterial phospholipase D reminiscent of purple acid phosphatases. J Biol Chem 278:13706–13711

Zunino H, Borie F, Aguilera S, Martin JP, Haider K (1982) Decomposition of ^{14}C-labelled glucose, plant and microbial products and phenols in volcanic ash-derived soils of Chile. Soil Biol Biochem 14:37–43

Chapter 18
Screening, Characterisation and Optimization of Microbial Pectinase

V. Suneetha and Zaved Ahmed Khan

18.1 Introduction

Enzymes are biocatalysts that possess extra ordinary specificity and remarkable catalysis power. They are superior to chemical catalysis due to their environment friendly character. As against traditional chemical methods, enzymatic processes generally yield products of improved quality and reduce the use of hazardous and polluting chemicals by their applications in industry with the advancement of scientific and industrial developments. The rapid growth of enzyme technology has resulted in the emergence of a new branch of science viz., Biotechnology (Chaplin and Bucke 1990).

Nature is the largest laboratory for the development of novel biocatalysts for industrial processes. Microorganisms in particular have been investigated as sources for new enzymes and other bioactive molecules. Pectinases are important industrial enzymes, used for cloud point stabilization in juices, to increase pulp extraction from fruits and vegetables, in cocoa bean fermentation and for soluble tea preparations. More recently, they have been used in the textile industries for the degumming of fiber crops, in wastewater treatment and in the paper industries (Kashyap et al. 2001). Actually, there are processes using a specific type of pectinase, such as the preparation of citrus and orange juices, where endopolygalacturonases are preferred to maintain the turbidity and the opaque aspect of juices (Manachini et al. 1988). Pectic substances, present in the primary cell wall and middle lamella of higher plants, contribute to the firmness and structure of plant tissues (Sathyanarayana and Panda 2003). Different pectinolytic enzymes are involved in the breakdown of pectin and are widely distributed in higher plants and microorganisms. They are important for plants as they help in cell wall

V. Suneetha (✉) and Z.A. Khan
School of Biotechnology, Chemical and Biomedical engineering, VIT University, Vellore 632014, Tamilnadu, India
e-mail: vsuneetha@vit.ac.in, zavedkhan@vit.ac.in, khan.zaved@gmail.com

G. Shukla and A. Varma (eds.), *Soil Enzymology*, Soil Biology 22,
DOI 10.1007/978-3-642-14225-3_18, © Springer-Verlag Berlin Heidelberg 2011

extension and fruit softening. They have a role in maintaining ecological balance by causing decomposition of plant material.

18.2 Structure of Pectin

Pectin is a heterogeneous grouping of acidic structural polysaccharides found in fruit and vegetables and mainly prepared from "waste" citrus peel and apple pomace. Pectins are complex branched heteropolysaccharides primarily containing an α-(1–4) polygalacturonic acid backbone which can be randomly acetylated and methylated. Three different pectins have been isolated from plant cell walls:

- Homogalacturonans are composed of the simple α-(1–4) polygalacturonic acid backbone.
- Substituted homogalacturonans are modifications of this backbone with β-D-xylose branching at C_3, or apiofuranose substitutions in the backbone with β-D-Apiosyl-(1,3′)-β-D-Apiose branching.
- Rhamnogalacturonan I contains alternating α-(1–4) galacturonosyl and α-(1–2) rhamnosyl residues, with primarily oligo α-(1–3) arabinose and oligo β-(1–4) galactose branching. Rhamnogalacturonan II is composed of the simple α-(1–4) polygalacturonic acid backbone with complex branching composed of up to 11 different monosaccharide types.

The structural unit of Pectin has a complex structure. Preparations consist of substructural entities that depend on their source and extraction methodology. Commercial extraction causes extensive degradation of the neutral sugar-containing side chains. The majority of the structure consists of homopolymeric partially methylated poly-α-(1→4)-D-galacturonic acid residues (Fig. 18.1) but there are substantial "hairy" non-gelling areas (Fig. 18.2) of alternating α-(1→2)-L-rhamnosyl-α-(1→4)-D-galacturonosyl sections containing branch-points with mostly neutral side chains (1–20 residues) of mainly L-arabinose and D-galactose (rhamnogalacturonan I). Pectins may also contain rhamnogalacturonan II sidechains containing other

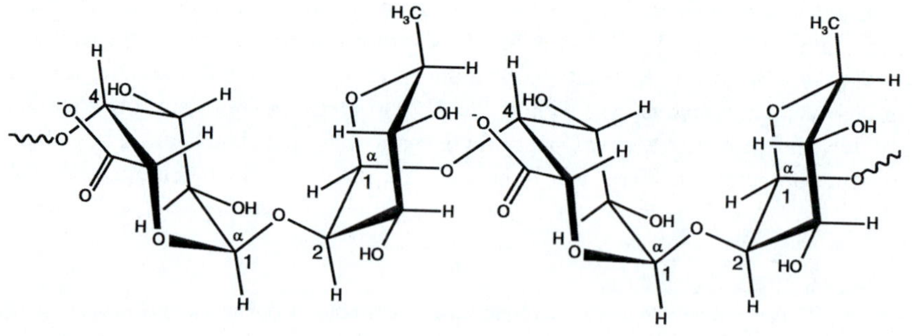

Fig. 18.1 Structure of pectin (From Schols et al. 1990)

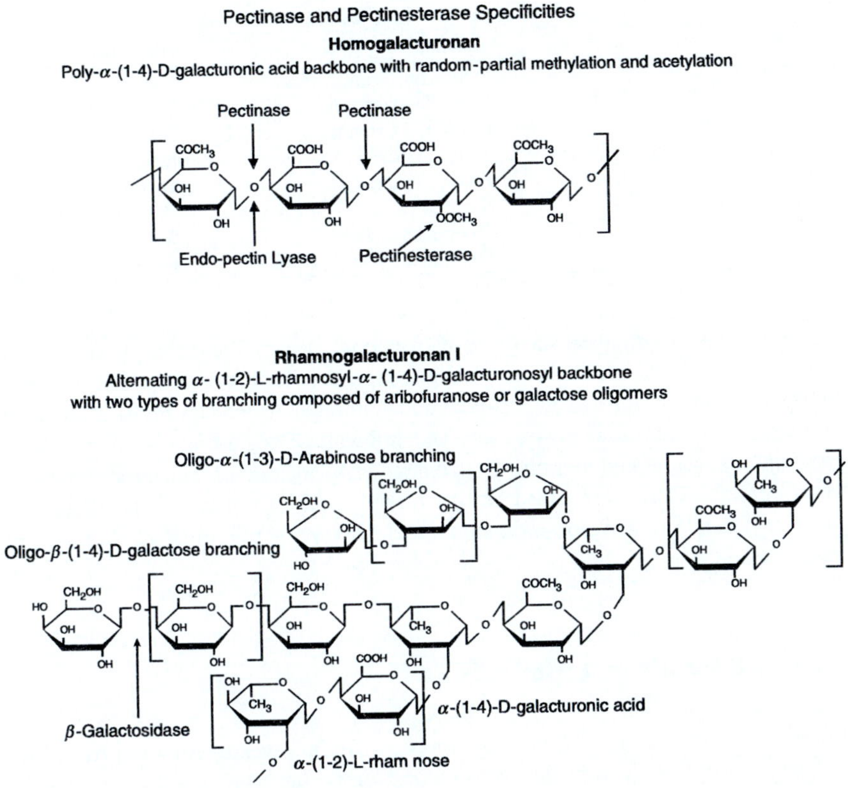

Fig. 18.2 Mode of action of pectinase

residues such as D-xylose, L-fucose, D-glucuronic acid, D-apiose, 3-deoxy-D-*manno*-2-octulosonic acid (Kdo) and 3-deoxy-D-*lyxo*-2-heptulosonic acid (Dha) attached to poly-α-(1→4)-D-galacturonic acid regions.

Pectinase was found to be the most efficient commercialized enzyme in degrading the fruit waste. Hydrolysis of pectic materials found in plants had an average of 53.6% decrease from the waste's original mass while cellulase had an average of 26.4% decrease from the original mass. In addition, as the length of time and the amount of concentration of the enzyme increased, those factors contributed in degrading a higher percentage of the fruit waste. Microbial pectinases account for 25% of the global food enzyme sale (Jayani et al. 2005). Applications of pectinases include fruit juice extraction and clarification, wine processing, oil extraction, coffee and tea leaf fermentation, retting and degumming of fibers, etc. (Kashyap et al. 2000).The commercial preparations of pectinases are produced mainly from fungi, especially *Aspergillus niger* (Torres et al. 2005).(Table 18.1).

Actinomycetes, the wonderful microorganisms are highly attractive as cell factories or bioreactors for applications in industrial, agricultural, environmental and pharmaceutical fields. Species of Actinomycetes have paved the way for

Table 18.1 Some commercial available pectinases (Kashyap et al. 2000)

Supplier	Location	Brand name
C.H. Boehringer Sohn	Ingelheim, West Germany	Panzym
Ciba-Geigy, A.G.	Basel, Switzerland	Ultrazyme
Grinsteelvaeket	Aarthus, Denmark	Pectolase
Kikkoman Syoyu, Co	Tokyo, Japan	Sclase
Schweizerische, Ferment, A.G	Basel, Switzerland	Pectinex
Societe Rapidase, S.A.	Seclin, France	Rapidase, Clarizyme
Wallerstein, Co.	Des plaines, USA	Klerzyme
Rohm, GmbH	DarmStadt, West Germany	Pectinol, Rohament

biochemical and structural analysis of important proteins and the production of such proteins as recombinants on a commercial scale. In this regard, there is a need for optimization of nutritional, physical and Computer Aided Design for maximum production of microbial pectinases. The knowledge regarding the intracellular environment, screening, identification, characterization and optimization (physical, nutritional and Computer Aided Design model (CAD) of microbial pectinase has made it possible to develop indigenous technology for microbial exploitation of Pectinase.

18.3 Materials and Methods

In our study we screened 100 soil samples for Actinomycetes isolation from fruit industry waste (taken from the surroundings of Tropic fruit products m/s Exotic fruits private ltd).

Media Employed:

Cultivation Media (g/100 ml)

- $K_2HPO_4 = 0.5$
- Casein = 3.0
- Maize Starch = 10.0
- Peptone = 1.0
- Yeast Extract = 1.0
- Malt Extract = 10.0
- Agar = 3.0

The pH of the medium was adjusted to 7.5 and 50 μg/ml of the antibiotic cyclohexamide was added to the media to inhibit the growth of fungi

The strain was grown for 24 h in 250 ml Erlenmeyer flask containing 100 ml of liquid medium. Adequate aeration was provided by agitation at 150 rpm at 30°C. The inoculum contained 2.5×10^{-4} CFU/ml.

Seed Media (in g/L):

- $KH_2PO_4 = 2$
- $K_2HPO_4 = 2$

- $(NH_4)_2SO_4 = 2$
- Yeast Extract = 3
- Pectin(SRL) = 5 (Aguilar and Huitron 1990)

The pH of the medium was adjusted to 7.6 using 1 M NaOH. Before production, we can inoculate the isolated cultures into the above seed media for innoculum development.

Production Media

Composition in (g/L):

- Pectin = 10
- Sucrose = 10
- Tryptone = 3
- Yeast extract = 2
- KCl = 0.5
- $MgCl_2.7H_2O = 0.5$
- $(NH_4)_2SO_4 = 2.0$

Supplemented with mineral salt solution of composition (g/100 ml)

- $CuSO_4.5H_2O = 0.04$
- $FeSO_4 = 0.08$
- $Na_2MoO_4 = 0.08$
- $ZnSO_4 = 0.8$
- $Na_2B_4O_7 = 0.004$
- MnSO4 = 0.008

Of 1 ml distilled water to make 1 L solution, pH should be maintained at 7.8. So an antifungal antibiotic (conc. 100 μg/ml) was added. Inoculated plates were incubated at 30°C for 5–7 days.

Maintenance Media

Yeast Extract Malt Extract Media

- Yeast extract = 1.0 gm
- Malt extract = 1.0 gm
- Casein = 0.1 gm
- Distilled water = 100 ml
- pH = 8.3

The strain was grown for 24 h in 250-ml Erlenmeyer flask containing 100 ml of liquid medium. Adequate aeration was provided by agitation at 175 rpm at 30°C. The inoculum contained 2.5×10^{-4} CFU/mL.

Designed Media

Composition in (g/L):

- Pectin = 10.0
- Starch = 8.0
- Tryptone = 3.0
- Yeast extract = 2.0

- KCl = 0.5
- $MgCl_2.7H_2O$ = 0.5
- Ground nut cake = 2.0

Supplemented with mineral salt solution of composition (g/100 ml)

- $CuSO_4.5H_2O$ = 0.04
- $FeSO_4$ = 0.08
- Na_2MoO_4 = 0.08
- $ZnSO_4$ = 0.8
- $Na_2B_4O_7$ = 0.004
- $MnSO_4$ = 0.008
- Distilled water-1 ml

The pH should be maintained at 7.8. There is the addition of antifungal antibiotic (conc. 50 µg/ml) to inhibit the growth of fungi and to facilitate the growth of Actinomycetes. The Inoculated plates were incubated at 50°C for 5–7 days.

18.4 Screening of Pectinolytic *Actinomycetes*

Hundred soil samples were screened for pectinolytic Actinomycetes. The soil samples were collected from different places of fruit industries present around our university campus.

- Screening was done by a Baiting technique. 3 of the 60 cultures were exhibiting digestion zones around the innoculum after 72 h of incubation, and then a rapid digestion was ensured .This was partly due to the slow growth of the remaining cultures.
- We repeated the technique several times for screening of pectinolytic *Actinomycetes* for confirmation.
- To avoid some of the misinterpretation and to see whether pectin stimulates the production of pectinase, the Actinomycetes were grown in shaker flasks containing pectin in addition to a small amount of hydrolyzed casein.

18.5 Assay of Pectinase

The principle of this assay depends upon measuring the amount of fermentation fluid or culture filtrate released from the Actinomycetes. Pectin (SRL) can be used as substrate for this assay.

- To make an extract of the fermentation, blend 2 cm^3 of water for every 1 g of pectin. Prepare at least 25 cm^3 of extract.

- Collect and label two boiling tubes and place 25 cm^3 of culture fluid in each of them.
- Add 25 cm^3 of extract to one of the tubes and 25 cm^3 of water to the other, to act as a control.
- Use a glass rod to mix thoroughly the contents of both tubes and then leave them in a boiling tube rack in a water bath at 35°C for at least 30 min.
- Take two similar sized funnels (funnel size about 100 cm^3 is suitable) and support them over two similar small (25 cm^3) measuring cylinders.
- After the incubation period, pour the contents of the two boiling tubes into the two funnels, and allow the filtrates to drain into the measuring cylinders. Allow at least 5 min for the draining to finish and note the readings at 280 nm using UV spectrophotometer (Shimadzu 260). The differences in the volume of culture filtrate between the two tubes give a measure of the pectinase activity in the extract.

18.6 Optimization of Pectinase

To optimize, the production of pectinase can be studied by taking both physical and chemical parameters such as Agitation, pH, temperature, carbon source and nitrogen source. CAD model regarding the optimization of pectinase was adopted.

1. Effect of agitation on pectinase:

To study the effect of agitation on fermentative production of pectinase, the fermentation flasks were incubated on a shaker adjusted to 50,100, and 150 rev/min at 30°C. Simultaneous flasks were also maintained under still condition, without shaking, for comparison.

2. Effect of pH

"pH" the negative logarithm of hydrogen ion concentration, has influence over all metabolisms of microorganisms and enzyme production. To study the effect of pH, the pH of the medium was adjusted to 4.5, 5.5, 6.5,7,8,9, and 10.The temperature of incubation was at 37°C and pectinase assay was done as described above.

3. Effect of Temperature

Temperature is one of the critical factors influencing the metabolism of micro organisms. A study was carried out by inoculating the organisms at 28, 37, 45, 50, 60, and 70°C .The pH of the medium was adjusted to 7–8 and the samples were assayed for pectinolytic activity.

4. Effect of carbon source on production of pectinase:

The different carbon sources were replaced with (1:1 w/v) to study the influence of carbon source on pectinase production. Sucrose, fructose, maltose, lactose, and starch were used as energy sources.

5. Effect of Nitrogen source on production of pectinase:

The different nitrogen sources were studied with (1:1 w/v) to study the influence of nitrogen source on pectinase production. The effect of nitrogen source on fermentation was determined by adding nitrogen sources like peptone, yeast extract, casein and GNC separately to the media.

6. Thermostability studies

Thermo stability of the enzyme activities of the crude Actinomycetes culture filtrates were determined by incubating the filtrates at 70, 60, and 50°C for various durations ranging from hours to weeks. The treated filtrates were then assayed for the respective enzyme activities by incubating an appropriately diluted aliquot of the treated sample with the assay substrate in citrate phosphate buffer at pH 5.0 and 60°C for 30 min, as described in the previous section.

18.7 Designing Model

We incorporated a new methodology to depict the pectin-pectinase interaction for various strains of Actinomycetes (S1, S2, S3), isolated with the help of 3-D modeling using sophisticated CAD Softwares.

18.8 Conclusion

Biotechnological solutions for environmental sustainability are recent innovations that help in the growth of the nation and are a boon for the welfare of human beings for the present and forthcoming generations. Pectinolytic enzymes or pectinases are a group of enzymes that hydrolyze the pectic substances, present mostly in plants. Enzymes are usually offered as "cocktails" of several activities rather than a single enzymatic activity. However, in many cases the enzyme activities can still act on the same composition, as the composition can have a complex chemical structure having various types of chemical bonds, requiring different enzyme activities for breakdown. An example of this is the enzyme cocktails offered as "pectinase". Such pectinase composition often contains one or more of the following activities: polygalacturonase, pectin lyase and pectin methyl esterase. Pectinase preparations are often used in fruit juice processing. It is preferred in the present invention that the enzyme preparation used contains at least one of these three activities mentioned, preferably two, more preferably all three. Pectinase has its largest application in fruit juice extraction and clarification process. In this study we have shown that addition of yeast extract, magnesium sulfate, manganese sulfate, and starch to the media formulation increased the activity of pectinase production. Pectin is a major component present in cell wall of the plants and the presence of pectin increases the viscosity of the juice. Pectinase has been used to increase the pressing efficiency of fruit juice. Pectinase has found commercial application in softening the peel of citrus and various other fruits. In future this may be used to replace hand cutting for the production of canned segments .The various screening methodologies for pectinase extraction, characterization and pectinase technology in juice industries will be economically worthy and helps in industrial development.

Acknowledgments The author Dr. V. Suneetha wants to acknowledge the DST (Department of Science and Technology) for awarding young scientist and financial help for carrying out to this research. Authors also want to express sincere thanks to VIT University for providing infrastructure facilities to carry out this research.

References

Aguilar G, Huitron C (1990) Constitutive exo-pectinase produced by *Aspergillus sp*. CH-4-1043 on different carbon source. Biotechnol Lett 12:655–660

Chaplin MF, Bucke K (1990) Enzyme technology. Cambridge University Press, NewYork, pp 1–67

Jayani RS, Saxena S, Gupta R (2005) Microbial pectinolytic enzymes: a review. Process Biochem 40:2931–2944

Kashyap DR, Vohra PK, Chopra S, Tewari R (2000) Applications of pectinases in the commercial sector: a review. Bioresour Technol 77:215–227

Kashyap DR, Vohra PK, Chopra S, Tewari R (2001) Applications of pectinases in the commercial sector: a review. Bioresour Technol 77:215–227

Manachini PL, Parani C, Fortina MG (1988) Pectic enzymes from *Aspergillus pullulans* LV10. Enzyme Microb Technol 10:682–685

Sathyanarayana NG, Panda T (2003) Purification and biochemical properties of microbial pectinases – a review. Process Biochem 38:987–996

Schols H, Geraeds C, Searle-Van-Leeuwen M, Komelink F, Voragen A (1990) Rhamnogalacturonase: a novel enzyme that degrades the hairy region of pectins. Carbohydrate Res 206:105–115

Torres EF, Aguilar C, Esquivel JCC, Gonzalez GV (2005) Pectinases. In: Pandey A, Webb C, Soccol CR, Larroche C (eds) Enzyme technology. Asiatech, New Delhi, India, pp 273–296

Chapter 19
Molecular Techniques to Study Polymorphism between Closely Related Microorganisms in Relation to Specific Protein Phosphatase

Rajani Malla, Utprekshya Pokharel, Ram Prasad, and Ajit Varma

19.1 Introduction to Sebacinales

Mycorrhizal taxa of *Sebacinaceae*, including mycobionts of ectomycorrhizas, orchid mycorrhizas (McCormick et al. 2004), ericoid mycorrhizas, and jungermannioid mycorrhizas, are distributed over two subgroups. One group contains species with microscopically visible basidiomes, whereas members of the other group probably lack basidiomes. *Sebacina* appears to be phylogenetic; current species concepts in *Sebacinaceae* are questionable. *S. vermifera* sensu consists of a broad complex of species possibly including mycobionts of jungermannioid and ericoid mycorrhizas. Extrapolating from the known rDNA sequences in *Sebacinaceae*, it is evident that there is a cosm of mycorrhizal biodiversity yet to be discovered in this group. Taxonomically, the *Sebacinaceae* recognized a new order, the *Sebacinales* (Weiß et al. 2004). The order primarily contains the genera *Sebacina, Tremelloscypha, Efibulobasidium, Craterocolla*, and *Piriformospora*. Proteomics and genomics data about *P. indica* fungus have recently been described (Peškan-Berghöfer et al. 2004; Shahollari et al. 2005; Kaldorf et al. 2005; Malla et al. 2004; Malla et al. 2007a, b).

Piriformospora indica and Sebacina vermifera (Fig 19.1) from *Sebacinales* are documented to function as biofertilizer, bioregulator, bioprotector, and supplement to the health of the plant and the soil (Verma et al. 1998; Varma et al. 1999; Malla et al. 2002; Malla et al. 2005). More recently, they are documented to act as an agent for biological hardening of tissue culture-raised plants. Despite their enormous potential, their biotechnological applications could not be exploited to the

R. Malla
Central Department of Biotechnology, Tribhuvan University, Kathmandu, Nepal

U. Pokharel
Department of Microbiology, Punjab University, Patiala, Chandigarh, Punjab

R. Prasad and A. Varma (✉)
Amity Institute of Microbial Technology, Amity University Uttar Pradesh, Sector 125, Noida 201303, India
e-mail: ajitvarma@aihmr.amity.edu

G. Shukla and A. Varma (eds.), *Soil Enzymology*, Soil Biology 22,
DOI 10.1007/978-3-642-14225-3_19, © Springer-Verlag Berlin Heidelberg 2011

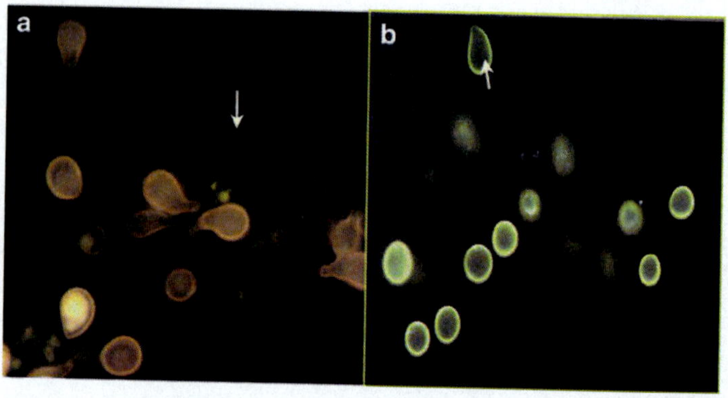

Fig. 19.1 Pear-shaped chlamydospores of (**a**) *Piriformospora indica* and (**b**) *Sebacina Vermifera*

level they deserve. The axenic cultivability of these fungi provided ample opportunity to study the comparative proteomic and genomic relationship (Malla et al. 2007a, b; Malla and Varma 2007) to establish variability in between these two fungi (Weiß et al. 2004; Malla and Varma 2005; Malla et al 2005, 2006a, b, 2007a, b).

Phosphatases are the enzymes of wide specificity which cleave phosphate ester bonds, and this plays an important role in the hydrolysis of polyphosphates and organic phosphates. Acid and alkaline phosphatases (ACPase and ALPase) are the two forms of intracellular phosphatase active at acidic and alkaline condition, respectively. ACPase was found to be mainly involved in the uptake of P by the fungal mycelia, and ALPase is linked with its assimilation (Fries et al. 1998).

Acid phosphatase (ACPase) in soil originates from both plants and fungi, while ALPase is believed to be of purely microbial origin (Gianinazzi-Pearson and Gianinazzi 1978; Tarafdar and Rao 1996). Indications such as ACPase of fungal origin has a higher hydrolyzing efficiency than enzymes of plant origin (Tarafdar et al. 2001) are found. Studying ACPase (Fig. 19.2) is difficult due to their multiform occurrence in organisms, their relative nonspecificity, their small quantity, and their instability in dilute solution. Their study is also complicated by wide variations in the activity and property of isozyme between species and between different stages in each plant's development (Alves et al. 1994).

Alkaline phosphatase has been proposed as a marker for analyzing the symbiotic efficiency of colonization (Tisserant et al. 1993). The argument for this was that ALPase is an important enzyme in metabolic processes, leading to P transfer to the host plant. The alkaline phosphatase activity is shown to be increased sharply prior to mycorrhizal stimulation of plant growth and then declined as the mycorrhizal colonization gets aged and P accumulates within the host. Arbuscule is speculated to be a site of nutrient exchange between the host plant and AM fungi (Cox et al. 1980). Phosphate efflux from the fungi to the host plant at arbuscules is supported by the recent discovery of novel plant Pi transporters that are localized around arbuscules and acquire Pi from the fungi (Rausch et al. 2001; Harrison et al. 2002; Paszkowski et al. 2002).

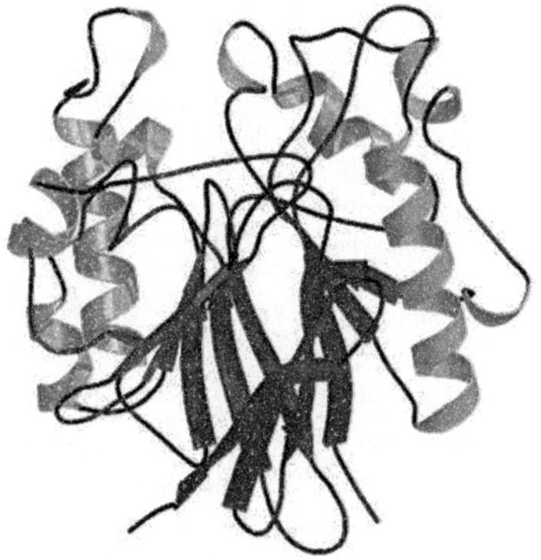

Fig. 19.2 A structural view of acid phosphatase

19.2 Techniques to Study Polymorphism

19.2.1 Cultivation of Fungi

19.2.1.1 Materials

Aspergillus medium

Constituents	Composition (g l^{-1})
Glucose	10
Peptone	2
Yeast extract	1
Casein amino acid	1
Vitamin stock solution	1 ml
Macroelements from stock	50 ml
Microelements from stock	1 ml
Agar	10
pH	6.5
Macroelements (major elements)	**Stock (g l^{-1})**
$NaNO_3$	120.0
KCl	10.4
$MgSO_4{\cdot}7H_2O$	10.4
KH_2PO_4	16.3
K_2HPO_4	20.9
Microelements (trace elements)	**Stock (g l^{-1})**
Zn $SO_4{\cdot}7H_2O$	22
H_3BO_3	11
$MnCl_2{\cdot}4H_2O$	5
$FeSO_4{\cdot}7H_2O$	5

(*continued*)

Microelements (trace elements)	Stock (g l^{-1})
$CoCl_2 \cdot 6H_2O$	1.6
$CuSO_4 \cdot 5H_2O$	1.6
$(NH_4)_6\ Mo_7O_{27} \cdot 4H_2O$	1.1
Na_2EDTA	60
Vitamins	**Percent**
Biotin	0.05
Nicotinamide	0.5
Pyridoxal phosphate	0.1
Amino benzoic acid	0.1
Riboflavin	0.25
pH 6.5	

The stocks were stored at 4°C and vitamin was stored at –20°C in aliquots. The stock of $FeSO_4 \cdot 7H_2O$ was prepared separately.

Transfer actively growing colonies of *P. indica* and *S. vermifera* sensu in modified *aspergillus* medium (Hill and Kaefer 2001) and incubate for 10 days at 28 ± 2°C in the dark with constant shaking at 120 rpm. The morphological features of the fungi can be studied with the aid of Leica microscope (Type 020-518.500, Germany).

19.2.2 *Enzyme Assay (Straker and Mitchell 1986)*

19.2.2.1 Extraction of Protein

Equipments

- Centrifuge
- Spectrophotometer

Reagents

- Disodium *p*-nitrophenyl phosphate (Sigma Chemical Co. N-2640)
- 0.05 M Sodium acetate buffer, adjusted to 5.3
- 0.05 M NaOH
- Phosphate buffer saline (PBS), pH 7.4

Protein Extraction Buffer (Rosendahl 1994)

Tris–HCl	10 mM
$NaHCO_3$	10 mM
$MgCl_2$	10 mM
Na_2EDTA	0.1 mM
β-mercaptoethanol	10 mM

(*continued*)

Sucrose	150 g l^{-1}
Triton X-100	1 ml l^{-1}
Protease inhibitors	From stock (−80°C)

Dissolve in distilled water and pH was adjusted to 8.0.

Procedure

Harvest and homogenize the biomass using extraction buffer of different according to need. Store the crude enzyme extract at −80°C in aliquots. Using those aliquots ALPase and ACPase activities can be determined spectrophotometrically with *P*-nitro phenyl phosphate (Sigma Chemical Co.) as a substrate.

19.2.2.2 Optimization of Physical Conditions

Equipment

- Water bath shaker
- Spectrophotometer

Reagent

- Disodium *p*-nitrophenyl phosphate (Sigma Chemical Co. N-2640) (2 mg *p*-nitrophenyl phosphate ml^{-1} in deionised water)
- 0.05 M Sodium acetate buffer, adjusted to pH 3.0, 3.5, 4.0, 4.5, 5.0, 5.5, 6.0, 6.5
- 0.05 M Tris–maleate buffers, adjusted to pH 7.0, 7.5, 8.0, 8.5, 9.0, 9.5, 10.0
- 0.05 M NaOH

Procedure

To 25 μl of crude enzyme extract, add 100 μl of *p*-nitrophenyl phosphate solution with 100 μl of buffer of different pH. Incubate the reaction mixture in a shaking water bath at 37°C for 30 min. Stop the reaction by adding 1 ml of 50 mM NaOH. Deactivate substrate suspension of each of the fungal treatments by heating to boiling in a microwave oven, and these can be used as substrate blanks to determine the background level of the substrate. The addition of distilled water instead of the substrate to the buffer solution can be taken as blanks to check the background of the *p*-nitrophenyl phosphate substrate. The absorbance is measured at 420 nm after filtration through a 0.45 μm (Millex, syringe driven filter unit, Millipore Co. MA 01730, USA). The amount of nitro phenol release can be estimated from a calibration curve of *p*-nitrophenol (spectrophotometer grade, Sigma, 104–8) ranging from 0 to 2 μmol ml^{-1}.

One unit of enzyme activity is defined as the amount of enzyme required to catalyze the formation of 1 μmol of *p*-nitrophenol per minute per ml under standard assay conditions, 37°C, shaking at 120 rpm in water bath shaker (CH-4103 BOTTMINGEN, GFL, Germany).

19.2.2.3 Calculation (see Fig. 19.3)

$$\text{Enzyme activity unit/mg} = \frac{\text{E}\Delta 410\ \text{nm/m} \times \text{Totalvolume} \times \text{D.F.}}{18.5\ (\text{Enz. Vol})\ \text{mg/ml}}.$$

$$\text{Specific Activity} = \frac{\text{Protein of interest (units) h}}{\text{Total Protein conc. Mg}}.$$

Equipment

- Water bath shaker
- Spectrophotometer

Reagents

- *p*-Nitrophenyl phosphate (Sigma Chemical Co.) (2 mg *p*-nitrophenyl phosphate per ml in deionized water)
- 50 mM Sodium acetate buffer, pH 5.3
- NaOH 50 mM

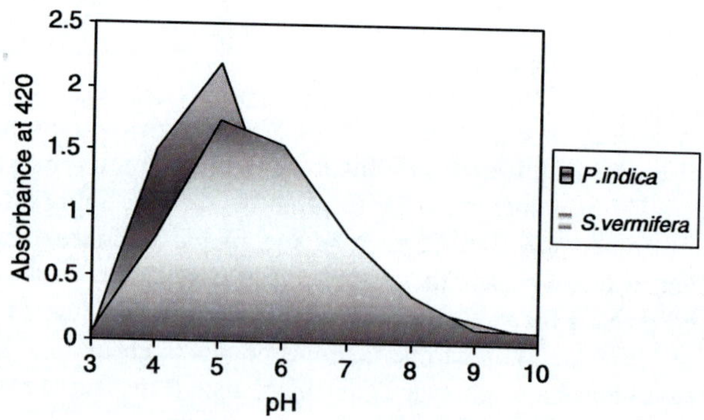

Fig. 19.3 Effect of pH on ACPase activity in *P. indica* and *S. vermifera* sensu

Procedure

The extracted crude enzyme can be assay for optimization of enzyme activity at different temperature ranging from 15 to 75°C under standard assay conditions of pH 5.3, shaking at 120 rpm for 30 min in water bath shaker. The absorbance is measure at 420 nm.

Calculation: Same as above (Fig. 19.4).

19.2.2.4 Determination of V_{max} and K_m value

Equipment

- Water bath shaker
- Spectrophotometer

Reagents

- *p*-Nitrophenyl phosphate (phosphomonoester) di-sodium salt prepared in concentration of 100–1,000 μM
- 50 mM Sodium acetate buffer, pH 5.3

Procedure

Enzyme kinetic studies can be performed at 37°C using *p*-nitrophenyl phosphate di-sodium salt as substrate in the concentration of 100–1,000 μM. The reaction mixture contains 200 μl of *p*-nitrophenyl phosphate solution with 200 μl of acetate buffer and 50 μl of protein (Fig. 19.5).

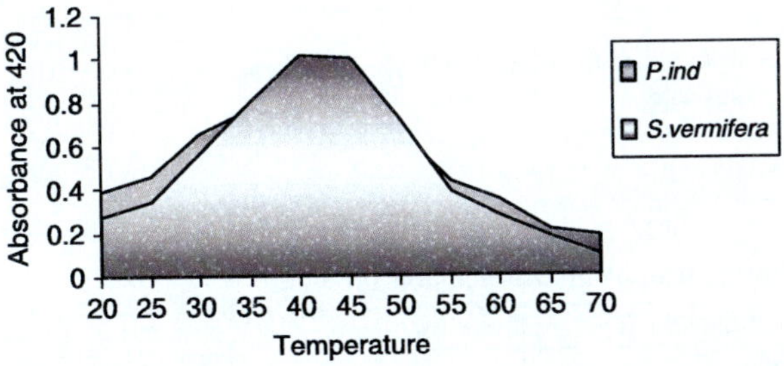

Fig. 19.4 Effect of temperature in ACPase activity in *P. indica* and *S. vermifera* sensu

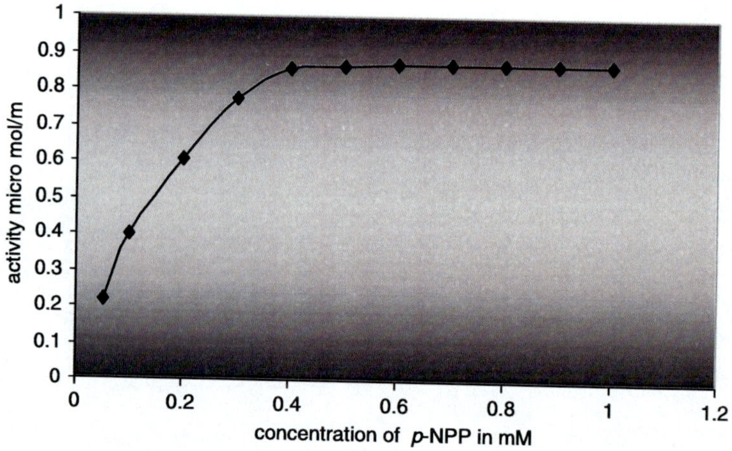

Fig. 19.5 Optimization of *p*-NPP (substrate) for activity of the enzyme

19.2.2.5 Two-Dimensional PAGE to Show Protein Polymorphism (Gravel and Golaz 1996)

Equipment

– Two-Dimensional PAGE apparatus Set (Bio-Rad, Hercules, CA)

Reagents

First dimension sample buffer

Urea	9 M
β-mercaptoethanol	0.5%
Biolyte 5–7 ampholyte	1.6%
Biolyte 3–10 ampholyte	0.4%
Bromophenol blue (w/v)	0.05%

Lysis solution

DTT	0.1 g
CHAPS (cholamidopropyldimethylhydroxypropane sulfonate	0.4 g
Urea	5.4 g
Biolyte 3/10 ampholyte	500 µl
Double-distilled water	6 ml

Preparation of first dimension tube gels

First dimension gel monomer solution	Concentration
Urea	9.2 M
Total monomer acrylamide/bisacrylamide stock	4.5%

(*continued*)

First dimension gel monomer solution	Concentration
Biolyte 5/7 ampholyte	0.1%
Biolyte 3/10 ampholyte	0.4%
CHAPS	1.5%
Nonidet P-40	0.5%

Running buffers for 2D

Cathodic buffer	Freshly prepared 20 mM NaOH in double-distilled water
Anodic buffer	Freshly prepared 10 mM H_3PO_4

Capillary gel equilibration buffer

0.5 M Tris–HCl, pH 8.8	40 ml
10.0% SDS stock solution	80 ml
0.05% Bromophenol blue stock solution	8 ml
Distilled water	150 ml

This solution can be stored at room temperature for 2–3 months.

Preparation of Second Dimension SDS gels

- Acrylamide and *N, N′*-methylene-bis-acrylamide stock: a stock solution containing acrylamide and *N, N′*-methylene-bis-acrylamide (29.2:0.8) was prepared in deionized warm water (to assist the dissolution of the bisacrylamide), pH 7, stored in dark bottle in room temperature
- Sodium dodecyl sulphate (SDS): 10% (w/v) stock solution prepared in deionized water, stored at room temperature
- Ammonium Persulfate (APS) Stock Solution
- Freshly prepared 10% w/v APS and stored at 4°C protected from light and made fresh every 2–3 weeks.
- *N, N, N′, N′*, tetra methylene ethylene diamine (TEMED) 0.04%

Resolving gel buffer stock: 1.5 M Tris (pH 8.8); the solution was filtered through Whatman No. 1 filter paper and stored at 4°C

Stacking gel buffer stock: 1 M Tris–HCl (pH 6.8); the solution was filtered through Whatman No. 1 filter paper and stored at 4°C

Running buffer stock (10×): 0.25 M Tris, 2.5 M glycine, 1% SDS (pH 8.3). The solution was stored at 4°C

Procedure

The first dimension is performed with the tube cell model 175 (Bio-Rad, Hercules, CA) and glass capillary tube (1.0–1.4 mm internal diameter and 210 mm long) as described by Gravel and Golaz 1996. Ampholyte pH 5.0–7.0 and 3.0–10.0 is from Bio-Rad. Load the samples on the top of the capillary (cathodic side); 80 μg of the sample was loaded using Hamilton syringe. Isoelectric Focusing (IEF) can be carried at 200 V constant voltage for 2 h, followed by 500 V constant voltage for 2 h and finally 800 V constant voltages for 16 h (overnight). Equilibrate the tube gels in equilibration buffer for ½ h at room temperature. The protean II chamber (Bio-Rad) is employed for second dimension. Cast the gel (160 × 200 × 1.5 mm) in the casting chamber (Bio-Rad). Transfer tube gels on the top of SDS polyacrylamide gel and separate at 90 V constant voltage in stacking gel and 120 V constant voltage in separating gel. Stain the proteins in the 2D gel with silver staining method.

19.2.2.6 Silver Staining (Horst 2000; Oakley et al. 1980)

Fixative

Methanol	40%
Glacial acetic acid	10%

Impregnating solution

$AgNO_3$	0.2% (w/v)
HCHO	75 μl of 37% solution

Developer

Na_2CO_3	2% (w/v)
HCHO	50 μl of 37% solution
NaS_2O_3	50 μl of 0.02% (w/v)

Terminator

- 10% glacial acetic acid

Procedure

Incubate the gel in fixative for 1 h, then wash thrice with milli Q water for 30 min each under constant gentle shaking conditions. Treat the gel with $Na_2S_2O_3$ (0.02% w/v) for 1 m and wash thrice in milli Q water for 15 s each. Incubate the gel in impregnating solution, rinse thrice in milli Q water for 15 s each. Subsequently, incubate the gel in developer under constant gentle shaking conditions till the desired intensity of protein bands develop. Rinse the gel with water and dip in 10% acetic acid to break the staining. The gel can be dried in gel dryer (Fig. 19.6).

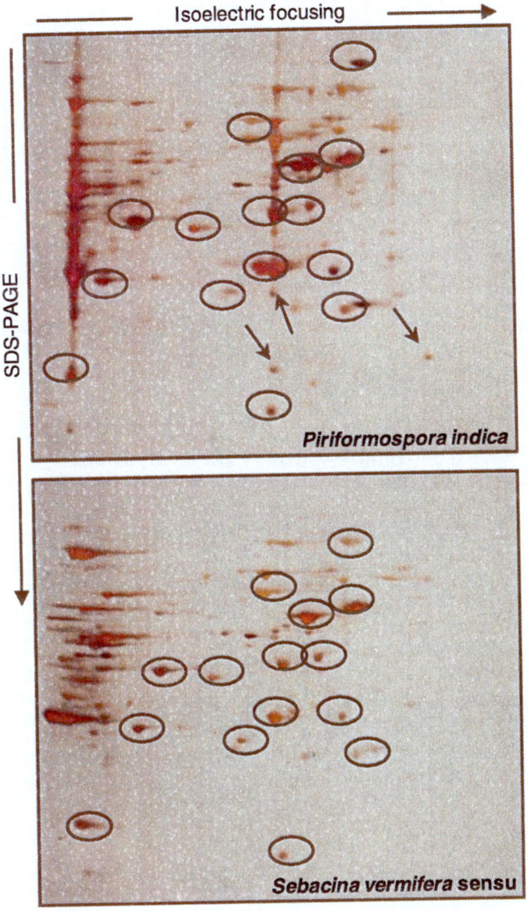

Fig. 19.6 Silver-stained two-dimensional maps of mycelial protein of *P. indica* and *S. vermifera* loaded with 80 μg of protein onto IEF gels. Separation in the horizontal dimension was achieved by IEF using carrier ampholyte in the pH range of 3–10 in the presence of 9.2 M urea and separation in the vertical dimension by 12% SDS-PAGE. The *arrow* in *P. indica* represents the bands absent in *S. vermifera*

19.2.3 Non-denatured Protein Polymorphism Study by Native Polyacrylamide Gel Electrophoresis (Walker 1994)

19.2.3.1 Equipments

- Vertical PAGE apparatus set
- Water Bath Shaker
- Cold room 4°C

19.2.3.2 Reagents

- 10% acrylamide solution from stock
- Separating gel buffer: 1.5 M Tris–HCl, pH 8.8

- Stacking gel buffer: 0.5 M Tris–HCl, pH 6.8
- APS 10%
- TEMED 0.04%
- Sample buffers 5×

A. Loading sample buffer, pH 6.8

Tris	2.5 g
β-mercaptoethanol	2.5 ml
Final volume	40 ml

B. Bromophenol blue glycerol solution

Bromophenol blue	5 mg
Glycerol	5.8 ml
Final volume	10 ml

19.2.3.3 Loading Buffer

Mix A and B in the ratio of 1:4

Running electrophoresis buffer

Tris	25 mM
Glycine	250 mM

19.2.3.4 Gel Staining Solution

Solution A

Distilled water	5 parts
Methanol	4 parts
Glacial acetic acid	1 part

Final staining solution

Coomassie brilliant blue G	0.1 g
Solution A	100 ml

Dissolved using stirrer, filtered through Whatman filter paper no. 1.

Gel De-staining solution

Glacial acetic acid	10 ml
Distilled water	60 ml

Note: All the stocks and solutions for Native PAGE are similar with SDS-PAGE stocks, excluding SDS in overall solutions in the process.

19.2.3.5 Procedure

Prepare the separating gel using 10% polyacrylamide gel solution. Allow the gel to polymerize. After polymerization of the separating gel, apply 4% stacking gel to the gel cassette. Place the well-forming comb into this solution and allow to polymerize. This preparation takes about 30 min.

Carefully remove the comb and spacer after the gel sets, and assemble the cassette in the electrophoresis tank. Fill the top and the bottom of the reservoir with electrophoresis buffer so that the buffers fully fill the sample loading wells. Mix 50 μg of protein with 5× sample loading buffer. Then centrifuge for 5 min at 5,000 rpm in micro-centrifuge. Load samples onto the gel well with the help of a Hamilton micro-syringe or gel loading tips. Slowly deliver the samples into the well. The dense sample settle at the bottom of the loading well. Connect the power pack to the apparatus and run the protein in stacking gel at constant voltage of 70 V and in separating gel at 120 V until the dye front reaches the bottom of the plate, 1 cm above the edge. The Native PAGE is run in cold room maintained at 4°C. After completion of electrophoresis, the gel is subjected to:

(a) ACPase enzyme assay
(b) Stain the gel for 60 min in Coomassie blue dye followed by destaining with the change of destain.

Observe the resulting bands and compare with bands in gel enzyme assay.

19.2.3.6 Gel Enzyme Assay (Walker 1996)

Reagents

- Sodium acetate buffer (50 mM), pH 5.3
- *p*-Nitrophenyl phosphate di-sodium salt (Sigma Chemical Co.) 2 mg/ml

Procedure

For the detection of protein for their biological activity, run duplicate samples in native gel. Stain one set of sample by Coomassie to obtain whole protein bands and the other set for phosphatase activity. Equilibrate the gel in 50 mM sodium acetate buffer pH 5.3 for 30 min at 4°C in cold room. Immerse the gel in solution containing 2 mg/ml concentration of the enzyme substrate (*p*-NPP) in shaking water bath till yellow color develops (Fig. 19.7).

Equipment

- Vertical PAGE apparatus set
- Sharp razor blade or dissecting scissors

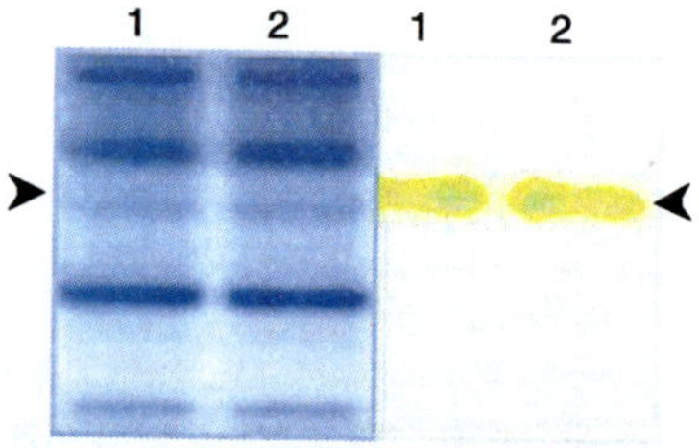

Fig. 19.7 Native PAGE stained with Coomassie blue (**a**) and gel assayed (**b**) using *p*-NPP as substrate. Lane 1. *P. indica* and Lane 2. *S. vermifera* sensu. Separation was done in 10% gel at 4°C for 6 h. Duplicate samples were run. One set of sample was stained for protein profile with Coomassie blue (**a**) and the other set for ACPase activity, washing the gel in 2 mg/ml substrate solution that gave *yellow colored p*-nitrophenol product at the site of enzyme. Represents the enzyme acid phosphatase, *light blue* in coomassie stained gel. Separation in this system depends on both the native charge on the protein and molecular mass

Enzyme Elution

The process is modified from that of Summers and Szewczyk (1996) "Elution of SDS-PAGE separated proteins from immobilon membranes for use as antigen"

Elution Buffer 1	Concentration
NH_4HCO_3	50 mM
SDS	0.4%
PMSF	2 mM
DTT	2 mM
TPCK	50 μM
Benzedene	50 μM
DTT	2 mM
Elution Buffer 2	**Concentration**
SDS	2%
Tritan X-100	1%
Tris–HCl	50 mM
pH 9.5	

Run Native polyacrylamide gel 6%. Assay for ACPase in presence of 2 mM *p*-NPP. Cut the yellow band thus formed with the help of sterile razor blade. Further cut the gel into small pieces and pass through different pore-sized needles with the help of plastic disposable syringe along with elution buffer 1. Transfer the gel to Falcon tube. Boil the mixture for 6 m, then keep at 60°C overnight in water bath. Centrifuge at 13,000 rpm in a spin filter. Collect the supernatant (Fig. 19.8).

SDS Polyacrylamide Gel Electrophoresis

All the stocks and solutions for SDS-PAGE are similar to those of Native PAGE stocks, except SDS; 10% SDS is used in sample buffer and in running buffer.

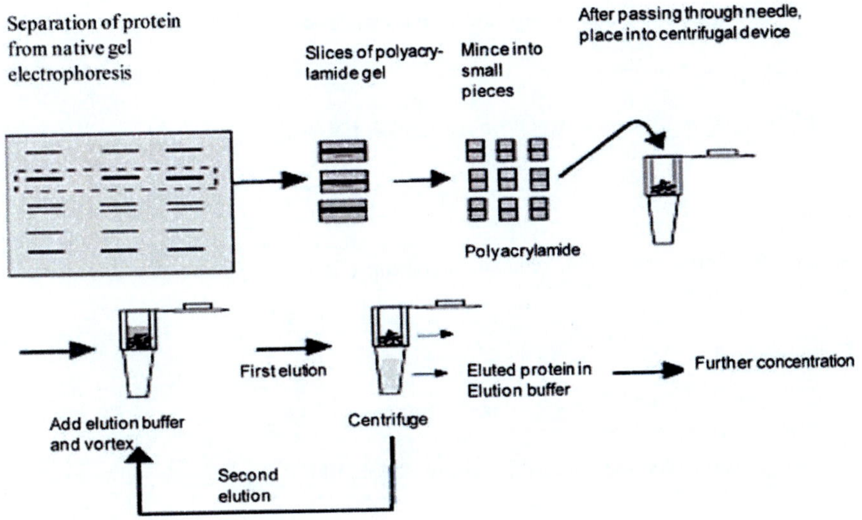

Fig. 19.8 Protocol for the elution of protein from Polyacrylamide gel electrophoresis

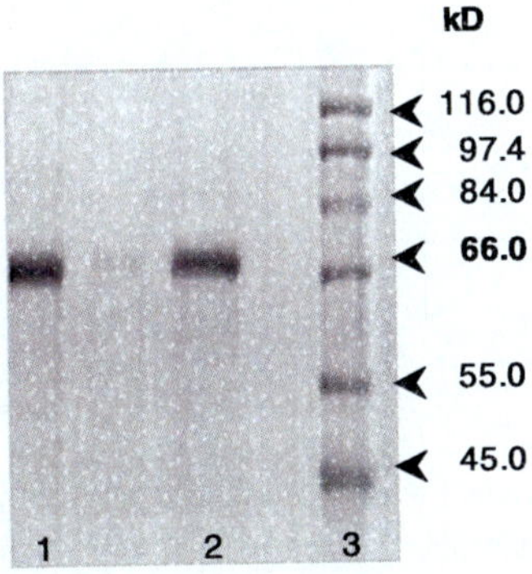

Fig. 19.9 Molecular mass determination of ACPase eluted from native gel. Lane 1. *P. indica*. Lane 2. *S. vermifera*. Lane 3. Molecular marker (Sigma wide range). The crude enzyme separated in 10% native PAGE and detected by assay using *p*-NPP. The eluted protein from native PAGE was separated by 12% SDS-PAGE along with wide range marker. The pure acid phosphatase showed 66 kD molecular mass

Similar as Native PAGE but the overall process is carried in room temperature. Along with eluted protein, run sigma marker (M-4038) in one of the wells of 12% PAGE. After the completion of the process, the gel can be stained with staining solution. Compare the thick band of ACPase with marker. Determine the molecular size of the ACPase (Fig. 19.9).

ELF 97 Endogenous Phosphatase Detection (van Aarle 2001)

Equipments

– Fluorescence Microscope (Olympus model, FV-300)

Reagents

– ELF-97 Endogenous Phosphatase Detection Kit

Component A

[2-(5′-chloro2′phosphoryloxyphenyl)-6-chloro-4-(^{3}H Quinazolinone (CPPCQ)]
20× concentrate ELF-97
Phosphatase substrate in 2 mM sodium Azide 500 μl

Component B

Detection Buffer = 10 ml

Component C

Mounting medium = 15 ml
PBS, pH 7.4.

Procedure

Fix the fungal cultures grown for 48 h in broth in 3.7% formaldehyde prepared in PBS for 1 h in 4°C. Permeabilize the samples in 0.2% Tween 20 in PBS buffer for 10 min at room temperature. Rinse the sample four times with PBS. Dilute the ELF 97 (Molecular Probes, Kit E-6610) phosphatase substrate (component A) 20-fold in detection buffer (component B) as already standardized. Filter the diluted substrate through 0.2 μm pore-sized spin filters (E-6606) just before applying to tissue sections. Add the samples to the spin filter and centrifuge in microcentrifuge for 2 min. Since the reaction occurs very fast, the reagent should add while the sample is on the microscope. Observe the sample using excitation filter and dichroic mirror from emission filter in the fluorescein set. The filter set provides the appropriate UV excitation and transmits wavelengths greater than 400 nm. Before applying the substrate solution, wick-off excess PBS from the sample, and then add 50 μl of substrate solution. Immediately, place the sample on the microscope and monitor the development of signal (Fig. 19.10).

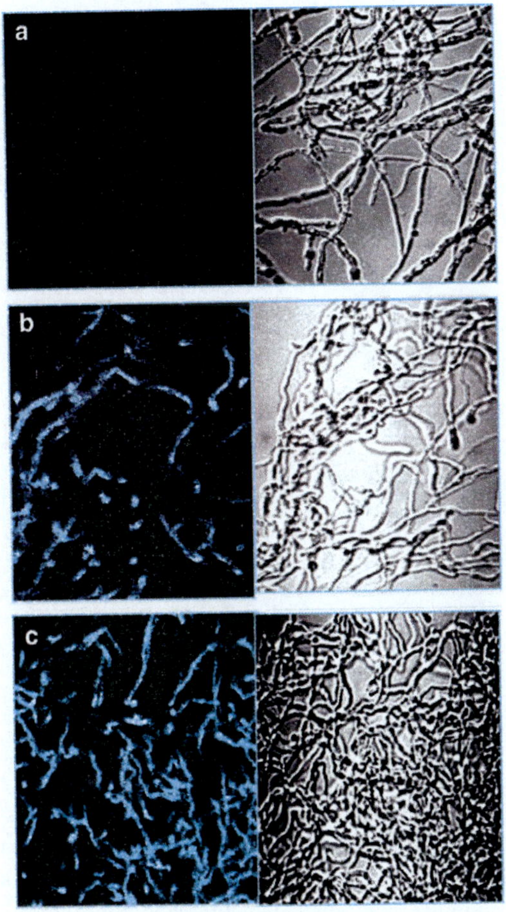

Fig. 19.10 ELF 97 Endogenous phosphatase detection (Molecular Probes, Kit E-6610) Detection of ACPase activity by enzyme labeled fluorescent substrate (ELF 97, Molecular Probes). (**a**) Control with out substrate (**b**) *P. indica* C. *S. vermifera* sensu. The result observed under Olympus standard microscope using excitation filter and dichroic mirror from the DAPI filter set and emission filter from the fluorescein set provides the appropriate UV excitation and transmits wavelengths greater than 400 nm. The results show uniform activity of phosphatase throughout the mycelium. None of the exatracellular activities is observed

Fast Garnet GBC Staining of Phosphatase Isozymes (Pasteur et al. 1988)

Equipments

- Vertical PAGE apparatus set
- Water Bath Shaker
- Cold room 4°C

Staining solution	
α-naphthyl phosphate	100 mg
Fast garnet GBC	100 mg
10% $MgCl_2$	200 μl
0.1 M Sodium acetate buffer, pH 5.3	100 ml

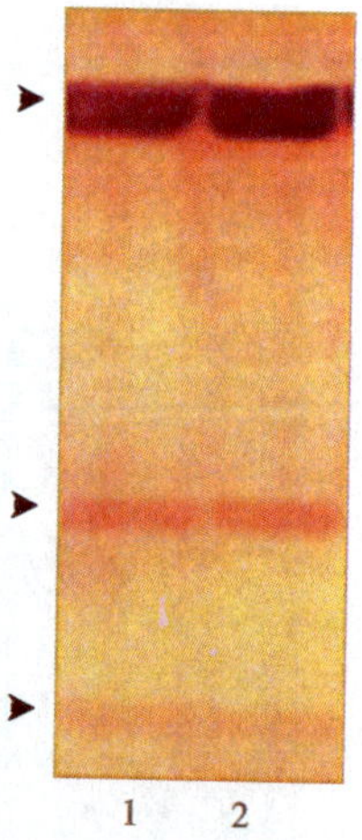

Fig. 19.11 Native PAGE Zymogram of acid phosphatase isoforms Fast garnet GBC staining of ACPase. The ACPase isoforms of *P. indica* and *S. vermifera* shows similar banding pattern. Lane 1. *P. indica* Lane 2. *S. vermifera* sensu. The native PAGE separated for 6 h at 4°C was neutralized with 50 mM sodium acetate buffer, stained with fast garnet GBC using α-naphthyl phosphate as substrate. The precisely localized band shows similar molecular mass and ionic strength of isoforms between *P. indica* and *S. vermifera*

De-staining solution

Acetic acid (v/v) = 0.1%

Procedure

Prepare horizontal electrophoretic PAGE as above in Native PAGE protocol. Run the PAGE at constant current of 70 and then 120 V for about 5 h. Neutralize the gel in 50 mM sodium acetate buffer for 30 min in cold room. Soak the gel in staining solution containing α-naphthyl acid phosphate to identify phosphatase isozyme in water bath shaker at 37°C. Destain with 1% acetic acid solution till the bands will clear (Fig. 19.11).

19.2.4 Polymorphism Based on Random Amplification of Polymorphic DNA (RAPD) Technique

19.2.4.1 Equipment

- Thermal Cycler
- Horizontal PAGE apparatus
- Centrifuge
- Refrigerator

19.2.4.2 Reagents

DNA isolation buffer (Moller et al. 1992)

CTAB (hexade-cyltrimethyl ammonium bromide)	2%
NaCl	1.4 M
EDTA	20 mM
Tris–HCl	100 mM

TE pH 8.0

Tris–HCl (pH 8.0)	10 mM
EDTA (pH 8.0)	1 mM

Sterilize by autoclaving at 15 lbs/sq. in. for 15 min.

Gel loading buffer

Bromophenol blue	0.25%
Sucrose in water	40% (w/v)

Store in small aliquots at 4°C.

Tris–borate EDTA (TBE)

Concentrated stock solutions (5×)		Working concentration	
Tris–Base	54.0 g/l	Tris–borate	0.089 M
Boric acid	27.5 g/l	Boric acid	0.089 M
0.5 M EDTA (pH 8.0)	20 ml	EDTA	0.002 M

Filter sterilize if necessary, but do not autoclave.

19.2.4.3 Ethidium Bromide (10 mg/ml)

Add 1 g of ethidium bromide (EtBr) to 100 ml of sterile water. Stir on magnetic stirrer for several hours to ensure that the dye has dissolved. Wrap the container in aluminum foil and store at room temperature in dark bottle.

Note: Ethidium bromide is carcinogenic

DNA Amplification Mixture for PCR (Operon Technologies Alameda, California)	(25 μl)
10× Buffer	2.5 μl
$MgCl_2$	2.5 μl
dNTPs 10 mM	0.8 μl
Primer (30 ng/μl)	1.0 μl
Taq Polymerage (3U/μl)	0.5 μl
Template DNA	1 μl
Milli Q water (Ultrapure)	16.7

RAPD amplification conditions

Initial denaturation cycle	95°C for 5 min	1 cycle
Denaturation cycle	94°C for 30 s	36 cycles
Annealing	36°C for 2 min	36 cycles
Extension	72°C for 2 min	36 cycles
Final extension	72°C for 5 min	1 cycle

19.2.4.4 Procedure

Isolate and purify the fungal DNA according to modified CTAB protocol of Moller et al. (1992). Carry DNA amplification in a total volume of 25 μl containing (μl): 2.5, Buffer (10× without $MgCl_2$), 2.5, $MgCl_2$, 0.8, dNTPs (10 mM), 1.0, primer (30 ng/μl), 0.5 *Taq* polymerase (3U/μl) and DNA concentration ranging from 5 to 25. Use random 10 bp oligonucleotide primers (Operon Technologies Alameda, California) to produce amplification. Amplify the DNA in PTC-200 Thermal Cycler (Techne make, UK). Carry Electrophoresis on 1.5% agrose gel in 1% TAE at 3.5 V/cm for 2 h and stain with ethidium bromide (Fig. 19.12).

Using the NTSYS-pc program (Rohlf 1992), perform statistical analysis. The degree of genetic relatedness or similarity can be estimated using the Jaccard coefficient. Clustering of similarity matrices can be done by UPGMA and projection by TREE program of NTSYS-pc (Fig. 19.13).

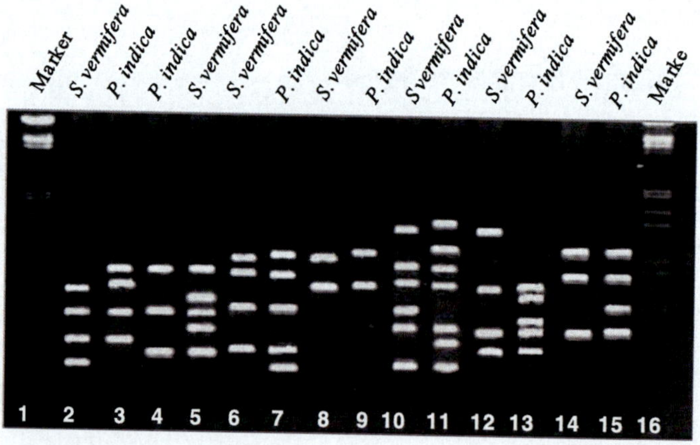

Fig. 19.12 Random amplified polymorphic DNA (RAPD) analysis of *Piriformospora indica* and *Sebacina vermifera* sensu. The RAPD analysis of *P. indica* and *S. vermifera* to show genetic variation between these two fungi. Out of seven primers used for amplification, six have given a productive polymorphism. Lanes 1 and 16. Marker. Lanes 2 and 3 Primer OPA10. Lanes 4 and 5 OPD01. Lanes 6 and 7. OPC06. Lanes 8 and 9. OPC10. Lanes 10 and 11. OPC 01. Lanes 12 and 13 OPI04. Lanes 14 and 15. OPI10. No polymorphism was observed when the genomic DNA was amplified with OPC10 Lanes 8 and 9

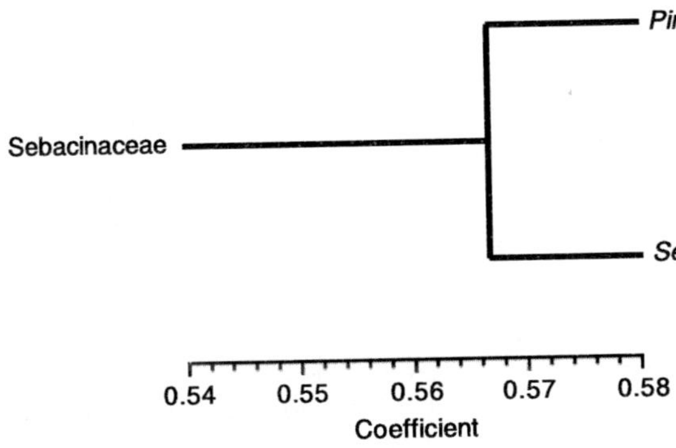

Fig. 19.13 Dendrogram showing phylogenetic relationship Phylogeny between *P. indica* and *S. vermifera* sensu. The NTSYS-pc (Numerical Taxonomy System, Applied Biostatistics) computer program was used for data analysis

19.3 Conclusion

The acid phosphatase in *Piriformospora indica* and *Sebacina vermifera* sensu were similar in their molecular mass. The optimum physical conditions such as pH and temperature are almost similar in both fungi, supporting strong relationship between these two. The protein separated in SDS-PAGE and Native PAGE of *P. indica* and *S. vermifera* is showed at precised location. The pattern of phosphatase isozymes in Native PAGE shows the evidences of strong relationship of these fungi. Almost similar observation was noticed in *P. indica* and *S. vermifera* sensu, showing closeness of these fungi to each other by applying ELF-97 substrate. Two-dimensional map of crude protein of these two fungi showed some differences in minor proteins.

Piriformospora indica and *Sebacina vermifera* sensu belonging to same taxonomic group show similar morphology, functions, and isozymes. However, they show distinct genetic variation based on the RAPD analysis. An average genetic similarity between both the fungi was 58% and can be considered to be placed in the species of the same ancestral roof.

The application of different techniques for characterization of ACPase in these two fungi has provided new insights into important aspects of this field. *Piriformospora indica* and *Sebacina vermifera* sensu belonging to the same taxonomic group show similar morphology, functions, protein profiles, and isozyme characterization along with close acid phosphatase relationships. However, they show distinct genetic polymorphism based on the RAPD analysis.

Acknowledgments Rajani Malla is thankful to Dr. Ashok K Chauhan, Founder President, Amity-RBEF, New Delhi for his encouragement and support. The author is also thankful to Pham Giang from International Center for Genetic Engineering and Biotechnology, Ram Prasad from Amity Institute, Dr. Upendra and Sweta from All India Institute of Medical Sciences.

References

Alves JM, Siachakr CD, Allot M, Tizroutine S, Mussio I, Servaes A (1994) Isozyme modifications and plant regeneration through somatic embryogenesis in sweet potato. Plant Cell Rep 13:437–441

Cox G, Moran KG, Sanders F, Nokolds C, Tinker PB (1980) Translocation and transfer of nutrients in vesicular arbuscular mycorrhizas. III. Polyphosphate granules and phosphorus translocation. New Phytol 84:649–659

Fries LLM, Pacovsky RS, Safir GR, Kaminski J (1998) Phosphorus effect on phosphatase activity in endomycorrhizal maize. Physiol Plantarum 103:162–171

Gianinazzi-Pearson V, Gianinazzi S (1978) Enzymatic studies on the metabolism of vesicular-arbuscular mycorrhiza. II. Soluble alkaline phosphatase specific to mycorrhizalinfection in onion roots. Physiol Plant Pathol 12:45–53

Gravel P, Golaz G (1996) Two-dimensional PAGE using carrier ampholyte pH gradients in the first dimension. In: Walker JM (ed) The protein protocols handbook. Humana, Totowa, New Jersey, pp 127–32

Harrison MJ, Dewbre GR, Liu JY (2002) A phosphate transporter from *Medicago truncatula* involved in the acquisition of phosphate released by arbuscular mycorrhizal fungi. Plant Cell 14:2413–2429

Horst MN (2000) Molecular approaches to glycobiology. pp 2–30

Hill TW, Kaefer E (2001) Improved protocols for aspergillus medium: trace elements and minimum medium salt stock solutions. Fungal Genet Newslett 48:20–21

van Aarle IM (2001) Microscopic detection of phosphatase activity of saprophytic and arbuscular mycorrhizal fungi using a fluorogenic substrate. Mycologia 93:17–24

Kaldorf M, Koch B, Rexer K-H, Kost G, Varma A (2005) Patterns of interaction between populus Esch5 and *Piriformospora indica*: a transition from mutualism to antagonism. Plant Biol 7:210–218

Malla R, Md Z, Yadav V, Suniti, Verma A, Rai M, A Varma A (2002) *Piriformospora indica* and plant growth promoting Rhizobacteria: An Appraisal. In: Rao GP, Bhat DJ, Lakhanpal TN, Manoharichari C (Eds). Frontiers of fungal diversity in India, pp 401–419

Malla R, Prasad R, Giang PH, Pokharel U, Oelmueller R, Varma A (2004) Characteristic features of symbiotic fungus *Piriformospora indica*. Endocytobiosis Cell Res 15:579–600

Malla R, Pokharel U, Varma A (2005) Random amplified polymorphic DNA of the two fungi from *Sebacinales*. Journal of Institute of Science and Technology 14:34–43

Malla R, Pokharel U, Varma A (2006a) Random amplified polymorphic DNA of the two fungi from members of Sebacinales. J Inst Sci Technol 14:34–43

Malla R, Pokharel U, Varma A (2006b) Immunological characterization of acid phosphatase in a endophytic fungus. Nepal J Sci Technol 7:77–84

Malla R, Pokharel U, Prasad R, Varma A (2007a) Proteomics and genomics approach to study plant-microbe cross communication. In: Chauhan AK, Harsha K, Varma A (eds) Microbes for human life, vol 4. IK International, India, pp 609–636

Malla R, Varma A (2005) Phosphatase (s) in microorganisms. Biotechnological applications of microbes. IK International, India, New York/Kluwer Academic, Holland, pp 125–50

Malla R, Varma A (2007) Use of short oligonucleotide primers in random amplified polymorphic DNA techniques for species identification. In: Oelmüller R, Varma A (eds) Advanced techniques in soil microbiology, vol 11. Springer, Germany, pp 235–244

Malla R, Pokharel U, Prasad R, Oelmüller R, Varma A (2007b) Immuno-technology for the localization of acid phosphatase using native gel bands in *Piriformospora indica* and other soil microorganism. In: Oelmüller R, Varma A (eds) Modern tools and techniques, vol 11. Springer-Verlag, Germany, pp 211–234

McCormick MK, Whigham DF, O'Neill J (2004) Mycorrhizal diversity in photosynthetic terrestrial orchid. New Phytol 163:425–438

Moller EM, Bahnweg G, Sandermann H, Geiger HH (1992) A simple and efficient protocol for isolation of high molecular weight DNA from filament fungi, fruit bodies and infected plant tissue. Nucleic Acids Res 20:6115–6116

Oakley BR, Kirsch DR, Morris NR (1980) A simplified ultrasensitive silver stain for detecting proteins in polyacrylamide gels. Anal Biochem 105:361–363

Pasteur N, Pasteur G, Bonhomme F, Catalan J, Davidson Brotton (1988) Practical isozyme genetics. Ellis-Horwood, Chichester

Paszkowski U, Kroken S, Roux C, Briggs SP (2002) Rice phosphate transporters include an evolutionarily divergent gene specifically activated in arbuscular mycorrhizal symbiosis. Proc Natl Acad Sci 99:13324–13329

Peškan-Berghöfer T, Shahollari B, Pham HG, Hehl S, Markent C, Blank V, Kost G, Varma A, Oelmueller R (2004) Association of *Piriformospora indica* with *Arabidopsis thaliana* roots represent a novel system to study beneficial plant-microbe interactions and involve in early plant protein modifications in the endocytoplasmic reticulum and in the plasma membrane. Physiol Plant 122:465–71

Rausch C, Daram P, Brunner S, Jansa J, Laloi M, Leggewie G, Amrhein N, Bucher M (2001) A phosphate transporter expressed in arbuscule containing cells in potato. Nature 414:462–466

Rohlf FJ, NTSYS-pc (1992) Numerical taxonomy and multivariate analysis system. Vers. 1.70. Exeter Software, Setauket, New York

Rosendahl S (1994) Isozyme analysis of mycorrhizal fungi and their mycorrhiza. In: Norris JR, Read D, Varma A (eds) Techniques for mycorrhizal research. Academic, London, pp 629–654

Shahollari B, Varma A, Oelmüller R (2005) Expression of a receptor kinase in *Arabidopsis* roots is stimulated by the basidiomycete *Piriformospora indica* andthe protein accumulates in Triton X-100 insoluble plasma membrane microdomains. J Plant Physiol (Holland) 162:945–958

Straker CJ, Mitchell DT (1986) The activity and characterization of acid phosphatases in endomycorrhizal fungi of the Ericaceae. New Phytol 104:243–256

Summers DF, Szewczyk B (1996) Elution of SDS-PAGE separated proteins from immobilon membranes for use as antigens. In: Walker JM (ed) The protein protocols handbook. Humana, Totowa, New Jersey, pp 699–702

Tarafdar JC, Rao AV (1996) Contribution of *Aspergillus* strains to acquisition of phosphorus by wheat (*Triticum aestivum* L.) and chick pea (*Cicer arietinum* L.) grown in a loamy soil. Appl Soil Ecol 3:190–214

Tarafdar JC, Yadav RS, Meena SC (2001) Comparative efficiency of acid phosphatase originated from plant and fungal sources. J Plant Nutr Soil Sci 164:279–282

Tisserant B, Gianinazzi-Pearson V, Gianinazzi S, Gollotte A (1993) In planta histochemical staining of fungal alkaline phosphatase activity for analysis of efficient arbuscular mycorrhizal infections. Mycol Res 97:245–250

Varma A, Verma S, Sudha, Sahay N, Britta B, Franken P (1999) *Piriformospora indica* – a cultivable plant growth promoting root endophyte with similarities to arbuscular mycorrhizal fungi. Appl and Environ Microbiol 65:2741–2744

Verma S, Varma A, Rexer KH, Hassel A, Kost G, Sarabhoy A, Bisen P, Bütenhorn B, Franken P (1998) *Piriformospora indica*, a new root colonizing fungus. Mycologia 90:896–903

Walker JM (1994) Basic protein and peptide protocols. Humana, Totowa, NJ, pp 86–187

Walker JM (1996) Non-denaturing polyacrylamide gel electrophoresis of protein. In: Walker JM (ed) The protein protocols handbook. Humana, NJ, pp 51–54

Weiß M, Selosse MA, Rexer KH, Urban A, Oberwinkler F (2004) *Sebacinales*: a hitherto overlooked cosm of heterobasidiomycetes with abroad mycorrhizal potential. Mycol Res 108:1–8

Chapter 20
Production of Ligninolytic Enzymes by White-rot Fungi during Bioremediation of Oil-contaminated Soil

Natalia N. Pozdnyakova, Ekaterina V. Dubrovskaya, Oleg E. Makarov, Valentina E. Nikitina, and Olga V. Turkovskaya

20.1 Introduction

Environmental pollution caused by various xenobiotics is becoming a global challenge. The most hazardous pollutants include oil, oil products, chlorinated compounds, synthetic dyes, and polycyclic aromatic hydrocarbons (PAHs) (Reddy 1995; Gramss 2000; Stanley et al. 2000). Industrial discharges or occasional releases of these compounds to the environment are very destructive, particularly when the biodegrading potential of the natural microflora is insufficient for their removal or detoxification.

Extensive studies are in progress to develop and improve methods for bioremediation of polluted soils and water. Mycoremediation is a process by which fungi degrade or transform hazardous organic contaminants to less toxic compounds (Sasek et al. 2003). White-rot fungi, which degrade predominantly wood in nature, are potential candidates for the treatment of contaminated soils because of their high capability of degrading a wide range of xenobiotics not only in liquid culture (Cripps et al. 1990; Morgan et al. 1991; Barr and Aust 1994; Reddy 1995; Bollag et al. 2003; Hou et al. 2004) but also in artificially contaminated soil (Lamar and Dietrich 1990; Morgan et al. 1991; Khadrani et al. 1999; Kubatova et al. 2001; Bhatt et al. 2002). Attempts have therefore been made to apply these fungi to the bioremediation of soils contaminated with compounds not sufficiently degradable by soil microorganisms (Lang et al. 1998).

In addition to lignin decomposition, the potential of white-rot fungi for degradation of various organopollutants in both sterile and nonsterile soil has been well

N.N. Pozdnyakova (✉)
Institute of Biochemistry and Physiology of Plants and Microorganisms Russian Academy of Sciences, 13 Prospekt Entuziastov, Saratov 410049, Russia
e-mail: nataliapozdnyakova@yahoo.com

E.V. Dubrovskaya, O.E. Makarov, V.E. Nikitina, and O.V. Turkovskaya
Institute of Biochemistry and Physiology of Plants and Microorganisms Russian Academy of Sciences, 13 Prospect Entuziastov, Saratov 410049, Russia

G. Shukla and A. Varma (eds.), *Soil Enzymology*, Soil Biology 22,
DOI 10.1007/978-3-642-14225-3_20, © Springer-Verlag Berlin Heidelberg 2011

documented (Lamar and Dietrich 1990; Boyle 1995; Eggen 1999; Bhatt et al. 2000; Marquez-Rocha et al. 2000). For example, *P. ostreatus* metabolized and mineralized soil-adsorbed PAHs and PAHs in creosote-contaminated soil (Eggen 1999; Novotny et al. 1999; Marquez-Rocha et al. 2000). *Irpex lacteus, Bjerkandera adusta*, and *Trametes versicolor* removed pentachlorophenol in nonsterile soil (Lestan and Lamar 1996).

In soil bioremediation, white-rot fungi are normally applied in the form of a mycelium pre-grown on wood chips, sawdust, chopped straw, or similar biological material that is mixed with the contaminated soil. Compared with the use of submerged cultures, this method increases the competitiveness of these fungi. For example, spent oyster mushroom (*P. ostreatus*) compost was applied to PAH degradation in creosote-contaminated soil (Eggen 1999). The spent mushroom compost of *P. pulmonarius* was used for bioremediation of PAH-contaminated samples (Lau et al. 2003). *Ph. chrysosporium*, *P. ostreatus*, and *C. versicolor* were pre-grown on shredded cardboard before being added to oil-contaminated soil (Yateem et al. 1998; Bhatt et al. 2000). Mycelia on solid substrates, such as straw or spent mushroom waste, can be used as economical sources of inoculum (Cohen et al. 2002).

If it is accepted that extracellular ligninolytic enzymes (laccase, lignin peroxidase, Mn-peroxidase, and versatile peroxidase) catalyze the initial reactions of pollutant transformation and degradation, the production and full activity of these enzymes in soil are two prerequisites for successful application of white-rot fungi in bioremediation. Oil, oil products, chlorinated compounds, and PAHs are usually poorly water soluble but must nevertheless be accessible to the extracellular enzymes in the soil matrix. Thus, the fungal hyphae can secrete the enzymes not only into their natural substrate, lignocellulose, but also into the contaminated soil, which is a completely different environment. The enzymes can also be produced and remain active in the presence of soil microorganisms (Lang et al. 1998).

Although many studies of xenobiotic degradation in contaminated soils by white-rot fungi have been performed, little attention has been paid to production of ligninolytic enzymes during mycoremediation. Most of the studies so far have focused on the model white-rot species *Phanerochaete chrysosporium*, *Trametes versicolor*, *Pleurotus ostreatus*, and several others. The ability to produce ligninolytic enzymes during cultivation in soil was found in different white-rot (Lang et al. 1998; Baldrian et al. 2000; Novotny et al. 2000; Snajdr and Baldrian 2006) and litter-decomposing fungi (Kahkonen et al. 2008).

Compared to wood, soil or litter is a more complex and heterogeneous environment, which may hamper the detection and estimation of enzyme activities. Laccase activity reflects the course of degradation of organic substances, and thus it varies with time. The activity of laccase also reflects the presence of fungal mycelia. Being the most abundant ligninolytic enzymes in soil, laccases participate in the transformation of lignin in forest litter. It is also generally presumed that laccases are able to react with soil humic substances that can be directly formed from lignin (Baldrian 2006).

Novotny et al. (1999) compared the abilities of *Ph. chrysosporium*, *P. ostreatus*, and *T. versicolor* to degrade PAHs and produce ligninolytic enzymes in soil. They found that colonization of sterilized soil by straw-grown inocula and degradation of anthracene, phenanthrene, and pyrene were the greatest with *P. ostreatus*. The production of Mn-peroxidase and laccase in soil was similar for *P. ostreatus* and *T. versicolor* but was extremely low for *P. chrysosporium*.

Production of ligninolytic enzymes by *Pleurotus* sp. and *Dichomitus squalens* in soil and on a lignocellulose substrate was studied by Lang et al. (1998). Both organisms produce laccase and Mn-peroxidase but not lignin peroxidase. The growth rate and the enzyme activities of *Pleurotus* sp. were not significantly influenced by the presence of soil microorganisms. In contrast, *D. squalens* did not penetrate nonsterile soil, and no enzyme activities could be detected in that soil (Lang et al. 1998). The extracellular activity of laccase and Mn-peroxidase during the growth of the ligninolytic fungus *P. ostreatus* in nonsterile soil with low and high carbon content available was determined. Addition of lignocellulose to soil increased the production of Mn-peroxidase by this fungus (Snajdr and Baldrian 2006). Well growing in soil, *P. ostreatus* produced Mn-peroxidase and laccase at levels lower than those observed on straw. *T. versicolor*, which produced high levels of Mn-peroxidase and laccase on straw, synthesized the two enzymes also in soil, in spite of its limited growth in this environment (Novotny et al. 1999). *Irpex lacteus* efficiently colonized sterile and nonsterile soil by mycelium growing from a wheat straw inoculum. Good colonization of nonsterile gasworks soil contaminated with PAHs and heavy metals was also observed. *I. lacteus* efficiently removed three- and four-ring PAHs, including anthracene, fluoranthene, and pyrene, from artificially spiked soil. Lignin peroxidase and laccase, but not Mn-peroxidase, were also detected when the fungus colonized the soil (Novotny et al. 2000).

The relationships between ligninolytic activity and pentachlorophenol biotransformation by *L. edodes* (Okeke et al. 1994) and between ligninolytic activity and PAH degradation by *P. ostreatus* (Eggen 1999) were demonstrated. In *P. ostreatus*, laccase and Mn-peroxidase were found, and their involvement in the removal of PAHs was possible, including the production of anthraquinone from anthracene (Novotny et al., 1999). A similar correlation was also reported for the expression of Mn-peroxidase and the removal of fluorene and chrysene by soil cultures of *P. chrysosporium* (Bogan et al. 1996).

Litter-decomposing fungi, represented by species inhabiting the natural environment of soil and decaying litter, are very promising candidates for the production of ligninolytic activities. There are very few reports describing the presence of ligninolytic activities in these species (Steffen et al. 2000, 2003). For example, litter-decomposing fungi, including *Agaricus bisporus*, *Agrocybe praecox*, *Gymnoporus peronatus*, *Gymnoporus sapineus*, *Mycena galericulata*, *Gymnopilus luteofolius*, *Stropharia aeruginosa*, and *Stropharia rugosoannulata*, produce laccase and Mn-peroxidase during cultivation in Pb-contaminated soil (Kahkonen et al. 2008).

A serious problem in the bioremediation technology is mixed pollution (e.g., oil and oil derivatives). Although there are numerous reports on the application of white-rot fungi to bioremediation of contaminated soil, only a few of them are addressed to mycoremediation of oil-polluted soil (Yateem et al. 1998; Ishikhuemhen et al. 2003; Pozdnyakova et al. 2008b).

Here, we investigate the ability of selected white-rot fungi, belonging to different ecological groups (white-rot and litter-decomposing fungi), to grow, decrease old oil-contamination, and produce extracellular ligninolytic enzymes in soil. The fungi chosen for study were the cultivated mushrooms *P. ostreatus*, *L. edodes*, and *Agaricus* sp., whose spent compost can be used for mycoremediation, and *Coriolus*, one of the most active producers of ligninolytic enzymes.

20.2 Materials and Method

Industrial soil with old oil-pollution was collected in the area around an oil-refining plant (Saratov, Russia). The concentration of total petroleum hydrocarbons (TPHs) was 36.2 mg/g dry soil, including, in mg/g dry soil: alkanes (11.4), naphthene (4.3), a fraction of low-molecular weight aromatic hydrocarbons (5.7), a fraction of high-molecular weight aromatic hydrocarbons (9.0), and tars (5.8). The total dry mass of the soil was 70% (pH 6.2). Soil was air-dried and sieved through a 2-mm mesh before use and analysis of its properties.

Experiments were carried out with the white-rot fungi *Pleurotus ostreatus* 336, *P. ostreatus* D1, *P. ostreatus* D2, *P. ostreatus* D/L, *P. ostreatus* D/o, *Lentinus edodes* F-249, *L. edodes* 0779, *L. edodes* 2T, *L. edodes* NY (obtained from the collection of the Laboratory of Microbiology and Mycology at IBPPM RAS), *Coriolus* sp. F-1, *Agaricus* sp. F-8, and *Agaricus* sp. F-17 (from the collection of the Environmental Biotechnology Laboratory at the same institute). Fungal cultures were maintained on Bezalel et al.'s (1997) basidiomycetes rich medium (pH 6.0), with our modifications. The composition of the medium was as follows (g/L): NH_4NO_3, 0.724; KH_2PO_4, 1.0; $MgSO_4 \times 7H_2O$, 1.0; KCl, 0.5; yeast extract, 0.5; $FeSO_4 \times 7H_2O$, 0.001; $ZnSO_4 \times 7H_2O$, 0.0028; $CaCl_2 \times 2H_2O$, 0.033; D-glucose, 10.0; peptone, 10.0; agar, 15.0.

To obtain submerged inocula, we grew the fungi at 29°C in basidiomycetes rich medium without agar. Solid-state fungal inocula were grown on sunflower-seed hulls. The hulls (5 g) were placed into flasks, moistened with 20 ml of tap water, autoclaved at 1 atm for 30 min, and supplemented with 5 ml of the submerged inoculum (ca. 200 mg wet weight). The cultures were incubated at 29°C for 7 days.

To study oil degradation, we placed industrial soil with old oil-pollution (40 g) into Petri dishes and autoclaved it three times one day apart. Fungal inocula were added to the soil as submerged (5 ml) or solid (2 g) cultures, to a final concentration of 50 mg/g. During the experiments, the soil moisture was about 30%. All experiments were performed in triplicate.

Before pH measurements, the soil was dried at 60°C to constant weight and was ground in a mortar. Subsequently, the soil (10 g) was thoroughly mixed with 25 ml of 1 N KCl (pH 6.0–6.5), and the pH was measured after 3 min. (Soils. GOST [State Standard] 26483–85.)

TPH content was determined by adsorption chromatography followed by gravimetric analysis. Soil samples were dried at 60°C to constant weight and were ground in a mortar. One gram of soil was loaded into a 10-ml Al_2O_3-containing column with pre-equilibrated with 10 ml of chloroform. TPHs were eluted with chloroform until the eluate was completely decolorized. The eluate was collected into weighed vials. The fractions were dried to constant weight, and the amounts of TPHs were calculated.

Fractional analysis of old oil-pollution in the soil was performed by adsorption chromatography followed by polarimetric and gravimetric analyses. Soil samples were prepared as above. Five grams of soil was loaded into a 56-ml Al_2O_3-containing column, equilibrated with 60 ml of hexane. Fractions were identified according to their refractive indices (n). The alkane fraction ($n = 1.41–1.43$) was eluted with 40 ml of hexane; the naphthene fraction, with 60 ml of hexane ($n = 1.45–1.49$); the fraction of low-molecular weight aromatic hydrocarbons, with 100 ml of 15% benzene in hexane ($n = 1.49–1.53$); the fraction of high-molecular weight aromatic hydrocarbons, with 100 ml of benzene ($n > 1.59$); and tars, with 80 ml of the ethyl alcohol–benzene mixture (1:1) (n not determinable). The fractions were dried until complete evaporation of the solvent and were weighed (Polunina and Kushik 1977).

For determination of enzyme activities, 5-g portions of soil were suspended in 10 ml of 50 mM Na/K-phosphate buffer (pH 6.0) and were extracted for 1 h with constant agitation. The suspension was centrifuged at 5,000 × g for 30 min. The supernatant liquid was tested for enzyme activities. Laccase production was assessed by enzymatic oxidation of ABTS at 436 nm ($\varepsilon = 29{,}300\ M^{-1}cm^{-1}$), according to Niku-Paavola et al. (1988). The reaction mixture contained 50 mM Na tartrate (pH 4.5) and 200 μM ABTS. Peroxidases were estimated by measurement of 2,6-dimethoxyphenol (DMOP) oxidation at 468 nm ($\varepsilon = 14{,}800\ M^{-1}cm^{-1}$). The reaction mixture also contained 50 mM Na tartrate (pH 4.5) and 200 μM H_2O_2, with or without 0.5 mM Mn^{2+} (Heinfling et al. 1998). Peroxidase activity was calculated as the difference between the values of DMOP oxidation with and without H_2O_2. Enzyme activities were expressed as μmol/min/g of dry soil (U/g).

To study the degradation of PAHs (anthracene, phenanthrene, fluorene, pyrene, fluoranthene, and chrysene), we obtained crude laccase from a solid-state culture of *Agaricus* sp. F-8. The fungus was grown by solid-phase fermentation on sunflower-seed hulls. Fermented hulls (5 g) were washed three times with 20 ml of distilled water. The water extracts were combined, the sediment was removed by centrifugation at 4,000 rpm for 30 min, and the supernatant liquid was used as crude laccase. Experiments with crude laccase (10 U/ml) were done in 50 mM phosphate buffer (pH 6.0) containing 1% (v/v) acetonitrile and 20 μM PAHs. The tubes were incubated for 10 days at 29°C. Residual PAHs were extracted from an acidified reaction mixture (pH 2.0) with ethyl acetate (equal volume, three times). The resulting extracts were evaporated to 500 μl. Controls were prepared identically, except that the enzymes had been inactivated by boiling for 20 min before addition of the PAHs. All experiments were run in triplicate. The PAHs were analyzed with an HPLC system (GPC, Laboratorni Přistroje Praha, Czech Republic) at isocratic elution (3 ml/min; acetonitrile:H_2O, 70:30, v/v), by using a UV detector at 254 nm.

A SupelcosilTM LC-PAH (5 cm × 4.6 mm, 3 μm) column was used. The sample volume was 20 μl. All experiments were performed in triplicate.

20.3 Salient Observations

The 12 strains of *Pleurotus*, *Lentinus*, *Coriolus*, and *Agaricus* were screened for their abilities to colonize the old oil–contaminated soil and to produce ligninolytic enzymes during the degradation. The submerged inocula of all studied fungi were unviable in old oil–contaminated soil, and no ligninolytic enzyme production was found. Yet, the solid-state cultures of all the fungi grew and produced ligninolytic enzymes in soil, but the rate and intensity of colonization were different. The growth rates and the mycelium densities in the soil decreased in the order *Agaricus* sp. > *P. ostreatus* > *L. edodes* > *Coriolus* sp.

The soil colonization by mycelia of *Pleurotus* and *Agaricus* was completed within the first week of the experiment (Fig. 10.1, leaders). The suitability of *Pleurotus* for soil remediation was found by different authors (Lang et al. 1998; Martens and Zadražil 1998; Baldrian et al. 2000). The fungi showed highly competitive saprophytic ability against soil microbiota in the soil–lignocellulose systems, grew with PAHs, and produced ligninolytic enzymes during the degradation of PAHs in soil (Baldrian et al. 2000). We also found that the *P. ostreatus* strains were similar in their abilities to colonize old oil–contaminated soil and to degrade the pollutant. The decrease in TPHs varied from 18 to 28% in sterile soil (Fig. 20.1a) and reached 64% in nonsterile soil (Fig. 20.1b). This was not

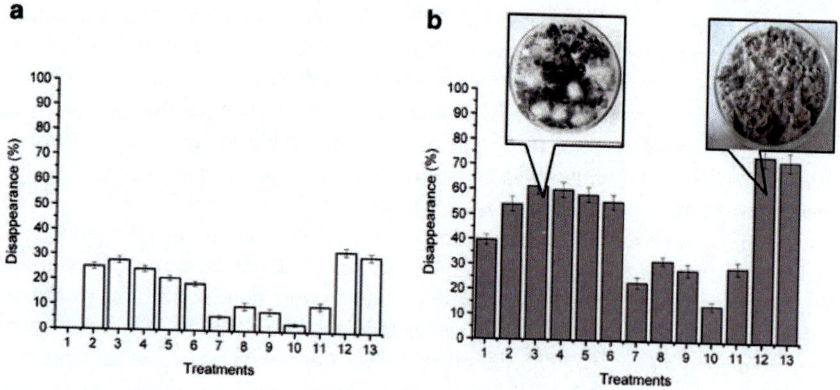

Fig. 20.1 Disappearance of TPHs during mycoremediation of sterile (**a**) and nonsterile (**b**) soil with old oil-pollution: (1) control (without fungi); (2) *P. ostreatus* 336; (3) *P. ostreatus* D1; (4) *P. ostreatus* D2; (5) *P. ostreatus* D/L; (6) *P. ostreatus* D/o; (7) *L. edodes* F-249; (8) *L. edodes* 0779; (9) *L. edodes* 2T; (10) *L. edodes* NY; (11) *Coriolus* sp. F-1; (12) *Agaricus* sp. F-8; (13) *Agaricus* sp. F-17; leaders – the growth of the white-rot fungi *P. ostreatus* D1 (3) and *Agaricus* sp. F-8 (12) in soil with old oil-pollution

unexpected because earlier we found that *P. ostreatus* D1 can actively degrade oil hydrocarbons during cultivation in soil (Pozdnyakova et al. 2008b).

White-rot fungi generally colonize compact wood and cannot compete in soil for a long time; therefore, their contribution to the removal of recalcitrant xenobiotics under natural conditions can be limited. There is, however, a second ecophysiological group of ligninolytic fungi, the litter-decomposing fungi, that are soil-inhabiting basidiomycetes (Steffen et al. 2002, 2003). They produce a ligninolytic enzyme system similar to that of the white-rot fungi (Bonnen et al. 1994; Heinzkill et al. 1998; Lankinen et al. 2001; Kahkonen et al. 2008) and are capable of metabolizing PAHs (Lambert et al. 1994; Lange et al. 1994; Steffen et al. 2002, 2003).

Two strains of the litter-decomposing fungus *Agaricus* sp. were collected by us from the oil-contaminated grounds of the Saratov petroleum refinery (Pozdnyakova et al. 2008a). In this work, we found that both *Agaricus* sp. strains actively colonized the soil (Fig. 20.1, leader) and decreased TPH content. They were very similar in these properties. The decrease in TPHs reached 32% in sterile and 75% in nonsterile soil. The presence of these fungi in the soil decreased old oil-pollution to about twofold, in comparison with the effect produced by the indigenous microflora alone (Fig. 20.1).

At the same time, the *Lentinus* fungi and *Coriolus* sp. F-1 grew very slowly, and complete colonization was not achieved. As a result, insignificant losses of TPHs in sterile soil occurred (Fig. 20.1a), but in nonsterile soil, the losses of TPHs did not exceed the control value (Fig. 20.1b).

The soil pH decreased from 6.2 to 5.0 at the end of the experiments, independently of the fungus used, and it was almost constant in the control.

We checked ligninolytic enzyme activities during 4 weeks of the experiments. The time courses of laccase activity are presented in Fig. 20.2. Enzyme activities in sterile and nonsterile soil were similar, but laccase activity tended to be higher in sterile soil. Laccase activity peaked during the first 2 weeks after sunflower-seed hulls colonized by the fungi had been added to the soil. Thereafter, laccase activity decreased and reached a minimal value, maintained in the course of the experiment. The trend revealed in TPH elimination was detected in these studies also. The fungi that were good colonizers of contaminated soil were good producers of laccase under the conditions used. The most active producers of laccase under these conditions were *Agaricus* sp. strains (Fig. 20.2e, f). All strains of *P. ostreatus* showed similar laccase activities in nonsterile soil, but in sterile soil, high activities were expressed by *P. ostreatus* D1 and D2 (Fig. 20.2a, b). Many authors showed the production of laccase by white-rot fungi during mycoremediation of contaminated soil. For example, Lang et al. (1998) found that the laccase activity of *Pleurotus* sp. was similar in sterile and nonsterile soil and in straw, but in that case, laccase activity tended to be higher in nonsterile soil.

The difference of our data from the findings of Lang et al. (1998) may have resulted from the use of another growth substrate (sunflower-seed hulls in our experiments and straw in Lang's) or old oil-pollutant in our experiments. Furthermore, we found that the weak colonizers of old oil–contaminated soil, including

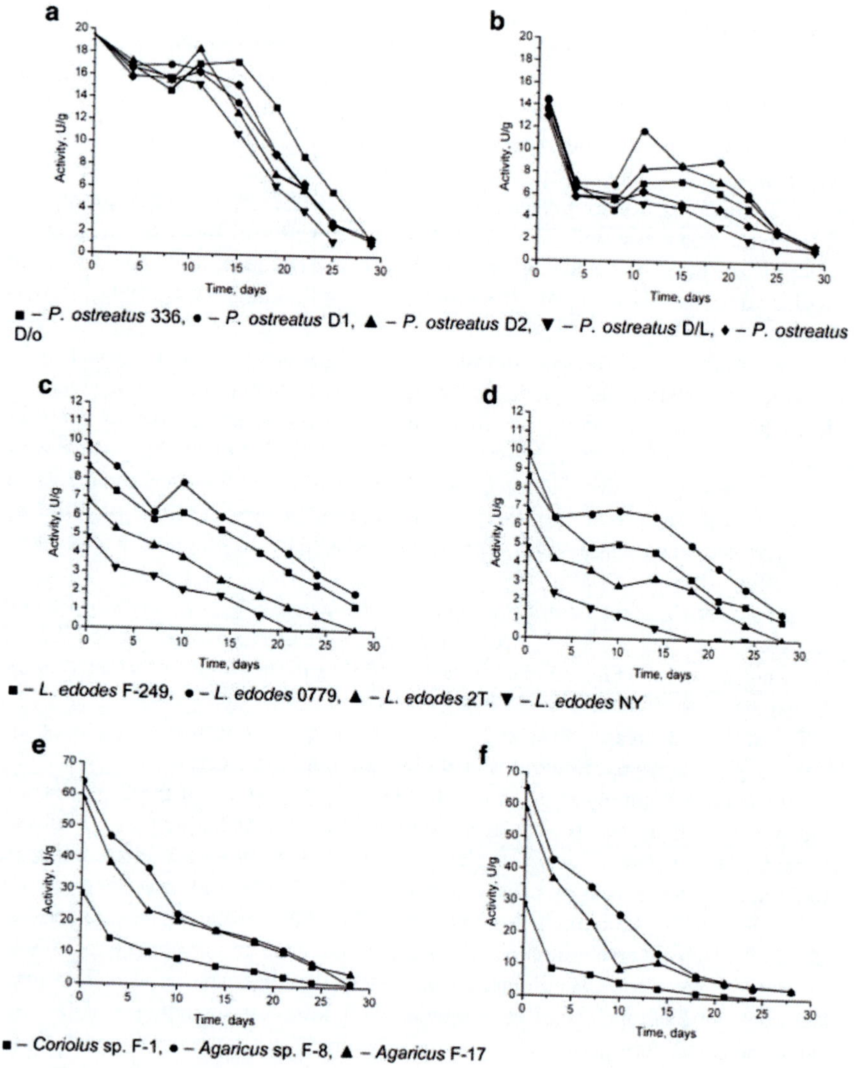

Fig. 20.2 Laccase production during mycoremediation of sterile (**a**, **c**, **e**) and nonsterile (**b**, **d**, **f**) soil with old oil pollution

L. edodes F-249, *L. edodes* 0779, *L. edodes* 2T, *L. edodes* NY, and *Coriolus* sp. F-1, were weak producers of laccase under these conditions (Fig. 20.2c, d).

Production of ligninolytic peroxidases during mycoremediation of old oil–contaminated soil was studied. Among the 12 strains of white-rot fungi used, only *Pleurotus* and *Agaricus* produced peroxidase under the conditions of which soil was studied. In either case, peroxidase activity was low and did not exceed

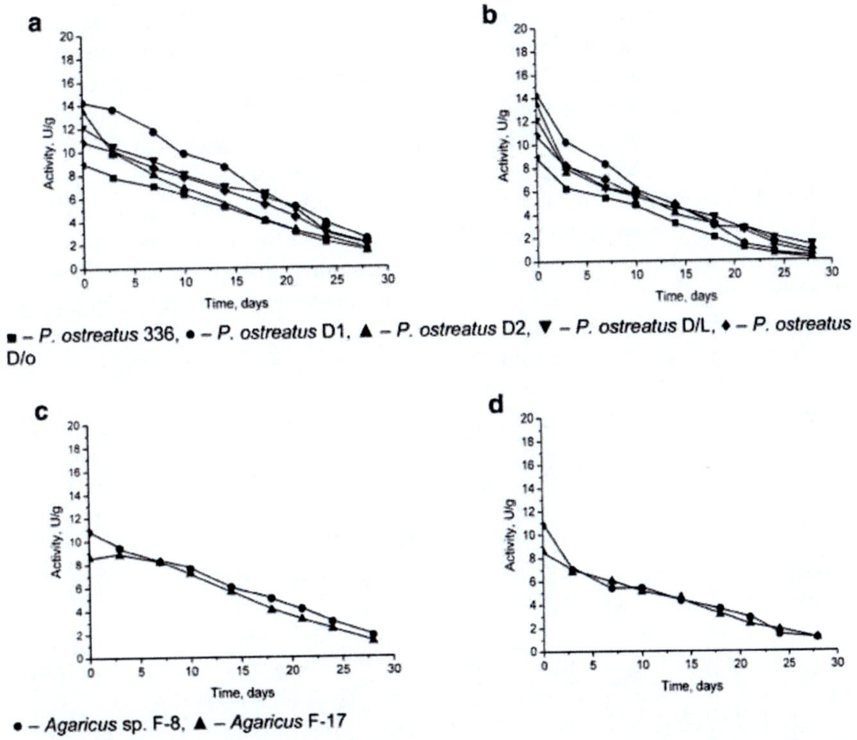

Fig. 20.3 Peroxidase production during mycoremediation of sterile (**a, c**) and nonsterile (**b, d**) soil with old oil pollution

14 U/g. As in the case of laccase activity, peroxidase activity was similar in sterile and nonsterile soil (Fig. 20.3).

Pleurotus and *Agaricus* fungi produce two ligninolytic peroxidases: Mn-peroxidase and versatile peroxidase. For example, the production of Mn-peroxidase by *Agaricus bisporus* ATCC 62459 was found in lignocellulose-containing culture (Lankinen et al. 2005). Lang et al. (1998) showed the production of Mn-peroxidase during growth of *Pleurotus* sp. in soil. Enzyme activity was similar in sterile and nonsterile soil. Mn-peroxidase activity increased during the first 4 weeks and remained constant for several weeks (Lang et al. 1998). A similar time course of enzyme activity was found by Baldrian et al. (2000), who studied the influence of heavy metals on the activities of ligninolytic enzymes during PAH degradation by *Pleurotus ostreatus* in soil. A weak affect of heavy metals on the activities of laccase and Mn-peroxidase produced by the fungus was found.

Unfortunately, we could not identify peroxidase activity in our studies. However, it should be noted that the peroxidase activities of the *P. ostreatus* strains detected in the presence of Mn^{2+} exceeded those without Mn^{2+} by about 20% only. The peroxidase activities of the *Agaricus* sp. strains could be detected only in the presence of Mn^{2+}.

White-rot fungi are good degraders of aromatic and polycyclic aromatic compounds in different environments, including soil (Baldrian et al. 2000; Bhatt et al. 2002; Marquez-Rocha et al. 2000). We studied changes in the fractional composition of oil during mycoremediation. The losses of low- and high-molecular weight aromatic hydrocarbons from sterile and nonsterile soil are presented in Fig. 20.4. In this case, the *P. ostreatus* and *Agaricus* sp. strains actively degraded both low- and high-molecular weight aromatic hydrocarbons. The decrease in the content of low-molecular weight aromatic compounds varied from 42 to 53% for the *P. ostreatus* strains and was about 60% for the *Agaricus* sp. strains in sterile soil (Fig. 20.4, Ia). In nonsterile soil, an important contribution to this decrease was made by the soil microflora (Fig. 20.4, Ib).

Similar data were obtained for the high-molecular weight aromatic hydrocarbons. This fraction was poorly available to the soil microflora, but all the white-rot fungi used by us could decrease its content in both sterile and nonsterile soil. In this

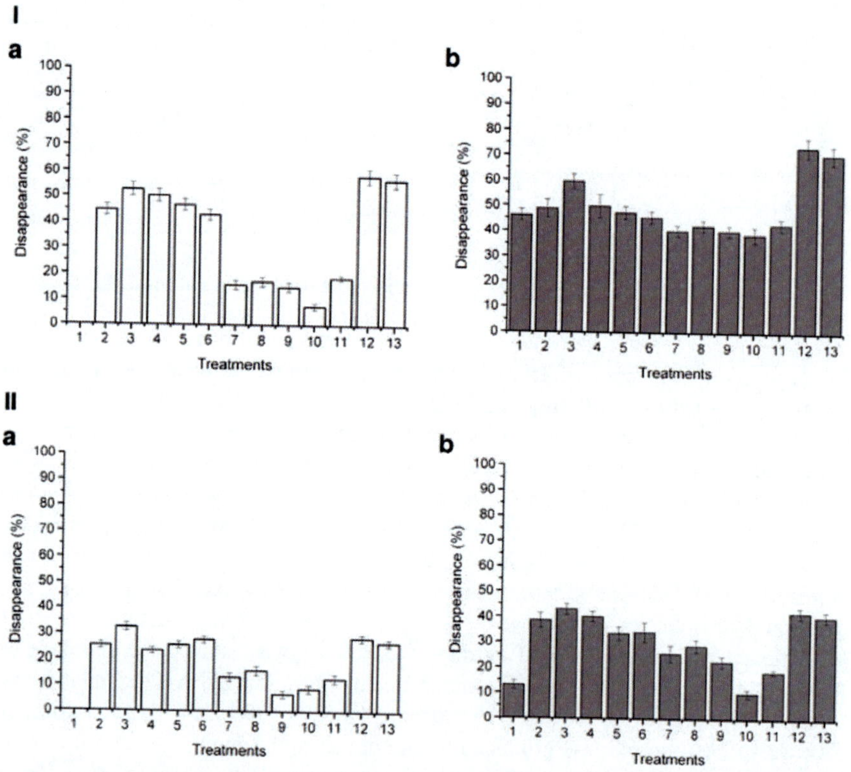

Fig. 20.4 Disappearance of low- (I) and high-molecular weight (II) aromatic hydrocarbons during mycoremediation of sterile (**a**) and nonsterile (**b**) soil with old oil pollution: (1) control (without the fungi); (2) *P. ostreatus* 336; (3) *P. ostreatus* D1; (4) *P. ostreatus* D2; (5) *P. ostreatus* D/L; (6) *P. ostreatus* D/o; (7) *L. edodes* F-249; (8) *L. edodes* 0779; (9) *L. edodes* 2T; (10) *L. edodes* NY; (11) *Coriolus* sp. F-1; (12) *Agaricus* sp. F-8; (13) *Agaricus* F-17

case, the more active ligninolytic enzyme producers (*P. ostreatus* and *Agaricus* sp.) actively decreased the content of high-molecular weight aromatic hydrocarbons in the soil (Fig. 20.4, II). In concordance with the obtained data, we propose that white-rot fungi make the main contribution to the decrease in the content of these compounds in nonsterile soil.

The strains of the litter-decomposing fungus *Agaricus* sp. were the most active. The suitability of *Agaricus* sp. F-8 for remediation of all components of old oil-pollution was estimated, and after 4-week cultivation of this fungus in contaminated soil, fractional analysis was made (Fig. 20.5).

We found that the indigenous microflora utilized 29% of alkanes. The loss of this fraction in the presence of *Agaricus* sp. F-8 in sterile soil was unexpected. Under these conditions, the content of alkanes decreased by 28% at the end of the experiment. The simultaneous presence of the fungus and the indigenous microflora resulted in a 42% removal of this fraction (Fig. 20.5).

The naphthene fraction was poorly available to *Agaricus* sp. F-8. At the same time, the indigenous microflora utilized about 40% of naphthene. In nonsterile soil, the fungus did not appreciably affect the decrease in the content of this fraction (Fig. 20.5).

The fraction of low-molecular weight aromatic hydrocarbons was available to the fungus and the indigenous microflora (60 and 45% losses, respectively). The simultaneous presence of the fungus and the indigenous microflora resulted in a 73% removal of this fraction (Fig. 20.5).

As mentioned above, *Agaricus* sp. F-8 actively decreased the content of the PAH fraction. The losses were about 30 and 45% in sterile and nonsterile soil, respectively (Fig. 20.5). Yateem et al. (1998) did not only make a fractional analysis of petroleum hydrocarbons in samples treated with *Ph. chrysosporium* but also find decreases in the PAH concentration (by 55%).

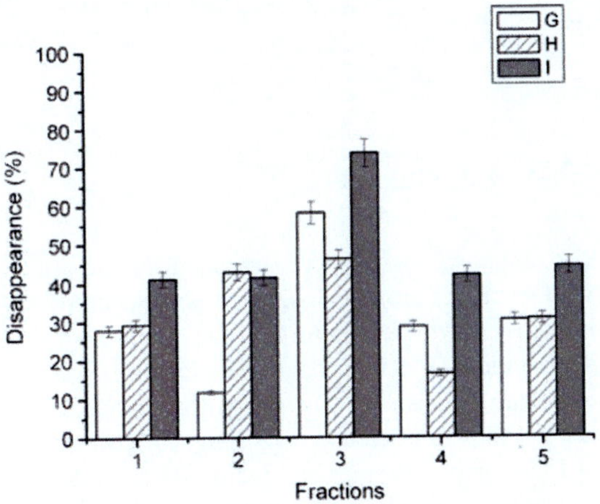

Fig. 20.5 Fractional analysis of old oil pollution in soil after treatment with *Agaricus* sp. F-8: *G*, sterile soil with F-8; *H*, nonsterile control (without the fungus); *I*, nonsterile soil with F-8; (1) alkanes; (2) naphthenes; (3) low-molecular weight aromatic hydrocarbons; (4) high-molecular weight aromatic hydrocarbons; (5) tars

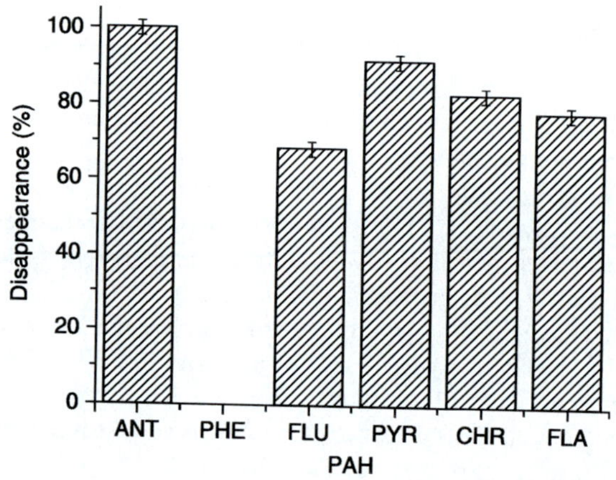

Fig. 20.6 PAH degradation by crude laccase from *Agaricus* sp. F-8: *ANT* anthracene, *PHE* phenanthrene, *FLU* fluorene, *PYR* pyrene, *CHR* chrysene, *FLA* fluoranthene

The decrease in the tar content varied from 26.5% in sterile to 48.9% in nonsterile soil with the fungus.

Litter-decomposing fungi produce a ligninolytic enzyme similar to that of the white-rot fungi (Bonnen et al. 1994; Heinzkill et al. 1998; Lankinen et al. 2001) and are capable of metabolizing PAHs (Lambert et al. 1994; Lange et al. 1994; Steffen et al. 2002, 2003). We also found the production of laccase and peroxidase during cultivation of *Agaricus* sp. F-8 in old oil–contaminated soil in the presence of sunflower-seed hulls. We obtained crude laccase from a solid-state culture of the fungus. The activity of this enzyme toward some individual compounds in the fraction of PAHs was studied. Among the tested three- and four-ring PAHs, only phenanthrene was not available to the enzyme. After 10 days of incubation, anthracene was degraded almost completely; fluorene, by 68.0 ± 3.6%; fluoranthene, by 78.2 ± 2.2%; chrysene, by 83.0 ± 1.2%; and pyrene by 91.9 ± 04% (Fig. 20.6).

20.4 Conclusions

The obtained results demonstrate different production of ligninolytic enzymes with respect to growth yields of various white-rot fungi growing in soil. The growth rates, the mycelium densities, the production of ligninolytic enzymes, and the degradation of old oil-contamination decreased in the order *Agaricus* sp. > *P. ostreatus* > *L. edodes* > *Coriolus* sp. The strains of the white-rot fungus *P. ostreatus* and the litter-decomposing fungus *Agaricus* sp. were the most active producers of ligninolytic enzymes and the most active degraders of old oil-contamination in soil. High-molecular weight aromatic hydrocarbons were poorly available to the soil microflora, but the more active ligninolytic enzyme producers (*P. ostreatus* and *Agaricus* sp.) actively decreased their content in soil. In concordance with the obtained data, we propose that white-rot fungi make the main contribution to the decrease in the content of these compounds in nonsterile soil.

Acknowledgments We are grateful to Dmitry N. Tychinin for his assistance in the preparation of the English text of this paper.

This work was supported by the federal target-oriented program "Issledovaniya i razrabotki po prioritetnym napravleniyam razvitiya nauchno-tekhnologicheskogo kompleksa Rossii na 2007–2012 gody" (Research and Developments on Priority Directions in the Development of Russia's Science and Technology Complex for 2007–2012), State contract no. 02.512.11.2210.

References

Baldrian P (2006) Fungal laccases – occurrence and properties. FEMS Microbiol Rev 30:215–242

Baldrian P, in der Wiesche C, Gabriel J, Nerud F, Zadrazil F (2000) Influence of cadmium and mercury on activities of ligninolytic enzymes and degradation of polycyclic aromatic hydrocarbons by *Pleurotus ostreatus* in soil. Appl Environ Microbiol 66:2471–2478

Barr DP, Aust SD (1994) Mechanisms white-rot fungi use to degrade pollutants. Environ Sci Technol 28:79A–87A

Bezalel L, Hadar Y, Cerniglia CE (1997) Enzymatic mechanisms involved in phenanthrene degradation by the white-rot fungus *Pleurotus ostreatus*. Appl Environ Microbiol 63:2495–2501

Bhatt M, Patel M, Rawal B, Novotny C, Molitoris HP, Sasek V (2000) Biological decolorization of the synthetic dye RBBR in contaminated soil. World J Microbiol Biotechnol 16:195–198

Bhatt M, Cajthaml T, Sasek V (2002) Mycoremediation of PAH-contaminated soil. Folia Microbiol 47:255–258

Bogan BW, Schoenike B, Lamar RT, Cullen D (1996) Manganese peroxidase mRNA and enzyme activity levels during bioremediation of polycyclic aromatic hydrocarbon contaminated soil with *Phanerochaete chrysosporium*. Appl Environ Microbiol 62:2381–2386

Bollag J-M, Chu H-L, Rao MA, Gianfreda L (2003) Enzymatic oxidative transformation of chlorophenol mixtures. J Environ Qual 32:63–69

Bonnen A, Anton L, Orth A (1994) Lignin-degrading enzymes of the commercial button mushroom, *Agaricus bisporus*. Appl Environ Microbiol 60:960–965

Boyle CD (1995) Development of a practical method for inducing white-rot fungi to grow into and degrade organopollutants in soil. Can J Microbiol 41:345–353

Cohen R, Persky L, Hadar Y (2002) Biotechnological applications and potential of wood-degrading mushrooms of the genus *Pleurotus*. Appl Microbiol Biotechnol 58:582–594

Cripps C, Bumpus JA, Aust SD (1990) Biodegradation of azo dyes by *Phanerochaete chrysosporium*. Appl Environ Microbiol 56:1114–1118

Eggen T (1999) Application of fungal substrate from commercial mushroom production – *Pleurotus ostreatus* – for bioremediation of creosote contaminated soil. Int Biodeterior Biodegradation 44:117–126

Soils. GOST 26483–85. (Soils. Preparation of salt extract and determination of its pH by SINAO Method) (in Russian)

Gramss G (2000) Degradation of aromatic xenobiotics in aerated soils by enzyme systems of microorganisms and plants. In: Wise DL, Trantolo DJ, Cichon EJ, Inyang HI, Stottmeister U (eds) Bioremediation of contaminated soils. Marcel Dekker, New York, pp 489–536

Heinfling A, Martinez MJ, Martinez AT, Bergbauer M, Szewzyk U (1998) Purification and characterization of peroxidases from dye-decolorizing fungus *Bjerkandera adusta*. FEMS Microbiol Lett 165:43–50

Heinzkill M, Bech L, Halkier T, Schneider P, Anke T (1998) Characterization of laccases and peroxidases from wood-rotting fungi (family *Coprinaceae*). Appl Environ Microbiol 64:1601–1606

Hou H, Zhou J, Wang J, Du C, Yan B (2004) Enhancement of laccase production by *Pleurotus ostreatus* and its use for the decolorization of anthraquinone dye. Process Biochem 39:1415–1419

Ishikhuemhen OS, Anoliefo GO, Oghale OI (2003) Bioremediation of crude oil polluted soil by the white-rot fungus, *Pleurotus tuberregium* (Fr.) Sing. Environ Sci Pollut Res 10:108–112

Kahkonen M, Lankinen P, Hatakka A (2008) Hydrolytic and ligninolytic enzyme activities in the Pb contaminated soil inoculated with litter-decomposing fungi. Chemosphere 72:708–714

Khadrani A, Siegile-Murandi F, Steiman R, Vrounsia T (1999) Degradation of three phenylurea herbicides (chlortorulon, isoproturon and diuron) by micromycetes isolated from soil. Chemosphere 38:3041–3050

Kubatova A, Erbanova P, Eichlerova I, Homolka L, Nerud F, Sasek V (2001) PCB congener selective biodegradation by the white-rot fungus *Pleurotus ostreatus* in contaminated soil. Chemosphere 43:207–215

Lamar R, Dietrich D (1990) In situ depletion of pentachlorophenol from contaminated soil by *Phanerochaete* spp. Appl Environ Microbiol 56:3093–3100

Lambert M, Kremer S, Sterner O, Anke H (1994) Metabolism of pyrene by the basidiomycete *Crinipellis stipitaria* and identification of pyrenequinones and their hydroxylated precursors in strain JK375. Appl Environ Microbiol 60:3597–3601

Lang E, Nerud F, Zadrazil F (1998) Production of ligninolytic enzymes by *Pleurotus* sp. and *Dichomitus squalens* in soil and lignocellulose substrate as influenced by soil microorganisms. FEMS Microbiol Lett 167:239–244

Lange B, Kremer S, Sterner O, Anke H (1994) Pyrene metabolism in *Crinipellis stipitaria*: identification of trans-4, 5-dihydro-4, 5-dihydroxypyrene and 1-pyrenylsulfate in strain JK364. Appl Environ Microbiol 60:3602–3607

Lankinen V, Bonnen A, Anton L, Wood D, Kalkkinen N, Hatakka A, Thurston C (2001) Characteristics and N-terminal amino acid sequence of manganese peroxidase from solid substrate cultures of *Agaricus bisporus*. Appl Microbiol Biotechnol 55:170–176

Lankinen P, Hilden K, Aro N, Salkinoja-Salonen M, Hatakka A (2005) Manganese peroxidase of *Agaricus bisporus*: grain bran-promoted production and gene characterization. Appl Microbiol Biotechnol 66:401–407

Lau K, Tsang Y, Chiu S (2003) Use of spent mushroom compost to bioremediate PAH-contaminated samples. Chemosphere 52:1539–1546

Lestan D, Lamar R (1996) Development of fungal inocula for bioaugmentation of contaminated soils. Appl Environ Microbiol 62:2045–2052

Marquez-Rocha F, Hernandez-Rodriguez V, Vazquez-Duhalt R (2000) Biodegradation of soil-adsorbed polycyclic aromatic hydrocarbons by the white-rot fungus *Pleurotus ostreatus*. Biotechnol Lett 22:469–472

Martens R, Zadražil F (1998) Screening of white-rot fungi for their ability to mineralize polycyclic aromatic hydrocarbons in soil. Folia Microbiol 43:97–103

Morgan P, Lewis T, Watkinson RJ (1991) Comparison of abilities of white-rot fungi to mineralize selected xenobiotic compounds. Appl Microbiol Biotechnol 34:693–696

Niku-Paavola M-L, Karhunen E, Salola P, Paunio V (1988) Ligninolytic enzymes of the white-rot fungus *Phlebia radiata*. Biochem J 254:877–883

Novotny C, Erbanova P, Šašek V, Kubatova A, Cajthaml T, Lang E, Krahl J, Zadražil F (1999) Extracellular oxidative enzyme production and PAH removal in soil by exploratory mycelium of white-rot fungi. Biodegradation 10:159–168

Novotny C, Erbanova P, Cajthaml T, Rothschild N, Dosoretz C, Sasek V (2000) *Irpex lacteus*, a white-rot fungus applicable to water and soil bioremediation. Appl Microbiol Biotechnol 54:850–853

Okeke B, Paterson A, Smith J, Watsoncraik I (1994) Relationships between ligninolytic activities of *Lentinula* spp. and biotransformation of pentachlorophenol in sterile soil. Lett Appl Microbiol 19:284–287

Polunina AG, Kushik GI (1977) Metody analiza organicheskogo veshchestva porod, nefti i gaza (Methods of Analyslis of Organic Matter in Rocks, Oil, and Gas). In: Ryl'kov AV (ed) Tyumen': Tr Zap-Sib. NIGNI, 122 (in Russian)

Pozdnyakova N, Dubrovskaya E, Makarov O, Turkovskaya O (2008a) Use of the complex "fungus *Agaricus* sp – indigenous microflora" in PAH degradation. In: Proceedings of the All-Russia conference "Basic and applied aspects of studies of symbiotic systems". p 30 (in Russian)

Pozdnyakova NN, Nikitina VE, Turovskaya OV (2008b) Bioremediation of oil-polluted soil with an association including the fungus *Pleurotus ostreatus* and soil microflora. Appl Biochem Microbiol 44:60–65

Reddy CA (1995) The potential for white-rot fungi in the treatment of pollutants. Curr Opin Biotechnol 6:320–328

Sasek V, Cajthaml T, Bhatt M (2003) Use of fungal technology in soil remediation: a case study. Water Air Soil Pollut Focus 3:5–14

Snajdr J, Baldrian P (2006) Production of lignocellulose-degrading enzymes and changes in soil bacterial communities during the growth of *Pleurotus ostreatus* in soil with different carbon content. Folia Microbiol 51:579–590

Stanley GA, Britz ML, Boonchan S, Juhasz A (2000) Detoxification of soil containing high-molecular weight polycyclic aromatic hydrocarbons by Gram-negative bacteria and bacterial-fungal cocultures. In: Wise DL, Trantglo DJ, Cichon EJ, Inyang HI, Stottmeister U (eds) Bioremediation of contaminated soils. Marcel Dekker, New York, pp 409–443

Steffen KT, Hofrichter M, Hatakka A (2000) Mineralization of ^{14}C-labelled synthetic lignin and ligninolytic enzyme activities of litter-decomposing basidiomycetous fungi. Appl Microbiol Biotechnol 54:819–825

Steffen K, Hatakka A, Hofrichter M (2002) Removal and mineralization of polycyclic aromatic hydrocarbons by litter-decomposing basidiomycetous fungi. Appl Microbiol Biotechnol 60:212–217

Steffen K, Hatakka A, Hofrichter M (2003) Degradation of benzo[a]pyrene by the litter-decomposing basidiomycete *Stropharia coronilla*: Role of manganese peroxidase. Appl Environ Microbiol 69:3957–3964

Yateem A, Balba MT, Al-Awadhi N, El-Nawawy AS (1998) White-rot fungi and their role in remediating oil-contaminated soil. Environ Int 24:181–187

Index

M

N

O

P